A TIME TRAVELER'S
GUIDE TO LIFE,
THE UNIVERSE,
AND EVERYTHING

JOURNEY BY STARLIGHT

IAN FLITCROFT
BRITT SPENCER

ONE PEACE
BOOKS

PUBLISHED BY ONE PEACE BOOKS

COVER DESIGN AND ILLUSTRATIONS BY BRITT SPENCER
ADDITIONAL INK WORK BY ALLIE JACHIMOWICZ, EMILY SPENCER, AND LOMAHO KRETZMANN
EDITING BY ERIN CANNING

ISBN: 978-1-935548-23-2

FIRST EDITION, FEBRUARY 2013

10 9 8 7 6 5 4 3 2 1

PRINTED IN CANADA

PHOTO (PAGE 75) COPYRIGHT © 2007 ISTOCKPHOTO LP/JAAP HART

DISTRIBUTION BY SCB DISTRIBUTORS
WWW.SCBDISTRIBUTORS.COM

FOR MORE INFORMATION, CONTACT:
ONE PEACE BOOKS
43-32 22ND STREET, SUITE #204
LONG ISLAND CITY, NY 11101
WWW.ONEPEACEBOOKS.COM

ABOUT THE AUTHOR AND ILLUSTRATOR

DR. IAN FLITCROFT IS A VISION SCIENTIST AND A CONSULTANT EYE SURGEON AT THE CHILDREN'S UNIVERSITY HOSPITAL, DUBLIN. AS A TEENAGER, HE MADE A CLOSE CALL BETWEEN CHOOSING MEDICINE OR ASTROPHYSICS AS A CAREER, BUT HE HOPES THIS BOOK WILL ENCOURAGE A NEW GENERATION OF ASTROPHYSICISTS, SCIENTISTS, AND, PERHAPS, EVEN A FEW EYE SURGEONS.

IAN STUDIED MEDICINE AT OXFORD UNIVERSITY, WHERE HE ALSO COMPLETED HIS DOCTORATE (D.PHIL) IN VISUAL PHYSIOLOGY BEFORE GOING ON TO COMPLETE HIS MEDICAL TRAINING AT ST MARY'S HOSPITAL, LONDON. AS A WRITER, HE HAS PUBLISHED MORE THAN 30 SCIENTIFIC PEER-REVIEWED PAPERS AND CHAPTERS IN MULTI-AUTHOR BOOKS. HIS WRITING HAS BEEN FEATURED IN *DO POLAR BEARS GET LONELY* (NEW SCIENTIST AND PROFILE BOOKS, 2008), AND HIS FIRST NOVEL, *THE RELUCTANT CANNIBALS*, WAS ONE OF THE WINNERS OF THE DUBLIN WRITERS CENTRE NOVEL COMPETITION 2012.

JOURNEY BY STARLIGHT IS BASED ON HIS AWARD-WINNING POPULAR SCIENCE BLOG (WWW.JOURNEYBYSTARLIGHT.COM), WHICH HAS BEEN VIEWED MORE THAN 700,000 TIMES. IAN LIVES IN DUBLIN WITH HIS WIFE AND THREE SONS.

BRITT SPENCER IS AN AWARD-WINNING ILLUSTRATOR WITH A BFA AND MFA FROM THE SAVANNAH COLLEGE OF ART AND DESIGN.

SINCE HIS ARRIVAL INTO THE COMMERCIAL ART WORLD, BRITT'S WORK HAS BEEN PUBLISHED INTERNATIONALLY AND BEEN RECOGNIZED BY THE DISTINGUISHED NEW YORK SOCIETY OF ILLUSTRATORS AND THE SOCIETY OF ILLUSTRATORS WEST. HE IS ALSO THE ILLUSTRATOR FOR SEVERAL CHILDREN'S BOOKS, INCLUDING PENGUIN/PHILOMEL BOOKS' *FLEAS!* (2008), *MAKE YOUR MARK, FRANKLIN ROOSEVELT* (2007), AND *ZARAFA: THE GIRAFFE WHO WALKED TO THE KING* (2009).

BRITT LIVES IN SAVANNAH, GEORGIA. YOU CAN SEE MORE OF HIS WORK AT HIS WEBSITE (WWW. BRITTSPENCER.COM).

TO THE SUN AND PLANETS OF MY OWN LITTLE SOLAR SYSTEM:
JEAN, CALLUM, MYLES, AND OLIVER.

TO THE FALLACY OF SUNK-COST.

CONTENTS

CHAPTER 1: THE POWER OF THOUGHT

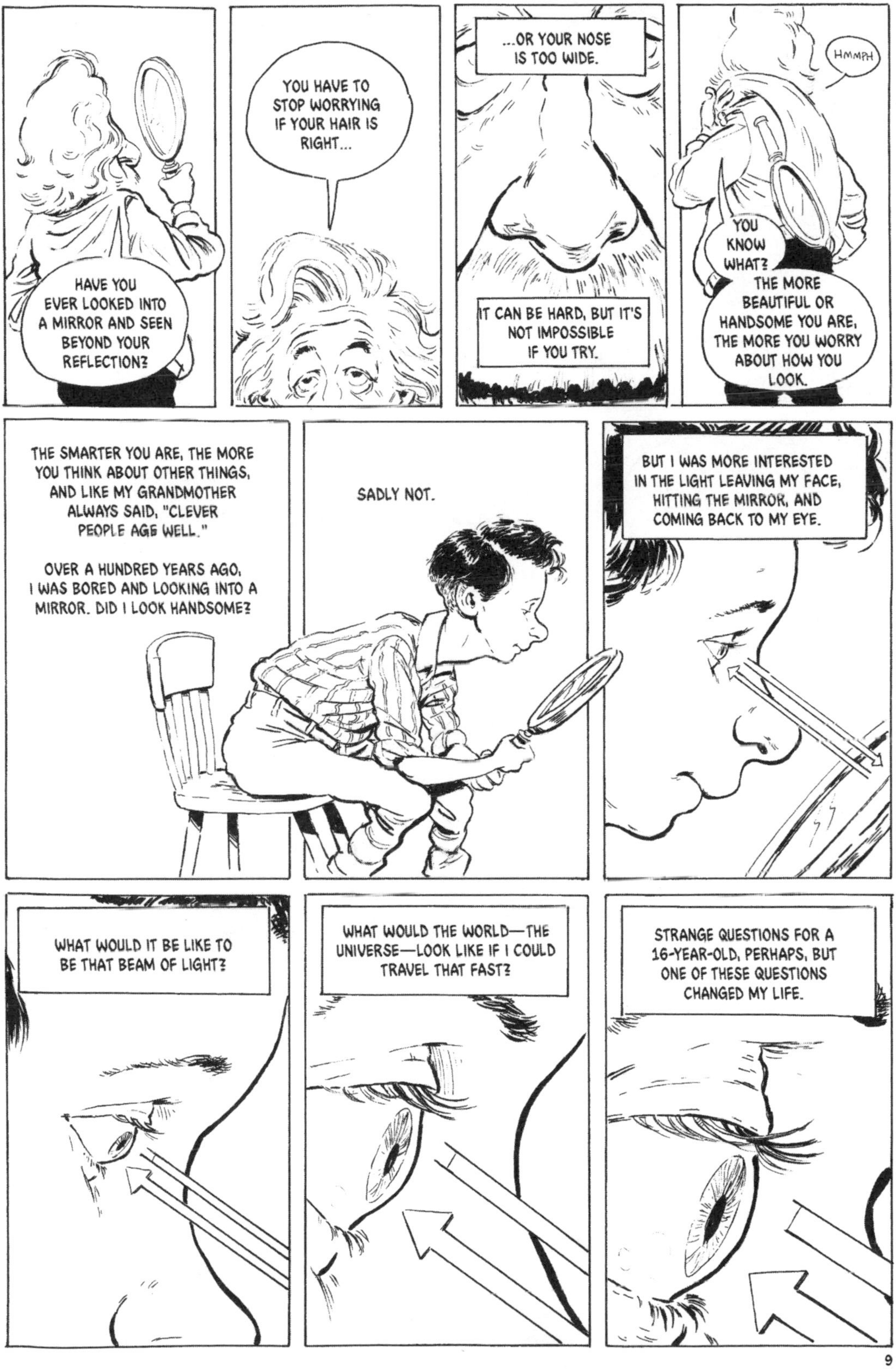

HAVE YOU EVER LOOKED INTO A MIRROR AND SEEN BEYOND YOUR REFLECTION?
YOU HAVE TO STOP WORRYING IF YOUR HAIR IS RIGHT...
...OR YOUR NOSE IS TOO WIDE.
IT CAN BE HARD, BUT IT'S NOT IMPOSSIBLE IF YOU TRY.
HMMPH
YOU KNOW WHAT?
THE MORE BEAUTIFUL OR HANDSOME YOU ARE, THE MORE YOU WORRY ABOUT HOW YOU LOOK.
THE SMARTER YOU ARE, THE MORE YOU THINK ABOUT OTHER THINGS, AND LIKE MY GRANDMOTHER ALWAYS SAID, "CLEVER PEOPLE AGE WELL."
OVER A HUNDRED YEARS AGO, I WAS BORED AND LOOKING INTO A MIRROR. DID I LOOK HANDSOME?
SADLY NOT.
BUT I WAS MORE INTERESTED IN THE LIGHT LEAVING MY FACE, HITTING THE MIRROR, AND COMING BACK TO MY EYE.
WHAT WOULD IT BE LIKE TO BE THAT BEAM OF LIGHT?
WHAT WOULD THE WORLD—THE UNIVERSE—LOOK LIKE IF I COULD TRAVEL THAT FAST?
STRANGE QUESTIONS FOR A 16-YEAR-OLD, PERHAPS, BUT ONE OF THESE QUESTIONS CHANGED MY LIFE.

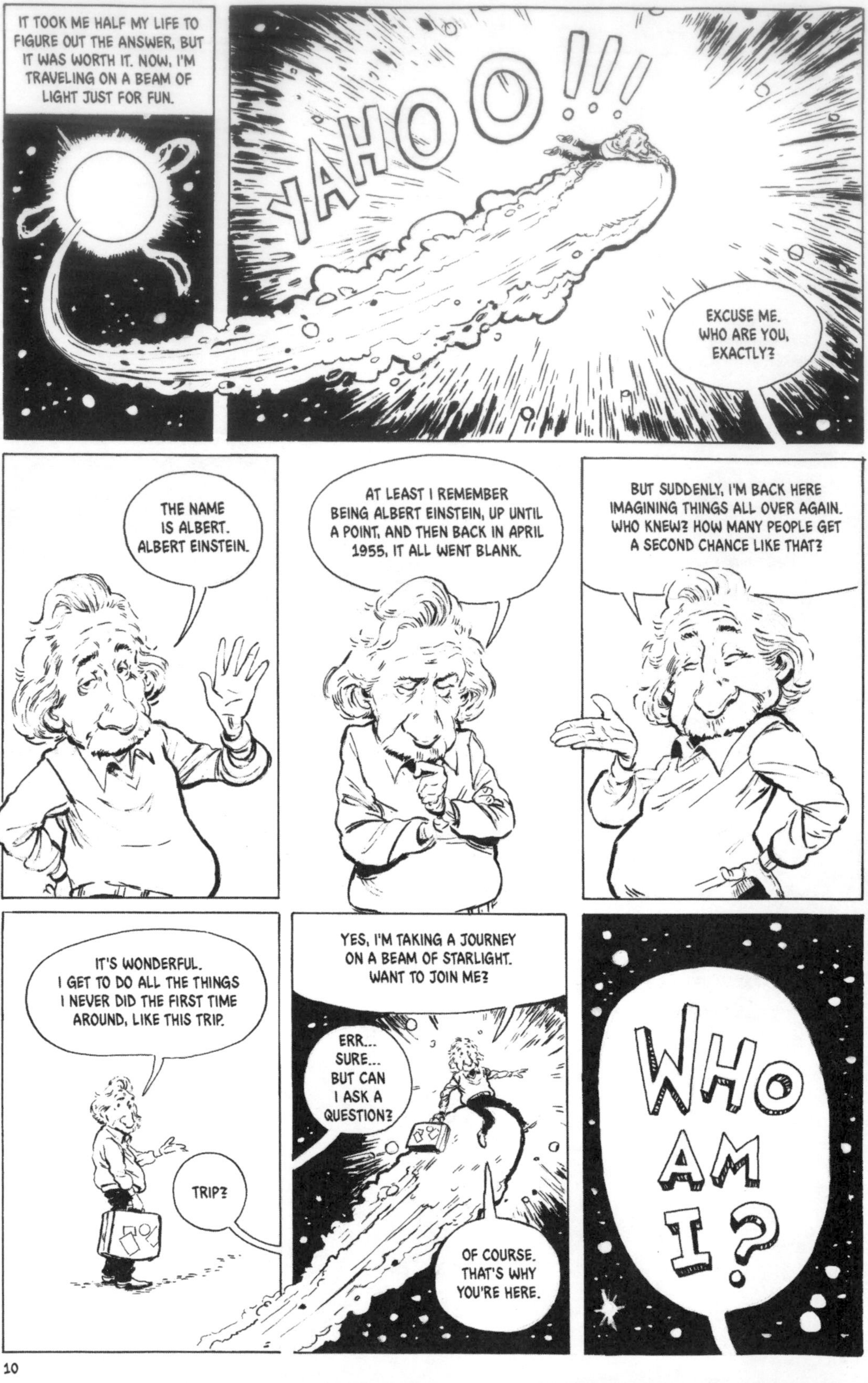

IT TOOK ME HALF MY LIFE TO FIGURE OUT THE ANSWER, BUT IT WAS WORTH IT. NOW, I'M TRAVELING ON A BEAM OF LIGHT JUST FOR FUN.
YAHOO!!!
EXCUSE ME. WHO ARE YOU, EXACTLY?
THE NAME IS ALBERT. ALBERT EINSTEIN.
AT LEAST I REMEMBER BEING ALBERT EINSTEIN, UP UNTIL A POINT, AND THEN BACK IN APRIL 1955, IT ALL WENT BLANK.
BUT SUDDENLY, I'M BACK HERE IMAGINING THINGS ALL OVER AGAIN. WHO KNEW? HOW MANY PEOPLE GET A SECOND CHANCE LIKE THAT?
IT'S WONDERFUL. I GET TO DO ALL THE THINGS I NEVER DID THE FIRST TIME AROUND, LIKE THIS TRIP.
TRIP?
YES, I'M TAKING A JOURNEY ON A BEAM OF STARLIGHT. WANT TO JOIN ME?
ERR... SURE... BUT CAN I ASK A QUESTION?
OF COURSE. THAT'S WHY YOU'RE HERE.
WHO AM I?

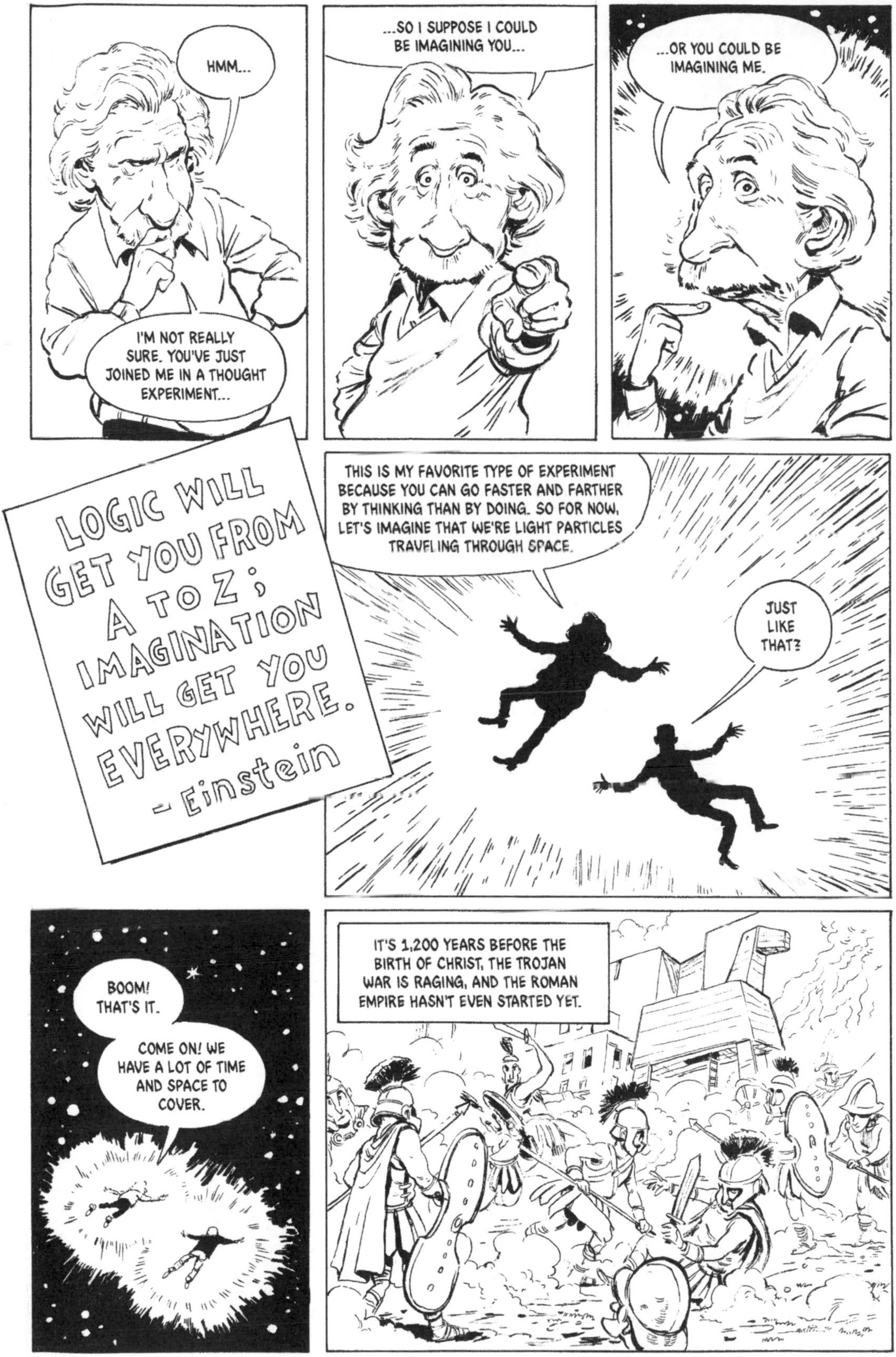
HMM...
I'M NOT REALLY SURE. YOU'VE JUST JOINED ME IN A THOUGHT EXPERIMENT...
...SO I SUPPOSE I COULD BE IMAGINING YOU...
...OR YOU COULD BE IMAGINING ME.
LOGIC WILL GET YOU FROM A TO Z; IMAGINATION WILL GET YOU EVERYWHERE. - Einstein
THIS IS MY FAVORITE TYPE OF EXPERIMENT BECAUSE YOU CAN GO FASTER AND FARTHER BY THINKING THAN BY DOING. SO FOR NOW, LET'S IMAGINE THAT WE'RE LIGHT PARTICLES TRAVELING THROUGH SPACE.
JUST LIKE THAT?
BOOM! THAT'S IT.
COME ON! WE HAVE A LOT OF TIME AND SPACE TO COVER.
IT'S 1,200 YEARS BEFORE THE BIRTH OF CHRIST, THE TROJAN WAR IS RAGING, AND THE ROMAN EMPIRE HASN'T EVEN STARTED YET.

DURING OUR JOURNEY, ALL THE GREAT ANCIENT CIVILIZATIONS WILL COME AND GO. BY THE END OF OUR JOURNEY, IN 3,200 YEARS, THE HUMAN RACE WILL HAVE PROGRESSED—IF THAT IS THE RIGHT WORD...

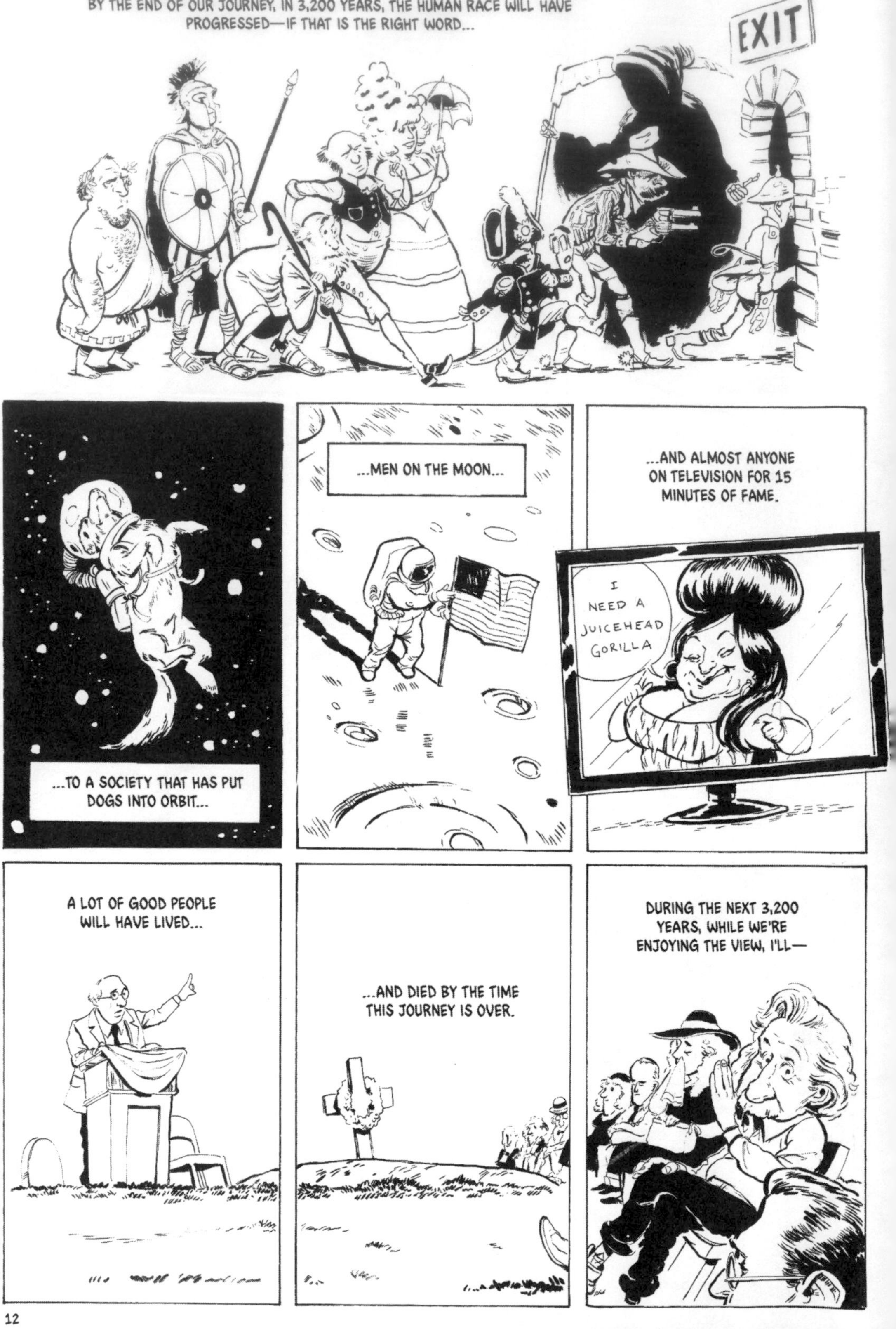
EXIT
...TO A SOCIETY THAT HAS PUT DOGS INTO ORBIT...
...MEN ON THE MOON...
...AND ALMOST ANYONE ON TELEVISION FOR 15 MINUTES OF FAME.
I NEED A JUICEHEAD GORILLA
A LOT OF GOOD PEOPLE WILL HAVE LIVED...
...AND DIED BY THE TIME THIS JOURNEY IS OVER.
DURING THE NEXT 3,200 YEARS, WHILE WE'RE ENJOYING THE VIEW, I'LL—

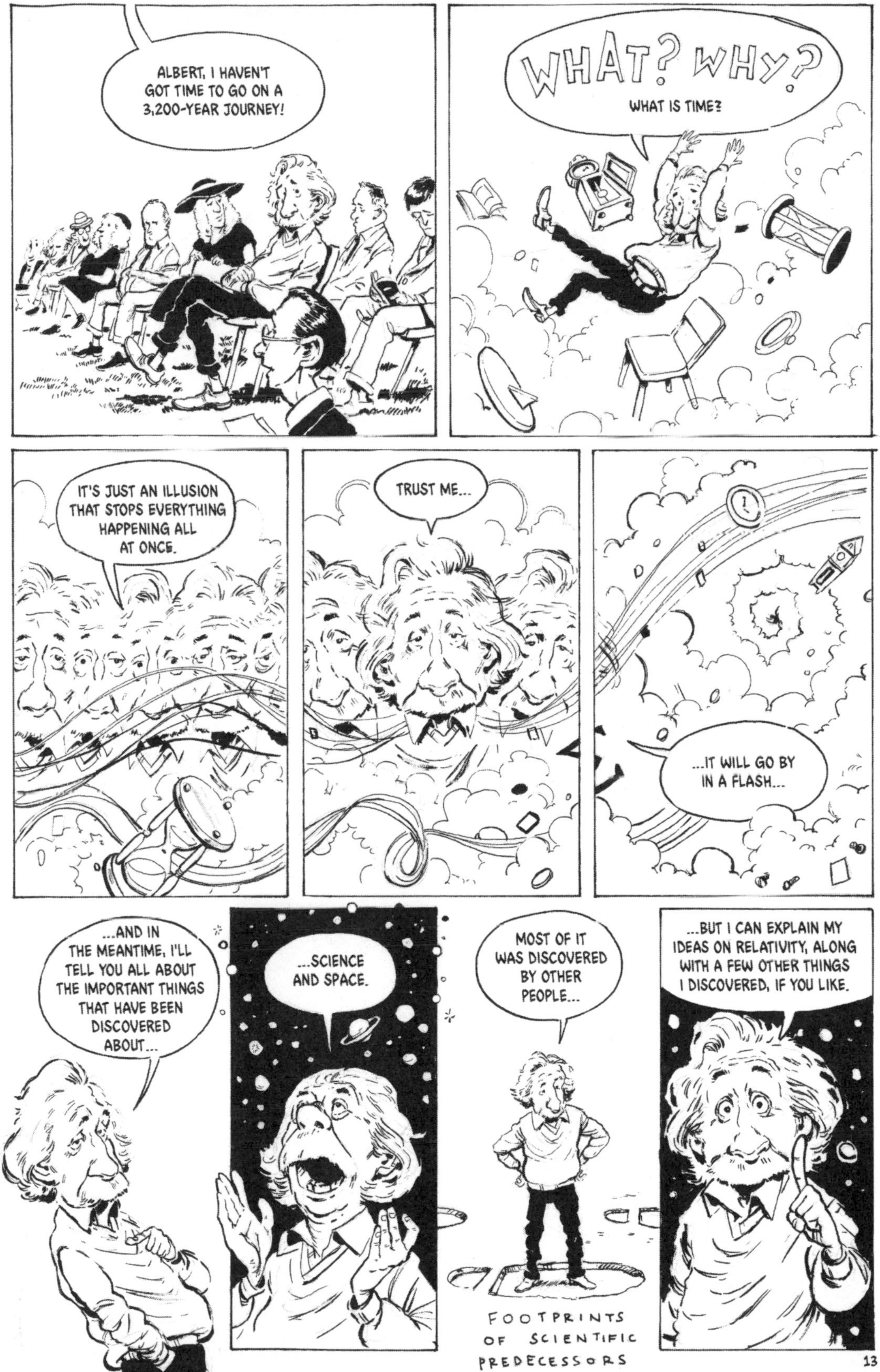
ALBERT, I HAVEN'T GOT TIME TO GO ON A 3,200-YEAR JOURNEY!
WHAT? WHY?
WHAT IS TIME?
IT'S JUST AN ILLUSION THAT STOPS EVERYTHING HAPPENING ALL AT ONCE.
TRUST ME...
...IT WILL GO BY IN A FLASH...
...AND IN THE MEANTIME, I'LL TELL YOU ALL ABOUT THE IMPORTANT THINGS THAT HAVE BEEN DISCOVERED ABOUT...
...SCIENCE AND SPACE.
MOST OF IT WAS DISCOVERED BY OTHER PEOPLE...
FOOTPRINTS OF SCIENTIFIC PREDECESSORS
...BUT I CAN EXPLAIN MY IDEAS ON RELATIVITY, ALONG WITH A FEW OTHER THINGS I DISCOVERED, IF YOU LIKE.

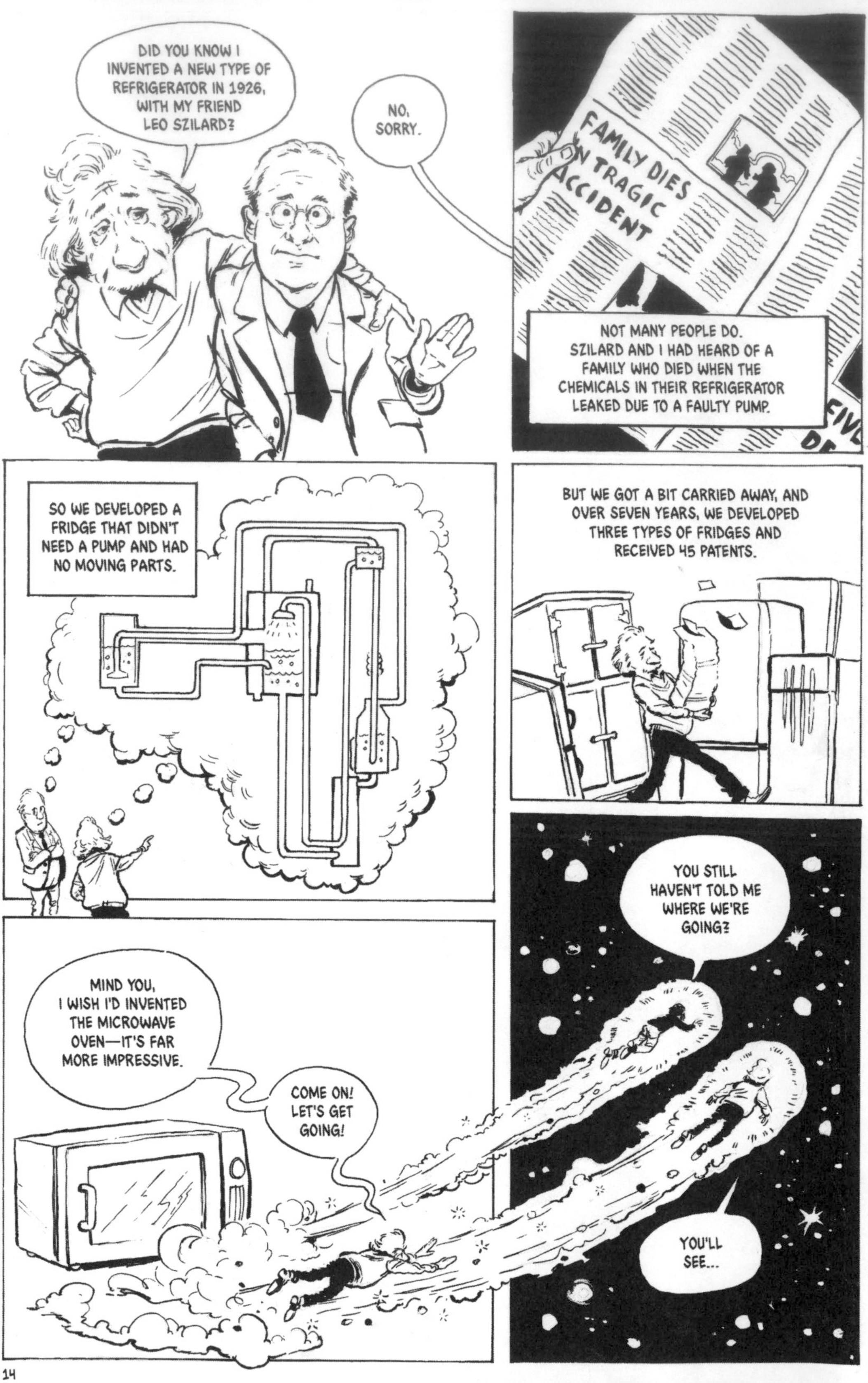
DID YOU KNOW I INVENTED A NEW TYPE OF REFRIGERATOR IN 1926, WITH MY FRIEND LEO SZILARD?
NO, SORRY.
FAMILY DIES
N TRAGIC
ACCIDENT
NOT MANY PEOPLE DO. SZILARD AND I HAD HEARD OF A FAMILY WHO DIED WHEN THE CHEMICALS IN THEIR REFRIGERATOR LEAKED DUE TO A FAULTY PUMP.
SO WE DEVELOPED A FRIDGE THAT DIDN'T NEED A PUMP AND HAD NO MOVING PARTS.
BUT WE GOT A BIT CARRIED AWAY, AND OVER SEVEN YEARS, WE DEVELOPED THREE TYPES OF FRIDGES AND RECEIVED 45 PATENTS.
MIND YOU, I WISH I'D INVENTED THE MICROWAVE OVEN—IT'S FAR MORE IMPRESSIVE.
COME ON! LET'S GET GOING!
YOU STILL HAVEN'T TOLD ME WHERE WE'RE GOING?
YOU'LL SEE...

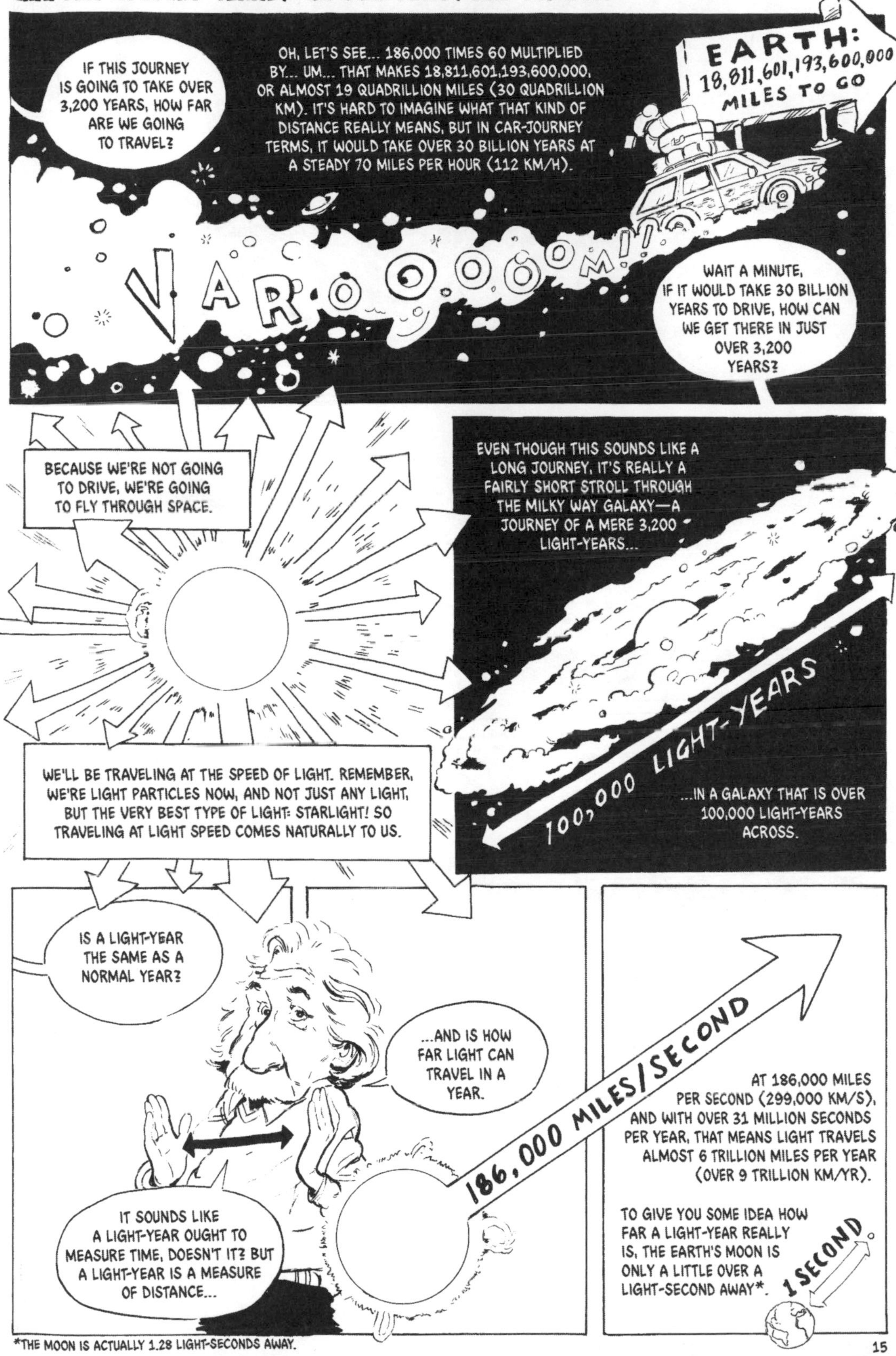
IF THIS JOURNEY IS GOING TO TAKE OVER 3,200 YEARS, HOW FAR ARE WE GOING TO TRAVEL?
OH, LET'S SEE... 186,000 TIMES 60 MULTIPLIED BY... UM... THAT MAKES 18,811,601,193,600,000, OR ALMOST 19 QUADRILLION MILES (30 QUADRILLION KM). IT'S HARD TO IMAGINE WHAT THAT KIND OF DISTANCE REALLY MEANS, BUT IN CAR-JOURNEY TERMS, IT WOULD TAKE OVER 30 BILLION YEARS AT A STEADY 70 MILES PER HOUR (112 KM/H).
EARTH: 18,811,601,193,600,000 MILES TO GO
VAROOOOOOM!!
WAIT A MINUTE, IF IT WOULD TAKE 30 BILLION YEARS TO DRIVE, HOW CAN WE GET THERE IN JUST OVER 3,200 YEARS?
BECAUSE WE'RE NOT GOING TO DRIVE, WE'RE GOING TO FLY THROUGH SPACE.
EVEN THOUGH THIS SOUNDS LIKE A LONG JOURNEY, IT'S REALLY A FAIRLY SHORT STROLL THROUGH THE MILKY WAY GALAXY—A JOURNEY OF A MERE 3,200 LIGHT-YEARS...
100,000 LIGHT-YEARS
...IN A GALAXY THAT IS OVER 100,000 LIGHT-YEARS ACROSS.
WE'LL BE TRAVELING AT THE SPEED OF LIGHT. REMEMBER, WE'RE LIGHT PARTICLES NOW, AND NOT JUST ANY LIGHT, BUT THE VERY BEST TYPE OF LIGHT: STARLIGHT! SO TRAVELING AT LIGHT SPEED COMES NATURALLY TO US.
IS A LIGHT-YEAR THE SAME AS A NORMAL YEAR?
IT SOUNDS LIKE A LIGHT-YEAR OUGHT TO MEASURE TIME, DOESN'T IT? BUT A LIGHT-YEAR IS A MEASURE OF DISTANCE...
...AND IS HOW FAR LIGHT CAN TRAVEL IN A YEAR.
186,000 MILES/SECOND
AT 186,000 MILES PER SECOND (299,000 KM/S), AND WITH OVER 31 MILLION SECONDS PER YEAR, THAT MEANS LIGHT TRAVELS ALMOST 6 TRILLION MILES PER YEAR (OVER 9 TRILLION KM/YR).
TO GIVE YOU SOME IDEA HOW FAR A LIGHT-YEAR REALLY IS, THE EARTH'S MOON IS ONLY A LITTLE OVER A LIGHT-SECOND AWAY*.
1 SECOND

*THE MOON IS ACTUALLY 1.28 LIGHT-SECONDS AWAY.

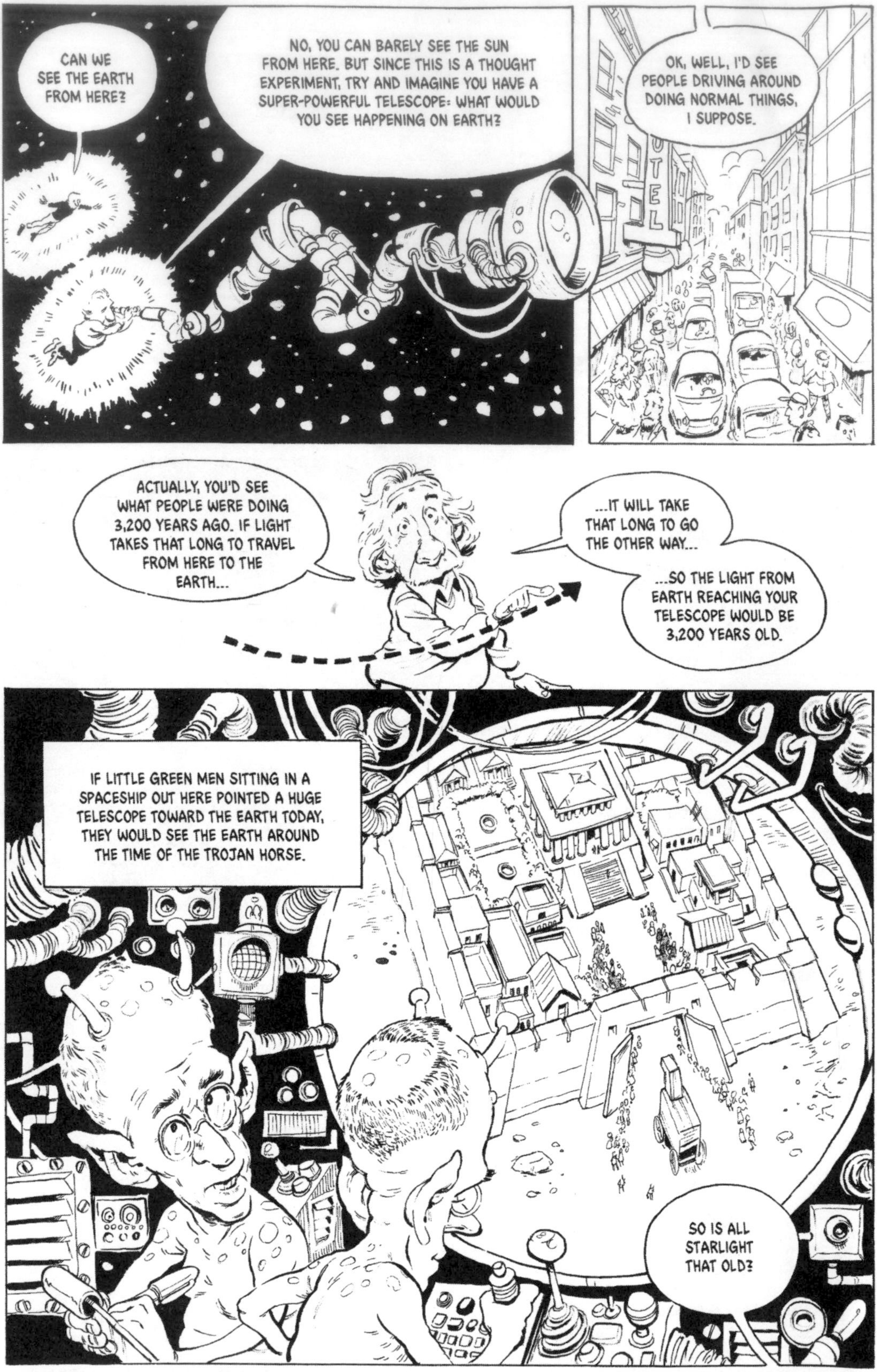
CAN WE SEE THE EARTH FROM HERE?
NO, YOU CAN BARELY SEE THE SUN FROM HERE. BUT SINCE THIS IS A THOUGHT EXPERIMENT, TRY AND IMAGINE YOU HAVE A SUPER-POWERFUL TELESCOPE: WHAT WOULD YOU SEE HAPPENING ON EARTH?
OK, WELL, I'D SEE PEOPLE DRIVING AROUND DOING NORMAL THINGS, I SUPPOSE.
ACTUALLY, YOU'D SEE WHAT PEOPLE WERE DOING 3,200 YEARS AGO. IF LIGHT TAKES THAT LONG TO TRAVEL FROM HERE TO THE EARTH...
...IT WILL TAKE THAT LONG TO GO THE OTHER WAY...
...SO THE LIGHT FROM EARTH REACHING YOUR TELESCOPE WOULD BE 3,200 YEARS OLD.
IF LITTLE GREEN MEN SITTING IN A SPACESHIP OUT HERE POINTED A HUGE TELESCOPE TOWARD THE EARTH TODAY, THEY WOULD SEE THE EARTH AROUND THE TIME OF THE TROJAN HORSE.
SO IS ALL STARLIGHT THAT OLD?

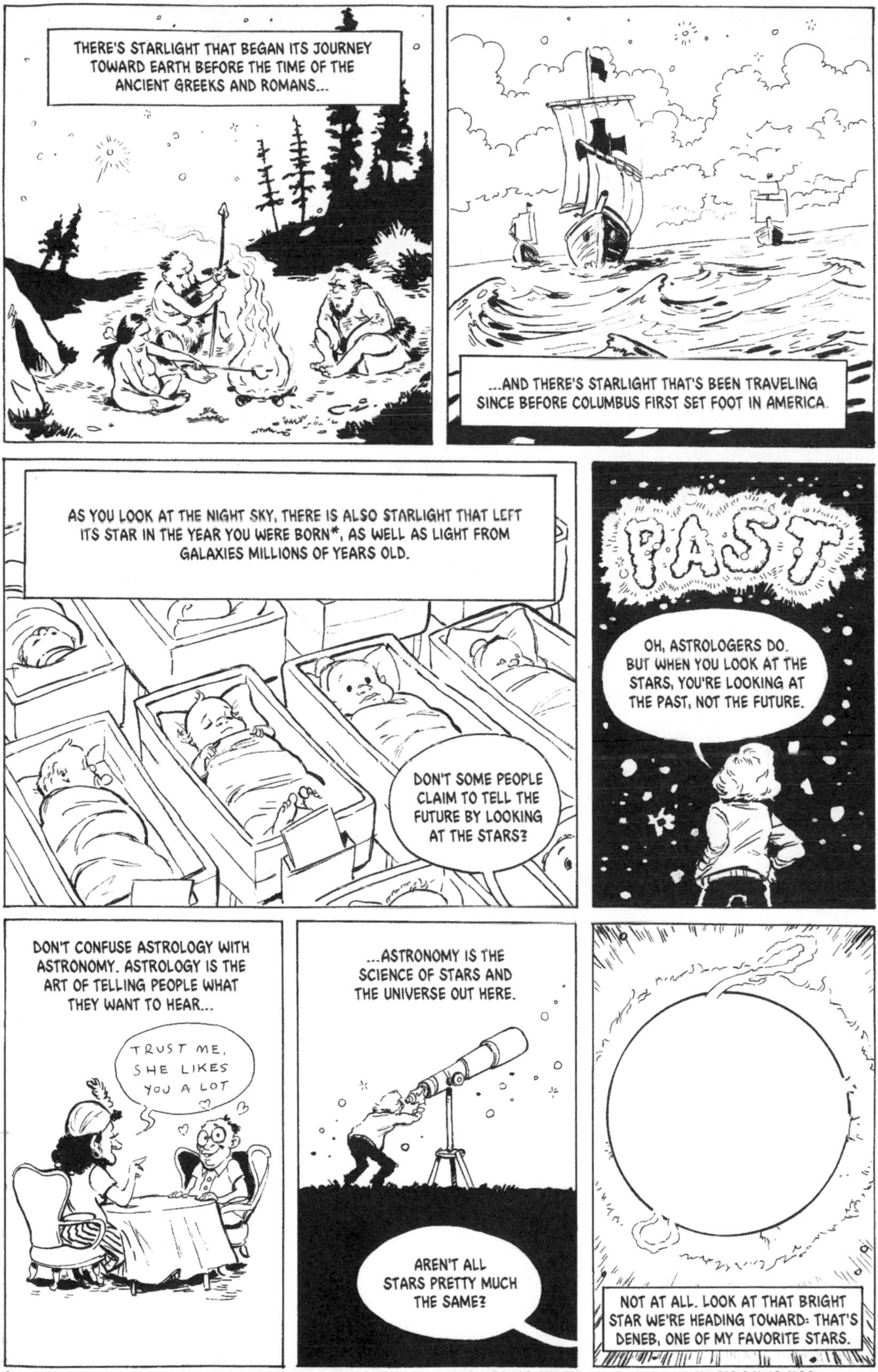
THERE'S STARLIGHT THAT BEGAN ITS JOURNEY TOWARD EARTH BEFORE THE TIME OF THE ANCIENT GREEKS AND ROMANS...
...AND THERE'S STARLIGHT THAT'S BEEN TRAVELING SINCE BEFORE COLUMBUS FIRST SET FOOT IN AMERICA.
AS YOU LOOK AT THE NIGHT SKY, THERE IS ALSO STARLIGHT THAT LEFT ITS STAR IN THE YEAR YOU WERE BORN*, AS WELL AS LIGHT FROM GALAXIES MILLIONS OF YEARS OLD.
DON'T SOME PEOPLE CLAIM TO TELL THE FUTURE BY LOOKING AT THE STARS?
PAST
OH, ASTROLOGERS DO. BUT WHEN YOU LOOK AT THE STARS, YOU'RE LOOKING AT THE PAST, NOT THE FUTURE.
DON'T CONFUSE ASTROLOGY WITH ASTRONOMY. ASTROLOGY IS THE ART OF TELLING PEOPLE WHAT THEY WANT TO HEAR...
TRUST ME, SHE LIKES YOU A LOT
...ASTRONOMY IS THE SCIENCE OF STARS AND THE UNIVERSE OUT HERE.
AREN'T ALL STARS PRETTY MUCH THE SAME?
NOT AT ALL. LOOK AT THAT BRIGHT STAR WE'RE HEADING TOWARD: THAT'S DENEB, ONE OF MY FAVORITE STARS.

*AS LONG AS YOU ARE AT LEAST 7 YEARS OLD. THERE ARE ALSO TWO STARS FOR 4-YEAR-OLDS AND ONE FOR 5-YEAR-OLDS, BUT NONE FOR 6-YEAR-OLDS.

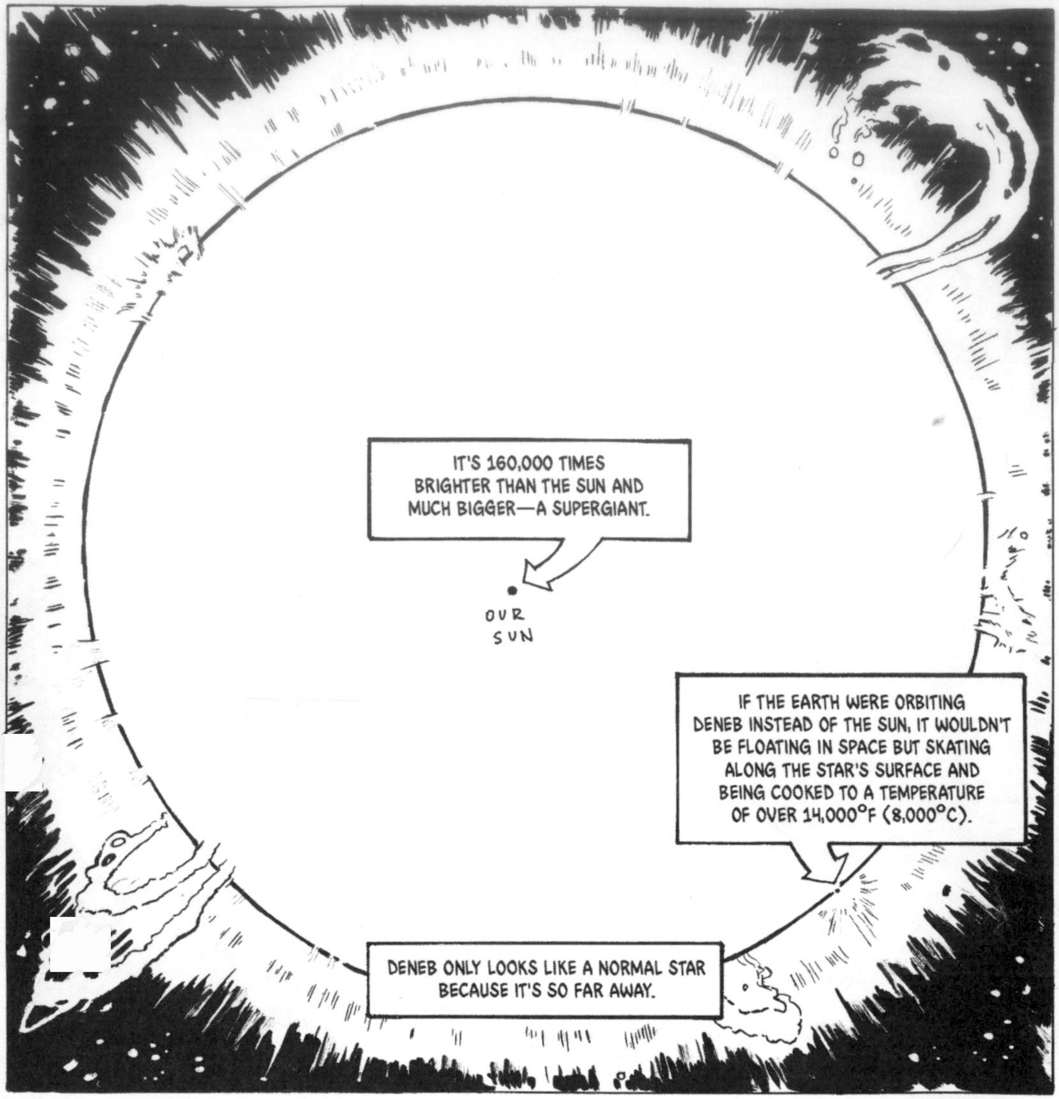
IT'S 160,000 TIMES BRIGHTER THAN THE SUN AND MUCH BIGGER—A SUPERGIANT.
OUR SUN
IF THE EARTH WERE ORBITING DENEB INSTEAD OF THE SUN, IT WOULDN'T BE FLOATING IN SPACE BUT SKATING ALONG THE STAR'S SURFACE AND BEING COOKED TO A TEMPERATURE OF OVER 14,000°F (8,000°C).
DENEB ONLY LOOKS LIKE A NORMAL STAR BECAUSE IT'S SO FAR AWAY.

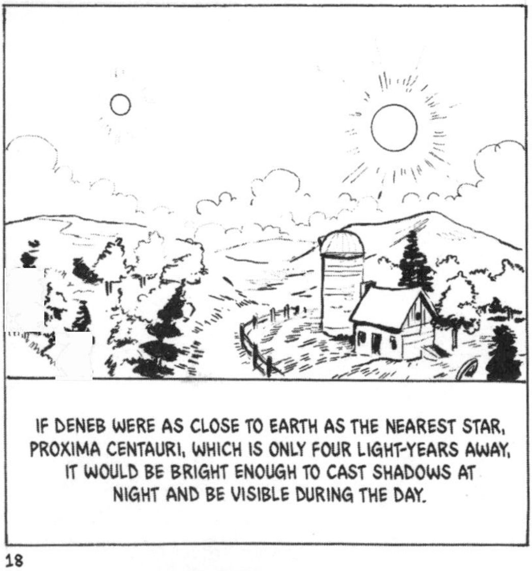
IF DENEB WERE AS CLOSE TO EARTH AS THE NEAREST STAR, PROXIMA CENTAURI, WHICH IS ONLY FOUR LIGHT-YEARS AWAY, IT WOULD BE BRIGHT ENOUGH TO CAST SHADOWS AT NIGHT AND BE VISIBLE DURING THE DAY.

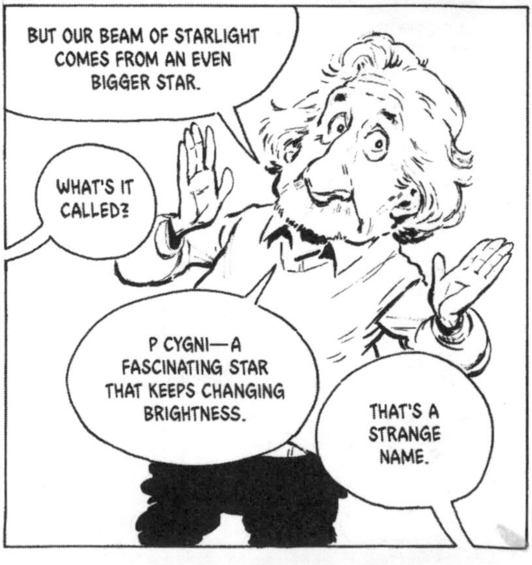
BUT OUR BEAM OF STARLIGHT COMES FROM AN EVEN BIGGER STAR.
WHAT'S IT CALLED?
P CYGNI—A FASCINATING STAR THAT KEEPS CHANGING BRIGHTNESS.
THAT'S A STRANGE NAME.

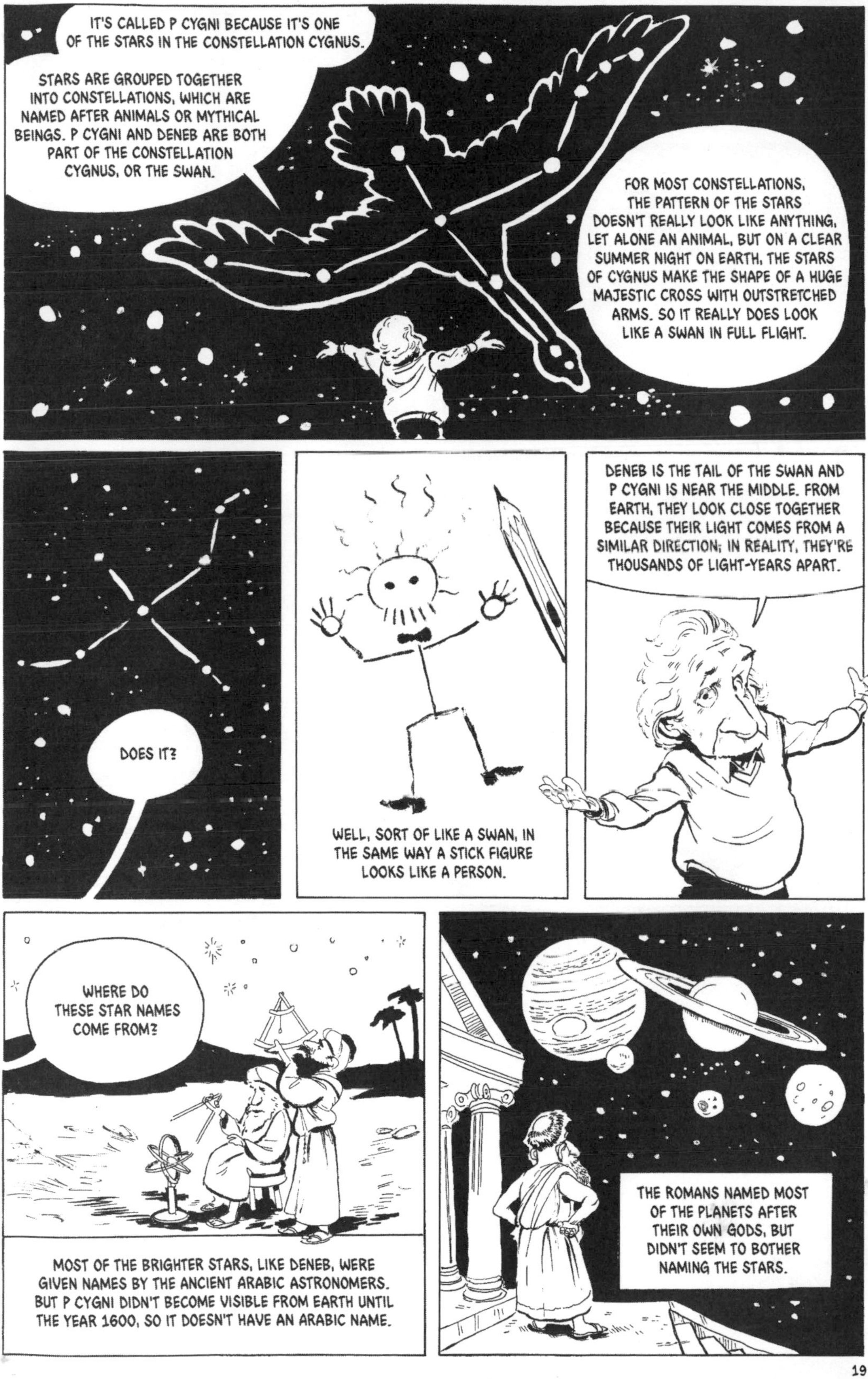
IT'S CALLED P CYGNI BECAUSE IT'S ONE OF THE STARS IN THE CONSTELLATION CYGNUS.
STARS ARE GROUPED TOGETHER INTO CONSTELLATIONS, WHICH ARE NAMED AFTER ANIMALS OR MYTHICAL BEINGS. P CYGNI AND DENEB ARE BOTH PART OF THE CONSTELLATION CYGNUS, OR THE SWAN.
FOR MOST CONSTELLATIONS, THE PATTERN OF THE STARS DOESN'T REALLY LOOK LIKE ANYTHING, LET ALONE AN ANIMAL, BUT ON A CLEAR SUMMER NIGHT ON EARTH, THE STARS OF CYGNUS MAKE THE SHAPE OF A HUGE MAJESTIC CROSS WITH OUTSTRETCHED ARMS. SO IT REALLY DOES LOOK LIKE A SWAN IN FULL FLIGHT.
DOES IT?
WELL, SORT OF LIKE A SWAN, IN THE SAME WAY A STICK FIGURE LOOKS LIKE A PERSON.
DENEB IS THE TAIL OF THE SWAN AND P CYGNI IS NEAR THE MIDDLE. FROM EARTH, THEY LOOK CLOSE TOGETHER BECAUSE THEIR LIGHT COMES FROM A SIMILAR DIRECTION; IN REALITY, THEY'RE THOUSANDS OF LIGHT-YEARS APART.
WHERE DO THESE STAR NAMES COME FROM?
MOST OF THE BRIGHTER STARS, LIKE DENEB, WERE GIVEN NAMES BY THE ANCIENT ARABIC ASTRONOMERS. BUT P CYGNI DIDN'T BECOME VISIBLE FROM EARTH UNTIL THE YEAR 1600, SO IT DOESN'T HAVE AN ARABIC NAME.
THE ROMANS NAMED MOST OF THE PLANETS AFTER THEIR OWN GODS, BUT DIDN'T SEEM TO BOTHER NAMING THE STARS.

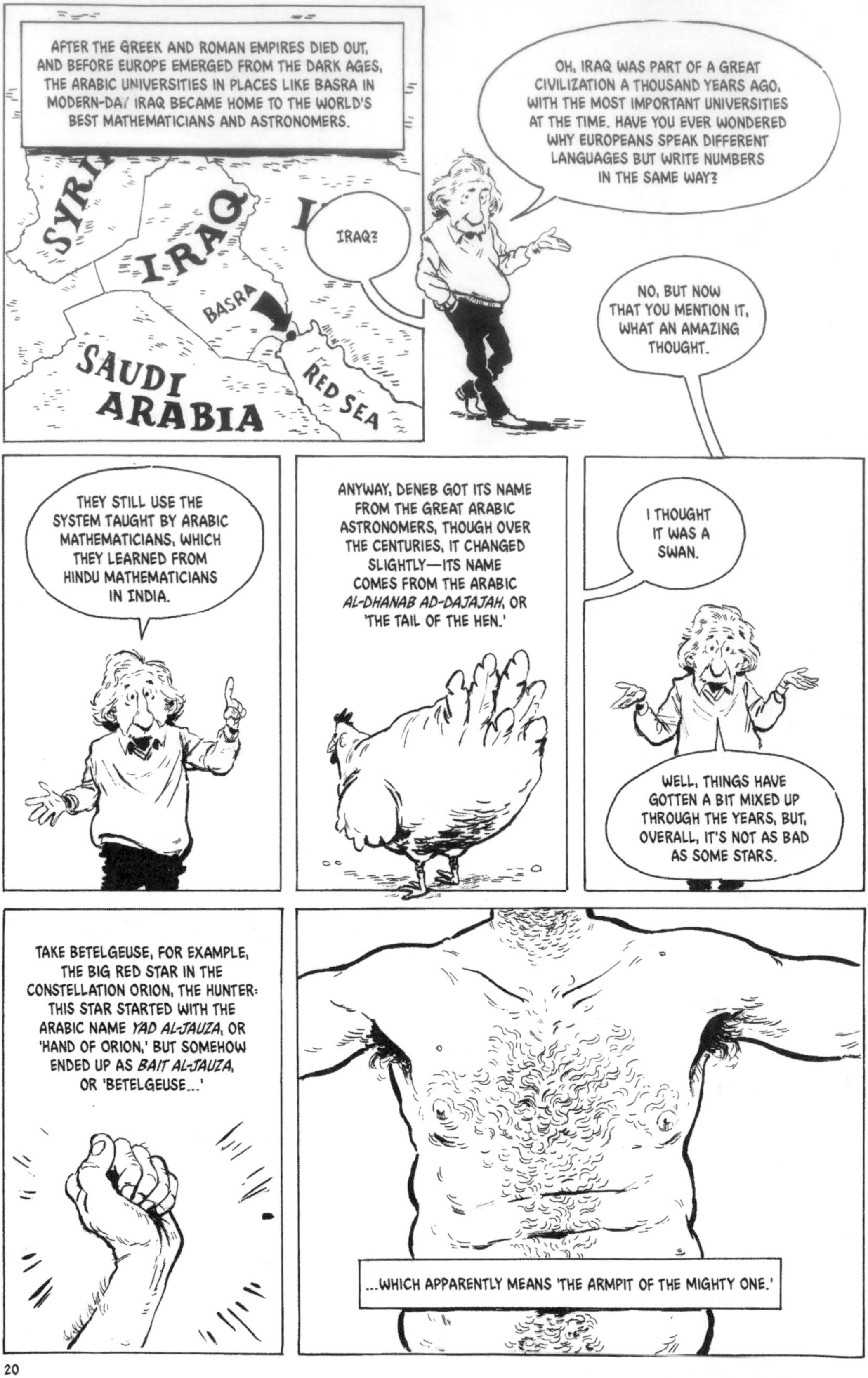
AFTER THE GREEK AND ROMAN EMPIRES DIED OUT, AND BEFORE EUROPE EMERGED FROM THE DARK AGES, THE ARABIC UNIVERSITIES IN PLACES LIKE BASRA IN MODERN-DAY IRAQ BECAME HOME TO THE WORLD'S BEST MATHEMATICIANS AND ASTRONOMERS.
SYRIA
IRAQ
BASRA
SAUDI ARABIA
RED SEA
IRAQ?
OH, IRAQ WAS PART OF A GREAT CIVILIZATION A THOUSAND YEARS AGO, WITH THE MOST IMPORTANT UNIVERSITIES AT THE TIME. HAVE YOU EVER WONDERED WHY EUROPEANS SPEAK DIFFERENT LANGUAGES BUT WRITE NUMBERS IN THE SAME WAY?
NO, BUT NOW THAT YOU MENTION IT, WHAT AN AMAZING THOUGHT.
THEY STILL USE THE SYSTEM TAUGHT BY ARABIC MATHEMATICIANS, WHICH THEY LEARNED FROM HINDU MATHEMATICIANS IN INDIA.
ANYWAY, DENEB GOT ITS NAME FROM THE GREAT ARABIC ASTRONOMERS, THOUGH OVER THE CENTURIES, IT CHANGED SLIGHTLY—ITS NAME COMES FROM THE ARABIC *AL-DHANAB AD-DAJAJAH*, OR 'THE TAIL OF THE HEN.'
I THOUGHT IT WAS A SWAN.
WELL, THINGS HAVE GOTTEN A BIT MIXED UP THROUGH THE YEARS, BUT, OVERALL, IT'S NOT AS BAD AS SOME STARS.
TAKE BETELGEUSE, FOR EXAMPLE, THE BIG RED STAR IN THE CONSTELLATION ORION, THE HUNTER: THIS STAR STARTED WITH THE ARABIC NAME *YAD AL-JAUZA*, OR 'HAND OF ORION,' BUT SOMEHOW ENDED UP AS *BAIT AL-JAUZA*, OR 'BETELGEUSE...'
...WHICH APPARENTLY MEANS 'THE ARMPIT OF THE MIGHTY ONE.'

CHAPTER 3: STARS, ATOMS, AND LEGO®

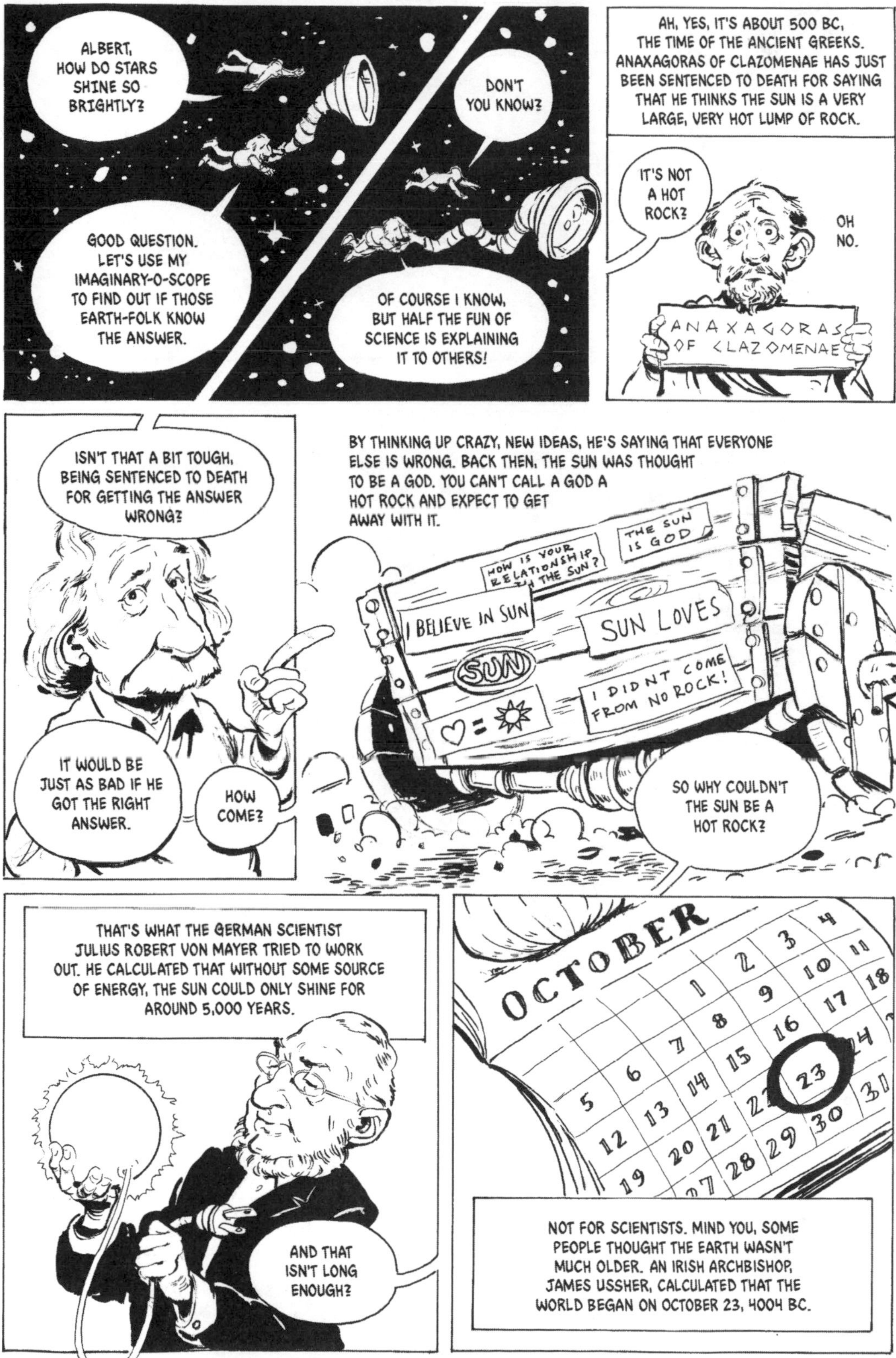
ALBERT, HOW DO STARS SHINE SO BRIGHTLY?
GOOD QUESTION. LET'S USE MY IMAGINARY-O-SCOPE TO FIND OUT IF THOSE EARTH-FOLK KNOW THE ANSWER.
DON'T YOU KNOW?
OF COURSE I KNOW, BUT HALF THE FUN OF SCIENCE IS EXPLAINING IT TO OTHERS!
AH, YES, IT'S ABOUT 500 BC, THE TIME OF THE ANCIENT GREEKS. ANAXAGORAS OF CLAZOMENAE HAS JUST BEEN SENTENCED TO DEATH FOR SAYING THAT HE THINKS THE SUN IS A VERY LARGE, VERY HOT LUMP OF ROCK.
IT'S NOT A HOT ROCK?
OH NO.
ANAXAGORAS OF CLAZOMENAE
ISN'T THAT A BIT TOUGH, BEING SENTENCED TO DEATH FOR GETTING THE ANSWER WRONG?
IT WOULD BE JUST AS BAD IF HE GOT THE RIGHT ANSWER.
HOW COME?
BY THINKING UP CRAZY, NEW IDEAS, HE'S SAYING THAT EVERYONE ELSE IS WRONG. BACK THEN, THE SUN WAS THOUGHT TO BE A GOD. YOU CAN'T CALL A GOD A HOT ROCK AND EXPECT TO GET AWAY WITH IT.
HOW IS YOUR RELATIONSHIP
THE SUN IS GOD
I BELIEVE IN SUN
SUN LOVES
SUN
I DIDNT COME FROM NO ROCK!
SO WHY COULDN'T THE SUN BE A HOT ROCK?
THAT'S WHAT THE GERMAN SCIENTIST JULIUS ROBERT VON MAYER TRIED TO WORK OUT. HE CALCULATED THAT WITHOUT SOME SOURCE OF ENERGY, THE SUN COULD ONLY SHINE FOR AROUND 5,000 YEARS.
AND THAT ISN'T LONG ENOUGH?
OCTOBER
NOT FOR SCIENTISTS. MIND YOU, SOME PEOPLE THOUGHT THE EARTH WASN'T MUCH OLDER. AN IRISH ARCHBISHOP, JAMES USSHER, CALCULATED THAT THE WORLD BEGAN ON OCTOBER 23, 4004 BC.

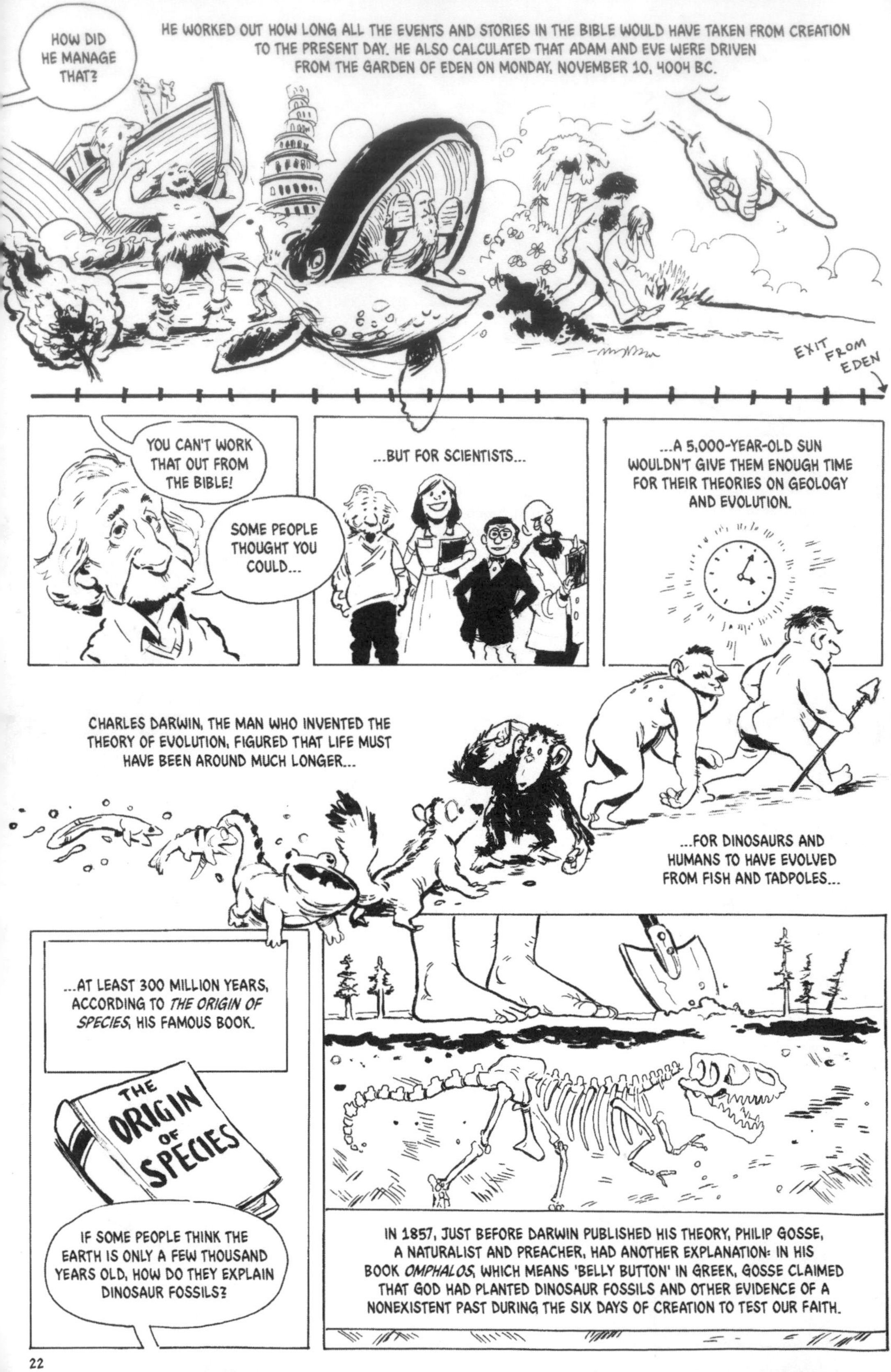
HOW DID HE MANAGE THAT?
HE WORKED OUT HOW LONG ALL THE EVENTS AND STORIES IN THE BIBLE WOULD HAVE TAKEN FROM CREATION TO THE PRESENT DAY. HE ALSO CALCULATED THAT ADAM AND EVE WERE DRIVEN FROM THE GARDEN OF EDEN ON MONDAY, NOVEMBER 10, 4004 BC.
EXIT FROM EDEN
YOU CAN'T WORK THAT OUT FROM THE BIBLE!
SOME PEOPLE THOUGHT YOU COULD...
...BUT FOR SCIENTISTS...
...A 5,000-YEAR-OLD SUN WOULDN'T GIVE THEM ENOUGH TIME FOR THEIR THEORIES ON GEOLOGY AND EVOLUTION.
CHARLES DARWIN, THE MAN WHO INVENTED THE THEORY OF EVOLUTION, FIGURED THAT LIFE MUST HAVE BEEN AROUND MUCH LONGER...
...FOR DINOSAURS AND HUMANS TO HAVE EVOLVED FROM FISH AND TADPOLES...
...AT LEAST 300 MILLION YEARS, ACCORDING TO *THE ORIGIN OF SPECIES*, HIS FAMOUS BOOK.
THE ORIGIN OF SPECIES
IF SOME PEOPLE THINK THE EARTH IS ONLY A FEW THOUSAND YEARS OLD, HOW DO THEY EXPLAIN DINOSAUR FOSSILS?
IN 1857, JUST BEFORE DARWIN PUBLISHED HIS THEORY, PHILIP GOSSE, A NATURALIST AND PREACHER, HAD ANOTHER EXPLANATION: IN HIS BOOK *OMPHALOS*, WHICH MEANS 'BELLY BUTTON' IN GREEK, GOSSE CLAIMED THAT GOD HAD PLANTED DINOSAUR FOSSILS AND OTHER EVIDENCE OF A NONEXISTENT PAST DURING THE SIX DAYS OF CREATION TO TEST OUR FAITH.

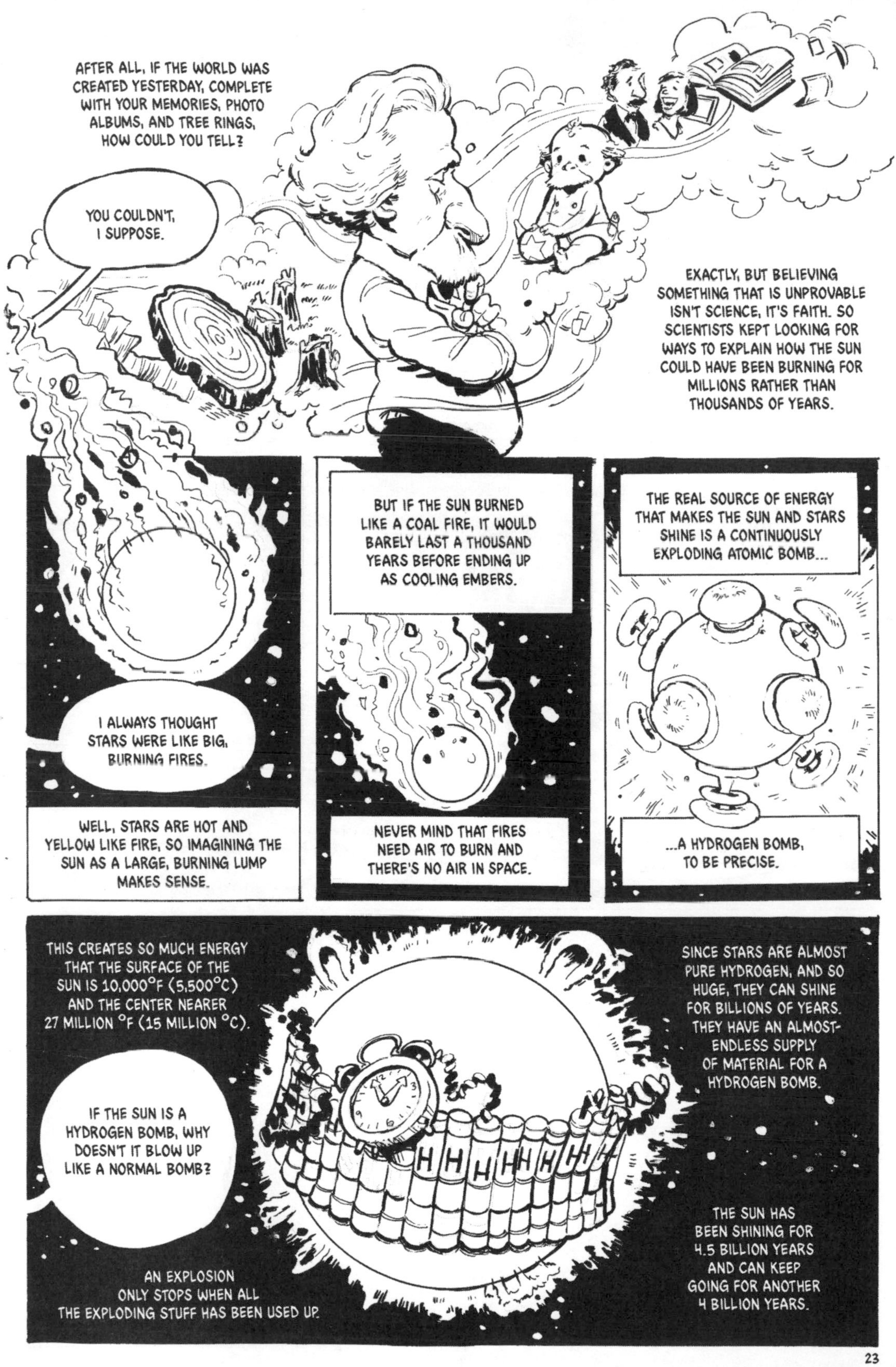
AFTER ALL, IF THE WORLD WAS CREATED YESTERDAY, COMPLETE WITH YOUR MEMORIES, PHOTO ALBUMS, AND TREE RINGS, HOW COULD YOU TELL?
YOU COULDN'T, I SUPPOSE.
EXACTLY, BUT BELIEVING SOMETHING THAT IS UNPROVABLE ISN'T SCIENCE, IT'S FAITH. SO SCIENTISTS KEPT LOOKING FOR WAYS TO EXPLAIN HOW THE SUN COULD HAVE BEEN BURNING FOR MILLIONS RATHER THAN THOUSANDS OF YEARS.
I ALWAYS THOUGHT STARS WERE LIKE BIG, BURNING FIRES.
WELL, STARS ARE HOT AND YELLOW LIKE FIRE, SO IMAGINING THE SUN AS A LARGE, BURNING LUMP MAKES SENSE.
BUT IF THE SUN BURNED LIKE A COAL FIRE, IT WOULD BARELY LAST A THOUSAND YEARS BEFORE ENDING UP AS COOLING EMBERS.
NEVER MIND THAT FIRES NEED AIR TO BURN AND THERE'S NO AIR IN SPACE.
THE REAL SOURCE OF ENERGY THAT MAKES THE SUN AND STARS SHINE IS A CONTINUOUSLY EXPLODING ATOMIC BOMB...
...A HYDROGEN BOMB, TO BE PRECISE.
THIS CREATES SO MUCH ENERGY THAT THE SURFACE OF THE SUN IS 10,000°F (5,500°C) AND THE CENTER NEARER 27 MILLION °F (15 MILLION °C).
SINCE STARS ARE ALMOST PURE HYDROGEN, AND SO HUGE, THEY CAN SHINE FOR BILLIONS OF YEARS. THEY HAVE AN ALMOST-ENDLESS SUPPLY OF MATERIAL FOR A HYDROGEN BOMB.
IF THE SUN IS A HYDROGEN BOMB, WHY DOESN'T IT BLOW UP LIKE A NORMAL BOMB?
AN EXPLOSION ONLY STOPS WHEN ALL THE EXPLODING STUFF HAS BEEN USED UP.
THE SUN HAS BEEN SHINING FOR 4.5 BILLION YEARS AND CAN KEEP GOING FOR ANOTHER 4 BILLION YEARS.

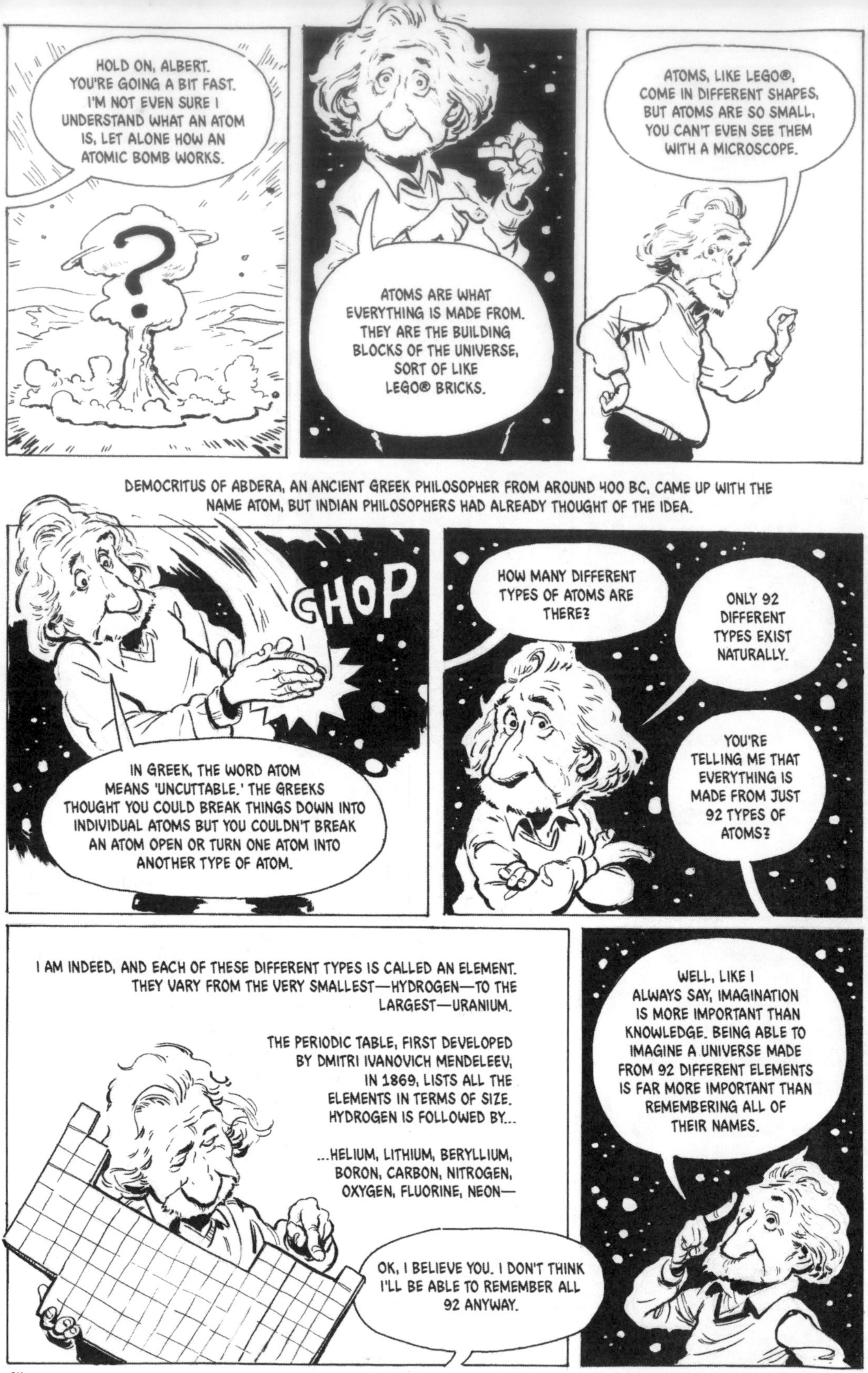
HOLD ON, ALBERT. YOU'RE GOING A BIT FAST. I'M NOT EVEN SURE I UNDERSTAND WHAT AN ATOM IS, LET ALONE HOW AN ATOMIC BOMB WORKS.
?
ATOMS ARE WHAT EVERYTHING IS MADE FROM. THEY ARE THE BUILDING BLOCKS OF THE UNIVERSE, SORT OF LIKE LEGO® BRICKS.
ATOMS, LIKE LEGO®, COME IN DIFFERENT SHAPES, BUT ATOMS ARE SO SMALL, YOU CAN'T EVEN SEE THEM WITH A MICROSCOPE.
DEMOCRITUS OF ABDERA, AN ANCIENT GREEK PHILOSOPHER FROM AROUND 400 BC, CAME UP WITH THE NAME ATOM, BUT INDIAN PHILOSOPHERS HAD ALREADY THOUGHT OF THE IDEA.
CHOP
IN GREEK, THE WORD ATOM MEANS 'UNCUTTABLE.' THE GREEKS THOUGHT YOU COULD BREAK THINGS DOWN INTO INDIVIDUAL ATOMS BUT YOU COULDN'T BREAK AN ATOM OPEN OR TURN ONE ATOM INTO ANOTHER TYPE OF ATOM.
HOW MANY DIFFERENT TYPES OF ATOMS ARE THERE?
ONLY 92 DIFFERENT TYPES EXIST NATURALLY.
YOU'RE TELLING ME THAT EVERYTHING IS MADE FROM JUST 92 TYPES OF ATOMS?
I AM INDEED, AND EACH OF THESE DIFFERENT TYPES IS CALLED AN ELEMENT. THEY VARY FROM THE VERY SMALLEST—HYDROGEN—TO THE LARGEST—URANIUM.
THE PERIODIC TABLE, FIRST DEVELOPED BY DMITRI IVANOVICH MENDELEEV, IN 1869, LISTS ALL THE ELEMENTS IN TERMS OF SIZE. HYDROGEN IS FOLLOWED BY...
...HELIUM, LITHIUM, BERYLLIUM, BORON, CARBON, NITROGEN, OXYGEN, FLUORINE, NEON—
OK, I BELIEVE YOU. I DON'T THINK I'LL BE ABLE TO REMEMBER ALL 92 ANYWAY.
WELL, LIKE I ALWAYS SAY, IMAGINATION IS MORE IMPORTANT THAN KNOWLEDGE. BEING ABLE TO IMAGINE A UNIVERSE MADE FROM 92 DIFFERENT ELEMENTS IS FAR MORE IMPORTANT THAN REMEMBERING ALL OF THEIR NAMES.

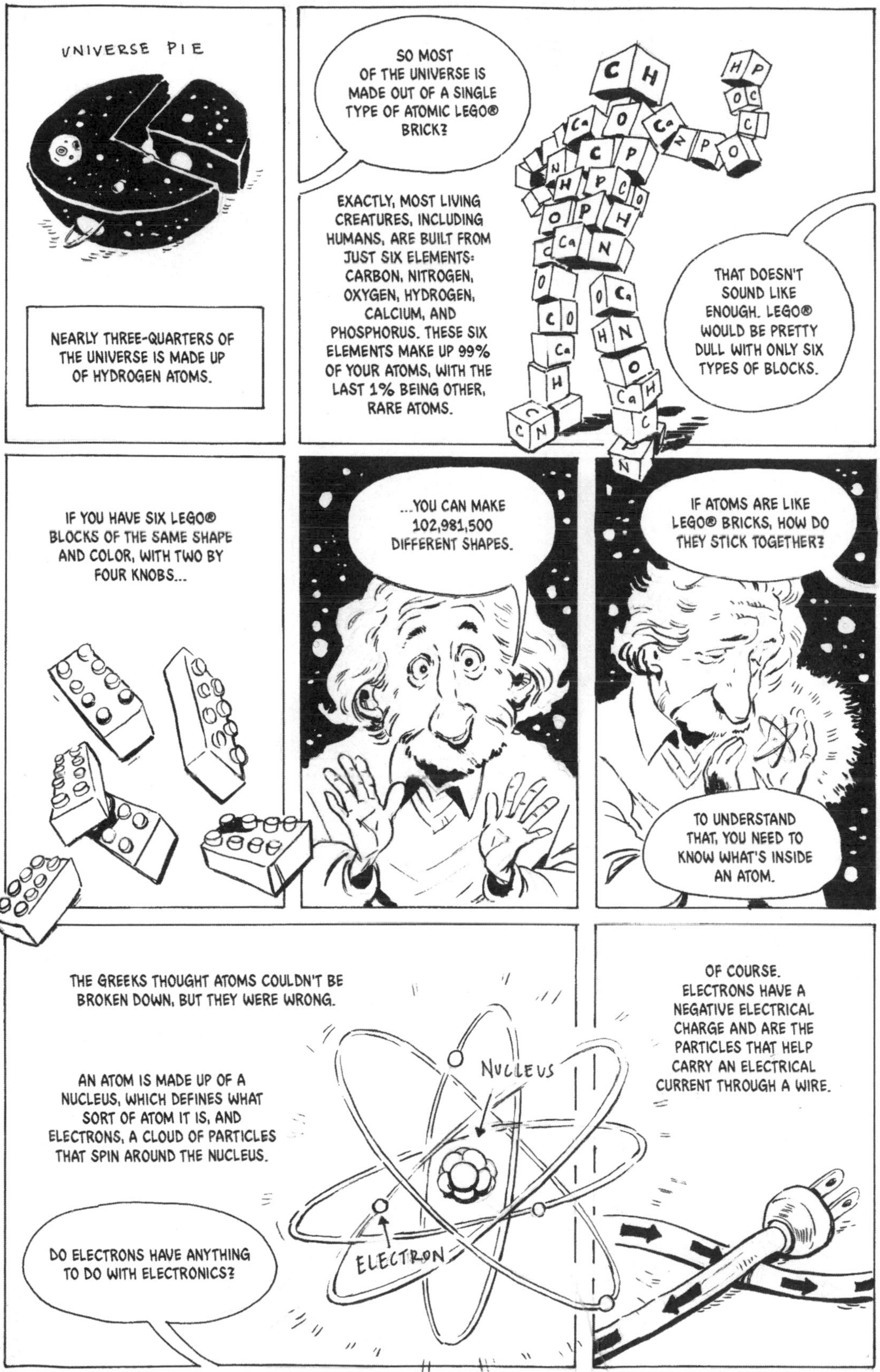

UNIVERSE PIE
NEARLY THREE-QUARTERS OF THE UNIVERSE IS MADE UP OF HYDROGEN ATOMS.
SO MOST OF THE UNIVERSE IS MADE OUT OF A SINGLE TYPE OF ATOMIC LEGO® BRICK?
EXACTLY, MOST LIVING CREATURES, INCLUDING HUMANS, ARE BUILT FROM JUST SIX ELEMENTS: CARBON, NITROGEN, OXYGEN, HYDROGEN, CALCIUM, AND PHOSPHORUS. THESE SIX ELEMENTS MAKE UP 99% OF YOUR ATOMS, WITH THE LAST 1% BEING OTHER, RARE ATOMS.
THAT DOESN'T SOUND LIKE ENOUGH. LEGO® WOULD BE PRETTY DULL WITH ONLY SIX TYPES OF BLOCKS.
IF YOU HAVE SIX LEGO® BLOCKS OF THE SAME SHAPE AND COLOR, WITH TWO BY FOUR KNOBS...
...YOU CAN MAKE 102,981,500 DIFFERENT SHAPES.
IF ATOMS ARE LIKE LEGO® BRICKS, HOW DO THEY STICK TOGETHER?
TO UNDERSTAND THAT, YOU NEED TO KNOW WHAT'S INSIDE AN ATOM.
THE GREEKS THOUGHT ATOMS COULDN'T BE BROKEN DOWN, BUT THEY WERE WRONG.
AN ATOM IS MADE UP OF A NUCLEUS, WHICH DEFINES WHAT SORT OF ATOM IT IS, AND ELECTRONS, A CLOUD OF PARTICLES THAT SPIN AROUND THE NUCLEUS.
NUCLEUS
ELECTRON
DO ELECTRONS HAVE ANYTHING TO DO WITH ELECTRONICS?
OF COURSE. ELECTRONS HAVE A NEGATIVE ELECTRICAL CHARGE AND ARE THE PARTICLES THAT HELP CARRY AN ELECTRICAL CURRENT THROUGH A WIRE.

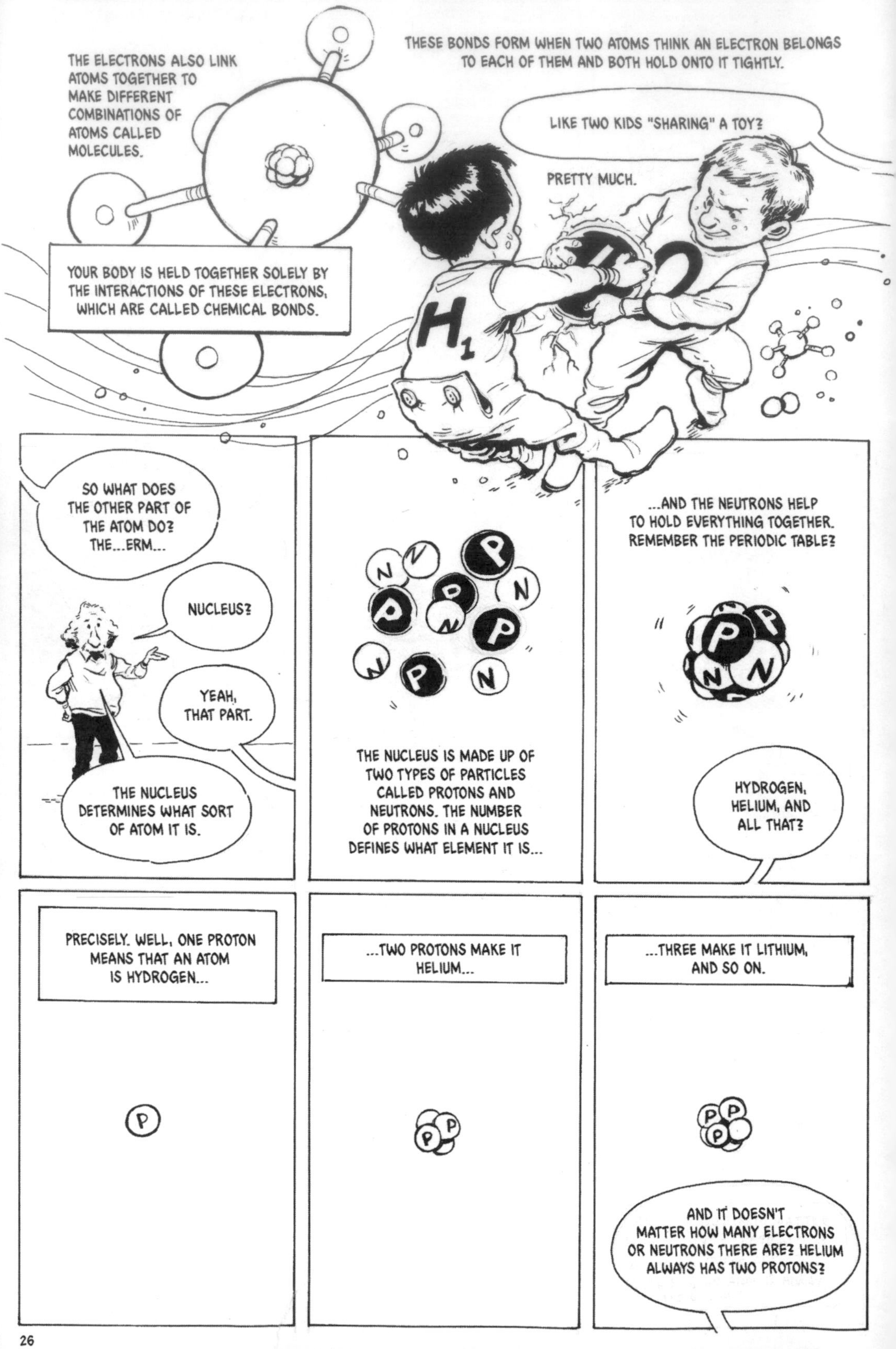
THE ELECTRONS ALSO LINK ATOMS TOGETHER TO MAKE DIFFERENT COMBINATIONS OF ATOMS CALLED MOLECULES.
YOUR BODY IS HELD TOGETHER SOLELY BY THE INTERACTIONS OF THESE ELECTRONS, WHICH ARE CALLED CHEMICAL BONDS.
THESE BONDS FORM WHEN TWO ATOMS THINK AN ELECTRON BELONGS TO EACH OF THEM AND BOTH HOLD ONTO IT TIGHTLY.
LIKE TWO KIDS "SHARING" A TOY?
PRETTY MUCH.
SO WHAT DOES THE OTHER PART OF THE ATOM DO? THE...ERM...
NUCLEUS?
YEAH, THAT PART.
THE NUCLEUS DETERMINES WHAT SORT OF ATOM IT IS.
THE NUCLEUS IS MADE UP OF TWO TYPES OF PARTICLES CALLED PROTONS AND NEUTRONS. THE NUMBER OF PROTONS IN A NUCLEUS DEFINES WHAT ELEMENT IT IS...
...AND THE NEUTRONS HELP TO HOLD EVERYTHING TOGETHER. REMEMBER THE PERIODIC TABLE?
HYDROGEN, HELIUM, AND ALL THAT?
PRECISELY. WELL, ONE PROTON MEANS THAT AN ATOM IS HYDROGEN...
...TWO PROTONS MAKE IT HELIUM...
...THREE MAKE IT LITHIUM, AND SO ON.
AND IT DOESN'T MATTER HOW MANY ELECTRONS OR NEUTRONS THERE ARE? HELIUM ALWAYS HAS TWO PROTONS?

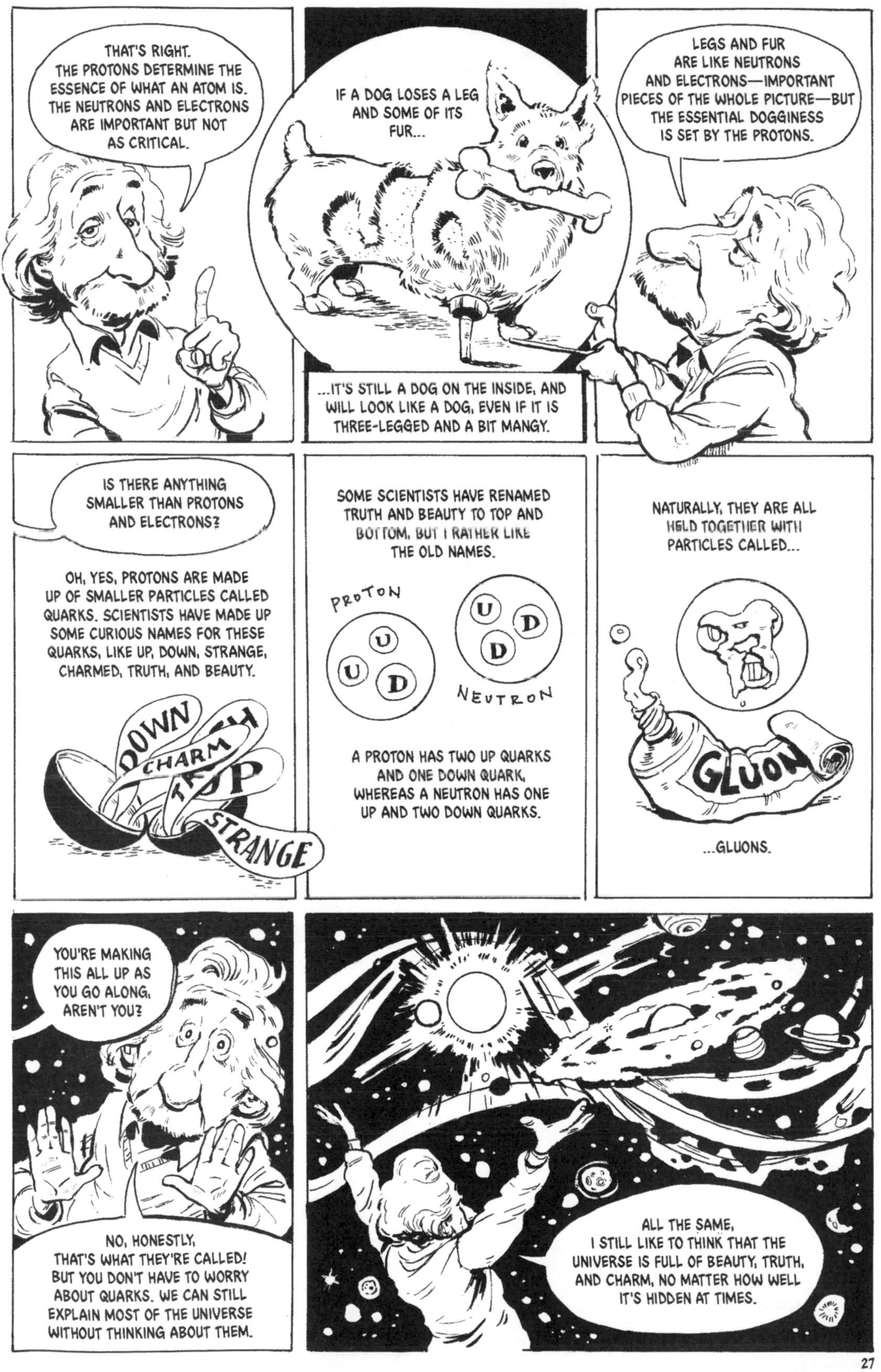
THAT'S RIGHT. THE PROTONS DETERMINE THE ESSENCE OF WHAT AN ATOM IS. THE NEUTRONS AND ELECTRONS ARE IMPORTANT BUT NOT AS CRITICAL.
IF A DOG LOSES A LEG AND SOME OF ITS FUR...
...IT'S STILL A DOG ON THE INSIDE, AND WILL LOOK LIKE A DOG, EVEN IF IT IS THREE-LEGGED AND A BIT MANGY.
LEGS AND FUR ARE LIKE NEUTRONS AND ELECTRONS—IMPORTANT PIECES OF THE WHOLE PICTURE—BUT THE ESSENTIAL DOGGINESS IS SET BY THE PROTONS.
IS THERE ANYTHING SMALLER THAN PROTONS AND ELECTRONS?
OH, YES, PROTONS ARE MADE UP OF SMALLER PARTICLES CALLED QUARKS. SCIENTISTS HAVE MADE UP SOME CURIOUS NAMES FOR THESE QUARKS, LIKE UP, DOWN, STRANGE, CHARMED, TRUTH, AND BEAUTY.
DOWN
CHARM
TOP
STRANGE
SOME SCIENTISTS HAVE RENAMED TRUTH AND BEAUTY TO TOP AND BOTTOM, BUT I RATHER LIKE THE OLD NAMES.
PROTON
U
U
D
NEUTRON
U
D
D
A PROTON HAS TWO UP QUARKS AND ONE DOWN QUARK, WHEREAS A NEUTRON HAS ONE UP AND TWO DOWN QUARKS.
NATURALLY, THEY ARE ALL HELD TOGETHER WITH PARTICLES CALLED...
GLUON
...GLUONS.
YOU'RE MAKING THIS ALL UP AS YOU GO ALONG, AREN'T YOU?
NO, HONESTLY, THAT'S WHAT THEY'RE CALLED! BUT YOU DON'T HAVE TO WORRY ABOUT QUARKS. WE CAN STILL EXPLAIN MOST OF THE UNIVERSE WITHOUT THINKING ABOUT THEM.
ALL THE SAME, I STILL LIKE TO THINK THAT THE UNIVERSE IS FULL OF BEAUTY, TRUTH, AND CHARM, NO MATTER HOW WELL IT'S HIDDEN AT TIMES.

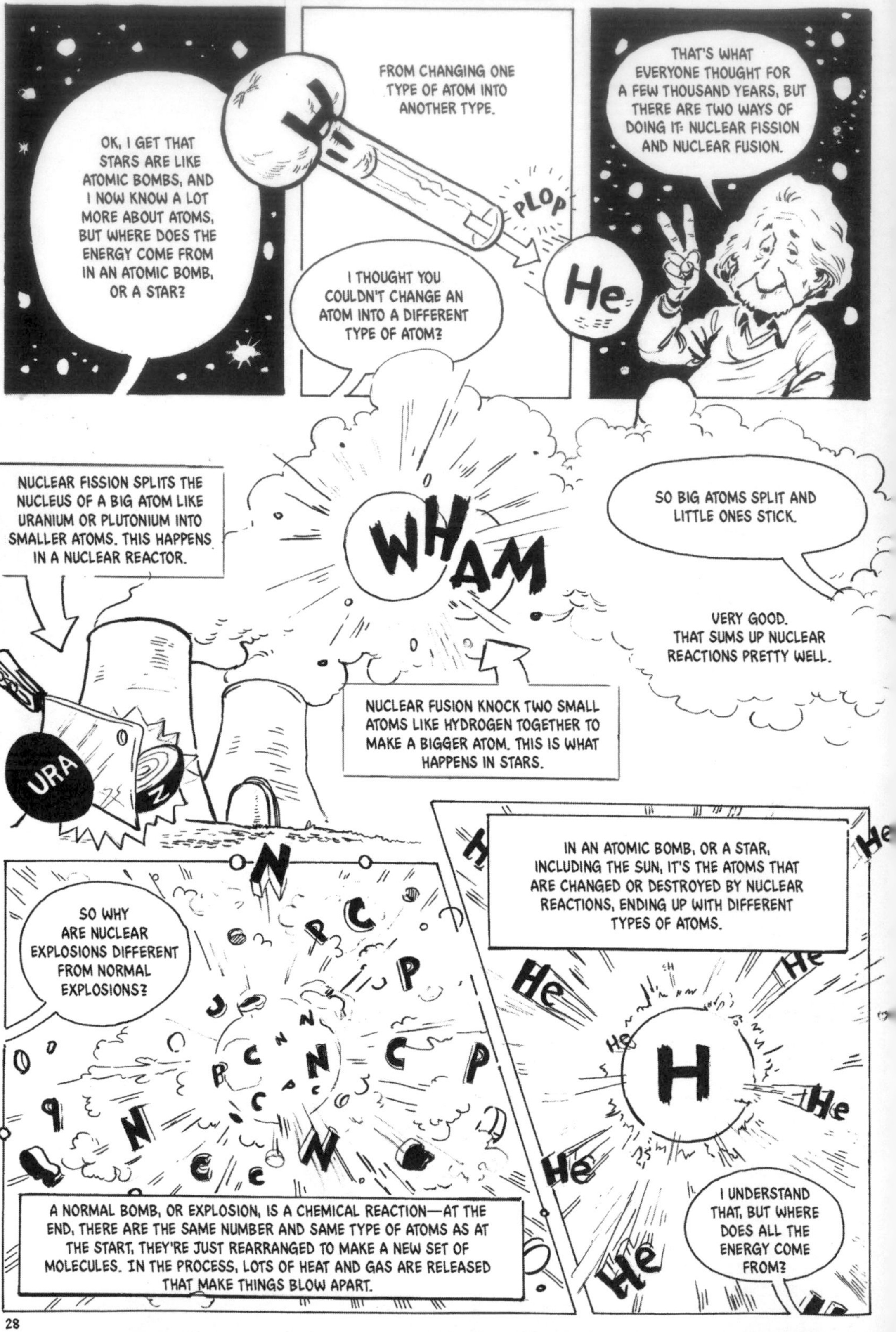
OK, I GET THAT STARS ARE LIKE ATOMIC BOMBS, AND I NOW KNOW A LOT MORE ABOUT ATOMS, BUT WHERE DOES THE ENERGY COME FROM IN AN ATOMIC BOMB, OR A STAR?
H
FROM CHANGING ONE TYPE OF ATOM INTO ANOTHER TYPE.
PLOP
I THOUGHT YOU COULDN'T CHANGE AN ATOM INTO A DIFFERENT TYPE OF ATOM?
He
THAT'S WHAT EVERYONE THOUGHT FOR A FEW THOUSAND YEARS, BUT THERE ARE TWO WAYS OF DOING IT: NUCLEAR FISSION AND NUCLEAR FUSION.
NUCLEAR FISSION SPLITS THE NUCLEUS OF A BIG ATOM LIKE URANIUM OR PLUTONIUM INTO SMALLER ATOMS. THIS HAPPENS IN A NUCLEAR REACTOR.
WHAM
SO BIG ATOMS SPLIT AND LITTLE ONES STICK.
VERY GOOD. THAT SUMS UP NUCLEAR REACTIONS PRETTY WELL.
URA
NUCLEAR FUSION KNOCK TWO SMALL ATOMS LIKE HYDROGEN TOGETHER TO MAKE A BIGGER ATOM. THIS IS WHAT HAPPENS IN STARS.
SO WHY ARE NUCLEAR EXPLOSIONS DIFFERENT FROM NORMAL EXPLOSIONS?
A NORMAL BOMB, OR EXPLOSION, IS A CHEMICAL REACTION—AT THE END, THERE ARE THE SAME NUMBER AND SAME TYPE OF ATOMS AS AT THE START, THEY'RE JUST REARRANGED TO MAKE A NEW SET OF MOLECULES. IN THE PROCESS, LOTS OF HEAT AND GAS ARE RELEASED THAT MAKE THINGS BLOW APART.
IN AN ATOMIC BOMB, OR A STAR, INCLUDING THE SUN, IT'S THE ATOMS THAT ARE CHANGED OR DESTROYED BY NUCLEAR REACTIONS, ENDING UP WITH DIFFERENT TYPES OF ATOMS.
H
I UNDERSTAND THAT, BUT WHERE DOES ALL THE ENERGY COME FROM?

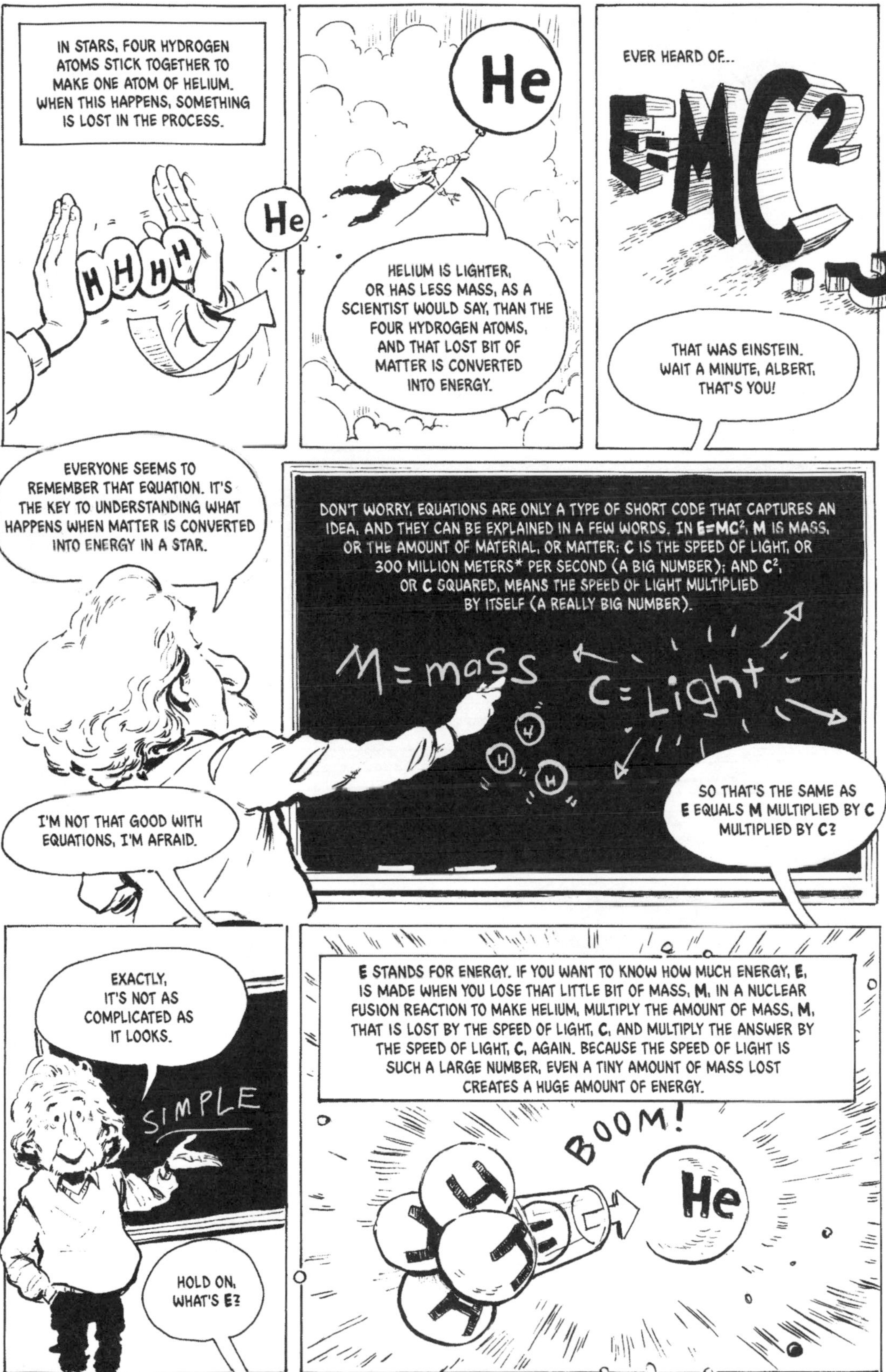

IN STARS, FOUR HYDROGEN ATOMS STICK TOGETHER TO MAKE ONE ATOM OF HELIUM. WHEN THIS HAPPENS, SOMETHING IS LOST IN THE PROCESS.
H H H H
He
He
HELIUM IS LIGHTER, OR HAS LESS MASS, AS A SCIENTIST WOULD SAY, THAN THE FOUR HYDROGEN ATOMS, AND THAT LOST BIT OF MATTER IS CONVERTED INTO ENERGY.
EVER HEARD OF...
E=MC²
THAT WAS EINSTEIN. WAIT A MINUTE, ALBERT, THAT'S YOU!
EVERYONE SEEMS TO REMEMBER THAT EQUATION. IT'S THE KEY TO UNDERSTANDING WHAT HAPPENS WHEN MATTER IS CONVERTED INTO ENERGY IN A STAR.
DON'T WORRY, EQUATIONS ARE ONLY A TYPE OF SHORT CODE THAT CAPTURES AN IDEA, AND THEY CAN BE EXPLAINED IN A FEW WORDS. IN E=MC², M IS MASS, OR THE AMOUNT OF MATERIAL, OR MATTER; C IS THE SPEED OF LIGHT, OR 300 MILLION METERS* PER SECOND (A BIG NUMBER); AND C², OR C SQUARED, MEANS THE SPEED OF LIGHT MULTIPLIED BY ITSELF (A REALLY BIG NUMBER).
M = mass
C = Light
I'M NOT THAT GOOD WITH EQUATIONS, I'M AFRAID.
SO THAT'S THE SAME AS E EQUALS M MULTIPLIED BY C MULTIPLIED BY C?
EXACTLY, IT'S NOT AS COMPLICATED AS IT LOOKS.
SIMPLE
HOLD ON, WHAT'S E?
E STANDS FOR ENERGY. IF YOU WANT TO KNOW HOW MUCH ENERGY, E, IS MADE WHEN YOU LOSE THAT LITTLE BIT OF MASS, M, IN A NUCLEAR FUSION REACTION TO MAKE HELIUM, MULTIPLY THE AMOUNT OF MASS, M, THAT IS LOST BY THE SPEED OF LIGHT, C, AND MULTIPLY THE ANSWER BY THE SPEED OF LIGHT, C, AGAIN. BECAUSE THE SPEED OF LIGHT IS SUCH A LARGE NUMBER, EVEN A TINY AMOUNT OF MASS LOST CREATES A HUGE AMOUNT OF ENERGY.
BOOM!
He

*OF COURSE, IT COULD BE MEASURED IN INCHES PER SECOND OR MILES PER HOUR, BUT THAT WOULD GIVE A DIFFERENT ANSWER, SO SCIENTISTS, LIKE ALBERT, HAVE ALL AGREED TO MEASURE THINGS IN THE SAME WAY—IN METERS, SECONDS, AND KILOGRAMS—SO THEY ALL GET THE SAME ANSWER!

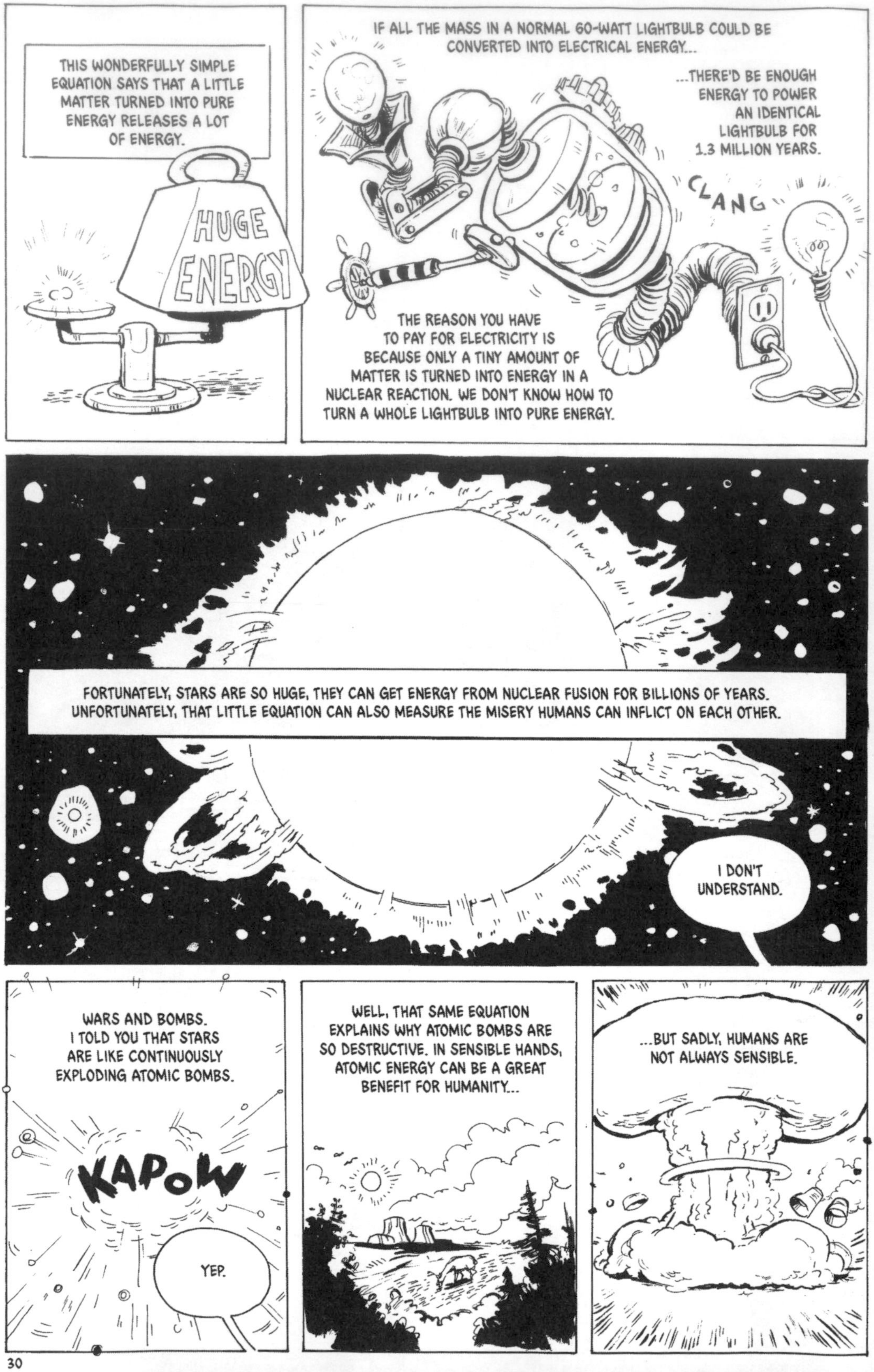
THIS WONDERFULLY SIMPLE EQUATION SAYS THAT A LITTLE MATTER TURNED INTO PURE ENERGY RELEASES A LOT OF ENERGY.
HUGE ENERGY
IF ALL THE MASS IN A NORMAL 60-WATT LIGHTBULB COULD BE CONVERTED INTO ELECTRICAL ENERGY...
...THERE'D BE ENOUGH ENERGY TO POWER AN IDENTICAL LIGHTBULB FOR 1.3 MILLION YEARS.
CLANG
THE REASON YOU HAVE TO PAY FOR ELECTRICITY IS BECAUSE ONLY A TINY AMOUNT OF MATTER IS TURNED INTO ENERGY IN A NUCLEAR REACTION. WE DON'T KNOW HOW TO TURN A WHOLE LIGHTBULB INTO PURE ENERGY.
FORTUNATELY, STARS ARE SO HUGE, THEY CAN GET ENERGY FROM NUCLEAR FUSION FOR BILLIONS OF YEARS. UNFORTUNATELY, THAT LITTLE EQUATION CAN ALSO MEASURE THE MISERY HUMANS CAN INFLICT ON EACH OTHER.
I DON'T UNDERSTAND.
WARS AND BOMBS. I TOLD YOU THAT STARS ARE LIKE CONTINUOUSLY EXPLODING ATOMIC BOMBS.
KAPOW
YEP.
WELL, THAT SAME EQUATION EXPLAINS WHY ATOMIC BOMBS ARE SO DESTRUCTIVE. IN SENSIBLE HANDS, ATOMIC ENERGY CAN BE A GREAT BENEFIT FOR HUMANITY...
...BUT SADLY, HUMANS ARE NOT ALWAYS SENSIBLE.

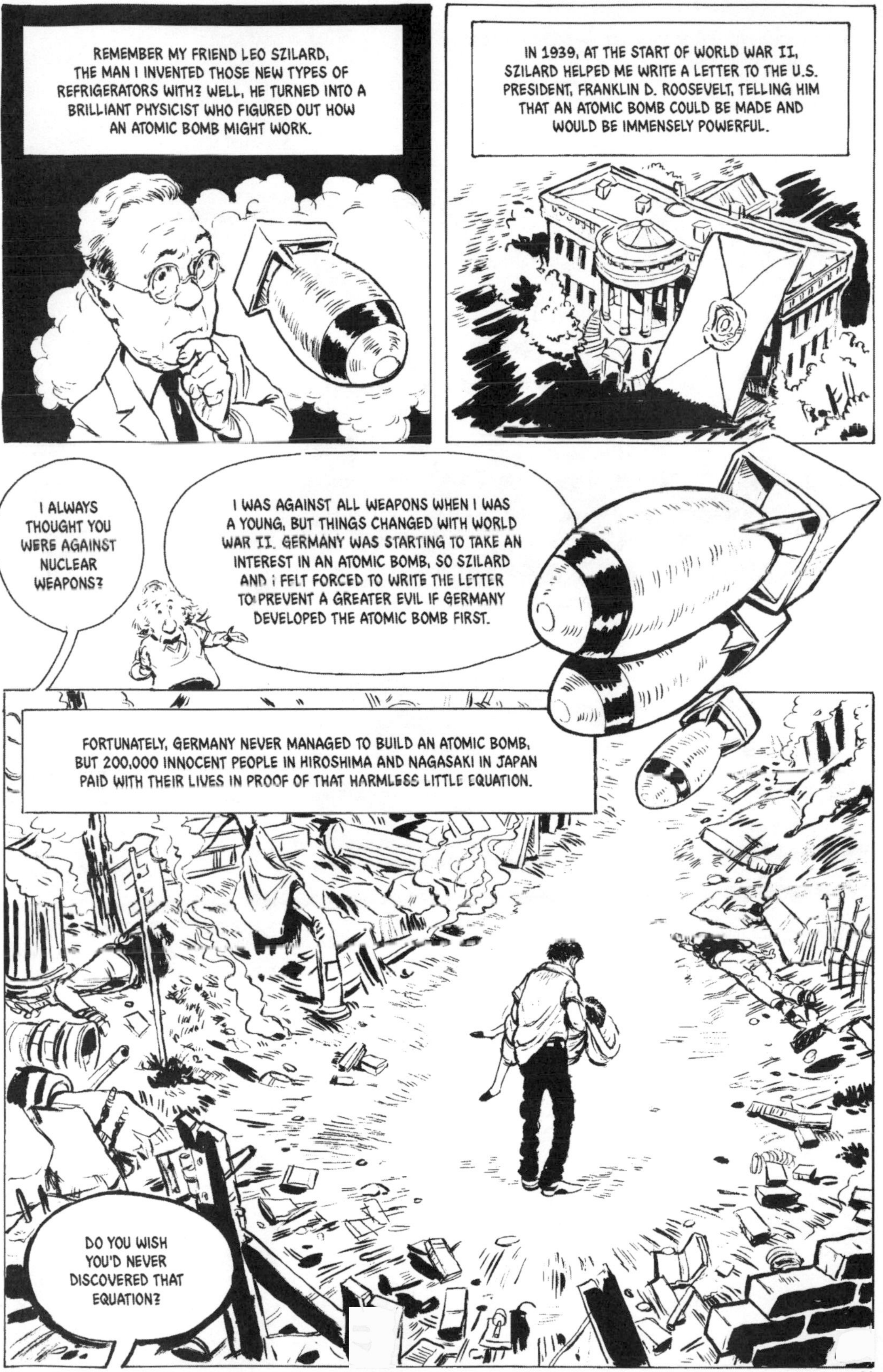
REMEMBER MY FRIEND LEO SZILARD, THE MAN I INVENTED THOSE NEW TYPES OF REFRIGERATORS WITH? WELL, HE TURNED INTO A BRILLIANT PHYSICIST WHO FIGURED OUT HOW AN ATOMIC BOMB MIGHT WORK.
IN 1939, AT THE START OF WORLD WAR II, SZILARD HELPED ME WRITE A LETTER TO THE U.S. PRESIDENT, FRANKLIN D. ROOSEVELT, TELLING HIM THAT AN ATOMIC BOMB COULD BE MADE AND WOULD BE IMMENSELY POWERFUL.
I ALWAYS THOUGHT YOU WERE AGAINST NUCLEAR WEAPONS?
I WAS AGAINST ALL WEAPONS WHEN I WAS A YOUNG, BUT THINGS CHANGED WITH WORLD WAR II. GERMANY WAS STARTING TO TAKE AN INTEREST IN AN ATOMIC BOMB, SO SZILARD AND I FELT FORCED TO WRITE THE LETTER TO PREVENT A GREATER EVIL IF GERMANY DEVELOPED THE ATOMIC BOMB FIRST.
FORTUNATELY, GERMANY NEVER MANAGED TO BUILD AN ATOMIC BOMB, BUT 200,000 INNOCENT PEOPLE IN HIROSHIMA AND NAGASAKI IN JAPAN PAID WITH THEIR LIVES IN PROOF OF THAT HARMLESS LITTLE EQUATION.
DO YOU WISH YOU'D NEVER DISCOVERED THAT EQUATION?

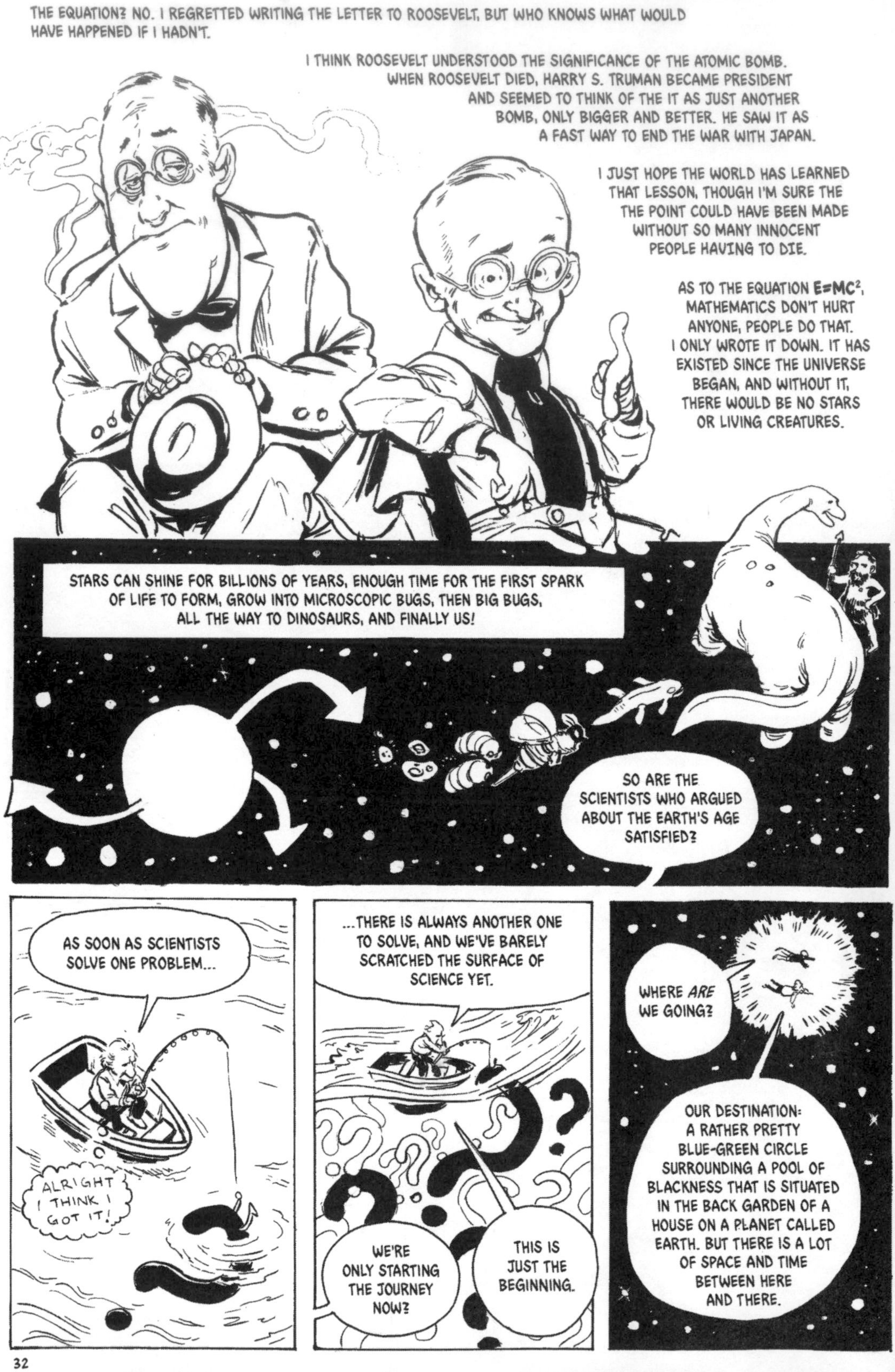
THE EQUATION? NO. I REGRETTED WRITING THE LETTER TO ROOSEVELT, BUT WHO KNOWS WHAT WOULD HAVE HAPPENED IF I HADN'T.
I THINK ROOSEVELT UNDERSTOOD THE SIGNIFICANCE OF THE ATOMIC BOMB. WHEN ROOSEVELT DIED, HARRY S. TRUMAN BECAME PRESIDENT AND SEEMED TO THINK OF THE IT AS JUST ANOTHER BOMB, ONLY BIGGER AND BETTER. HE SAW IT AS A FAST WAY TO END THE WAR WITH JAPAN.
I JUST HOPE THE WORLD HAS LEARNED THAT LESSON, THOUGH I'M SURE THE THE POINT COULD HAVE BEEN MADE WITHOUT SO MANY INNOCENT PEOPLE HAVING TO DIE.
AS TO THE EQUATION $E=MC^2$, MATHEMATICS DON'T HURT ANYONE, PEOPLE DO THAT. I ONLY WROTE IT DOWN. IT HAS EXISTED SINCE THE UNIVERSE BEGAN, AND WITHOUT IT, THERE WOULD BE NO STARS OR LIVING CREATURES.
STARS CAN SHINE FOR BILLIONS OF YEARS, ENOUGH TIME FOR THE FIRST SPARK OF LIFE TO FORM, GROW INTO MICROSCOPIC BUGS, THEN BIG BUGS, ALL THE WAY TO DINOSAURS, AND FINALLY US!
SO ARE THE SCIENTISTS WHO ARGUED ABOUT THE EARTH'S AGE SATISFIED?
AS SOON AS SCIENTISTS SOLVE ONE PROBLEM...
ALRIGHT I THINK I GOT IT!
...THERE IS ALWAYS ANOTHER ONE TO SOLVE, AND WE'VE BARELY SCRATCHED THE SURFACE OF SCIENCE YET.
WE'RE ONLY STARTING THE JOURNEY NOW?
THIS IS JUST THE BEGINNING.
WHERE *ARE* WE GOING?
OUR DESTINATION: A RATHER PRETTY BLUE-GREEN CIRCLE SURROUNDING A POOL OF BLACKNESS THAT IS SITUATED IN THE BACK GARDEN OF A HOUSE ON A PLANET CALLED EARTH. BUT THERE IS A LOT OF SPACE AND TIME BETWEEN HERE AND THERE.

CHAPTER 5: LET THERE BE LIGHT!

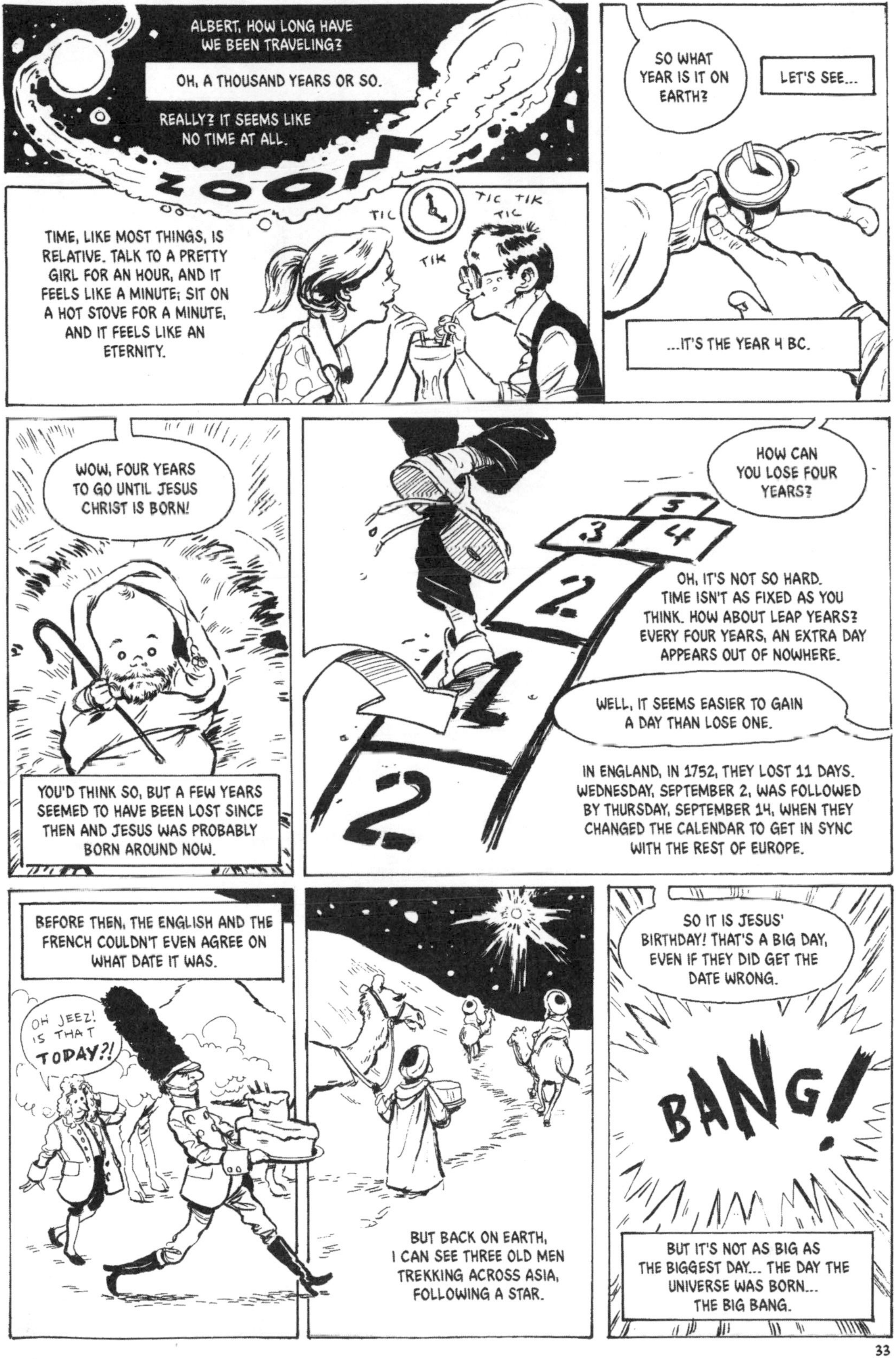
ALBERT, HOW LONG HAVE WE BEEN TRAVELING?
OH, A THOUSAND YEARS OR SO.
REALLY? IT SEEMS LIKE NO TIME AT ALL.
ZOOM
TIME, LIKE MOST THINGS, IS RELATIVE. TALK TO A PRETTY GIRL FOR AN HOUR, AND IT FEELS LIKE A MINUTE; SIT ON A HOT STOVE FOR A MINUTE, AND IT FEELS LIKE AN ETERNITY.
TIC TIK TIC
TIC
TIK
SO WHAT YEAR IS IT ON EARTH?
LET'S SEE...
...IT'S THE YEAR 4 BC.
WOW, FOUR YEARS TO GO UNTIL JESUS CHRIST IS BORN!
YOU'D THINK SO, BUT A FEW YEARS SEEMED TO HAVE BEEN LOST SINCE THEN AND JESUS WAS PROBABLY BORN AROUND NOW.
HOW CAN YOU LOSE FOUR YEARS?
OH, IT'S NOT SO HARD. TIME ISN'T AS FIXED AS YOU THINK. HOW ABOUT LEAP YEARS? EVERY FOUR YEARS, AN EXTRA DAY APPEARS OUT OF NOWHERE.
WELL, IT SEEMS EASIER TO GAIN A DAY THAN LOSE ONE.
IN ENGLAND, IN 1752, THEY LOST 11 DAYS. WEDNESDAY, SEPTEMBER 2, WAS FOLLOWED BY THURSDAY, SEPTEMBER 14, WHEN THEY CHANGED THE CALENDAR TO GET IN SYNC WITH THE REST OF EUROPE.
BEFORE THEN, THE ENGLISH AND THE FRENCH COULDN'T EVEN AGREE ON WHAT DATE IT WAS.
OH JEEZ! IS THAT TODAY?!
BUT BACK ON EARTH, I CAN SEE THREE OLD MEN TREKKING ACROSS ASIA, FOLLOWING A STAR.
SO IT IS JESUS' BIRTHDAY! THAT'S A BIG DAY, EVEN IF THEY DID GET THE DATE WRONG.
BANG!
BUT IT'S NOT AS BIG AS THE BIGGEST DAY... THE DAY THE UNIVERSE WAS BORN... THE BIG BANG.

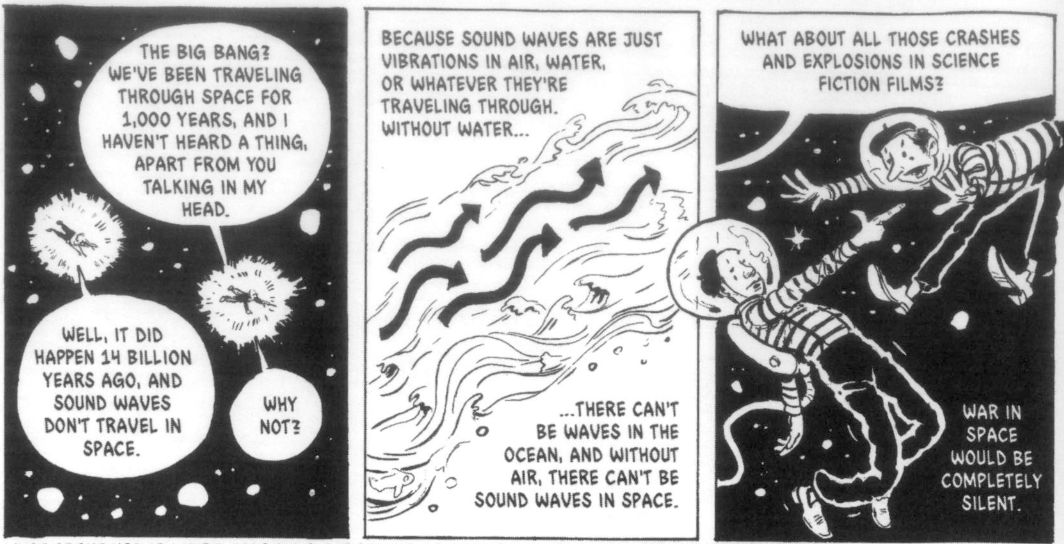
THE BIG BANG? WE'VE BEEN TRAVELING THROUGH SPACE FOR 1,000 YEARS, AND I HAVEN'T HEARD A THING, APART FROM YOU TALKING IN MY HEAD.
WELL, IT DID HAPPEN 14 BILLION YEARS AGO, AND SOUND WAVES DON'T TRAVEL IN SPACE.
WHY NOT?
BECAUSE SOUND WAVES ARE JUST VIBRATIONS IN AIR, WATER, OR WHATEVER THEY'RE TRAVELING THROUGH. WITHOUT WATER...
...THERE CAN'T BE WAVES IN THE OCEAN, AND WITHOUT AIR, THERE CAN'T BE SOUND WAVES IN SPACE.
WHAT ABOUT ALL THOSE CRASHES AND EXPLOSIONS IN SCIENCE FICTION FILMS?
WAR IN SPACE WOULD BE COMPLETELY SILENT.

WHAT ABOUT US? IF WE'RE IMAGINING OURSELVES AS BEAMS OF LIGHT, HOW CAN WE TRAVEL THROUGH SPACE?

LIGHT ISN'T JUST A WAVE, LIKE SOUND, IT'S MADE UP OF PHOTONS, OR LIGHT PARTICLES. THAT'S WHY WE CAN TRAVEL THROUGH A VACUUM LIKE SPACE.

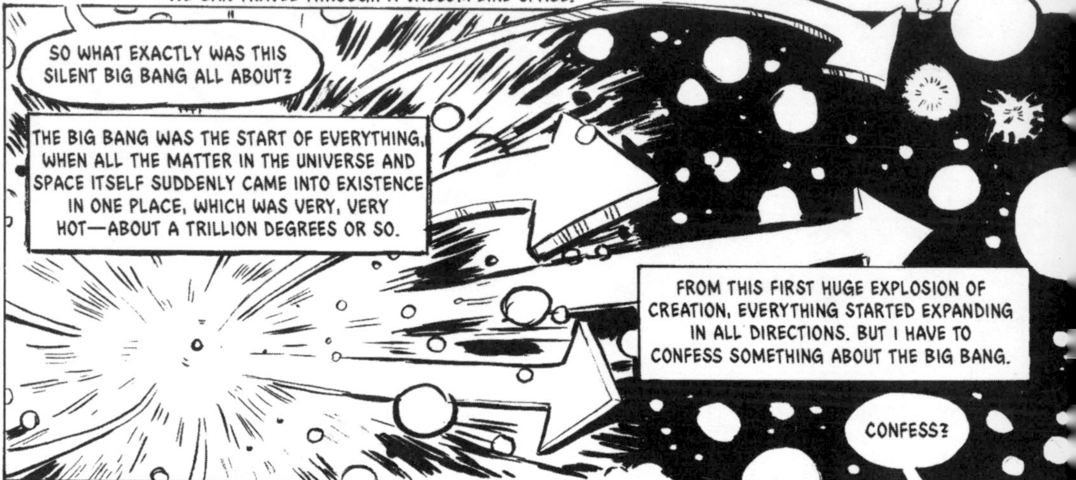
SO WHAT EXACTLY WAS THIS SILENT BIG BANG ALL ABOUT?
THE BIG BANG WAS THE START OF EVERYTHING, WHEN ALL THE MATTER IN THE UNIVERSE AND SPACE ITSELF SUDDENLY CAME INTO EXISTENCE IN ONE PLACE, WHICH WAS VERY, VERY HOT—ABOUT A TRILLION DEGREES OR SO.
FROM THIS FIRST HUGE EXPLOSION OF CREATION, EVERYTHING STARTED EXPANDING IN ALL DIRECTIONS. BUT I HAVE TO CONFESS SOMETHING ABOUT THE BIG BANG.
CONFESS?

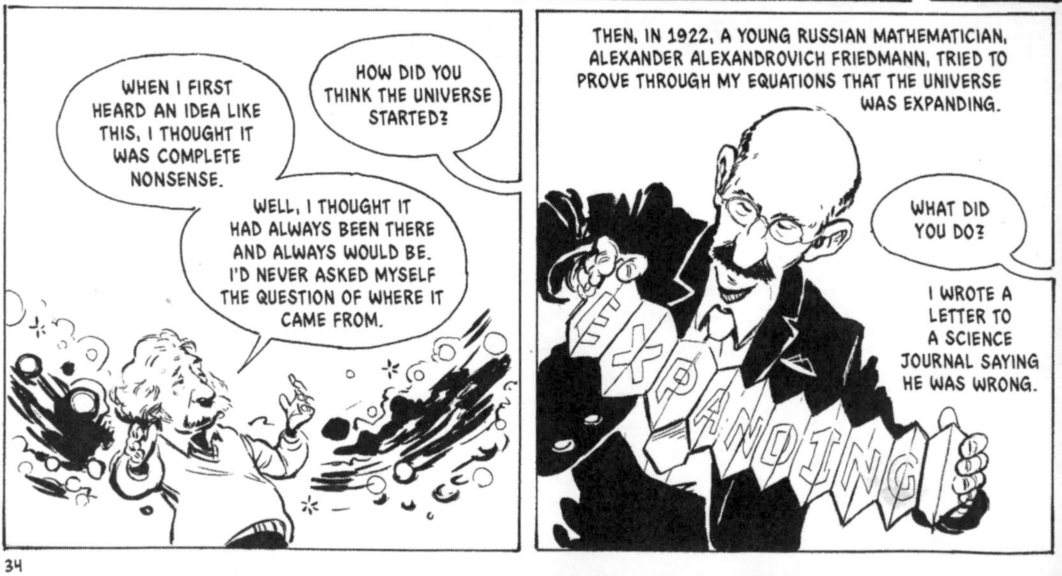
WHEN I FIRST HEARD AN IDEA LIKE THIS, I THOUGHT IT WAS COMPLETE NONSENSE.
HOW DID YOU THINK THE UNIVERSE STARTED?
WELL, I THOUGHT IT HAD ALWAYS BEEN THERE AND ALWAYS WOULD BE. I'D NEVER ASKED MYSELF THE QUESTION OF WHERE IT CAME FROM.
THEN, IN 1922, A YOUNG RUSSIAN MATHEMATICIAN, ALEXANDER ALEXANDROVICH FRIEDMANN, TRIED TO PROVE THROUGH MY EQUATIONS THAT THE UNIVERSE WAS EXPANDING.
WHAT DID YOU DO?
I WROTE A LETTER TO A SCIENCE JOURNAL SAYING HE WAS WRONG.
EXPANDING

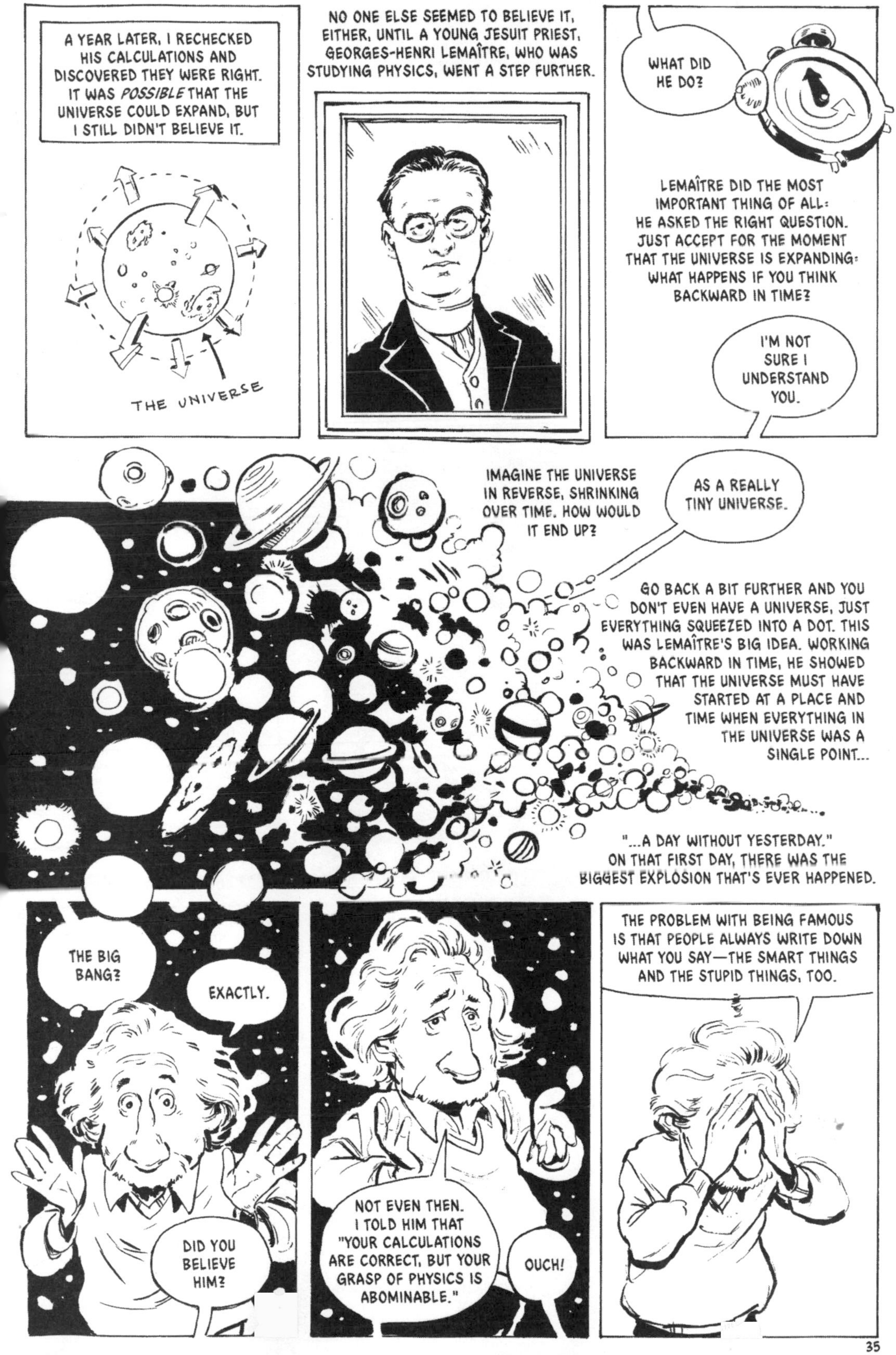
A YEAR LATER, I RECHECKED HIS CALCULATIONS AND DISCOVERED THEY WERE RIGHT. IT WAS *POSSIBLE* THAT THE UNIVERSE COULD EXPAND, BUT I STILL DIDN'T BELIEVE IT.
THE UNIVERSE
NO ONE ELSE SEEMED TO BELIEVE IT, EITHER, UNTIL A YOUNG JESUIT PRIEST, GEORGES-HENRI LEMAÎTRE, WHO WAS STUDYING PHYSICS, WENT A STEP FURTHER.
WHAT DID HE DO?
LEMAÎTRE DID THE MOST IMPORTANT THING OF ALL: HE ASKED THE RIGHT QUESTION. JUST ACCEPT FOR THE MOMENT THAT THE UNIVERSE IS EXPANDING: WHAT HAPPENS IF YOU THINK BACKWARD IN TIME?
I'M NOT SURE I UNDERSTAND YOU.
IMAGINE THE UNIVERSE IN REVERSE, SHRINKING OVER TIME. HOW WOULD IT END UP?
AS A REALLY TINY UNIVERSE.
GO BACK A BIT FURTHER AND YOU DON'T EVEN HAVE A UNIVERSE, JUST EVERYTHING SQUEEZED INTO A DOT. THIS WAS LEMAÎTRE'S BIG IDEA. WORKING BACKWARD IN TIME, HE SHOWED THAT THE UNIVERSE MUST HAVE STARTED AT A PLACE AND TIME WHEN EVERYTHING IN THE UNIVERSE WAS A SINGLE POINT...
"...A DAY WITHOUT YESTERDAY." ON THAT FIRST DAY, THERE WAS THE BIGGEST EXPLOSION THAT'S EVER HAPPENED.
THE BIG BANG?
EXACTLY.
DID YOU BELIEVE HIM?
NOT EVEN THEN. I TOLD HIM THAT "YOUR CALCULATIONS ARE CORRECT, BUT YOUR GRASP OF PHYSICS IS ABOMINABLE."
OUCH!
THE PROBLEM WITH BEING FAMOUS IS THAT PEOPLE ALWAYS WRITE DOWN WHAT YOU SAY—THE SMART THINGS AND THE STUPID THINGS, TOO.

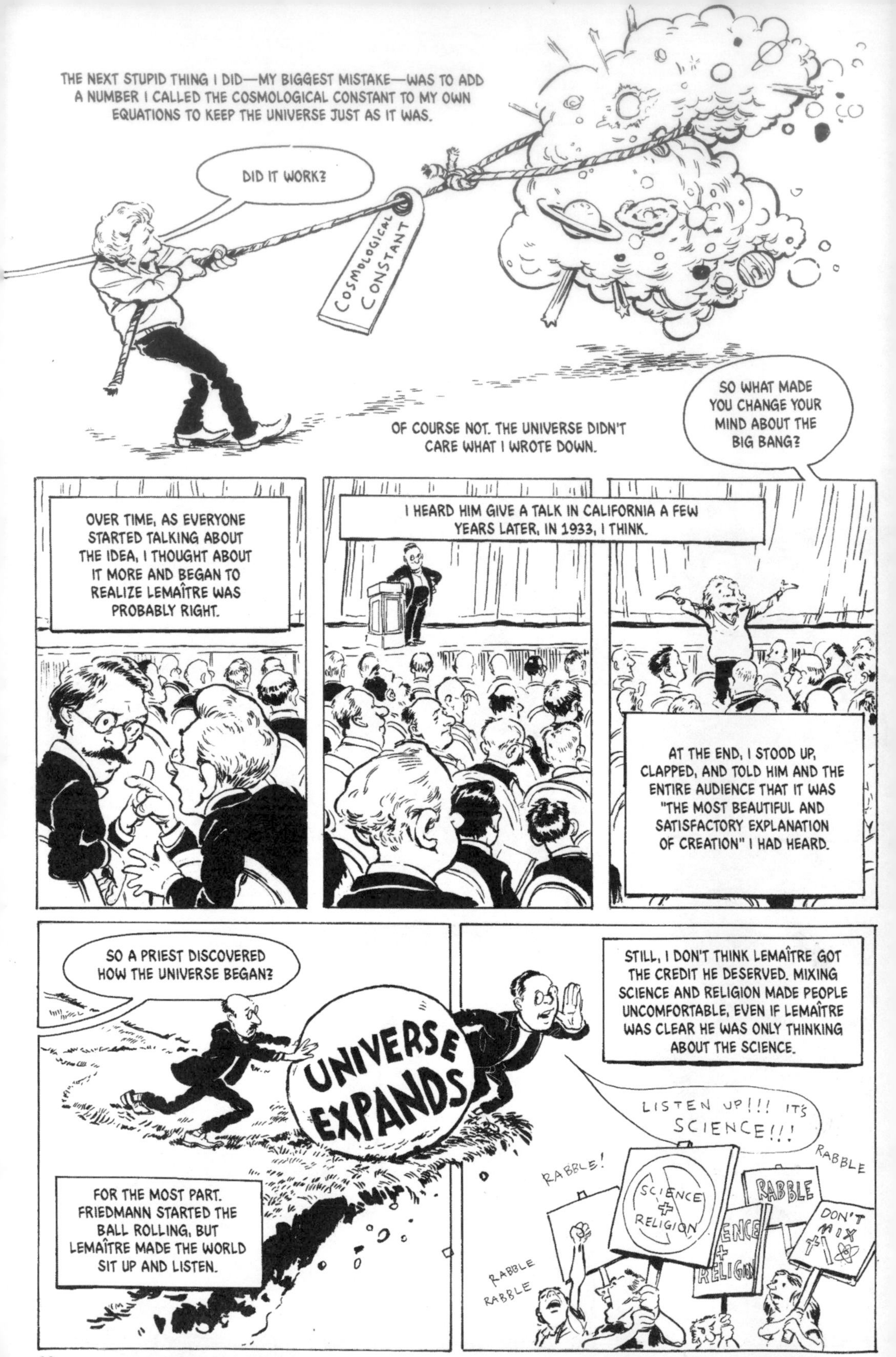
THE NEXT STUPID THING I DID—MY BIGGEST MISTAKE—WAS TO ADD A NUMBER I CALLED THE COSMOLOGICAL CONSTANT TO MY OWN EQUATIONS TO KEEP THE UNIVERSE JUST AS IT WAS.
DID IT WORK?
COSMOLOGICAL CONSTANT
OF COURSE NOT. THE UNIVERSE DIDN'T CARE WHAT I WROTE DOWN.
SO WHAT MADE YOU CHANGE YOUR MIND ABOUT THE BIG BANG?
OVER TIME, AS EVERYONE STARTED TALKING ABOUT THE IDEA, I THOUGHT ABOUT IT MORE AND BEGAN TO REALIZE LEMAÎTRE WAS PROBABLY RIGHT.
I HEARD HIM GIVE A TALK IN CALIFORNIA A FEW YEARS LATER, IN 1933, I THINK.
AT THE END, I STOOD UP, CLAPPED, AND TOLD HIM AND THE ENTIRE AUDIENCE THAT IT WAS "THE MOST BEAUTIFUL AND SATISFACTORY EXPLANATION OF CREATION" I HAD HEARD.
SO A PRIEST DISCOVERED HOW THE UNIVERSE BEGAN?
UNIVERSE EXPANDS
FOR THE MOST PART. FRIEDMANN STARTED THE BALL ROLLING, BUT LEMAÎTRE MADE THE WORLD SIT UP AND LISTEN.
STILL, I DON'T THINK LEMAÎTRE GOT THE CREDIT HE DESERVED. MIXING SCIENCE AND RELIGION MADE PEOPLE UNCOMFORTABLE, EVEN IF LEMAÎTRE WAS CLEAR HE WAS ONLY THINKING ABOUT THE SCIENCE.
LISTEN UP!!! IT'S SCIENCE!!!
RABBLE!
RABBLE
SCIENCE + RELIGION
RABBLE
DON'T MIX
RABBLE RABBLE

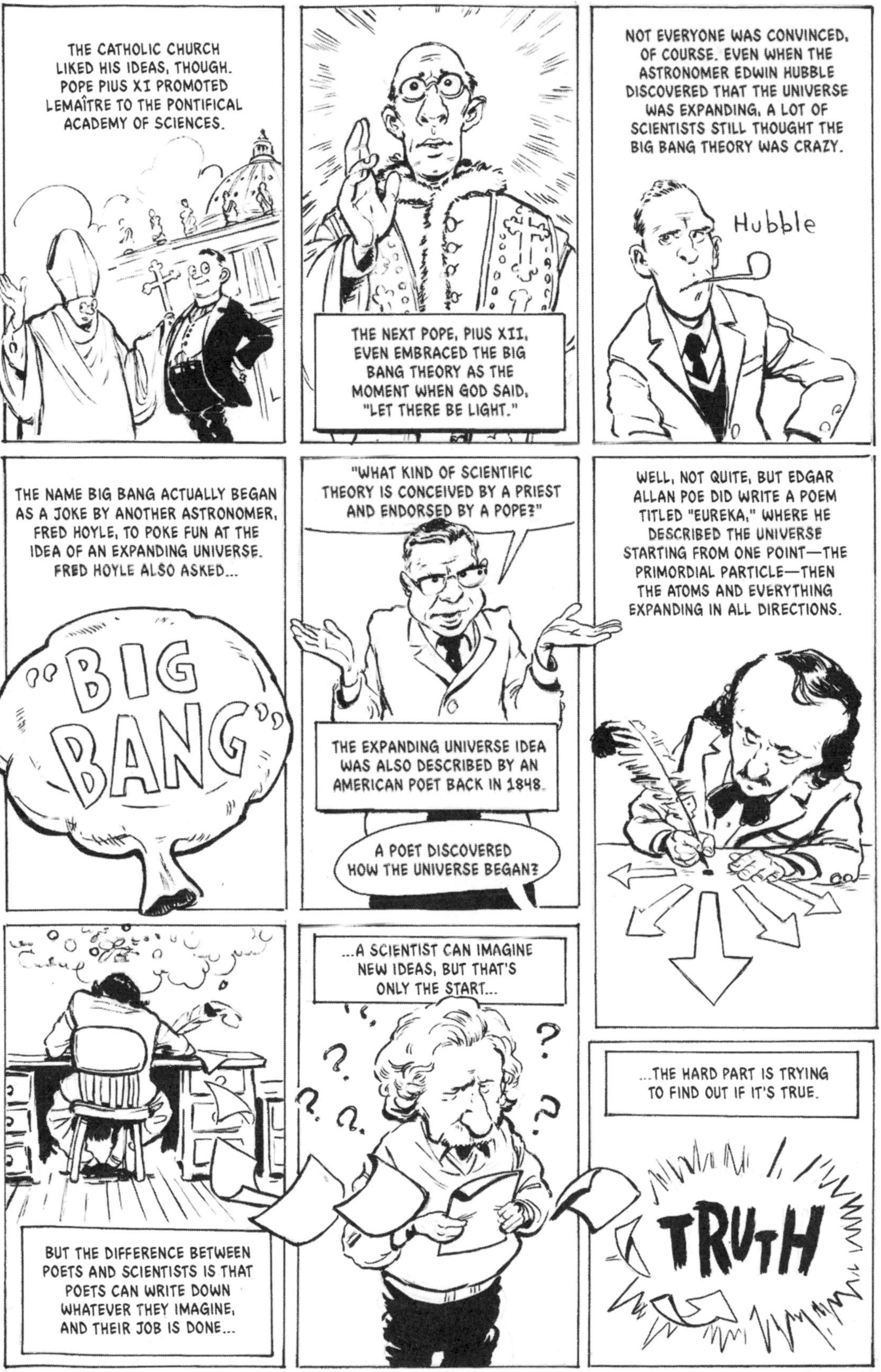
THE CATHOLIC CHURCH LIKED HIS IDEAS, THOUGH. POPE PIUS XI PROMOTED LEMAÎTRE TO THE PONTIFICAL ACADEMY OF SCIENCES.
THE NEXT POPE, PIUS XII, EVEN EMBRACED THE BIG BANG THEORY AS THE MOMENT WHEN GOD SAID, "LET THERE BE LIGHT."
NOT EVERYONE WAS CONVINCED, OF COURSE. EVEN WHEN THE ASTRONOMER EDWIN HUBBLE DISCOVERED THAT THE UNIVERSE WAS EXPANDING, A LOT OF SCIENTISTS STILL THOUGHT THE BIG BANG THEORY WAS CRAZY.
Hubble
THE NAME BIG BANG ACTUALLY BEGAN AS A JOKE BY ANOTHER ASTRONOMER, FRED HOYLE, TO POKE FUN AT THE IDEA OF AN EXPANDING UNIVERSE. FRED HOYLE ALSO ASKED...
"BIG BANG"
"WHAT KIND OF SCIENTIFIC THEORY IS CONCEIVED BY A PRIEST AND ENDORSED BY A POPE?"
THE EXPANDING UNIVERSE IDEA WAS ALSO DESCRIBED BY AN AMERICAN POET BACK IN 1848.
A POET DISCOVERED HOW THE UNIVERSE BEGAN?
WELL, NOT QUITE, BUT EDGAR ALLAN POE DID WRITE A POEM TITLED "EUREKA," WHERE HE DESCRIBED THE UNIVERSE STARTING FROM ONE POINT—THE PRIMORDIAL PARTICLE—THEN THE ATOMS AND EVERYTHING EXPANDING IN ALL DIRECTIONS.
BUT THE DIFFERENCE BETWEEN POETS AND SCIENTISTS IS THAT POETS CAN WRITE DOWN WHATEVER THEY IMAGINE, AND THEIR JOB IS DONE...
...A SCIENTIST CAN IMAGINE NEW IDEAS, BUT THAT'S ONLY THE START...
...THE HARD PART IS TRYING TO FIND OUT IF IT'S TRUE.
TRUTH

SO HOW CAN YOU TELL THE UNIVERSE IS EXPANDING?

NO MATTER WHICH DIRECTION WE LOOK, EVERYTHING IS MOVING AWAY FROM US.

YES, BUT HOW DO WE KNOW THAT?

IN THE SAME WAY A BLIND MAN CAN TELL WHETHER A FIRE ENGINE IS COMING OR GOING. THE SIREN FROM A FIRE ENGINE THAT'S TRAVELING TOWARD YOU IS SLIGHTLY HIGHER PITCHED THAN THE SOUND OF THE SAME SIREN WHEN IT'S TRAVELING AWAY FROM YOU. THIS IS AN EFFECT THAT CHRISTIAN DOPPLER, AN AUSTRIAN ASTRONOMER AND MATHEMATICIAN, DISCOVERED IN 1842 THAT IS NOW CALLED THE DOPPLER EFFECT.

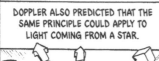

DOPPLER ALSO PREDICTED THAT THE SAME PRINCIPLE COULD APPLY TO LIGHT COMING FROM A STAR.

RATHER THAN BECOMING LOWER PITCHED WHEN MOVING AWAY, THE LIGHT SHIFTS VERY SLIGHTLY IN COLOR, BECOMING REDDER. THE FASTER THE MOVEMENT, THE BIGGER THE SHIFT.

THE EFFECT IS CALLED THE RED SHIFT AND WAS DISCOVERED BY EDWIN HUBBLE IN 1929. FROM LOOKING AT LIGHT FROM DISTANT GALAXIES, IT APPEARS THAT ALL THE GALAXIES ARE MOVING APART. THE FARTHER AWAY A GALAXY IS, THE FASTER IT SEEMS TO BE MOVING.

SO WHY DON'T FIRE ENGINES LOOK REDDER WHEN THEY'RE MOVING AWAY?

LIGHT

SOUND

GALAXIES ARE A LOT FASTER THAN FIRE ENGINES AND CAN TRAVEL AT MILLIONS OF MILES PER HOUR, SO THEIR LIGHT APPEARS SLIGHTLY REDDER THAN IT IS.

THE DOPPLER EFFECT ONLY WORKS WITH THE SIRENS BECAUSE FIRE ENGINES CAN TRAVEL AT A REASONABLE FRACTION OF THE SPEED OF SOUND, WHICH IS ABOUT 760 MILES PER HOUR (AROUND 1,200 KM/H). LIGHT TRAVELS ALMOST A MILLION TIMES FASTER, AT 670 MILLION MILES PER HOUR (1.1 BILLION KM/H).

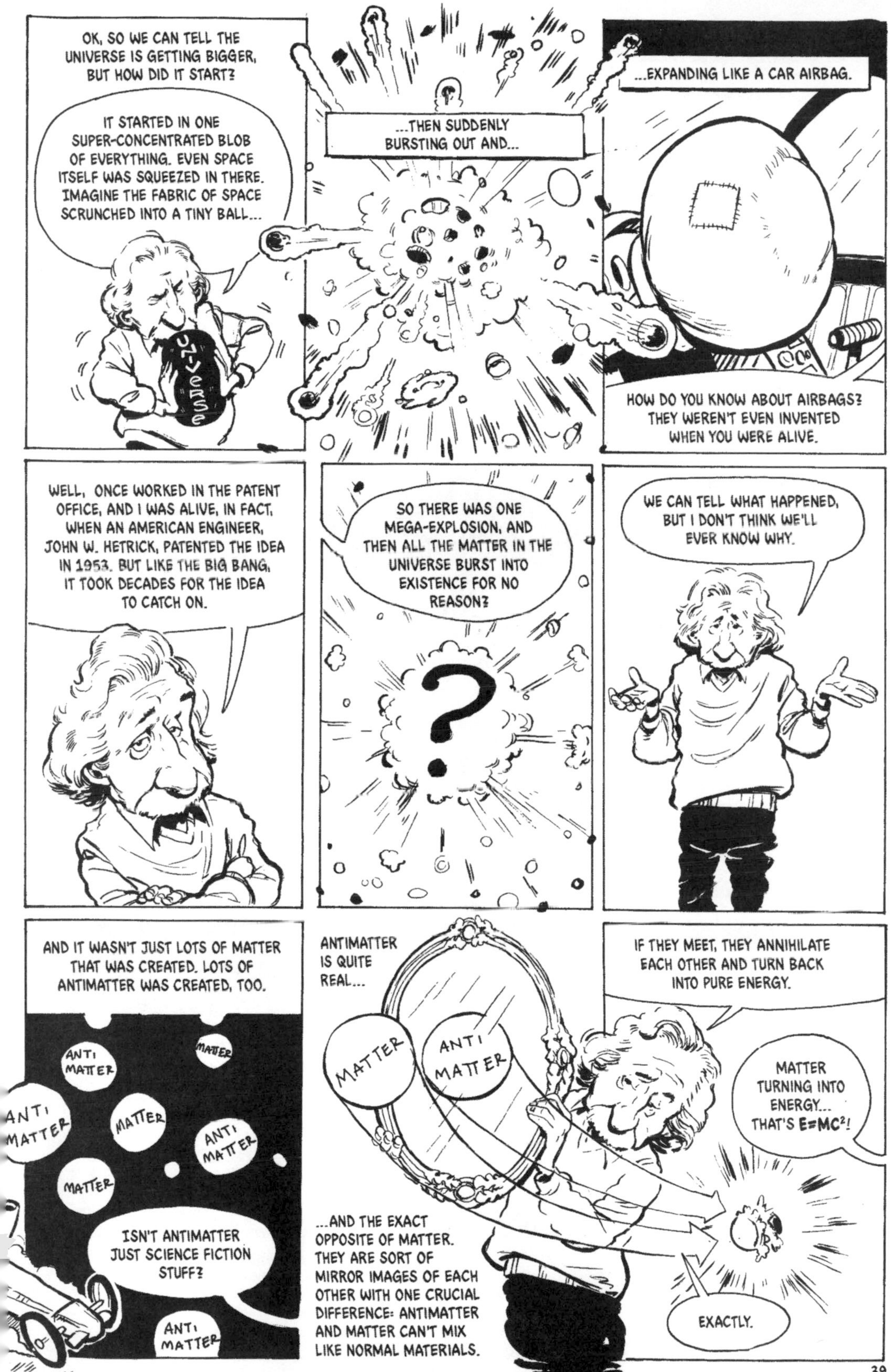
OK, SO WE CAN TELL THE UNIVERSE IS GETTING BIGGER, BUT HOW DID IT START?
IT STARTED IN ONE SUPER-CONCENTRATED BLOB OF EVERYTHING. EVEN SPACE ITSELF WAS SQUEEZED IN THERE. IMAGINE THE FABRIC OF SPACE SCRUNCHED INTO A TINY BALL...
UNIVERSE
...THEN SUDDENLY BURSTING OUT AND...
...EXPANDING LIKE A CAR AIRBAG.
HOW DO YOU KNOW ABOUT AIRBAGS? THEY WEREN'T EVEN INVENTED WHEN YOU WERE ALIVE.
WELL, ONCE WORKED IN THE PATENT OFFICE, AND I WAS ALIVE, IN FACT, WHEN AN AMERICAN ENGINEER, JOHN W. HETRICK, PATENTED THE IDEA IN 1953. BUT LIKE THE BIG BANG, IT TOOK DECADES FOR THE IDEA TO CATCH ON.
SO THERE WAS ONE MEGA-EXPLOSION, AND THEN ALL THE MATTER IN THE UNIVERSE BURST INTO EXISTENCE FOR NO REASON?
?
WE CAN TELL WHAT HAPPENED, BUT I DON'T THINK WE'LL EVER KNOW WHY.
AND IT WASN'T JUST LOTS OF MATTER THAT WAS CREATED. LOTS OF ANTIMATTER WAS CREATED, TOO.
ANTI MATTER
MATTER
ANTI MATTER
MATTER
ANTI MATTER
MATTER
ISN'T ANTIMATTER JUST SCIENCE FICTION STUFF?
ANTI MATTER
ANTIMATTER IS QUITE REAL...
MATTER
ANTI MATTER
...AND THE EXACT OPPOSITE OF MATTER. THEY ARE SORT OF MIRROR IMAGES OF EACH OTHER WITH ONE CRUCIAL DIFFERENCE: ANTIMATTER AND MATTER CAN'T MIX LIKE NORMAL MATERIALS.
IF THEY MEET, THEY ANNIHILATE EACH OTHER AND TURN BACK INTO PURE ENERGY.
MATTER TURNING INTO ENERGY... THAT'S E=MC²!
EXACTLY.

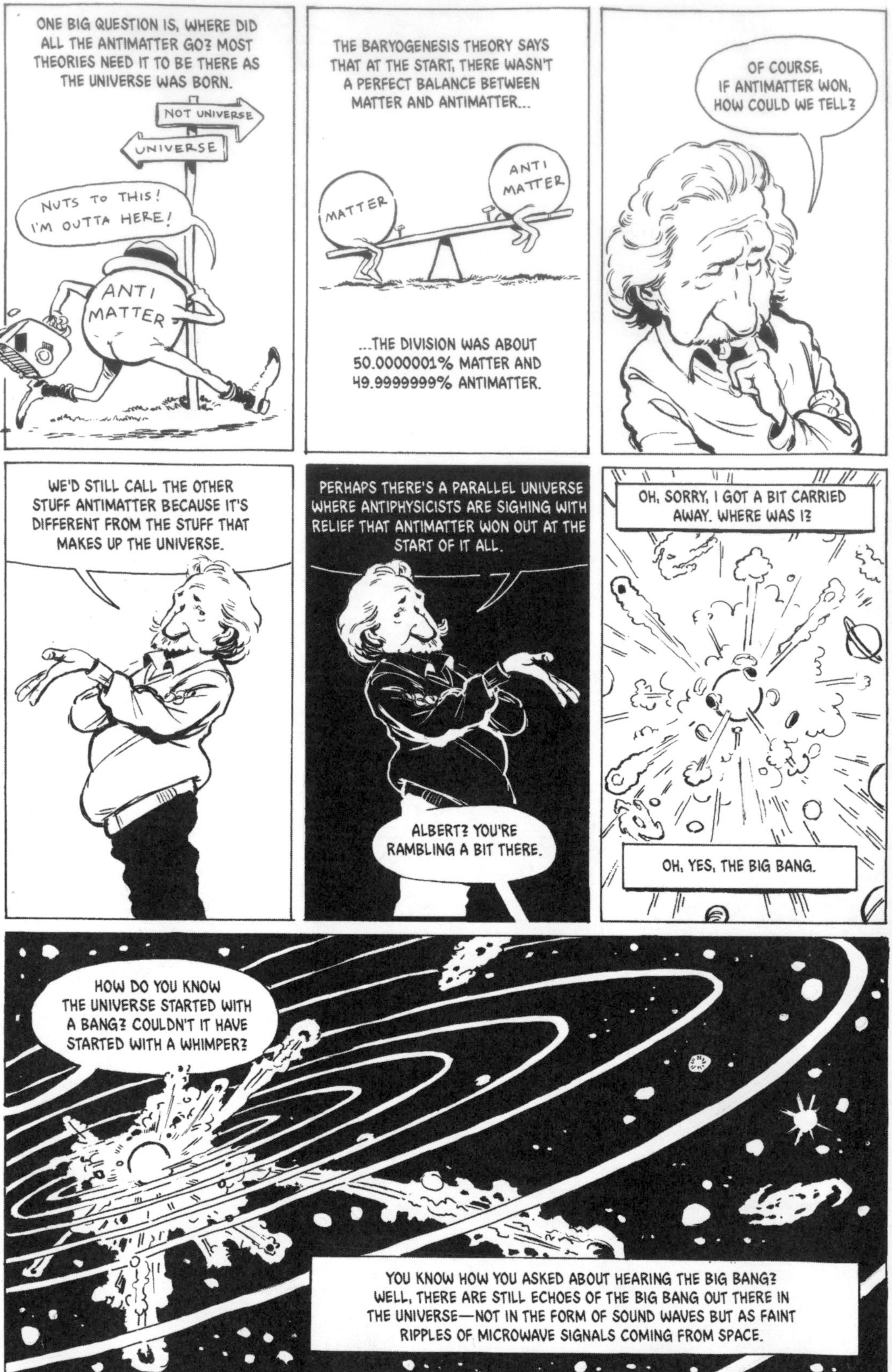
ONE BIG QUESTION IS, WHERE DID ALL THE ANTIMATTER GO? MOST THEORIES NEED IT TO BE THERE AS THE UNIVERSE WAS BORN.
NOT UNIVERSE
UNIVERSE
NUTS TO THIS! I'M OUTTA HERE!
ANTI MATTER
THE BARYOGENESIS THEORY SAYS THAT AT THE START, THERE WASN'T A PERFECT BALANCE BETWEEN MATTER AND ANTIMATTER...
MATTER
ANTI MATTER
...THE DIVISION WAS ABOUT 50.0000001% MATTER AND 49.9999999% ANTIMATTER.
OF COURSE, IF ANTIMATTER WON, HOW COULD WE TELL?
WE'D STILL CALL THE OTHER STUFF ANTIMATTER BECAUSE IT'S DIFFERENT FROM THE STUFF THAT MAKES UP THE UNIVERSE.
PERHAPS THERE'S A PARALLEL UNIVERSE WHERE ANTIPHYSICISTS ARE SIGHING WITH RELIEF THAT ANTIMATTER WON OUT AT THE START OF IT ALL.
ALBERT? YOU'RE RAMBLING A BIT THERE.
OH, SORRY, I GOT A BIT CARRIED AWAY. WHERE WAS I?
OH, YES, THE BIG BANG.
HOW DO YOU KNOW THE UNIVERSE STARTED WITH A BANG? COULDN'T IT HAVE STARTED WITH A WHIMPER?
YOU KNOW HOW YOU ASKED ABOUT HEARING THE BIG BANG? WELL, THERE ARE STILL ECHOES OF THE BIG BANG OUT THERE IN THE UNIVERSE—NOT IN THE FORM OF SOUND WAVES BUT AS FAINT RIPPLES OF MICROWAVE SIGNALS COMING FROM SPACE.

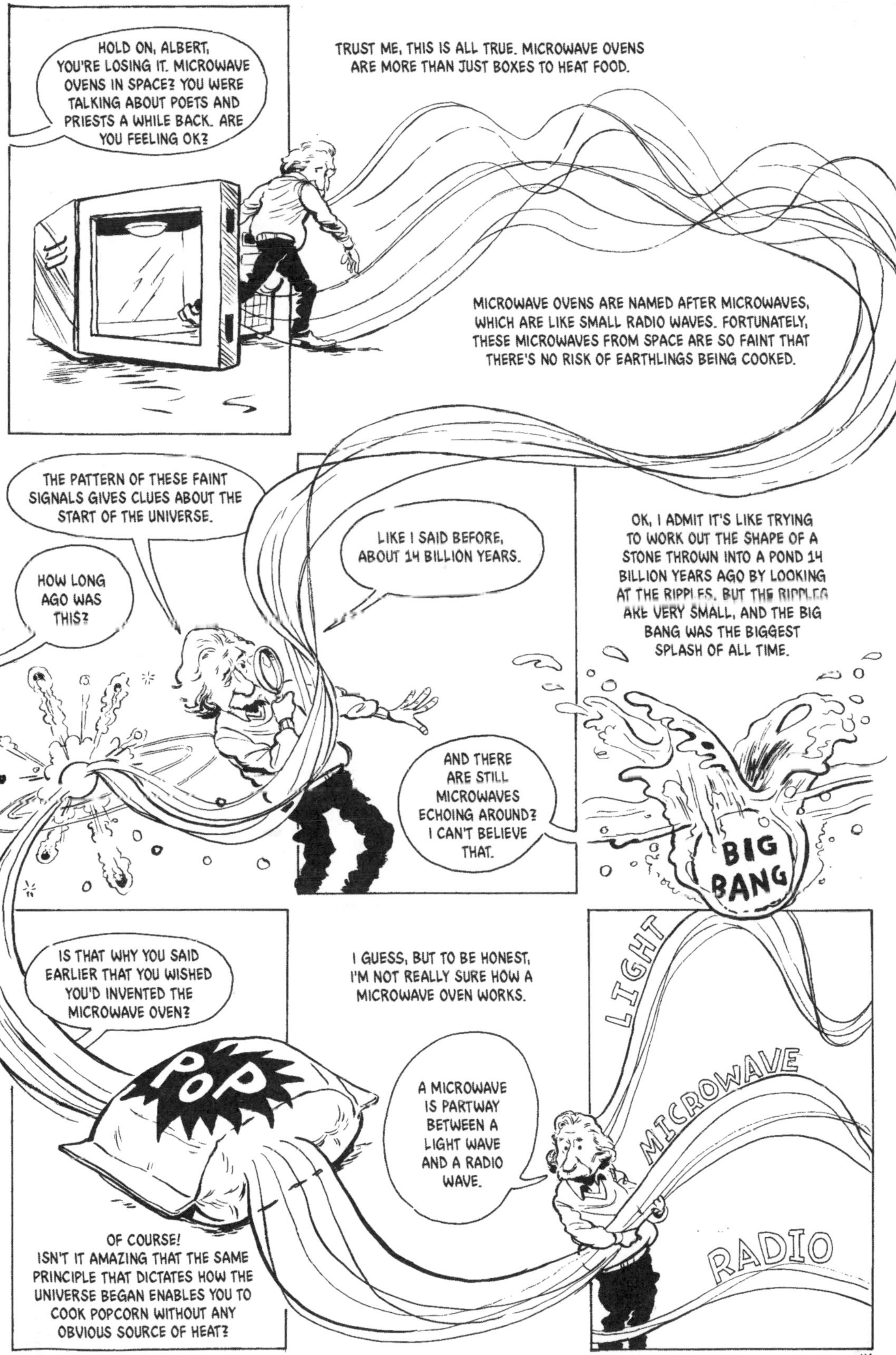
HOLD ON, ALBERT, YOU'RE LOSING IT. MICROWAVE OVENS IN SPACE? YOU WERE TALKING ABOUT POETS AND PRIESTS A WHILE BACK. ARE YOU FEELING OK?
TRUST ME, THIS IS ALL TRUE. MICROWAVE OVENS ARE MORE THAN JUST BOXES TO HEAT FOOD.
MICROWAVE OVENS ARE NAMED AFTER MICROWAVES, WHICH ARE LIKE SMALL RADIO WAVES. FORTUNATELY, THESE MICROWAVES FROM SPACE ARE SO FAINT THAT THERE'S NO RISK OF EARTHLINGS BEING COOKED.
THE PATTERN OF THESE FAINT SIGNALS GIVES CLUES ABOUT THE START OF THE UNIVERSE.
HOW LONG AGO WAS THIS?
LIKE I SAID BEFORE, ABOUT 14 BILLION YEARS.
AND THERE ARE STILL MICROWAVES ECHOING AROUND? I CAN'T BELIEVE THAT.
OK, I ADMIT IT'S LIKE TRYING TO WORK OUT THE SHAPE OF A STONE THROWN INTO A POND 14 BILLION YEARS AGO BY LOOKING AT THE RIPPLES, BUT THE RIPPLES ARE VERY SMALL, AND THE BIG BANG WAS THE BIGGEST SPLASH OF ALL TIME.
BIG BANG
IS THAT WHY YOU SAID EARLIER THAT YOU WISHED YOU'D INVENTED THE MICROWAVE OVEN?
I GUESS, BUT TO BE HONEST, I'M NOT REALLY SURE HOW A MICROWAVE OVEN WORKS.
LIGHT
MICROWAVE
POP
A MICROWAVE IS PARTWAY BETWEEN A LIGHT WAVE AND A RADIO WAVE.
RADIO
OF COURSE! ISN'T IT AMAZING THAT THE SAME PRINCIPLE THAT DICTATES HOW THE UNIVERSE BEGAN ENABLES YOU TO COOK POPCORN WITHOUT ANY OBVIOUS SOURCE OF HEAT?

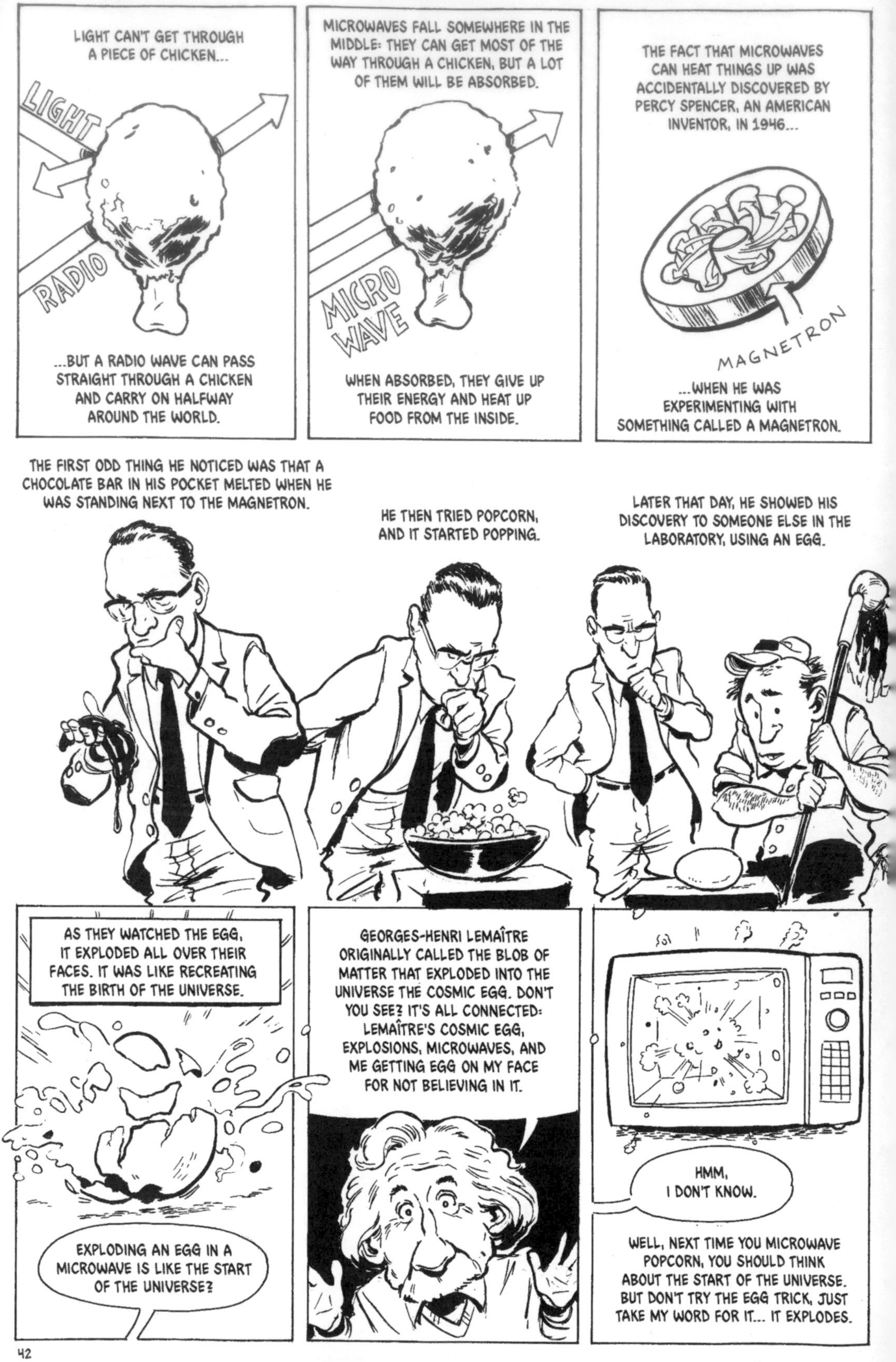
LIGHT CAN'T GET THROUGH A PIECE OF CHICKEN...
LIGHT
RADIO
...BUT A RADIO WAVE CAN PASS STRAIGHT THROUGH A CHICKEN AND CARRY ON HALFWAY AROUND THE WORLD.
MICROWAVES FALL SOMEWHERE IN THE MIDDLE: THEY CAN GET MOST OF THE WAY THROUGH A CHICKEN, BUT A LOT OF THEM WILL BE ABSORBED.
MICRO WAVE
WHEN ABSORBED, THEY GIVE UP THEIR ENERGY AND HEAT UP FOOD FROM THE INSIDE.
THE FACT THAT MICROWAVES CAN HEAT THINGS UP WAS ACCIDENTALLY DISCOVERED BY PERCY SPENCER, AN AMERICAN INVENTOR, IN 1946...
MAGNETRON
...WHEN HE WAS EXPERIMENTING WITH SOMETHING CALLED A MAGNETRON.
THE FIRST ODD THING HE NOTICED WAS THAT A CHOCOLATE BAR IN HIS POCKET MELTED WHEN HE WAS STANDING NEXT TO THE MAGNETRON.
HE THEN TRIED POPCORN, AND IT STARTED POPPING.
LATER THAT DAY, HE SHOWED HIS DISCOVERY TO SOMEONE ELSE IN THE LABORATORY, USING AN EGG.
AS THEY WATCHED THE EGG, IT EXPLODED ALL OVER THEIR FACES. IT WAS LIKE RECREATING THE BIRTH OF THE UNIVERSE.
EXPLODING AN EGG IN A MICROWAVE IS LIKE THE START OF THE UNIVERSE?
GEORGES-HENRI LEMAÎTRE ORIGINALLY CALLED THE BLOB OF MATTER THAT EXPLODED INTO THE UNIVERSE THE COSMIC EGG. DON'T YOU SEE? IT'S ALL CONNECTED: LEMAÎTRE'S COSMIC EGG, EXPLOSIONS, MICROWAVES, AND ME GETTING EGG ON MY FACE FOR NOT BELIEVING IN IT.
HMM, I DON'T KNOW.
WELL, NEXT TIME YOU MICROWAVE POPCORN, YOU SHOULD THINK ABOUT THE START OF THE UNIVERSE. BUT DON'T TRY THE EGG TRICK, JUST TAKE MY WORD FOR IT... IT EXPLODES.

CHAPTER 7: DARK MATTER AND HOW WIMPS MAY SAVE THE UNIVERSE

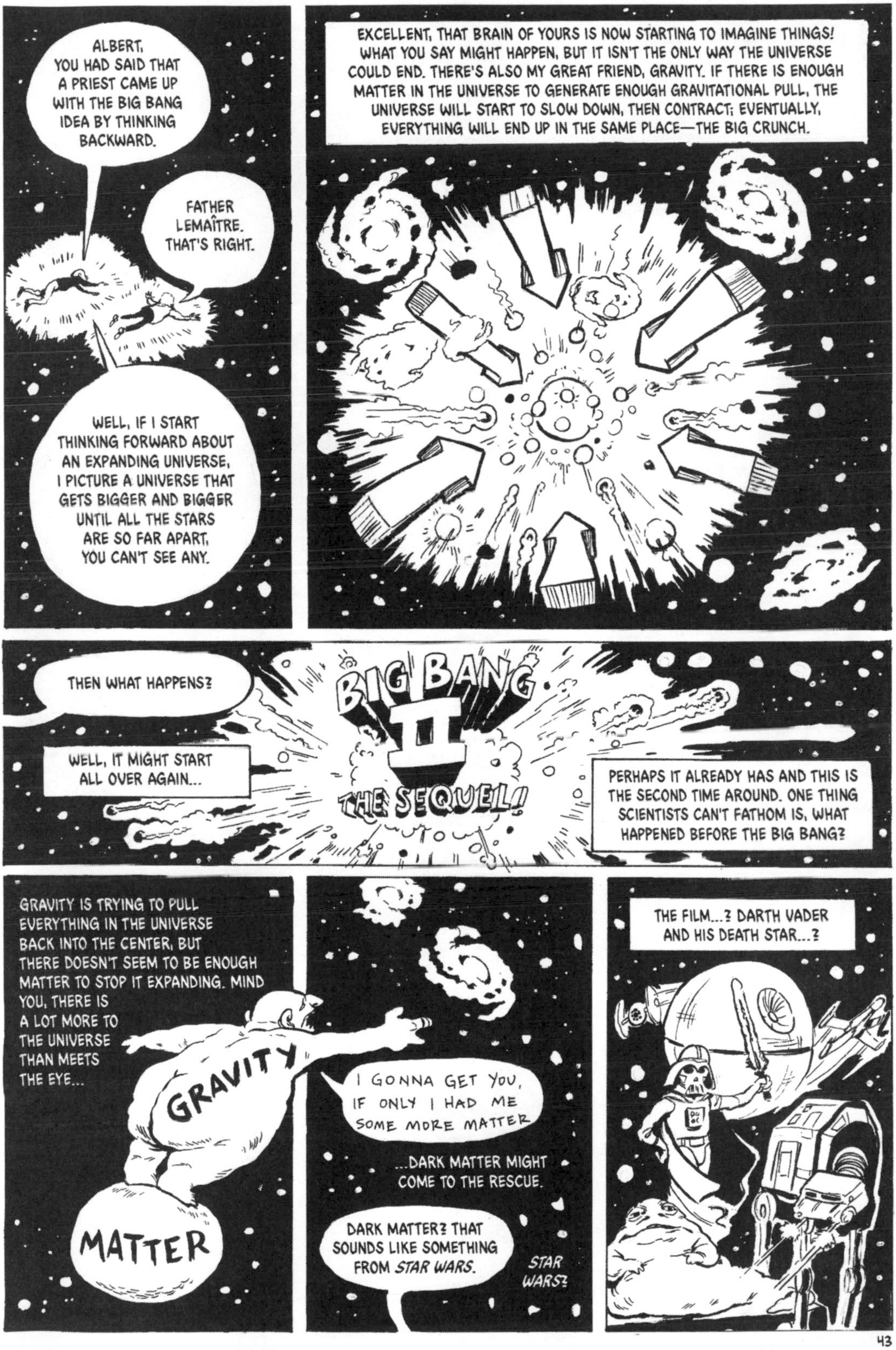
ALBERT, YOU HAD SAID THAT A PRIEST CAME UP WITH THE BIG BANG IDEA BY THINKING BACKWARD.
FATHER LEMAÎTRE. THAT'S RIGHT.
WELL, IF I START THINKING FORWARD ABOUT AN EXPANDING UNIVERSE, I PICTURE A UNIVERSE THAT GETS BIGGER AND BIGGER UNTIL ALL THE STARS ARE SO FAR APART, YOU CAN'T SEE ANY.
EXCELLENT, THAT BRAIN OF YOURS IS NOW STARTING TO IMAGINE THINGS! WHAT YOU SAY MIGHT HAPPEN, BUT IT ISN'T THE ONLY WAY THE UNIVERSE COULD END. THERE'S ALSO MY GREAT FRIEND, GRAVITY. IF THERE IS ENOUGH MATTER IN THE UNIVERSE TO GENERATE ENOUGH GRAVITATIONAL PULL, THE UNIVERSE WILL START TO SLOW DOWN, THEN CONTRACT; EVENTUALLY, EVERYTHING WILL END UP IN THE SAME PLACE—THE BIG CRUNCH.
THEN WHAT HAPPENS?
WELL, IT MIGHT START ALL OVER AGAIN...
BIG BANG II
THE SEQUEL!
PERHAPS IT ALREADY HAS AND THIS IS THE SECOND TIME AROUND. ONE THING SCIENTISTS CAN'T FATHOM IS, WHAT HAPPENED BEFORE THE BIG BANG?
GRAVITY IS TRYING TO PULL EVERYTHING IN THE UNIVERSE BACK INTO THE CENTER, BUT THERE DOESN'T SEEM TO BE ENOUGH MATTER TO STOP IT EXPANDING. MIND YOU, THERE IS A LOT MORE TO THE UNIVERSE THAN MEETS THE EYE...
GRAVITY
MATTER
I GONNA GET YOU, IF ONLY I HAD ME SOME MORE MATTER
...DARK MATTER MIGHT COME TO THE RESCUE.
DARK MATTER? THAT SOUNDS LIKE SOMETHING FROM *STAR WARS*.
STAR WARS?
THE FILM...? DARTH VADER AND HIS DEATH STAR...?

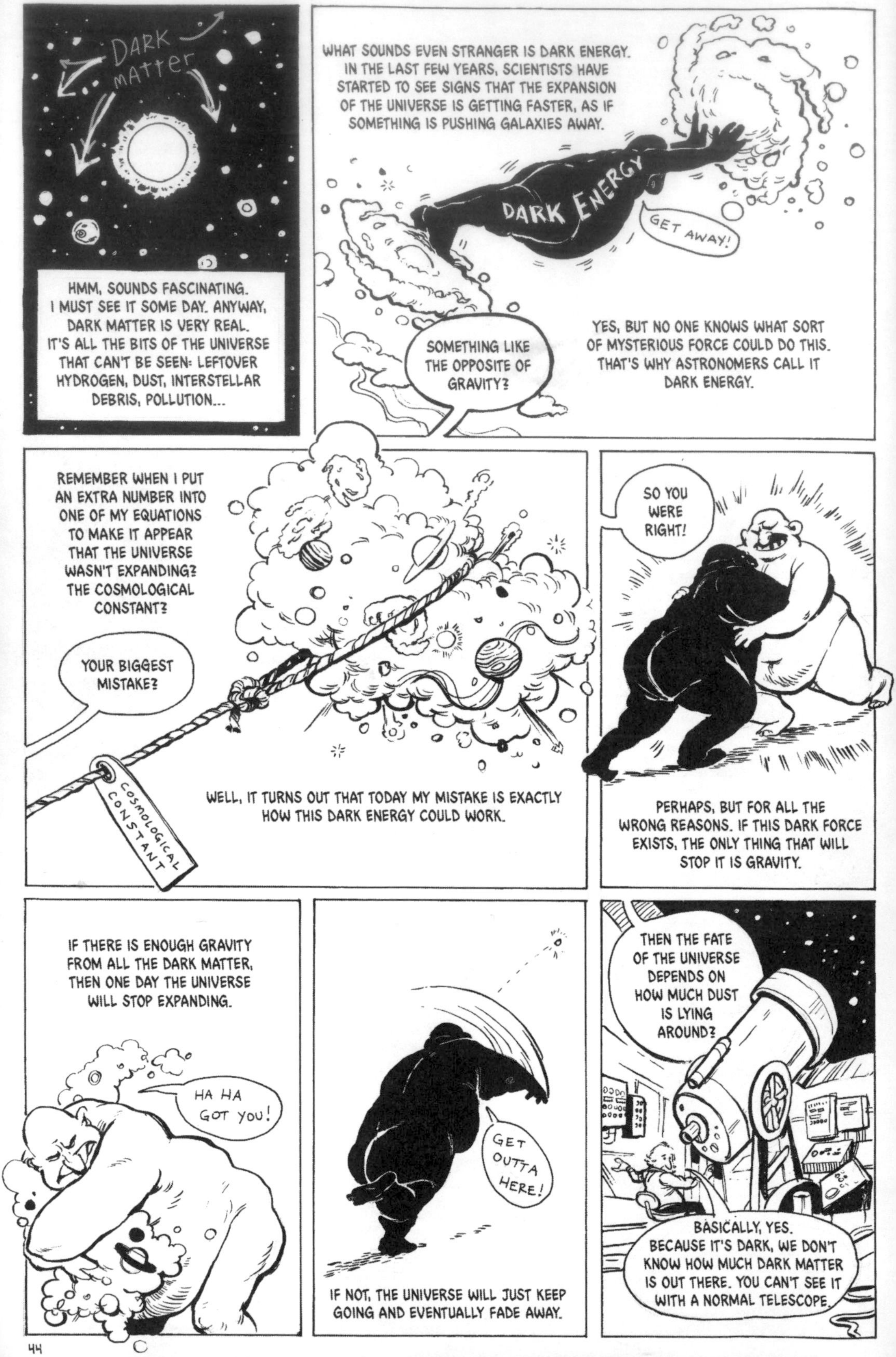
DARK MATTER
HMM, SOUNDS FASCINATING. I MUST SEE IT SOME DAY. ANYWAY, DARK MATTER IS VERY REAL. IT'S ALL THE BITS OF THE UNIVERSE THAT CAN'T BE SEEN: LEFTOVER HYDROGEN, DUST, INTERSTELLAR DEBRIS, POLLUTION...
WHAT SOUNDS EVEN STRANGER IS DARK ENERGY. IN THE LAST FEW YEARS, SCIENTISTS HAVE STARTED TO SEE SIGNS THAT THE EXPANSION OF THE UNIVERSE IS GETTING FASTER, AS IF SOMETHING IS PUSHING GALAXIES AWAY.
DARK ENERGY
GET AWAY!
SOMETHING LIKE THE OPPOSITE OF GRAVITY?
YES, BUT NO ONE KNOWS WHAT SORT OF MYSTERIOUS FORCE COULD DO THIS. THAT'S WHY ASTRONOMERS CALL IT DARK ENERGY.
REMEMBER WHEN I PUT AN EXTRA NUMBER INTO ONE OF MY EQUATIONS TO MAKE IT APPEAR THAT THE UNIVERSE WASN'T EXPANDING? THE COSMOLOGICAL CONSTANT?
YOUR BIGGEST MISTAKE?
COSMOLOGICAL CONSTANT
WELL, IT TURNS OUT THAT TODAY MY MISTAKE IS EXACTLY HOW THIS DARK ENERGY COULD WORK.
SO YOU WERE RIGHT!
PERHAPS, BUT FOR ALL THE WRONG REASONS. IF THIS DARK FORCE EXISTS, THE ONLY THING THAT WILL STOP IT IS GRAVITY.
IF THERE IS ENOUGH GRAVITY FROM ALL THE DARK MATTER, THEN ONE DAY THE UNIVERSE WILL STOP EXPANDING.
HA HA GOT YOU!
GET OUTTA HERE!
IF NOT, THE UNIVERSE WILL JUST KEEP GOING AND EVENTUALLY FADE AWAY.
THEN THE FATE OF THE UNIVERSE DEPENDS ON HOW MUCH DUST IS LYING AROUND?
BASICALLY, YES. BECAUSE IT'S DARK, WE DON'T KNOW HOW MUCH DARK MATTER IS OUT THERE. YOU CAN'T SEE IT WITH A NORMAL TELESCOPE.

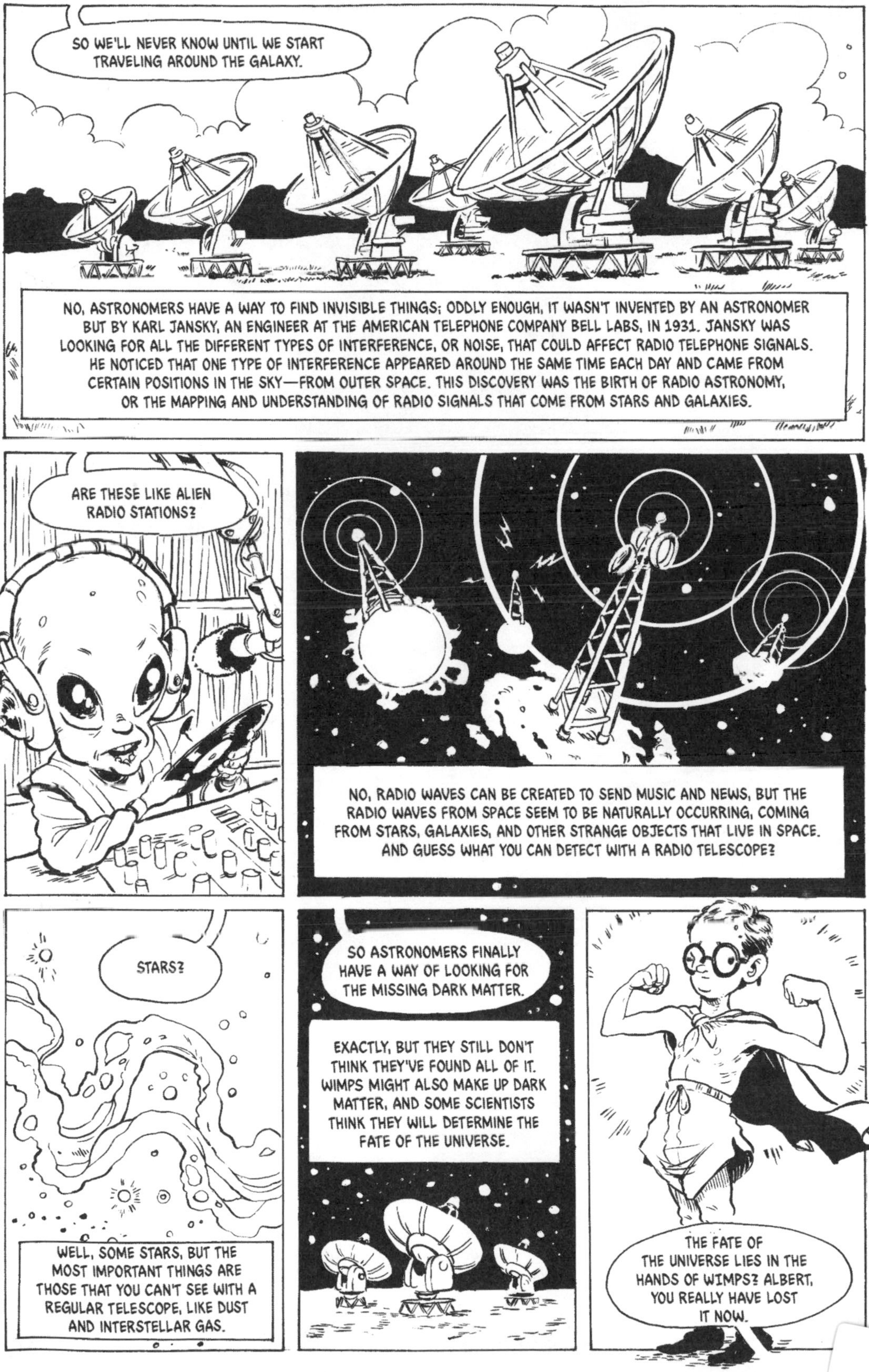
SO WE'LL NEVER KNOW UNTIL WE START TRAVELING AROUND THE GALAXY.
NO, ASTRONOMERS HAVE A WAY TO FIND INVISIBLE THINGS; ODDLY ENOUGH, IT WASN'T INVENTED BY AN ASTRONOMER BUT BY KARL JANSKY, AN ENGINEER AT THE AMERICAN TELEPHONE COMPANY BELL LABS, IN 1931. JANSKY WAS LOOKING FOR ALL THE DIFFERENT TYPES OF INTERFERENCE, OR NOISE, THAT COULD AFFECT RADIO TELEPHONE SIGNALS. HE NOTICED THAT ONE TYPE OF INTERFERENCE APPEARED AROUND THE SAME TIME EACH DAY AND CAME FROM CERTAIN POSITIONS IN THE SKY—FROM OUTER SPACE. THIS DISCOVERY WAS THE BIRTH OF RADIO ASTRONOMY, OR THE MAPPING AND UNDERSTANDING OF RADIO SIGNALS THAT COME FROM STARS AND GALAXIES.
ARE THESE LIKE ALIEN RADIO STATIONS?
NO, RADIO WAVES CAN BE CREATED TO SEND MUSIC AND NEWS, BUT THE RADIO WAVES FROM SPACE SEEM TO BE NATURALLY OCCURRING, COMING FROM STARS, GALAXIES, AND OTHER STRANGE OBJECTS THAT LIVE IN SPACE. AND GUESS WHAT YOU CAN DETECT WITH A RADIO TELESCOPE?
STARS?
WELL, SOME STARS, BUT THE MOST IMPORTANT THINGS ARE THOSE THAT YOU CAN'T SEE WITH A REGULAR TELESCOPE, LIKE DUST AND INTERSTELLAR GAS.
SO ASTRONOMERS FINALLY HAVE A WAY OF LOOKING FOR THE MISSING DARK MATTER.
EXACTLY, BUT THEY STILL DON'T THINK THEY'VE FOUND ALL OF IT. WIMPS MIGHT ALSO MAKE UP DARK MATTER, AND SOME SCIENTISTS THINK THEY WILL DETERMINE THE FATE OF THE UNIVERSE.
THE FATE OF THE UNIVERSE LIES IN THE HANDS OF WIMPS? ALBERT, YOU REALLY HAVE LOST IT NOW.

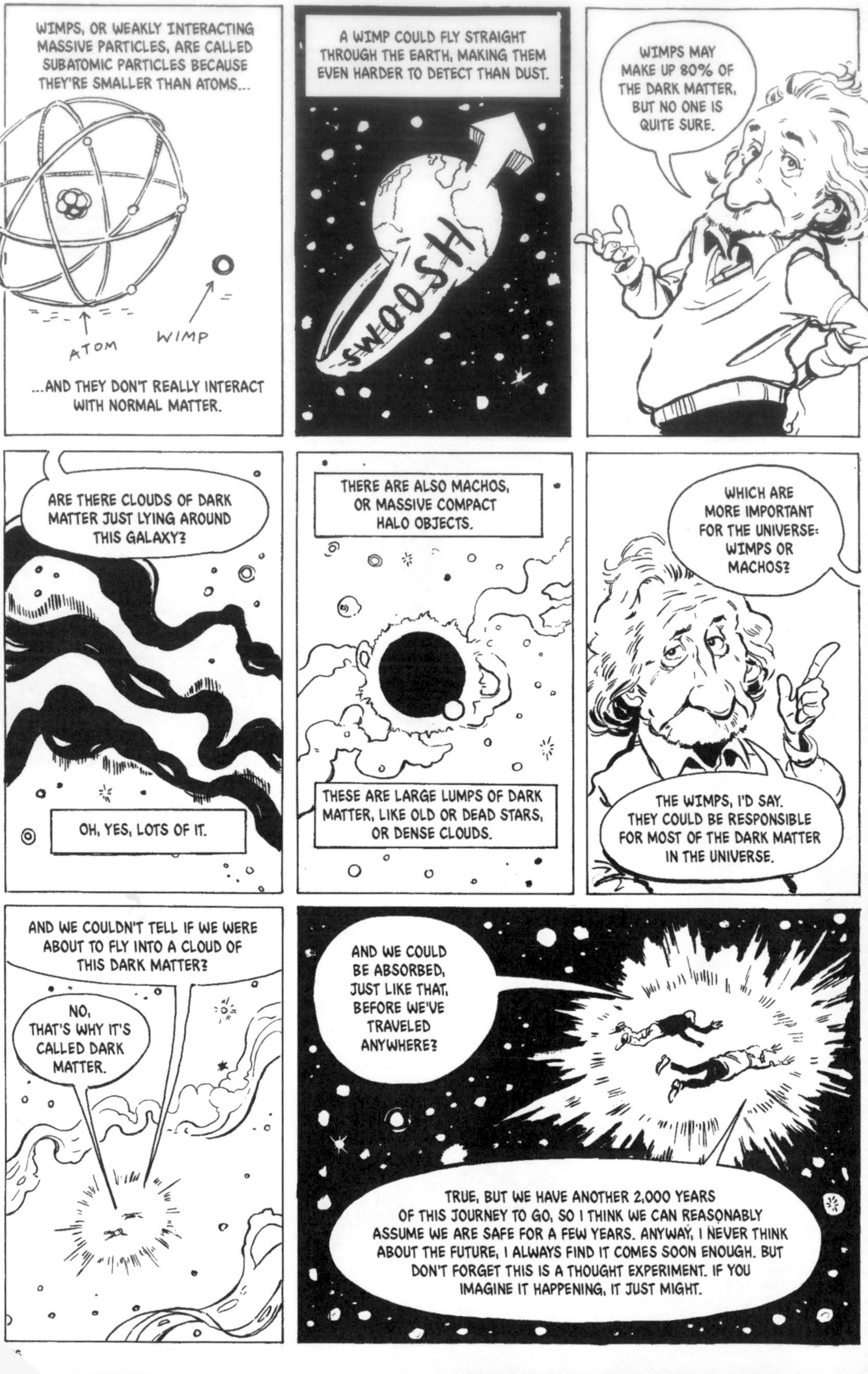
WIMPS, OR WEAKLY INTERACTING MASSIVE PARTICLES, ARE CALLED SUBATOMIC PARTICLES BECAUSE THEY'RE SMALLER THAN ATOMS...
ATOM
WIMP
...AND THEY DON'T REALLY INTERACT WITH NORMAL MATTER.
A WIMP COULD FLY STRAIGHT THROUGH THE EARTH, MAKING THEM EVEN HARDER TO DETECT THAN DUST.
SWOOSH
WIMPS MAY MAKE UP 80% OF THE DARK MATTER, BUT NO ONE IS QUITE SURE.
ARE THERE CLOUDS OF DARK MATTER JUST LYING AROUND THIS GALAXY?
OH, YES, LOTS OF IT.
THERE ARE ALSO MACHOS, OR MASSIVE COMPACT HALO OBJECTS.
THESE ARE LARGE LUMPS OF DARK MATTER, LIKE OLD OR DEAD STARS, OR DENSE CLOUDS.
WHICH ARE MORE IMPORTANT FOR THE UNIVERSE: WIMPS OR MACHOS?
THE WIMPS, I'D SAY. THEY COULD BE RESPONSIBLE FOR MOST OF THE DARK MATTER IN THE UNIVERSE.
AND WE COULDN'T TELL IF WE WERE ABOUT TO FLY INTO A CLOUD OF THIS DARK MATTER?
NO, THAT'S WHY IT'S CALLED DARK MATTER.
AND WE COULD BE ABSORBED, JUST LIKE THAT, BEFORE WE'VE TRAVELED ANYWHERE?
TRUE, BUT WE HAVE ANOTHER 2,000 YEARS OF THIS JOURNEY TO GO, SO I THINK WE CAN REASONABLY ASSUME WE ARE SAFE FOR A FEW YEARS. ANYWAY, I NEVER THINK ABOUT THE FUTURE, I ALWAYS FIND IT COMES SOON ENOUGH. BUT DON'T FORGET THIS IS A THOUGHT EXPERIMENT. IF YOU IMAGINE IT HAPPENING, IT JUST MIGHT.

CHAPTER 8:
THE STORY OF GRAVITY: THE WEAKLING, THE PLAGUE, AND THE APPLE

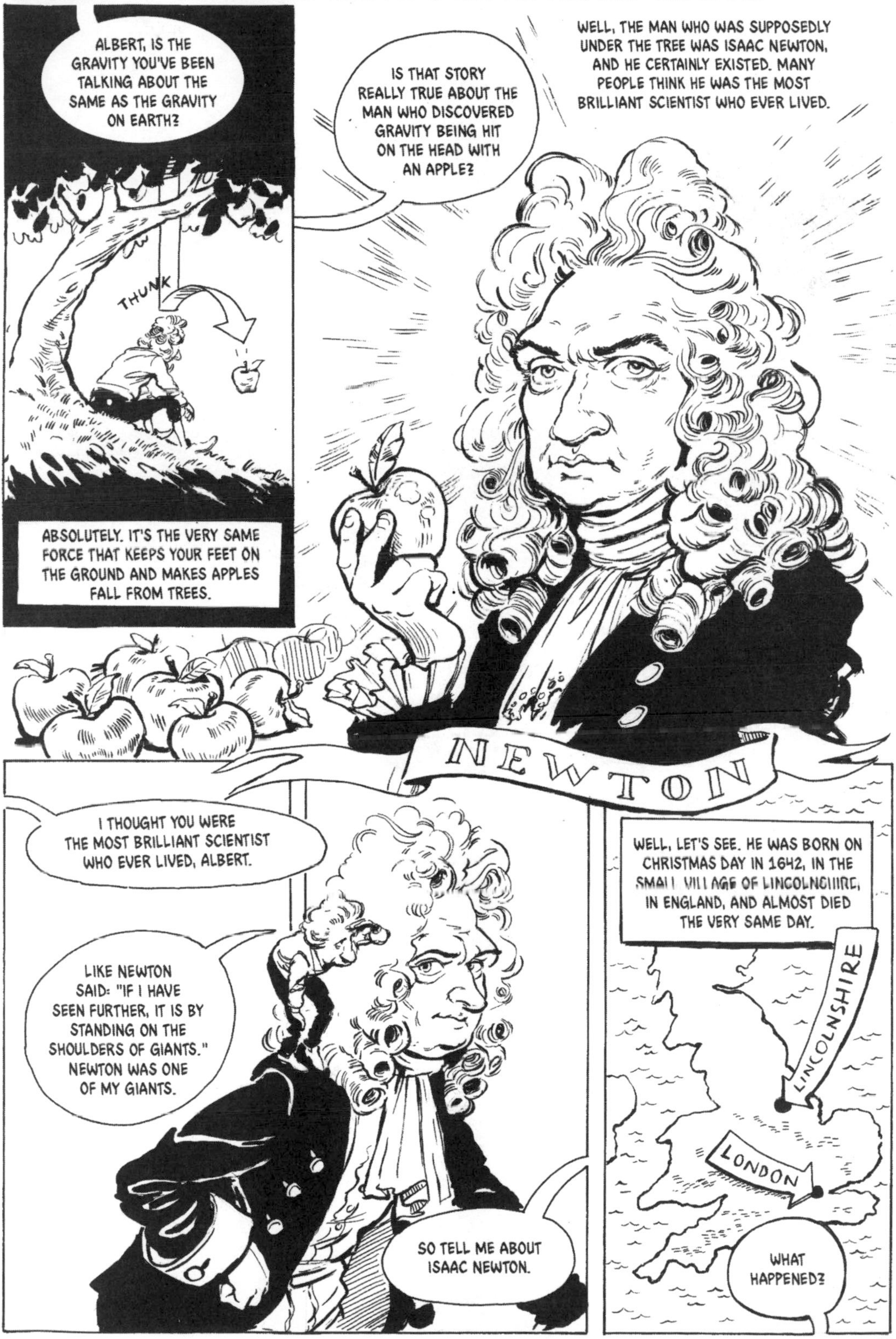
ALBERT, IS THE GRAVITY YOU'VE BEEN TALKING ABOUT THE SAME AS THE GRAVITY ON EARTH?
THUNK
ABSOLUTELY. IT'S THE VERY SAME FORCE THAT KEEPS YOUR FEET ON THE GROUND AND MAKES APPLES FALL FROM TREES.
IS THAT STORY REALLY TRUE ABOUT THE MAN WHO DISCOVERED GRAVITY BEING HIT ON THE HEAD WITH AN APPLE?
WELL, THE MAN WHO WAS SUPPOSEDLY UNDER THE TREE WAS ISAAC NEWTON, AND HE CERTAINLY EXISTED. MANY PEOPLE THINK HE WAS THE MOST BRILLIANT SCIENTIST WHO EVER LIVED.
NEWTON
I THOUGHT YOU WERE THE MOST BRILLIANT SCIENTIST WHO EVER LIVED, ALBERT.
LIKE NEWTON SAID: "IF I HAVE SEEN FURTHER, IT IS BY STANDING ON THE SHOULDERS OF GIANTS." NEWTON WAS ONE OF MY GIANTS.
SO TELL ME ABOUT ISAAC NEWTON.
WELL, LET'S SEE. HE WAS BORN ON CHRISTMAS DAY IN 1642, IN THE SMALL VILLAGE OF LINCOLNSHIRE, IN ENGLAND, AND ALMOST DIED THE VERY SAME DAY.
LINCOLNSHIRE
LONDON
WHAT HAPPENED?

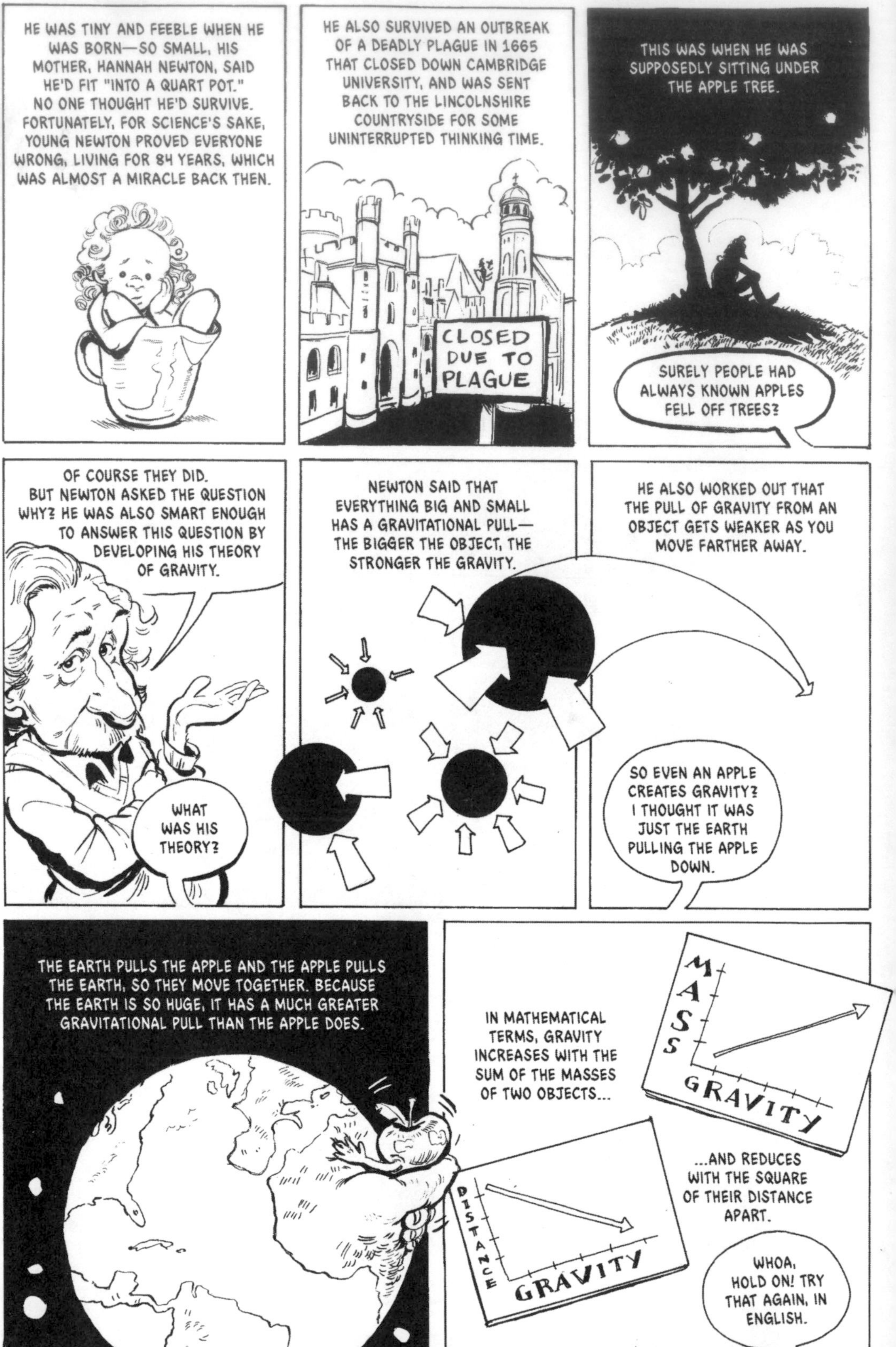
HE WAS TINY AND FEEBLE WHEN HE WAS BORN—SO SMALL, HIS MOTHER, HANNAH NEWTON, SAID HE'D FIT "INTO A QUART POT." NO ONE THOUGHT HE'D SURVIVE. FORTUNATELY, FOR SCIENCE'S SAKE, YOUNG NEWTON PROVED EVERYONE WRONG, LIVING FOR 84 YEARS, WHICH WAS ALMOST A MIRACLE BACK THEN.
HE ALSO SURVIVED AN OUTBREAK OF A DEADLY PLAGUE IN 1665 THAT CLOSED DOWN CAMBRIDGE UNIVERSITY, AND WAS SENT BACK TO THE LINCOLNSHIRE COUNTRYSIDE FOR SOME UNINTERRUPTED THINKING TIME.
CLOSED DUE TO PLAGUE
THIS WAS WHEN HE WAS SUPPOSEDLY SITTING UNDER THE APPLE TREE.
SURELY PEOPLE HAD ALWAYS KNOWN APPLES FELL OFF TREES?
OF COURSE THEY DID. BUT NEWTON ASKED THE QUESTION WHY? HE WAS ALSO SMART ENOUGH TO ANSWER THIS QUESTION BY DEVELOPING HIS THEORY OF GRAVITY.
WHAT WAS HIS THEORY?
NEWTON SAID THAT EVERYTHING BIG AND SMALL HAS A GRAVITATIONAL PULL—THE BIGGER THE OBJECT, THE STRONGER THE GRAVITY.
HE ALSO WORKED OUT THAT THE PULL OF GRAVITY FROM AN OBJECT GETS WEAKER AS YOU MOVE FARTHER AWAY.
SO EVEN AN APPLE CREATES GRAVITY? I THOUGHT IT WAS JUST THE EARTH PULLING THE APPLE DOWN.
THE EARTH PULLS THE APPLE AND THE APPLE PULLS THE EARTH, SO THEY MOVE TOGETHER. BECAUSE THE EARTH IS SO HUGE, IT HAS A MUCH GREATER GRAVITATIONAL PULL THAN THE APPLE DOES.
IN MATHEMATICAL TERMS, GRAVITY INCREASES WITH THE SUM OF THE MASSES OF TWO OBJECTS...
MASS
GRAVITY
...AND REDUCES WITH THE SQUARE OF THEIR DISTANCE APART.
DISTANCE
GRAVITY
WHOA, HOLD ON! TRY THAT AGAIN, IN ENGLISH.

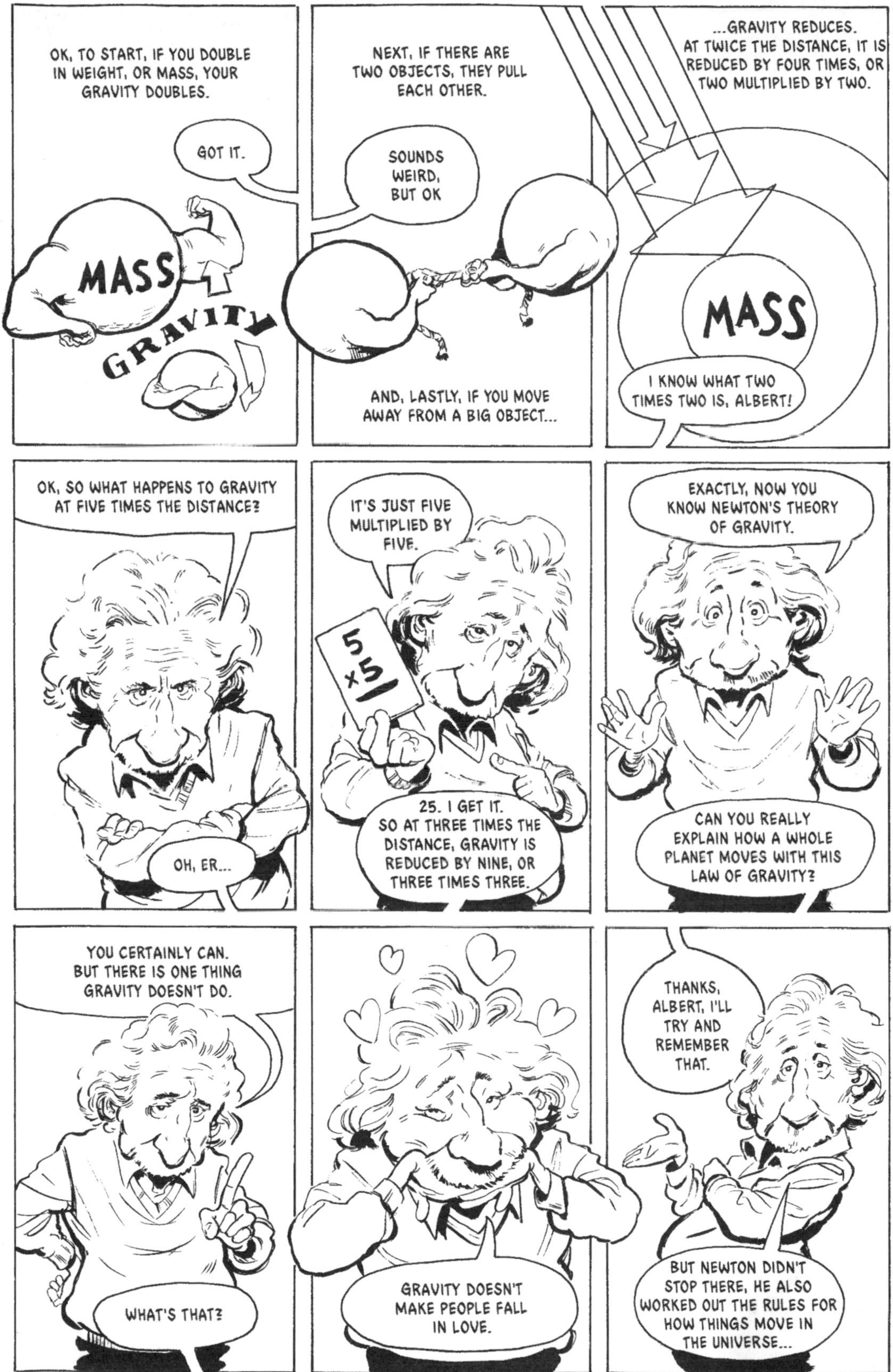
OK, TO START, IF YOU DOUBLE IN WEIGHT, OR MASS, YOUR GRAVITY DOUBLES.
GOT IT.
MASS
GRAVITY
NEXT, IF THERE ARE TWO OBJECTS, THEY PULL EACH OTHER.
SOUNDS WEIRD, BUT OK
AND, LASTLY, IF YOU MOVE AWAY FROM A BIG OBJECT...
...GRAVITY REDUCES. AT TWICE THE DISTANCE, IT IS REDUCED BY FOUR TIMES, OR TWO MULTIPLIED BY TWO.
MASS
I KNOW WHAT TWO TIMES TWO IS, ALBERT!
OK, SO WHAT HAPPENS TO GRAVITY AT FIVE TIMES THE DISTANCE?
OH, ER...
IT'S JUST FIVE MULTIPLIED BY FIVE.
5 x5
25. I GET IT. SO AT THREE TIMES THE DISTANCE, GRAVITY IS REDUCED BY NINE, OR THREE TIMES THREE.
EXACTLY, NOW YOU KNOW NEWTON'S THEORY OF GRAVITY.
CAN YOU REALLY EXPLAIN HOW A WHOLE PLANET MOVES WITH THIS LAW OF GRAVITY?
YOU CERTAINLY CAN. BUT THERE IS ONE THING GRAVITY DOESN'T DO.
WHAT'S THAT?
GRAVITY DOESN'T MAKE PEOPLE FALL IN LOVE.
THANKS, ALBERT, I'LL TRY AND REMEMBER THAT.
BUT NEWTON DIDN'T STOP THERE, HE ALSO WORKED OUT THE RULES FOR HOW THINGS MOVE IN THE UNIVERSE...

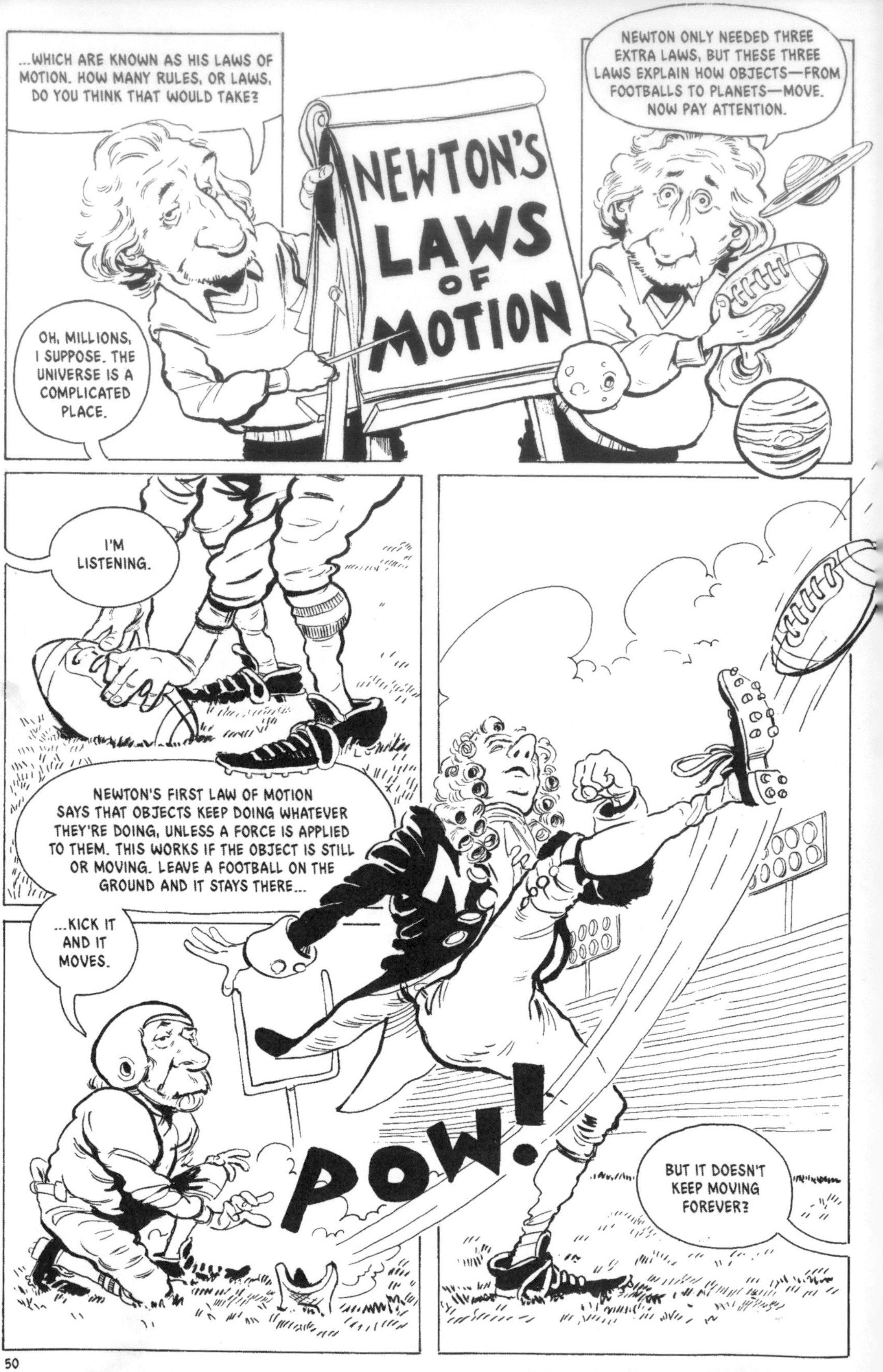
...WHICH ARE KNOWN AS HIS LAWS OF MOTION. HOW MANY RULES, OR LAWS, DO YOU THINK THAT WOULD TAKE?
NEWTON'S LAWS OF MOTION
OH, MILLIONS, I SUPPOSE. THE UNIVERSE IS A COMPLICATED PLACE.
NEWTON ONLY NEEDED THREE EXTRA LAWS, BUT THESE THREE LAWS EXPLAIN HOW OBJECTS—FROM FOOTBALLS TO PLANETS—MOVE. NOW PAY ATTENTION.
I'M LISTENING.
NEWTON'S FIRST LAW OF MOTION SAYS THAT OBJECTS KEEP DOING WHATEVER THEY'RE DOING, UNLESS A FORCE IS APPLIED TO THEM. THIS WORKS IF THE OBJECT IS STILL OR MOVING. LEAVE A FOOTBALL ON THE GROUND AND IT STAYS THERE...
...KICK IT AND IT MOVES.
POW!
BUT IT DOESN'T KEEP MOVING FOREVER?

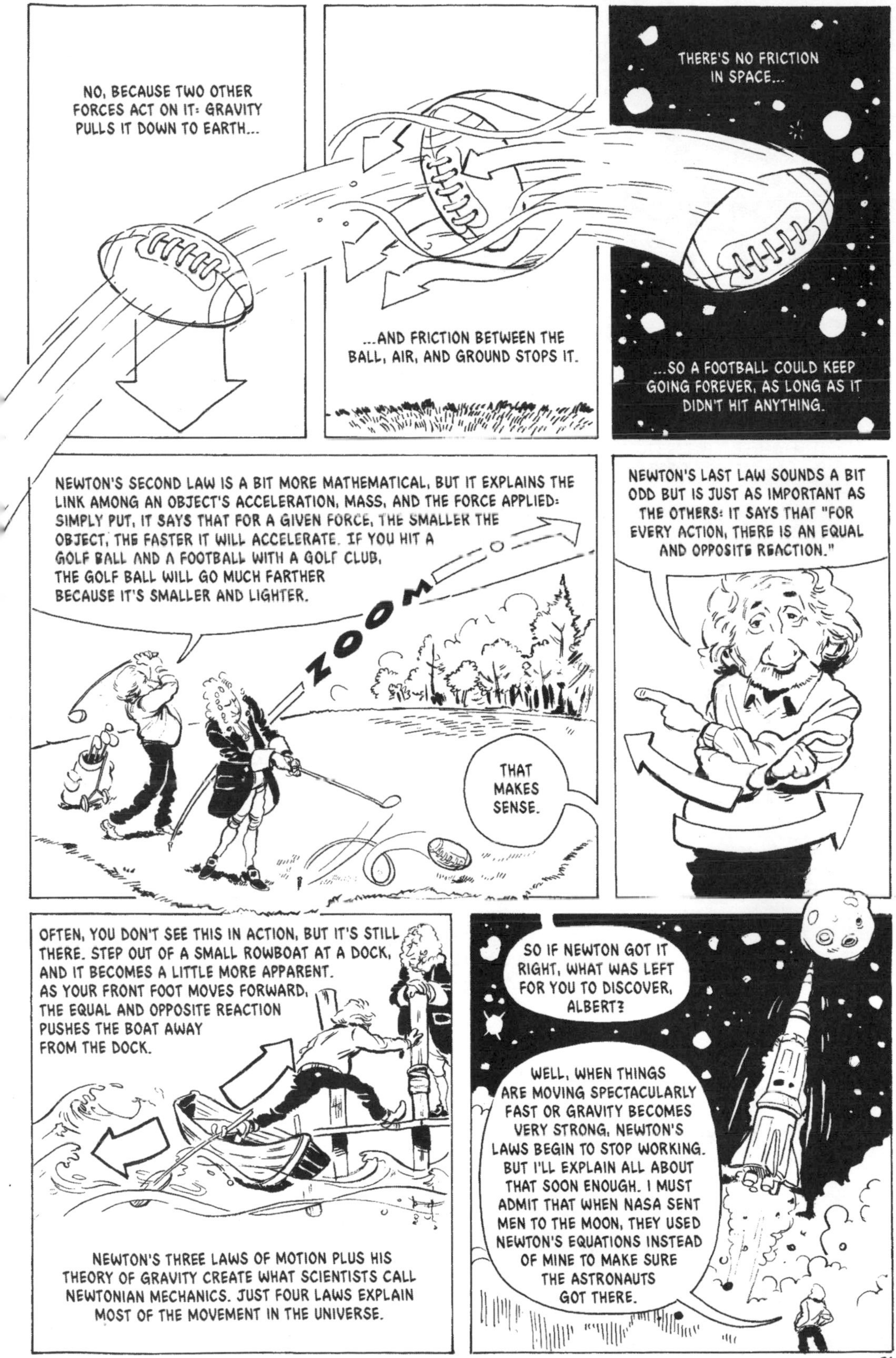
NO, BECAUSE TWO OTHER FORCES ACT ON IT: GRAVITY PULLS IT DOWN TO EARTH...
...AND FRICTION BETWEEN THE BALL, AIR, AND GROUND STOPS IT.
THERE'S NO FRICTION IN SPACE...
...SO A FOOTBALL COULD KEEP GOING FOREVER, AS LONG AS IT DIDN'T HIT ANYTHING.
NEWTON'S SECOND LAW IS A BIT MORE MATHEMATICAL, BUT IT EXPLAINS THE LINK AMONG AN OBJECT'S ACCELERATION, MASS, AND THE FORCE APPLIED: SIMPLY PUT, IT SAYS THAT FOR A GIVEN FORCE, THE SMALLER THE OBJECT, THE FASTER IT WILL ACCELERATE. IF YOU HIT A GOLF BALL AND A FOOTBALL WITH A GOLF CLUB, THE GOLF BALL WILL GO MUCH FARTHER BECAUSE IT'S SMALLER AND LIGHTER.
ZOOM
THAT MAKES SENSE.
NEWTON'S LAST LAW SOUNDS A BIT ODD BUT IS JUST AS IMPORTANT AS THE OTHERS: IT SAYS THAT "FOR EVERY ACTION, THERE IS AN EQUAL AND OPPOSITE REACTION."
OFTEN, YOU DON'T SEE THIS IN ACTION, BUT IT'S STILL THERE. STEP OUT OF A SMALL ROWBOAT AT A DOCK, AND IT BECOMES A LITTLE MORE APPARENT. AS YOUR FRONT FOOT MOVES FORWARD, THE EQUAL AND OPPOSITE REACTION PUSHES THE BOAT AWAY FROM THE DOCK.
NEWTON'S THREE LAWS OF MOTION PLUS HIS THEORY OF GRAVITY CREATE WHAT SCIENTISTS CALL NEWTONIAN MECHANICS. JUST FOUR LAWS EXPLAIN MOST OF THE MOVEMENT IN THE UNIVERSE.
SO IF NEWTON GOT IT RIGHT, WHAT WAS LEFT FOR YOU TO DISCOVER, ALBERT?
WELL, WHEN THINGS ARE MOVING SPECTACULARLY FAST OR GRAVITY BECOMES VERY STRONG, NEWTON'S LAWS BEGIN TO STOP WORKING. BUT I'LL EXPLAIN ALL ABOUT THAT SOON ENOUGH. I MUST ADMIT THAT WHEN NASA SENT MEN TO THE MOON, THEY USED NEWTON'S EQUATIONS INSTEAD OF MINE TO MAKE SURE THE ASTRONAUTS GOT THERE.

CHAPTER 9: ISAAC NEWTON'S LINK TO HARRY POTTER AND HOW TO WEIGH A PLANET

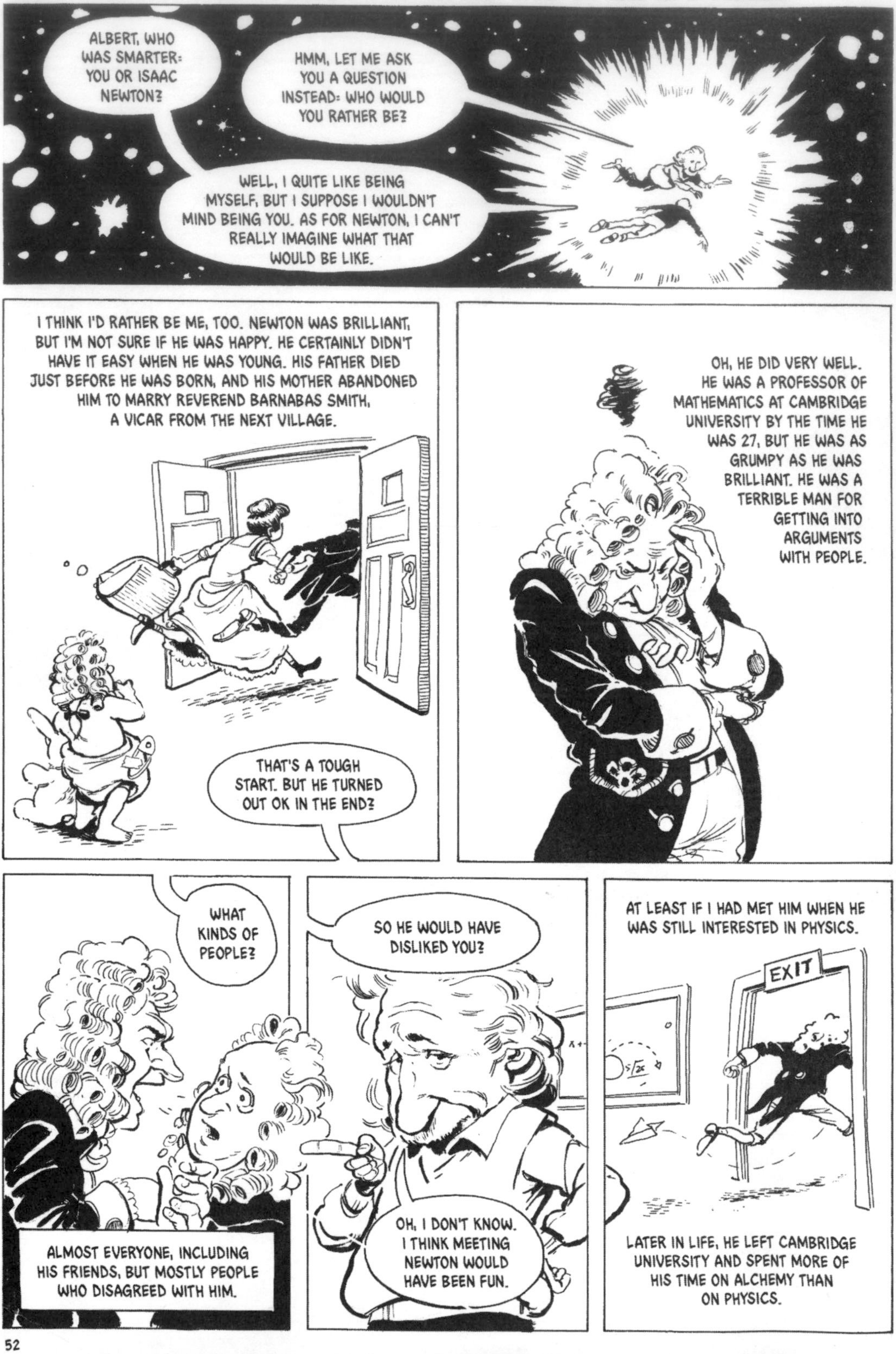

ALBERT, WHO WAS SMARTER: YOU OR ISAAC NEWTON?
HMM, LET ME ASK YOU A QUESTION INSTEAD: WHO WOULD YOU RATHER BE?
WELL, I QUITE LIKE BEING MYSELF, BUT I SUPPOSE I WOULDN'T MIND BEING YOU. AS FOR NEWTON, I CAN'T REALLY IMAGINE WHAT THAT WOULD BE LIKE.
I THINK I'D RATHER BE ME, TOO. NEWTON WAS BRILLIANT, BUT I'M NOT SURE IF HE WAS HAPPY. HE CERTAINLY DIDN'T HAVE IT EASY WHEN HE WAS YOUNG. HIS FATHER DIED JUST BEFORE HE WAS BORN, AND HIS MOTHER ABANDONED HIM TO MARRY REVEREND BARNABAS SMITH, A VICAR FROM THE NEXT VILLAGE.
THAT'S A TOUGH START. BUT HE TURNED OUT OK IN THE END?
OH, HE DID VERY WELL. HE WAS A PROFESSOR OF MATHEMATICS AT CAMBRIDGE UNIVERSITY BY THE TIME HE WAS 27, BUT HE WAS AS GRUMPY AS HE WAS BRILLIANT. HE WAS A TERRIBLE MAN FOR GETTING INTO ARGUMENTS WITH PEOPLE.
WHAT KINDS OF PEOPLE?
ALMOST EVERYONE, INCLUDING HIS FRIENDS, BUT MOSTLY PEOPLE WHO DISAGREED WITH HIM.
SO HE WOULD HAVE DISLIKED YOU?
OH, I DON'T KNOW. I THINK MEETING NEWTON WOULD HAVE BEEN FUN.
AT LEAST IF I HAD MET HIM WHEN HE WAS STILL INTERESTED IN PHYSICS.
EXIT
LATER IN LIFE, HE LEFT CAMBRIDGE UNIVERSITY AND SPENT MORE OF HIS TIME ON ALCHEMY THAN ON PHYSICS.

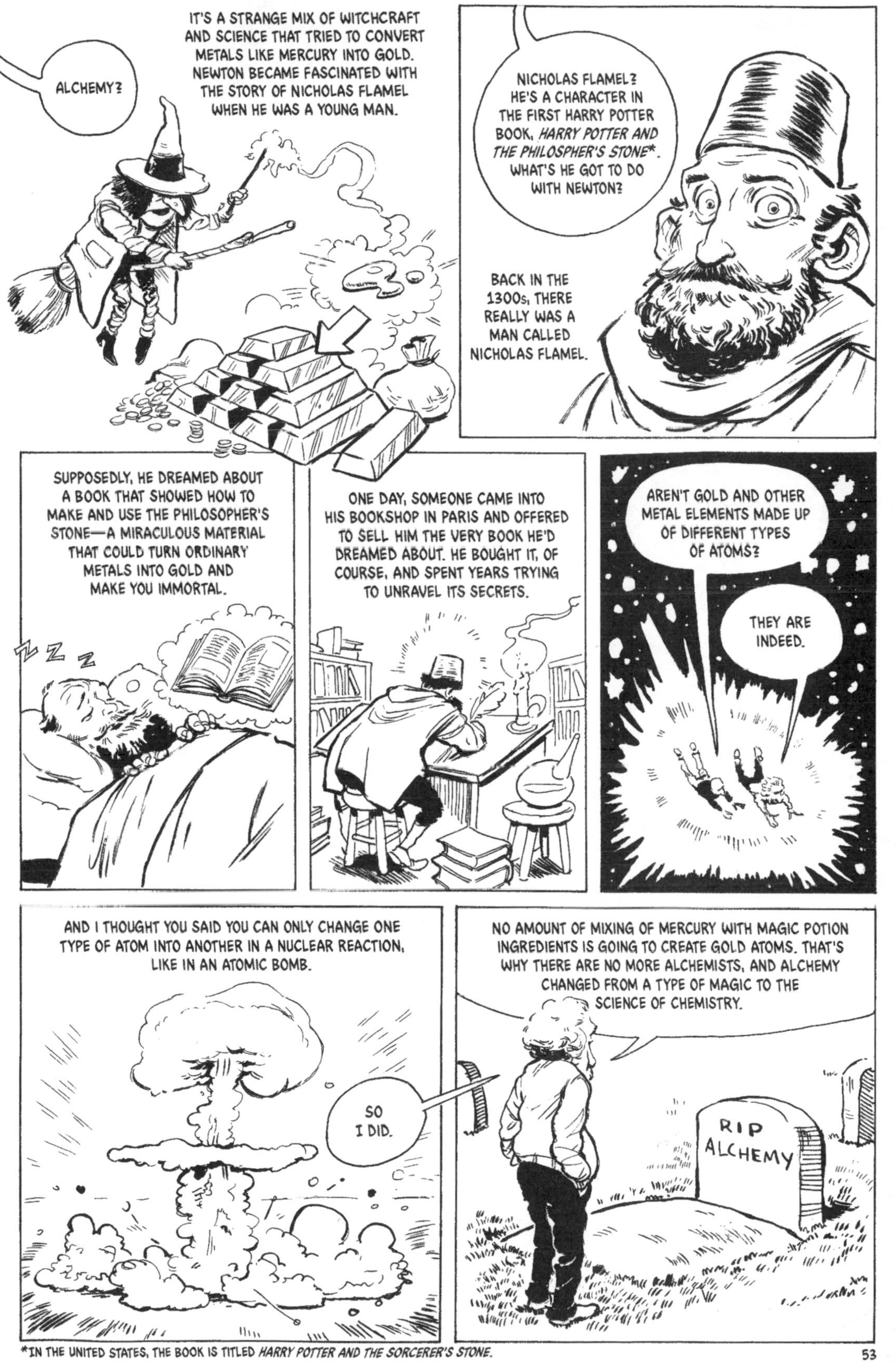
ALCHEMY?
IT'S A STRANGE MIX OF WITCHCRAFT AND SCIENCE THAT TRIED TO CONVERT METALS LIKE MERCURY INTO GOLD. NEWTON BECAME FASCINATED WITH THE STORY OF NICHOLAS FLAMEL WHEN HE WAS A YOUNG MAN.
NICHOLAS FLAMEL? HE'S A CHARACTER IN THE FIRST HARRY POTTER BOOK, *HARRY POTTER AND THE PHILOSPHER'S STONE**. WHAT'S HE GOT TO DO WITH NEWTON?
BACK IN THE 1300s, THERE REALLY WAS A MAN CALLED NICHOLAS FLAMEL.
SUPPOSEDLY, HE DREAMED ABOUT A BOOK THAT SHOWED HOW TO MAKE AND USE THE PHILOSOPHER'S STONE—A MIRACULOUS MATERIAL THAT COULD TURN ORDINARY METALS INTO GOLD AND MAKE YOU IMMORTAL.
Z Z Z
ONE DAY, SOMEONE CAME INTO HIS BOOKSHOP IN PARIS AND OFFERED TO SELL HIM THE VERY BOOK HE'D DREAMED ABOUT. HE BOUGHT IT, OF COURSE, AND SPENT YEARS TRYING TO UNRAVEL ITS SECRETS.
AREN'T GOLD AND OTHER METAL ELEMENTS MADE UP OF DIFFERENT TYPES OF ATOMS?
THEY ARE INDEED.
AND I THOUGHT YOU SAID YOU CAN ONLY CHANGE ONE TYPE OF ATOM INTO ANOTHER IN A NUCLEAR REACTION, LIKE IN AN ATOMIC BOMB.
SO I DID.
NO AMOUNT OF MIXING OF MERCURY WITH MAGIC POTION INGREDIENTS IS GOING TO CREATE GOLD ATOMS. THAT'S WHY THERE ARE NO MORE ALCHEMISTS, AND ALCHEMY CHANGED FROM A TYPE OF MAGIC TO THE SCIENCE OF CHEMISTRY.
RIP ALCHEMY

*IN THE UNITED STATES, THE BOOK IS TITLED *HARRY POTTER AND THE SORCERER'S STONE.*

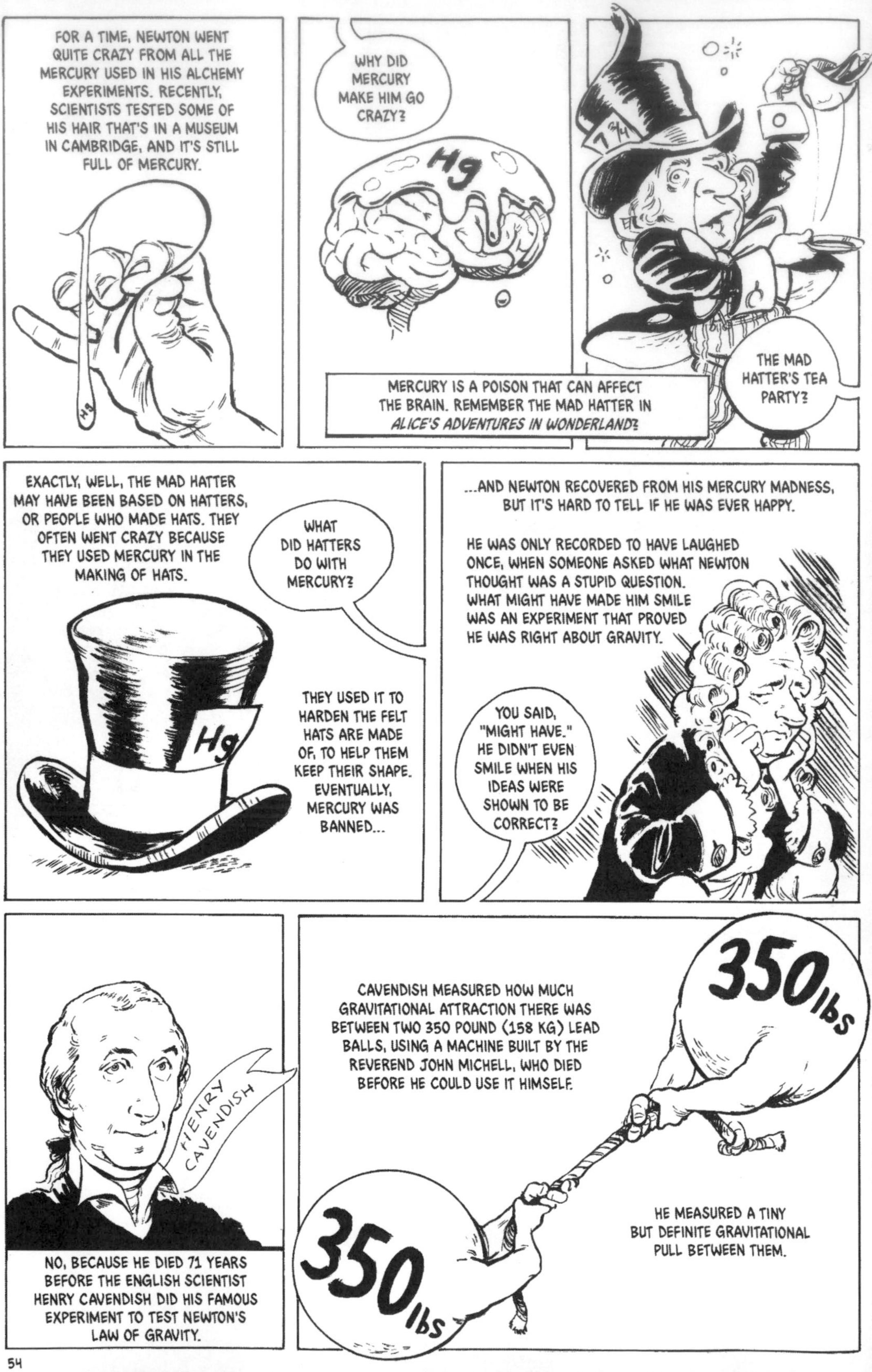
FOR A TIME, NEWTON WENT QUITE CRAZY FROM ALL THE MERCURY USED IN HIS ALCHEMY EXPERIMENTS. RECENTLY, SCIENTISTS TESTED SOME OF HIS HAIR THAT'S IN A MUSEUM IN CAMBRIDGE, AND IT'S STILL FULL OF MERCURY.
Hg
WHY DID MERCURY MAKE HIM GO CRAZY?
Hg
MERCURY IS A POISON THAT CAN AFFECT THE BRAIN. REMEMBER THE MAD HATTER IN *ALICE'S ADVENTURES IN WONDERLAND*?
THE MAD HATTER'S TEA PARTY?
EXACTLY, WELL, THE MAD HATTER MAY HAVE BEEN BASED ON HATTERS, OR PEOPLE WHO MADE HATS. THEY OFTEN WENT CRAZY BECAUSE THEY USED MERCURY IN THE MAKING OF HATS.
WHAT DID HATTERS DO WITH MERCURY?
Hg
THEY USED IT TO HARDEN THE FELT HATS ARE MADE OF, TO HELP THEM KEEP THEIR SHAPE. EVENTUALLY, MERCURY WAS BANNED...
...AND NEWTON RECOVERED FROM HIS MERCURY MADNESS, BUT IT'S HARD TO TELL IF HE WAS EVER HAPPY.
HE WAS ONLY RECORDED TO HAVE LAUGHED ONCE, WHEN SOMEONE ASKED WHAT NEWTON THOUGHT WAS A STUPID QUESTION. WHAT MIGHT HAVE MADE HIM SMILE WAS AN EXPERIMENT THAT PROVED HE WAS RIGHT ABOUT GRAVITY.
YOU SAID, "MIGHT HAVE." HE DIDN'T EVEN SMILE WHEN HIS IDEAS WERE SHOWN TO BE CORRECT?
HENRY CAVENDISH
NO, BECAUSE HE DIED 71 YEARS BEFORE THE ENGLISH SCIENTIST HENRY CAVENDISH DID HIS FAMOUS EXPERIMENT TO TEST NEWTON'S LAW OF GRAVITY.
CAVENDISH MEASURED HOW MUCH GRAVITATIONAL ATTRACTION THERE WAS BETWEEN TWO 350 POUND (158 KG) LEAD BALLS, USING A MACHINE BUILT BY THE REVEREND JOHN MICHELL, WHO DIED BEFORE HE COULD USE IT HIMSELF.
350 lbs
HE MEASURED A TINY BUT DEFINITE GRAVITATIONAL PULL BETWEEN THEM.
350 lbs

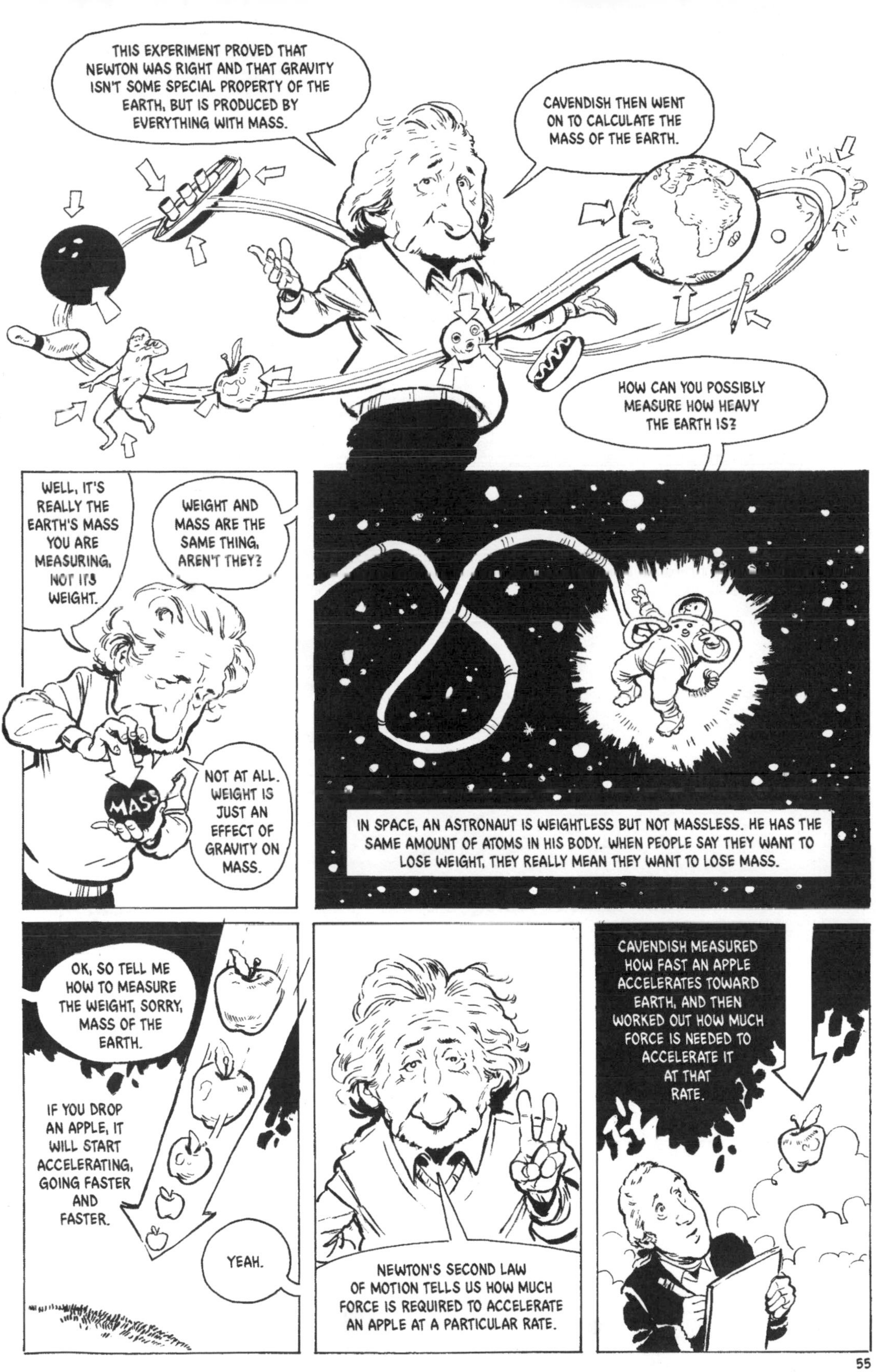
THIS EXPERIMENT PROVED THAT NEWTON WAS RIGHT AND THAT GRAVITY ISN'T SOME SPECIAL PROPERTY OF THE EARTH, BUT IS PRODUCED BY EVERYTHING WITH MASS.
CAVENDISH THEN WENT ON TO CALCULATE THE MASS OF THE EARTH.
HOW CAN YOU POSSIBLY MEASURE HOW HEAVY THE EARTH IS?
WELL, IT'S REALLY THE EARTH'S MASS YOU ARE MEASURING, NOT ITS WEIGHT.
WEIGHT AND MASS ARE THE SAME THING, AREN'T THEY?
MASS
NOT AT ALL. WEIGHT IS JUST AN EFFECT OF GRAVITY ON MASS.
IN SPACE, AN ASTRONAUT IS WEIGHTLESS BUT NOT MASSLESS. HE HAS THE SAME AMOUNT OF ATOMS IN HIS BODY. WHEN PEOPLE SAY THEY WANT TO LOSE WEIGHT, THEY REALLY MEAN THEY WANT TO LOSE MASS.
OK, SO TELL ME HOW TO MEASURE THE WEIGHT, SORRY, MASS OF THE EARTH.
IF YOU DROP AN APPLE, IT WILL START ACCELERATING, GOING FASTER AND FASTER.
YEAH.
NEWTON'S SECOND LAW OF MOTION TELLS US HOW MUCH FORCE IS REQUIRED TO ACCELERATE AN APPLE AT A PARTICULAR RATE.
CAVENDISH MEASURED HOW FAST AN APPLE ACCELERATES TOWARD EARTH, AND THEN WORKED OUT HOW MUCH FORCE IS NEEDED TO ACCELERATE IT AT THAT RATE.

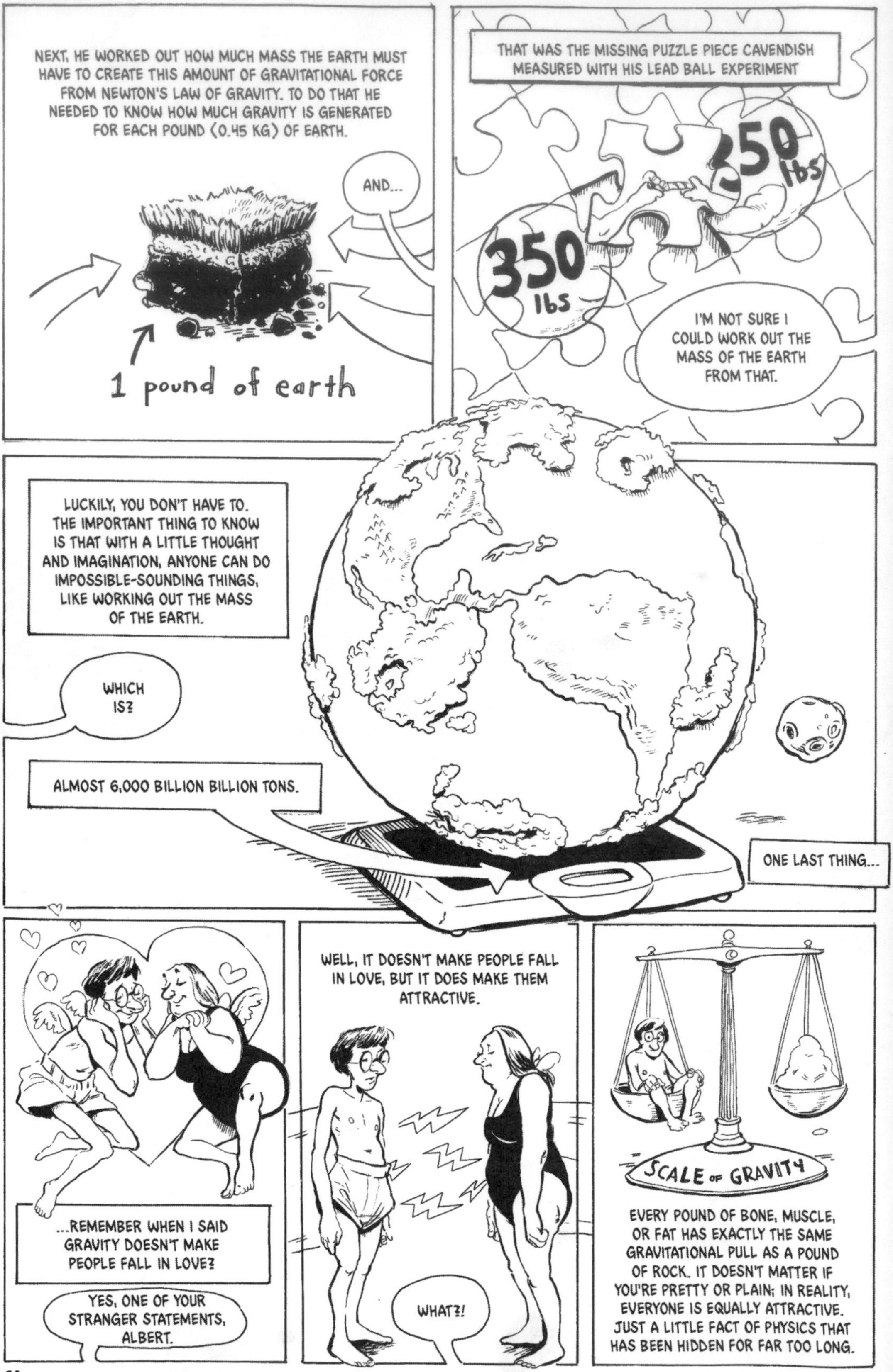

NEXT, HE WORKED OUT HOW MUCH MASS THE EARTH MUST HAVE TO CREATE THIS AMOUNT OF GRAVITATIONAL FORCE FROM NEWTON'S LAW OF GRAVITY. TO DO THAT HE NEEDED TO KNOW HOW MUCH GRAVITY IS GENERATED FOR EACH POUND (0.45 KG) OF EARTH.
AND...
1 pound of earth
THAT WAS THE MISSING PUZZLE PIECE CAVENDISH MEASURED WITH HIS LEAD BALL EXPERIMENT
350 lbs
350 lbs
I'M NOT SURE I COULD WORK OUT THE MASS OF THE EARTH FROM THAT.
LUCKILY, YOU DON'T HAVE TO. THE IMPORTANT THING TO KNOW IS THAT WITH A LITTLE THOUGHT AND IMAGINATION, ANYONE CAN DO IMPOSSIBLE-SOUNDING THINGS, LIKE WORKING OUT THE MASS OF THE EARTH.
WHICH IS?
ALMOST 6,000 BILLION BILLION TONS.
ONE LAST THING...
...REMEMBER WHEN I SAID GRAVITY DOESN'T MAKE PEOPLE FALL IN LOVE?
YES, ONE OF YOUR STRANGER STATEMENTS, ALBERT.
WELL, IT DOESN'T MAKE PEOPLE FALL IN LOVE, BUT IT DOES MAKE THEM ATTRACTIVE.
WHAT?!
SCALE OF GRAVITY
EVERY POUND OF BONE, MUSCLE, OR FAT HAS EXACTLY THE SAME GRAVITATIONAL PULL AS A POUND OF ROCK. IT DOESN'T MATTER IF YOU'RE PRETTY OR PLAIN; IN REALITY, EVERYONE IS EQUALLY ATTRACTIVE. JUST A LITTLE FACT OF PHYSICS THAT HAS BEEN HIDDEN FOR FAR TOO LONG.

CHAPTER 10: EXPLODING STARS AND WHY EVERYONE HAS A LITTLE STAR QUALITY

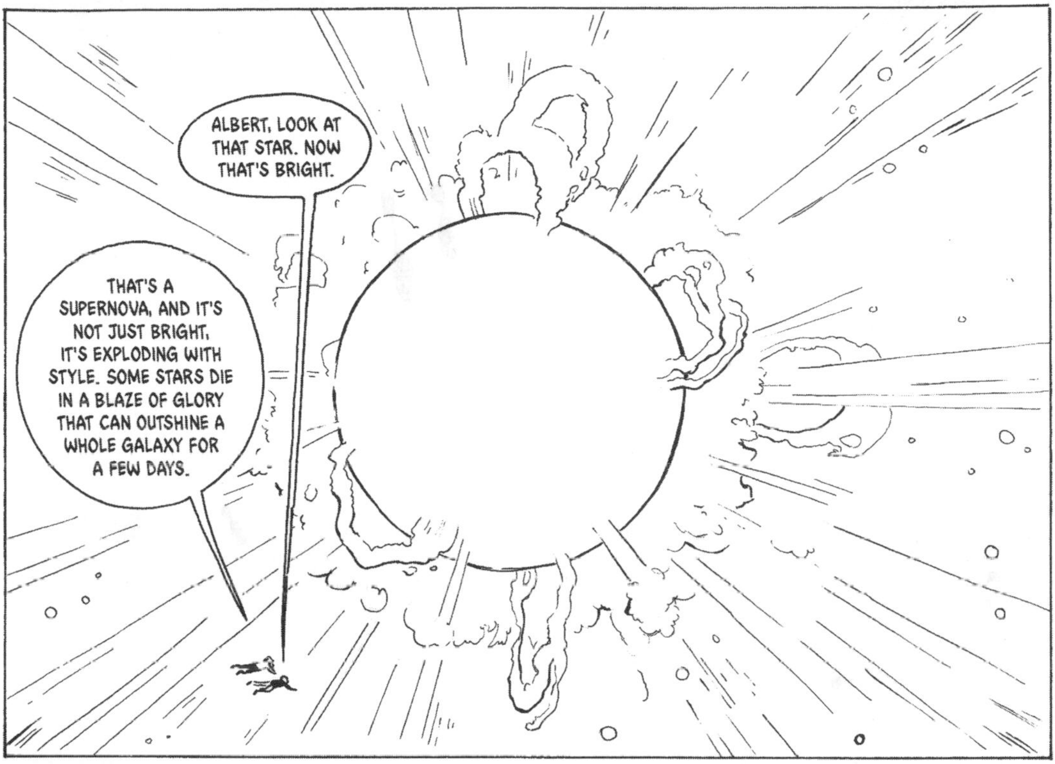
ALBERT, LOOK AT THAT STAR. NOW THAT'S BRIGHT.
THAT'S A SUPERNOVA, AND IT'S NOT JUST BRIGHT, IT'S EXPLODING WITH STYLE. SOME STARS DIE IN A BLAZE OF GLORY THAT CAN OUTSHINE A WHOLE GALAXY FOR A FEW DAYS.

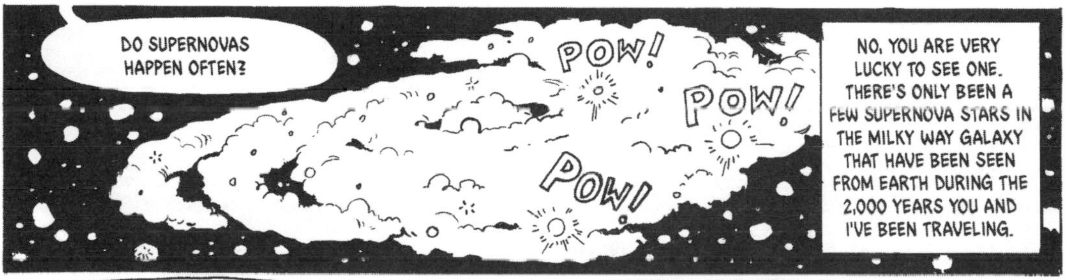
DO SUPERNOVAS HAPPEN OFTEN?
POW!
POW!
POW!
NO, YOU ARE VERY LUCKY TO SEE ONE. THERE'S ONLY BEEN A FEW SUPERNOVA STARS IN THE MILKY WAY GALAXY THAT HAVE BEEN SEEN FROM EARTH DURING THE 2,000 YEARS YOU AND I'VE BEEN TRAVELING.

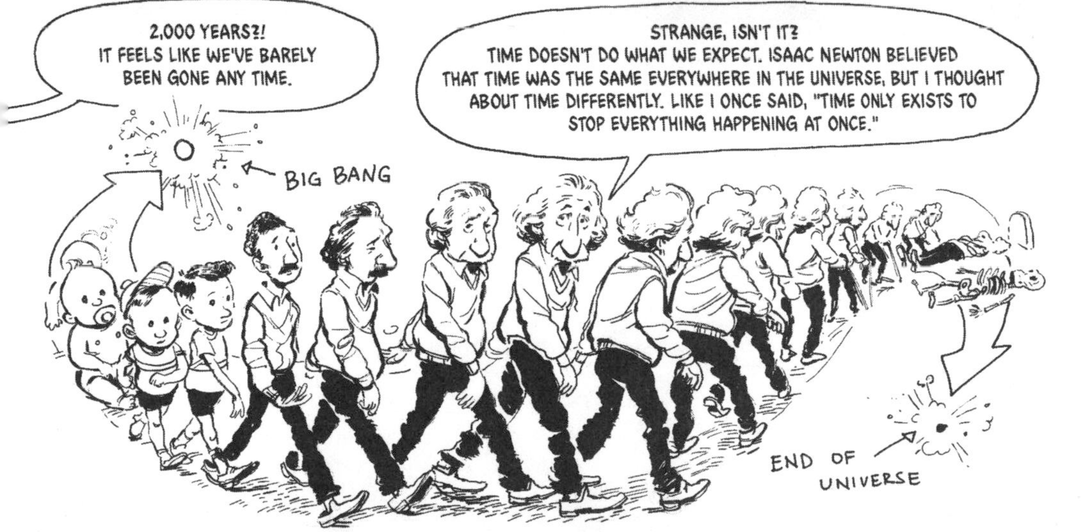
2,000 YEARS?! IT FEELS LIKE WE'VE BARELY BEEN GONE ANY TIME.
STRANGE, ISN'T IT? TIME DOESN'T DO WHAT WE EXPECT. ISAAC NEWTON BELIEVED THAT TIME WAS THE SAME EVERYWHERE IN THE UNIVERSE, BUT I THOUGHT ABOUT TIME DIFFERENTLY. LIKE I ONCE SAID, "TIME ONLY EXISTS TO STOP EVERYTHING HAPPENING AT ONCE."
BIG BANG
END OF UNIVERSE

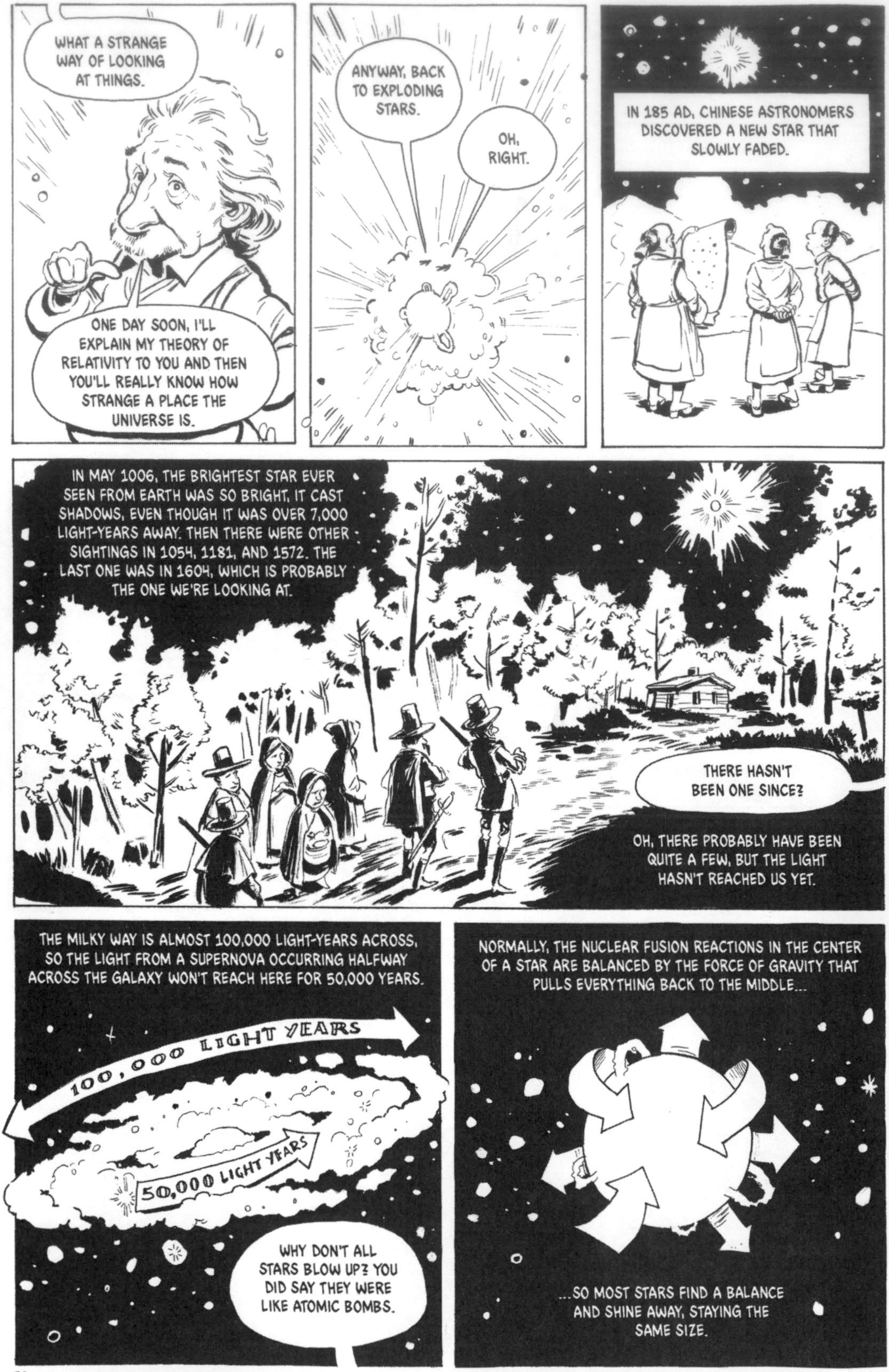

WHAT A STRANGE WAY OF LOOKING AT THINGS.
ONE DAY SOON, I'LL EXPLAIN MY THEORY OF RELATIVITY TO YOU AND THEN YOU'LL REALLY KNOW HOW STRANGE A PLACE THE UNIVERSE IS.
ANYWAY, BACK TO EXPLODING STARS.
OH, RIGHT.
IN 185 AD, CHINESE ASTRONOMERS DISCOVERED A NEW STAR THAT SLOWLY FADED.
IN MAY 1006, THE BRIGHTEST STAR EVER SEEN FROM EARTH WAS SO BRIGHT, IT CAST SHADOWS, EVEN THOUGH IT WAS OVER 7,000 LIGHT-YEARS AWAY. THEN THERE WERE OTHER SIGHTINGS IN 1054, 1181, AND 1572. THE LAST ONE WAS IN 1604, WHICH IS PROBABLY THE ONE WE'RE LOOKING AT.
THERE HASN'T BEEN ONE SINCE?
OH, THERE PROBABLY HAVE BEEN QUITE A FEW, BUT THE LIGHT HASN'T REACHED US YET.
THE MILKY WAY IS ALMOST 100,000 LIGHT-YEARS ACROSS, SO THE LIGHT FROM A SUPERNOVA OCCURRING HALFWAY ACROSS THE GALAXY WON'T REACH HERE FOR 50,000 YEARS.
100,000 LIGHT YEARS
50,000 LIGHT YEARS
WHY DON'T ALL STARS BLOW UP? YOU DID SAY THEY WERE LIKE ATOMIC BOMBS.
NORMALLY, THE NUCLEAR FUSION REACTIONS IN THE CENTER OF A STAR ARE BALANCED BY THE FORCE OF GRAVITY THAT PULLS EVERYTHING BACK TO THE MIDDLE...
...SO MOST STARS FIND A BALANCE AND SHINE AWAY, STAYING THE SAME SIZE.

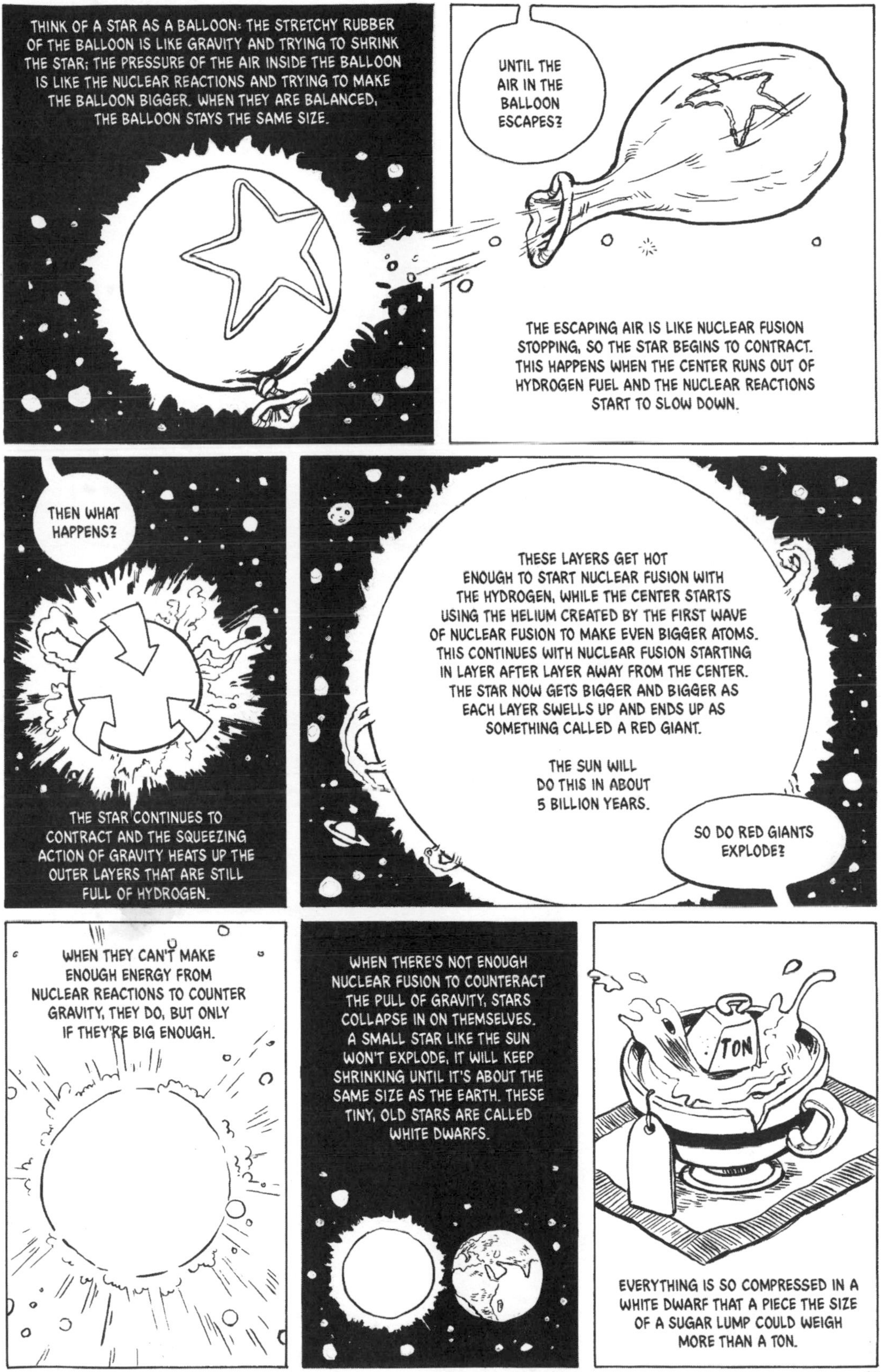
THINK OF A STAR AS A BALLOON: THE STRETCHY RUBBER OF THE BALLOON IS LIKE GRAVITY AND TRYING TO SHRINK THE STAR; THE PRESSURE OF THE AIR INSIDE THE BALLOON IS LIKE THE NUCLEAR REACTIONS AND TRYING TO MAKE THE BALLOON BIGGER. WHEN THEY ARE BALANCED, THE BALLOON STAYS THE SAME SIZE.
UNTIL THE AIR IN THE BALLOON ESCAPES?
THE ESCAPING AIR IS LIKE NUCLEAR FUSION STOPPING, SO THE STAR BEGINS TO CONTRACT. THIS HAPPENS WHEN THE CENTER RUNS OUT OF HYDROGEN FUEL AND THE NUCLEAR REACTIONS START TO SLOW DOWN.
THEN WHAT HAPPENS?
THE STAR CONTINUES TO CONTRACT AND THE SQUEEZING ACTION OF GRAVITY HEATS UP THE OUTER LAYERS THAT ARE STILL FULL OF HYDROGEN.
THESE LAYERS GET HOT ENOUGH TO START NUCLEAR FUSION WITH THE HYDROGEN, WHILE THE CENTER STARTS USING THE HELIUM CREATED BY THE FIRST WAVE OF NUCLEAR FUSION TO MAKE EVEN BIGGER ATOMS. THIS CONTINUES WITH NUCLEAR FUSION STARTING IN LAYER AFTER LAYER AWAY FROM THE CENTER. THE STAR NOW GETS BIGGER AND BIGGER AS EACH LAYER SWELLS UP AND ENDS UP AS SOMETHING CALLED A RED GIANT.
THE SUN WILL DO THIS IN ABOUT 5 BILLION YEARS.
SO DO RED GIANTS EXPLODE?
WHEN THEY CAN'T MAKE ENOUGH ENERGY FROM NUCLEAR REACTIONS TO COUNTER GRAVITY, THEY DO, BUT ONLY IF THEY'RE BIG ENOUGH.
WHEN THERE'S NOT ENOUGH NUCLEAR FUSION TO COUNTERACT THE PULL OF GRAVITY, STARS COLLAPSE IN ON THEMSELVES. A SMALL STAR LIKE THE SUN WON'T EXPLODE, IT WILL KEEP SHRINKING UNTIL IT'S ABOUT THE SAME SIZE AS THE EARTH. THESE TINY, OLD STARS ARE CALLED WHITE DWARFS.
TON
EVERYTHING IS SO COMPRESSED IN A WHITE DWARF THAT A PIECE THE SIZE OF A SUGAR LUMP COULD WEIGH MORE THAN A TON.

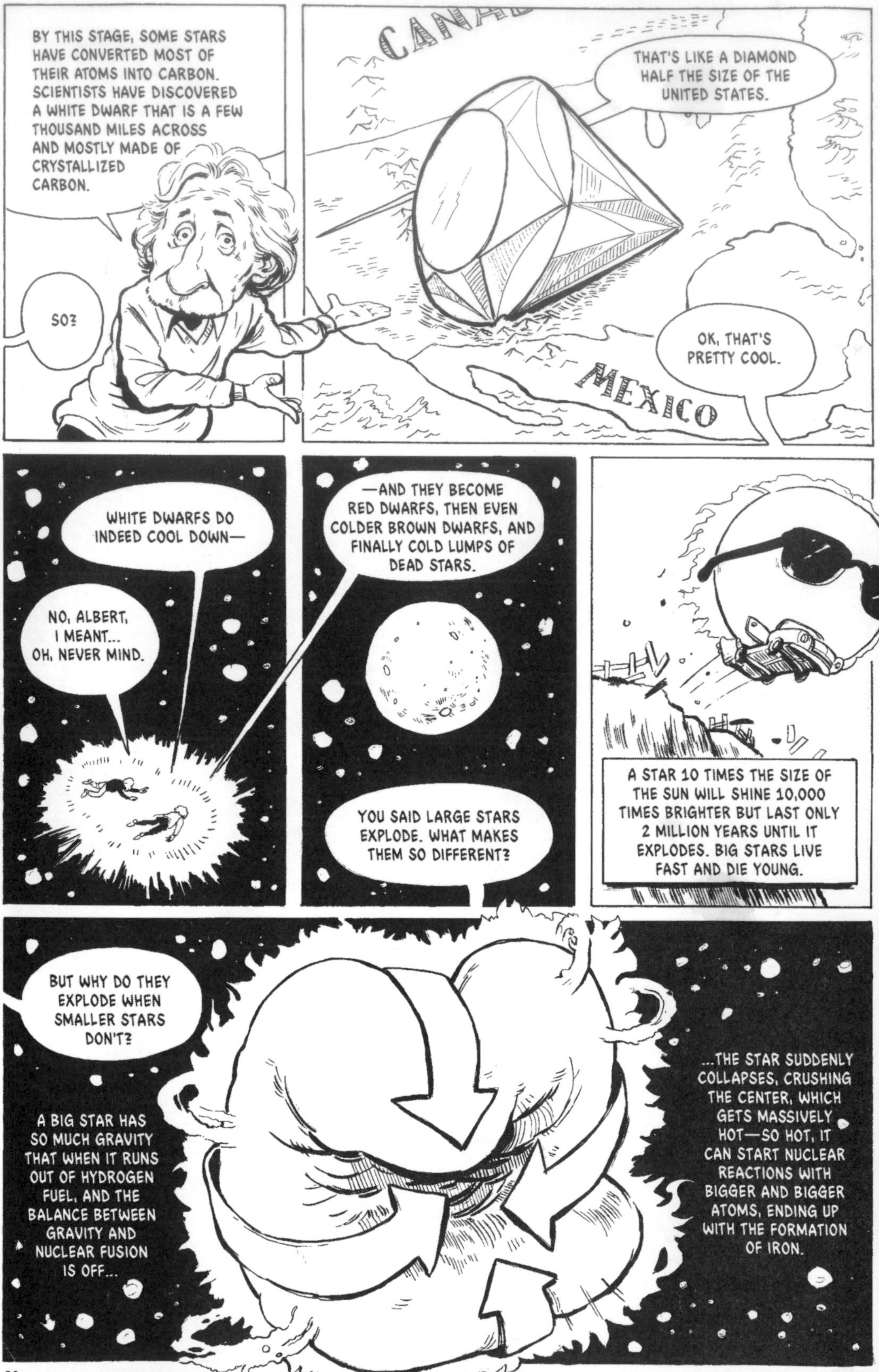
BY THIS STAGE, SOME STARS HAVE CONVERTED MOST OF THEIR ATOMS INTO CARBON. SCIENTISTS HAVE DISCOVERED A WHITE DWARF THAT IS A FEW THOUSAND MILES ACROSS AND MOSTLY MADE OF CRYSTALLIZED CARBON.
SO?
CANAD
THAT'S LIKE A DIAMOND HALF THE SIZE OF THE UNITED STATES.
OK, THAT'S PRETTY COOL.
MEXICO
WHITE DWARFS DO INDEED COOL DOWN—
NO, ALBERT, I MEANT... OH, NEVER MIND.
—AND THEY BECOME RED DWARFS, THEN EVEN COLDER BROWN DWARFS, AND FINALLY COLD LUMPS OF DEAD STARS.
YOU SAID LARGE STARS EXPLODE. WHAT MAKES THEM SO DIFFERENT?
A STAR 10 TIMES THE SIZE OF THE SUN WILL SHINE 10,000 TIMES BRIGHTER BUT LAST ONLY 2 MILLION YEARS UNTIL IT EXPLODES. BIG STARS LIVE FAST AND DIE YOUNG.
BUT WHY DO THEY EXPLODE WHEN SMALLER STARS DON'T?
A BIG STAR HAS SO MUCH GRAVITY THAT WHEN IT RUNS OUT OF HYDROGEN FUEL, AND THE BALANCE BETWEEN GRAVITY AND NUCLEAR FUSION IS OFF...
...THE STAR SUDDENLY COLLAPSES, CRUSHING THE CENTER, WHICH GETS MASSIVELY HOT—SO HOT, IT CAN START NUCLEAR REACTIONS WITH BIGGER AND BIGGER ATOMS, ENDING UP WITH THE FORMATION OF IRON.

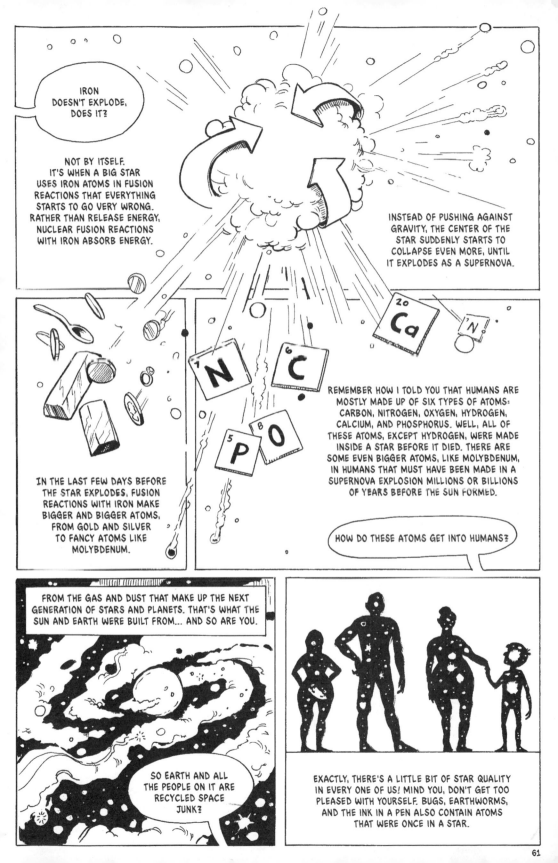

IRON DOESN'T EXPLODE, DOES IT?

NOT BY ITSELF. IT'S WHEN A BIG STAR USES IRON ATOMS IN FUSION REACTIONS THAT EVERYTHING STARTS TO GO VERY WRONG. RATHER THAN RELEASE ENERGY, NUCLEAR FUSION REACTIONS WITH IRON ABSORB ENERGY.

INSTEAD OF PUSHING AGAINST GRAVITY, THE CENTER OF THE STAR SUDDENLY STARTS TO COLLAPSE EVEN MORE, UNTIL IT EXPLODES AS A SUPERNOVA.

IN THE LAST FEW DAYS BEFORE THE STAR EXPLODES, FUSION REACTIONS WITH IRON MAKE BIGGER AND BIGGER ATOMS, FROM GOLD AND SILVER TO FANCY ATOMS LIKE MOLYBDENUM.

REMEMBER HOW I TOLD YOU THAT HUMANS ARE MOSTLY MADE UP OF SIX TYPES OF ATOMS: CARBON, NITROGEN, OXYGEN, HYDROGEN, CALCIUM, AND PHOSPHORUS. WELL, ALL OF THESE ATOMS, EXCEPT HYDROGEN, WERE MADE INSIDE A STAR BEFORE IT DIED. THERE ARE SOME EVEN BIGGER ATOMS, LIKE MOLYBDENUM, IN HUMANS THAT MUST HAVE BEEN MADE IN A SUPERNOVA EXPLOSION MILLIONS OR BILLIONS OF YEARS BEFORE THE SUN FORMED.

HOW DO THESE ATOMS GET INTO HUMANS?

FROM THE GAS AND DUST THAT MAKE UP THE NEXT GENERATION OF STARS AND PLANETS. THAT'S WHAT THE SUN AND EARTH WERE BUILT FROM... AND SO ARE YOU.

SO EARTH AND ALL THE PEOPLE ON IT ARE RECYCLED SPACE JUNK?

EXACTLY, THERE'S A LITTLE BIT OF STAR QUALITY IN EVERY ONE OF US! MIND YOU, DON'T GET TOO PLEASED WITH YOURSELF. BUGS, EARTHWORMS, AND THE INK IN A PEN ALSO CONTAIN ATOMS THAT WERE ONCE IN A STAR.

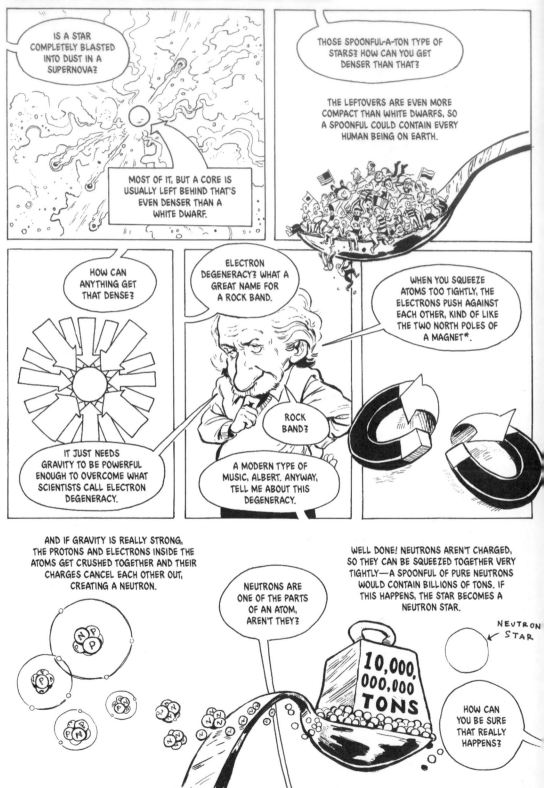

IS A STAR COMPLETELY BLASTED INTO DUST IN A SUPERNOVA?

MOST OF IT, BUT A CORE IS USUALLY LEFT BEHIND THAT'S EVEN DENSER THAN A WHITE DWARF.

THOSE SPOONFUL-A-TON TYPE OF STARS? HOW CAN YOU GET DENSER THAN THAT?

THE LEFTOVERS ARE EVEN MORE COMPACT THAN WHITE DWARFS, SO A SPOONFUL COULD CONTAIN EVERY HUMAN BEING ON EARTH.

HOW CAN ANYTHING GET THAT DENSE?

ELECTRON DEGENERACY? WHAT A GREAT NAME FOR A ROCK BAND.

WHEN YOU SQUEEZE ATOMS TOO TIGHTLY, THE ELECTRONS PUSH AGAINST EACH OTHER, KIND OF LIKE THE TWO NORTH POLES OF A MAGNET*.

IT JUST NEEDS GRAVITY TO BE POWERFUL ENOUGH TO OVERCOME WHAT SCIENTISTS CALL ELECTRON DEGENERACY.

ROCK BAND?

A MODERN TYPE OF MUSIC, ALBERT. ANYWAY, TELL ME ABOUT THIS DEGENERACY.

AND IF GRAVITY IS REALLY STRONG, THE PROTONS AND ELECTRONS INSIDE THE ATOMS GET CRUSHED TOGETHER AND THEIR CHARGES CANCEL EACH OTHER OUT, CREATING A NEUTRON.

NEUTRONS ARE ONE OF THE PARTS OF AN ATOM, AREN'T THEY?

WELL DONE! NEUTRONS AREN'T CHARGED, SO THEY CAN BE SQUEEZED TOGETHER VERY TIGHTLY—A SPOONFUL OF PURE NEUTRONS WOULD CONTAIN BILLIONS OF TONS. IF THIS HAPPENS, THE STAR BECOMES A NEUTRON STAR.

NEUTRON STAR

10,000, 000,000 TONS

HOW CAN YOU BE SURE THAT REALLY HAPPENS?

*THIS REALLY HAS TO DO WITH QUANTUM MECHANICS RATHER THAN MAGNETISM, BUT WE'LL GET TO THAT SOON.

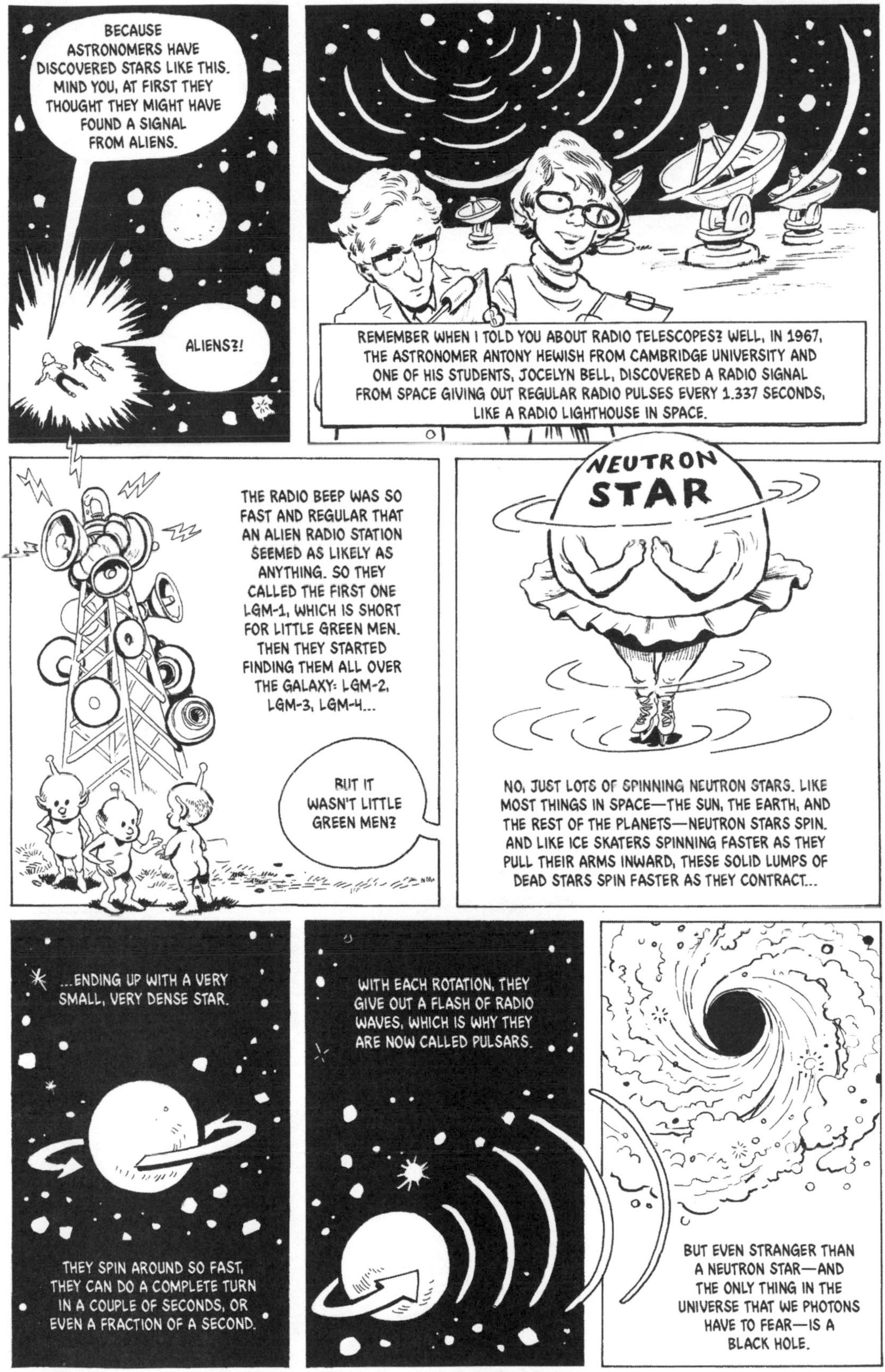

BECAUSE ASTRONOMERS HAVE DISCOVERED STARS LIKE THIS. MIND YOU, AT FIRST THEY THOUGHT THEY MIGHT HAVE FOUND A SIGNAL FROM ALIENS.
ALIENS?!
REMEMBER WHEN I TOLD YOU ABOUT RADIO TELESCOPES? WELL, IN 1967, THE ASTRONOMER ANTONY HEWISH FROM CAMBRIDGE UNIVERSITY AND ONE OF HIS STUDENTS, JOCELYN BELL, DISCOVERED A RADIO SIGNAL FROM SPACE GIVING OUT REGULAR RADIO PULSES EVERY 1.337 SECONDS, LIKE A RADIO LIGHTHOUSE IN SPACE.
THE RADIO BEEP WAS SO FAST AND REGULAR THAT AN ALIEN RADIO STATION SEEMED AS LIKELY AS ANYTHING. SO THEY CALLED THE FIRST ONE LGM-1, WHICH IS SHORT FOR LITTLE GREEN MEN. THEN THEY STARTED FINDING THEM ALL OVER THE GALAXY: LGM-2, LGM-3, LGM-4...
BUT IT WASN'T LITTLE GREEN MEN?
NEUTRON STAR
NO, JUST LOTS OF SPINNING NEUTRON STARS. LIKE MOST THINGS IN SPACE—THE SUN, THE EARTH, AND THE REST OF THE PLANETS—NEUTRON STARS SPIN. AND LIKE ICE SKATERS SPINNING FASTER AS THEY PULL THEIR ARMS INWARD, THESE SOLID LUMPS OF DEAD STARS SPIN FASTER AS THEY CONTRACT...
...ENDING UP WITH A VERY SMALL, VERY DENSE STAR.
THEY SPIN AROUND SO FAST, THEY CAN DO A COMPLETE TURN IN A COUPLE OF SECONDS, OR EVEN A FRACTION OF A SECOND.
WITH EACH ROTATION, THEY GIVE OUT A FLASH OF RADIO WAVES, WHICH IS WHY THEY ARE NOW CALLED PULSARS.
BUT EVEN STRANGER THAN A NEUTRON STAR—AND THE ONLY THING IN THE UNIVERSE THAT WE PHOTONS HAVE TO FEAR—IS A BLACK HOLE.

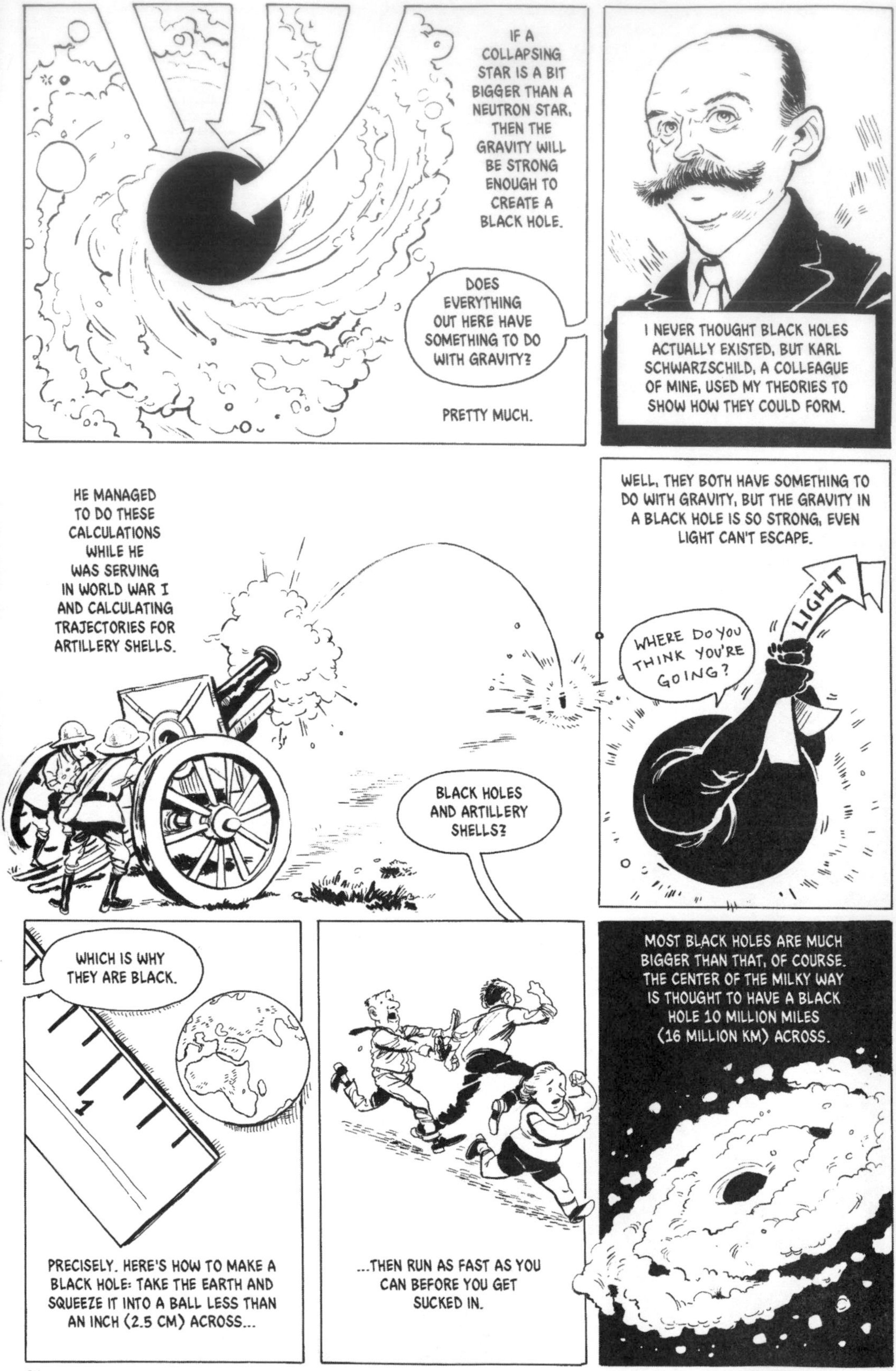
IF A COLLAPSING STAR IS A BIT BIGGER THAN A NEUTRON STAR, THEN THE GRAVITY WILL BE STRONG ENOUGH TO CREATE A BLACK HOLE.
DOES EVERYTHING OUT HERE HAVE SOMETHING TO DO WITH GRAVITY?
PRETTY MUCH.
I NEVER THOUGHT BLACK HOLES ACTUALLY EXISTED, BUT KARL SCHWARZSCHILD, A COLLEAGUE OF MINE, USED MY THEORIES TO SHOW HOW THEY COULD FORM.
HE MANAGED TO DO THESE CALCULATIONS WHILE HE WAS SERVING IN WORLD WAR I AND CALCULATING TRAJECTORIES FOR ARTILLERY SHELLS.
BLACK HOLES AND ARTILLERY SHELLS?
WELL, THEY BOTH HAVE SOMETHING TO DO WITH GRAVITY, BUT THE GRAVITY IN A BLACK HOLE IS SO STRONG, EVEN LIGHT CAN'T ESCAPE.
LIGHT
WHERE DO YOU THINK YOU'RE GOING?
WHICH IS WHY THEY ARE BLACK.
1
PRECISELY. HERE'S HOW TO MAKE A BLACK HOLE: TAKE THE EARTH AND SQUEEZE IT INTO A BALL LESS THAN AN INCH (2.5 CM) ACROSS...
...THEN RUN AS FAST AS YOU CAN BEFORE YOU GET SUCKED IN.
MOST BLACK HOLES ARE MUCH BIGGER THAN THAT, OF COURSE. THE CENTER OF THE MILKY WAY IS THOUGHT TO HAVE A BLACK HOLE 10 MILLION MILES (16 MILLION KM) ACROSS.

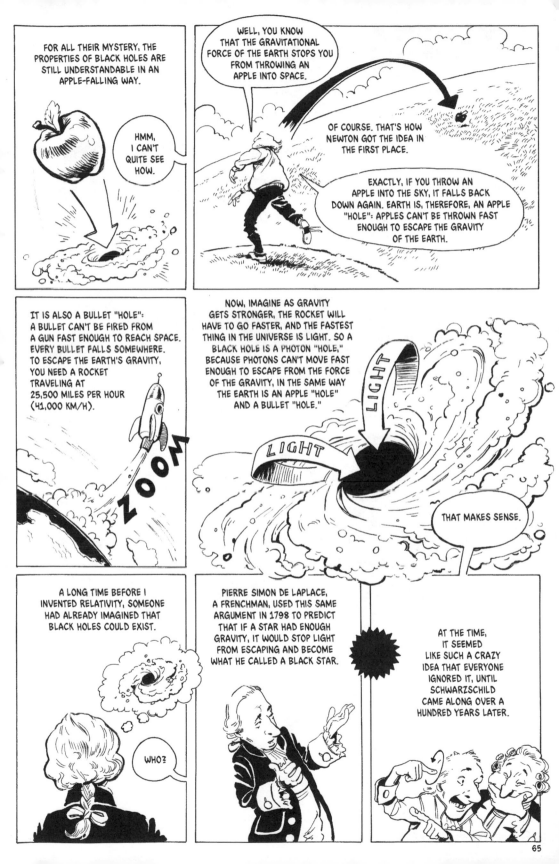

FOR ALL THEIR MYSTERY, THE PROPERTIES OF BLACK HOLES ARE STILL UNDERSTANDABLE IN AN APPLE-FALLING WAY.

HMM, I CAN'T QUITE SEE HOW.

WELL, YOU KNOW THAT THE GRAVITATIONAL FORCE OF THE EARTH STOPS YOU FROM THROWING AN APPLE INTO SPACE.

OF COURSE. THAT'S HOW NEWTON GOT THE IDEA IN THE FIRST PLACE.

EXACTLY, IF YOU THROW AN APPLE INTO THE SKY, IT FALLS BACK DOWN AGAIN. EARTH IS, THEREFORE, AN APPLE "HOLE": APPLES CAN'T BE THROWN FAST ENOUGH TO ESCAPE THE GRAVITY OF THE EARTH.

IT IS ALSO A BULLET "HOLE": A BULLET CAN'T BE FIRED FROM A GUN FAST ENOUGH TO REACH SPACE. EVERY BULLET FALLS SOMEWHERE. TO ESCAPE THE EARTH'S GRAVITY, YOU NEED A ROCKET TRAVELING AT 25,500 MILES PER HOUR (41,000 KM/H).

ZOOM

NOW, IMAGINE AS GRAVITY GETS STRONGER, THE ROCKET WILL HAVE TO GO FASTER, AND THE FASTEST THING IN THE UNIVERSE IS LIGHT. SO A BLACK HOLE IS A PHOTON "HOLE," BECAUSE PHOTONS CAN'T MOVE FAST ENOUGH TO ESCAPE FROM THE FORCE OF THE GRAVITY, IN THE SAME WAY THE EARTH IS AN APPLE "HOLE" AND A BULLET "HOLE."

LIGHT

LIGHT

THAT MAKES SENSE.

A LONG TIME BEFORE I INVENTED RELATIVITY, SOMEONE HAD ALREADY IMAGINED THAT BLACK HOLES COULD EXIST.

WHO?

PIERRE SIMON DE LAPLACE, A FRENCHMAN, USED THIS SAME ARGUMENT IN 1798 TO PREDICT THAT IF A STAR HAD ENOUGH GRAVITY, IT WOULD STOP LIGHT FROM ESCAPING AND BECOME WHAT HE CALLED A BLACK STAR.

AT THE TIME, IT SEEMED LIKE SUCH A CRAZY IDEA THAT EVERYONE IGNORED IT, UNTIL SCHWARZSCHILD CAME ALONG OVER A HUNDRED YEARS LATER.

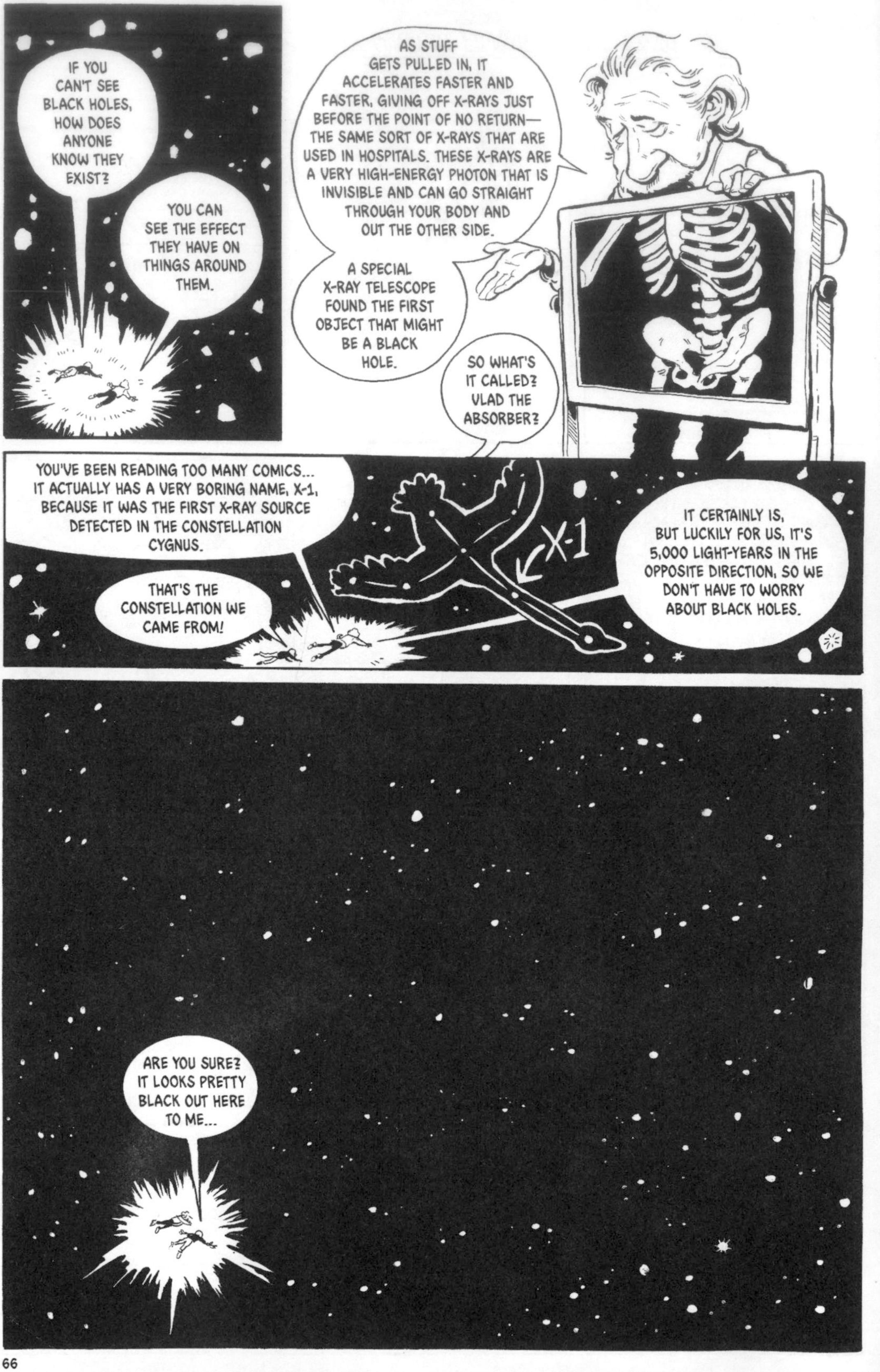
IF YOU CAN'T SEE BLACK HOLES, HOW DOES ANYONE KNOW THEY EXIST?
YOU CAN SEE THE EFFECT THEY HAVE ON THINGS AROUND THEM.
AS STUFF GETS PULLED IN, IT ACCELERATES FASTER AND FASTER, GIVING OFF X-RAYS JUST BEFORE THE POINT OF NO RETURN—THE SAME SORT OF X-RAYS THAT ARE USED IN HOSPITALS. THESE X-RAYS ARE A VERY HIGH-ENERGY PHOTON THAT IS INVISIBLE AND CAN GO STRAIGHT THROUGH YOUR BODY AND OUT THE OTHER SIDE.
A SPECIAL X-RAY TELESCOPE FOUND THE FIRST OBJECT THAT MIGHT BE A BLACK HOLE.
SO WHAT'S IT CALLED? VLAD THE ABSORBER?
YOU'VE BEEN READING TOO MANY COMICS... IT ACTUALLY HAS A VERY BORING NAME, X-1, BECAUSE IT WAS THE FIRST X-RAY SOURCE DETECTED IN THE CONSTELLATION CYGNUS.
THAT'S THE CONSTELLATION WE CAME FROM!
X-1
IT CERTAINLY IS, BUT LUCKILY FOR US, IT'S 5,000 LIGHT-YEARS IN THE OPPOSITE DIRECTION, SO WE DON'T HAVE TO WORRY ABOUT BLACK HOLES.
ARE YOU SURE? IT LOOKS PRETTY BLACK OUT HERE TO ME...

CHAPTER 12:
THE END OF THE DARK AGES AND THE INVENTION OF FROZEN CHICKEN

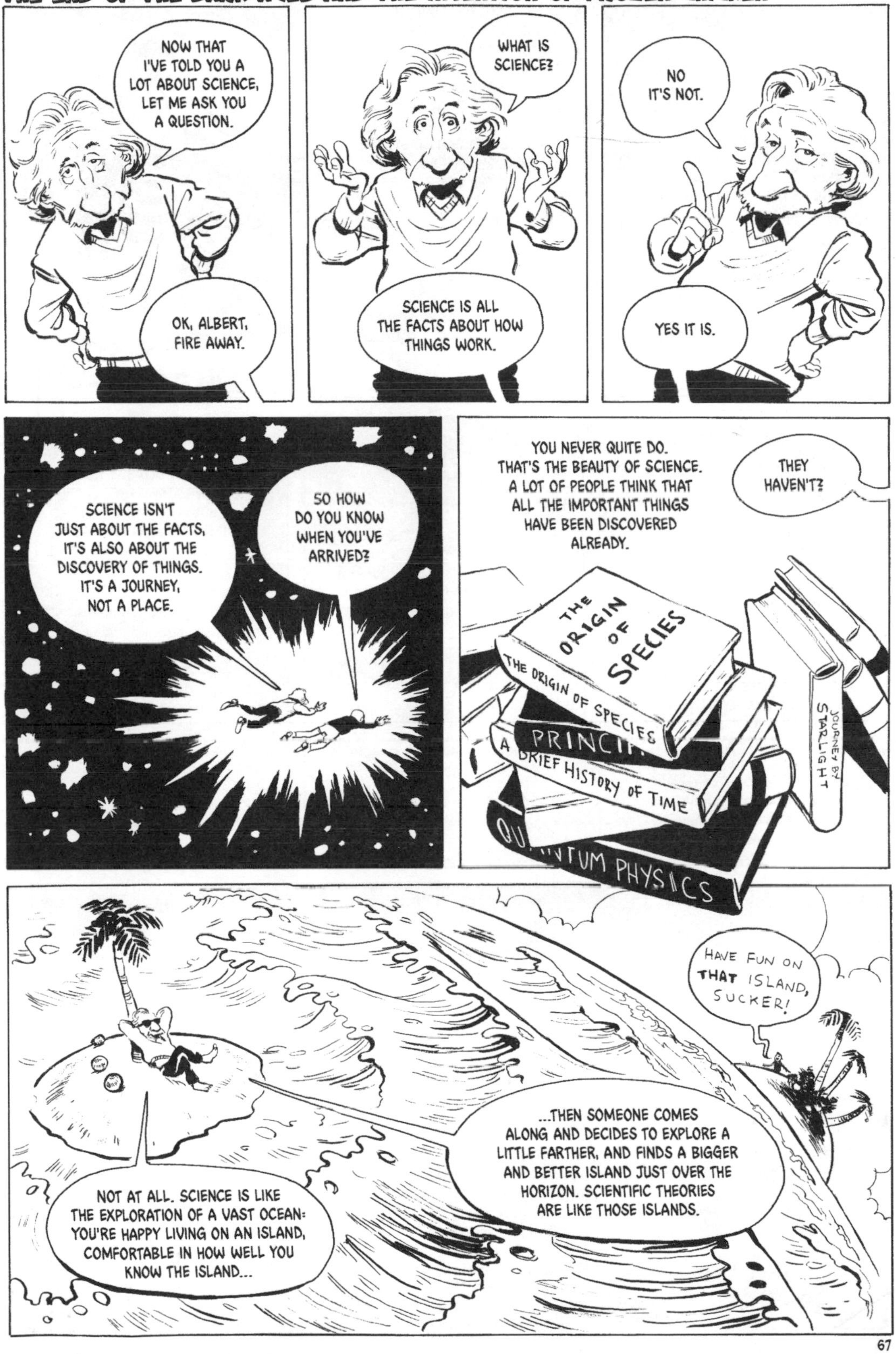

NOW THAT I'VE TOLD YOU A LOT ABOUT SCIENCE, LET ME ASK YOU A QUESTION.
OK, ALBERT, FIRE AWAY.
WHAT IS SCIENCE?
SCIENCE IS ALL THE FACTS ABOUT HOW THINGS WORK.
NO IT'S NOT.
YES IT IS.
SCIENCE ISN'T JUST ABOUT THE FACTS, IT'S ALSO ABOUT THE DISCOVERY OF THINGS. IT'S A JOURNEY, NOT A PLACE.
SO HOW DO YOU KNOW WHEN YOU'VE ARRIVED?
YOU NEVER QUITE DO. THAT'S THE BEAUTY OF SCIENCE. A LOT OF PEOPLE THINK THAT ALL THE IMPORTANT THINGS HAVE BEEN DISCOVERED ALREADY.
THEY HAVEN'T?
THE ORIGIN OF SPECIES
THE ORIGIN OF SPECIES
A BRIEF HISTORY OF TIME
JOURNEY BY STARLIGHT
NOT AT ALL. SCIENCE IS LIKE THE EXPLORATION OF A VAST OCEAN: YOU'RE HAPPY LIVING ON AN ISLAND, COMFORTABLE IN HOW WELL YOU KNOW THE ISLAND...
...THEN SOMEONE COMES ALONG AND DECIDES TO EXPLORE A LITTLE FARTHER, AND FINDS A BIGGER AND BETTER ISLAND JUST OVER THE HORIZON. SCIENTIFIC THEORIES ARE LIKE THOSE ISLANDS.
HAVE FUN ON THAT ISLAND, SUCKER!

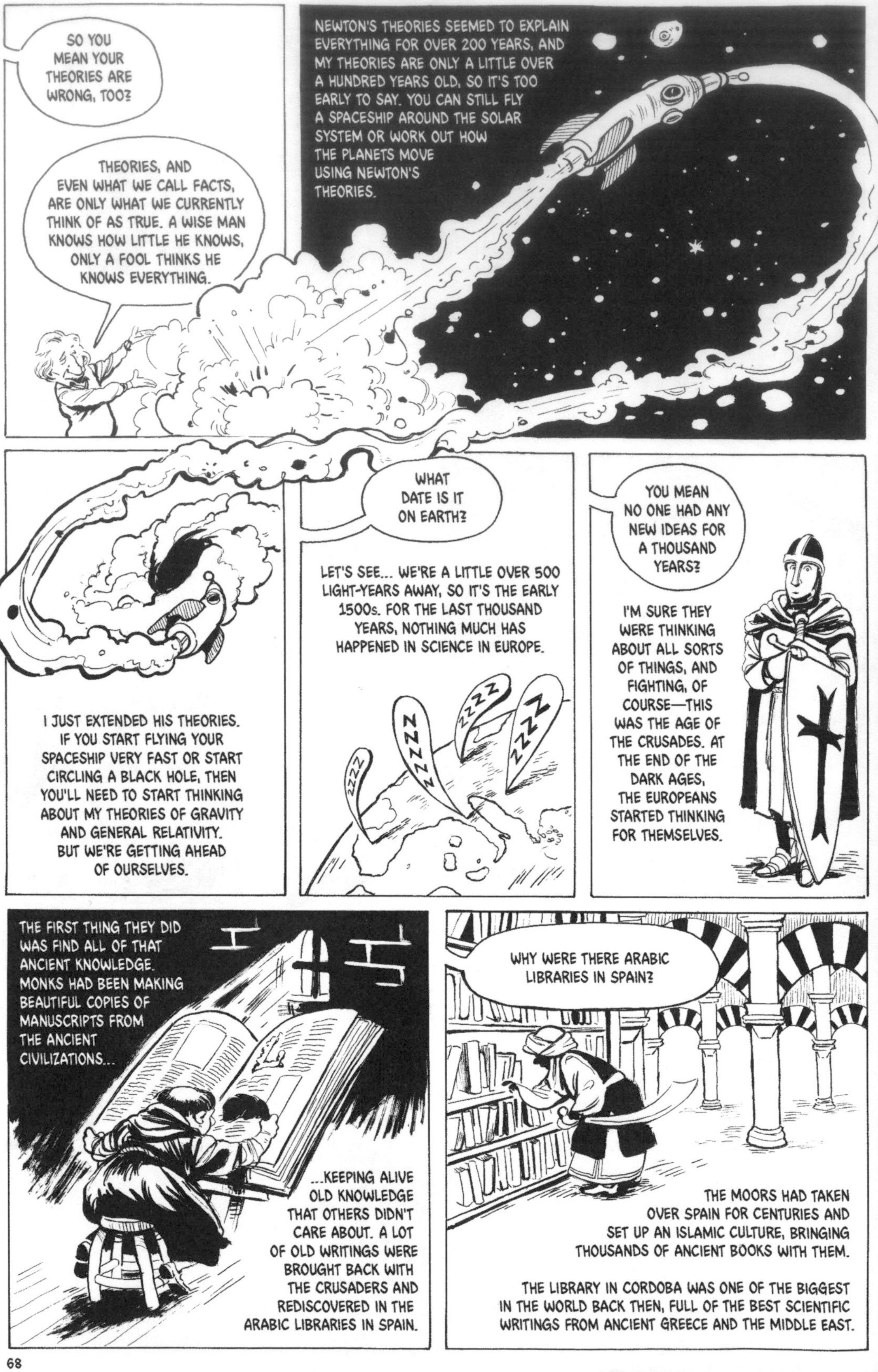
SO YOU MEAN YOUR THEORIES ARE WRONG, TOO?
THEORIES, AND EVEN WHAT WE CALL FACTS, ARE ONLY WHAT WE CURRENTLY THINK OF AS TRUE. A WISE MAN KNOWS HOW LITTLE HE KNOWS, ONLY A FOOL THINKS HE KNOWS EVERYTHING.
NEWTON'S THEORIES SEEMED TO EXPLAIN EVERYTHING FOR OVER 200 YEARS, AND MY THEORIES ARE ONLY A LITTLE OVER A HUNDRED YEARS OLD, SO IT'S TOO EARLY TO SAY. YOU CAN STILL FLY A SPACESHIP AROUND THE SOLAR SYSTEM OR WORK OUT HOW THE PLANETS MOVE USING NEWTON'S THEORIES.
I JUST EXTENDED HIS THEORIES. IF YOU START FLYING YOUR SPACESHIP VERY FAST OR START CIRCLING A BLACK HOLE, THEN YOU'LL NEED TO START THINKING ABOUT MY THEORIES OF GRAVITY AND GENERAL RELATIVITY. BUT WE'RE GETTING AHEAD OF OURSELVES.
WHAT DATE IS IT ON EARTH?
LET'S SEE... WE'RE A LITTLE OVER 500 LIGHT-YEARS AWAY, SO IT'S THE EARLY 1500s. FOR THE LAST THOUSAND YEARS, NOTHING MUCH HAS HAPPENED IN SCIENCE IN EUROPE.
ZZZZ
ZZZZZ
ZZZZ
ZZZZZ
YOU MEAN NO ONE HAD ANY NEW IDEAS FOR A THOUSAND YEARS?
I'M SURE THEY WERE THINKING ABOUT ALL SORTS OF THINGS, AND FIGHTING, OF COURSE—THIS WAS THE AGE OF THE CRUSADES. AT THE END OF THE DARK AGES, THE EUROPEANS STARTED THINKING FOR THEMSELVES.
THE FIRST THING THEY DID WAS FIND ALL OF THAT ANCIENT KNOWLEDGE. MONKS HAD BEEN MAKING BEAUTIFUL COPIES OF MANUSCRIPTS FROM THE ANCIENT CIVILIZATIONS...
...KEEPING ALIVE OLD KNOWLEDGE THAT OTHERS DIDN'T CARE ABOUT. A LOT OF OLD WRITINGS WERE BROUGHT BACK WITH THE CRUSADERS AND REDISCOVERED IN THE ARABIC LIBRARIES IN SPAIN.
WHY WERE THERE ARABIC LIBRARIES IN SPAIN?
THE MOORS HAD TAKEN OVER SPAIN FOR CENTURIES AND SET UP AN ISLAMIC CULTURE, BRINGING THOUSANDS OF ANCIENT BOOKS WITH THEM.
THE LIBRARY IN CORDOBA WAS ONE OF THE BIGGEST IN THE WORLD BACK THEN, FULL OF THE BEST SCIENTIFIC WRITINGS FROM ANCIENT GREECE AND THE MIDDLE EAST.

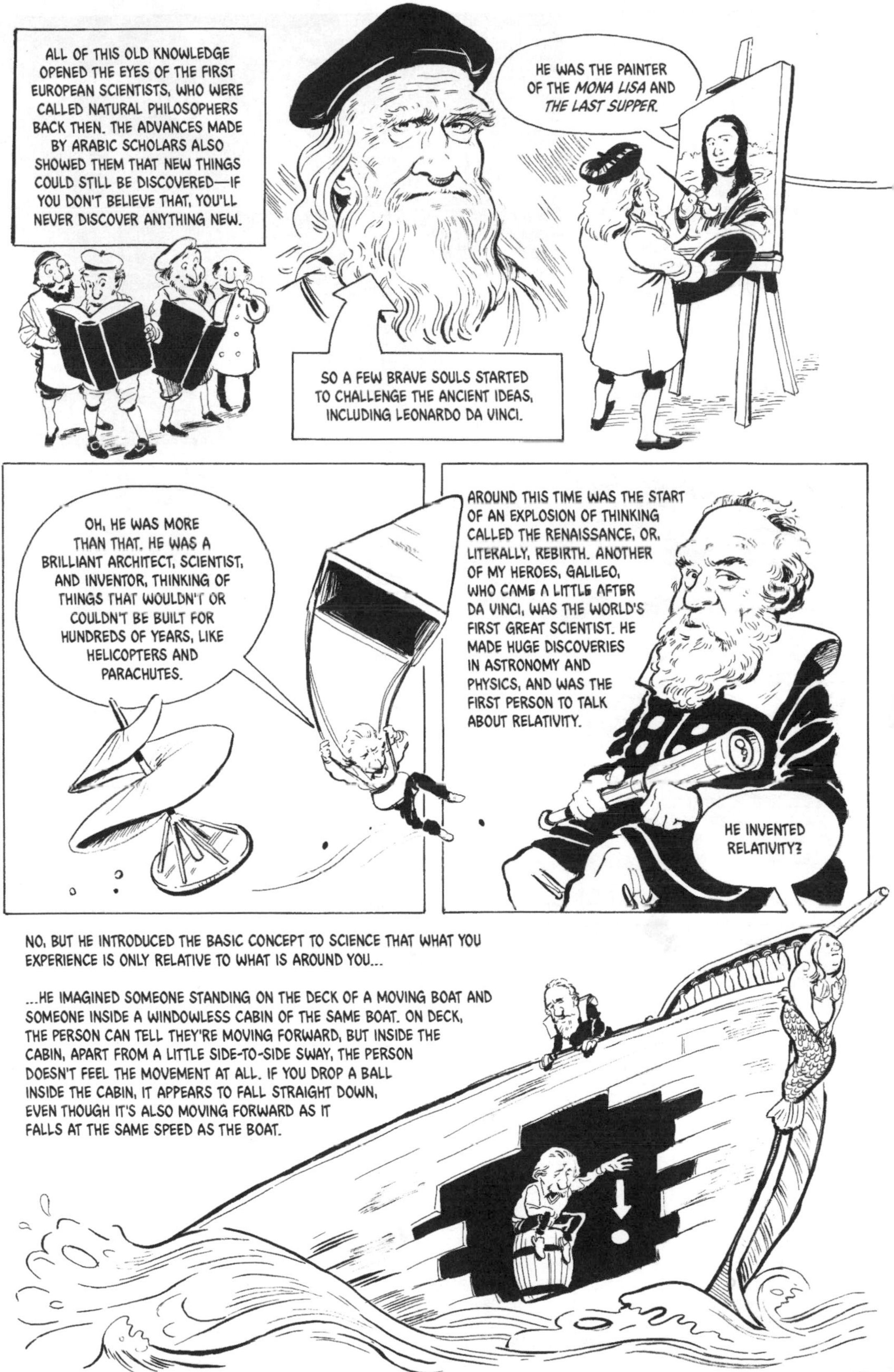
ALL OF THIS OLD KNOWLEDGE OPENED THE EYES OF THE FIRST EUROPEAN SCIENTISTS, WHO WERE CALLED NATURAL PHILOSOPHERS BACK THEN. THE ADVANCES MADE BY ARABIC SCHOLARS ALSO SHOWED THEM THAT NEW THINGS COULD STILL BE DISCOVERED—IF YOU DON'T BELIEVE THAT, YOU'LL NEVER DISCOVER ANYTHING NEW.
SO A FEW BRAVE SOULS STARTED TO CHALLENGE THE ANCIENT IDEAS, INCLUDING LEONARDO DA VINCI.
HE WAS THE PAINTER OF THE *MONA LISA* AND *THE LAST SUPPER.*
OH, HE WAS MORE THAN THAT. HE WAS A BRILLIANT ARCHITECT, SCIENTIST, AND INVENTOR, THINKING OF THINGS THAT WOULDN'T OR COULDN'T BE BUILT FOR HUNDREDS OF YEARS, LIKE HELICOPTERS AND PARACHUTES.
AROUND THIS TIME WAS THE START OF AN EXPLOSION OF THINKING CALLED THE RENAISSANCE, OR, LITERALLY, REBIRTH. ANOTHER OF MY HEROES, GALILEO, WHO CAME A LITTLE AFTER DA VINCI, WAS THE WORLD'S FIRST GREAT SCIENTIST. HE MADE HUGE DISCOVERIES IN ASTRONOMY AND PHYSICS, AND WAS THE FIRST PERSON TO TALK ABOUT RELATIVITY.
HE INVENTED RELATIVITY?
NO, BUT HE INTRODUCED THE BASIC CONCEPT TO SCIENCE THAT WHAT YOU EXPERIENCE IS ONLY RELATIVE TO WHAT IS AROUND YOU...
...HE IMAGINED SOMEONE STANDING ON THE DECK OF A MOVING BOAT AND SOMEONE INSIDE A WINDOWLESS CABIN OF THE SAME BOAT. ON DECK, THE PERSON CAN TELL THEY'RE MOVING FORWARD, BUT INSIDE THE CABIN, APART FROM A LITTLE SIDE-TO-SIDE SWAY, THE PERSON DOESN'T FEEL THE MOVEMENT AT ALL. IF YOU DROP A BALL INSIDE THE CABIN, IT APPEARS TO FALL STRAIGHT DOWN, EVEN THOUGH IT'S ALSO MOVING FORWARD AS IT FALLS AT THE SAME SPEED AS THE BOAT.

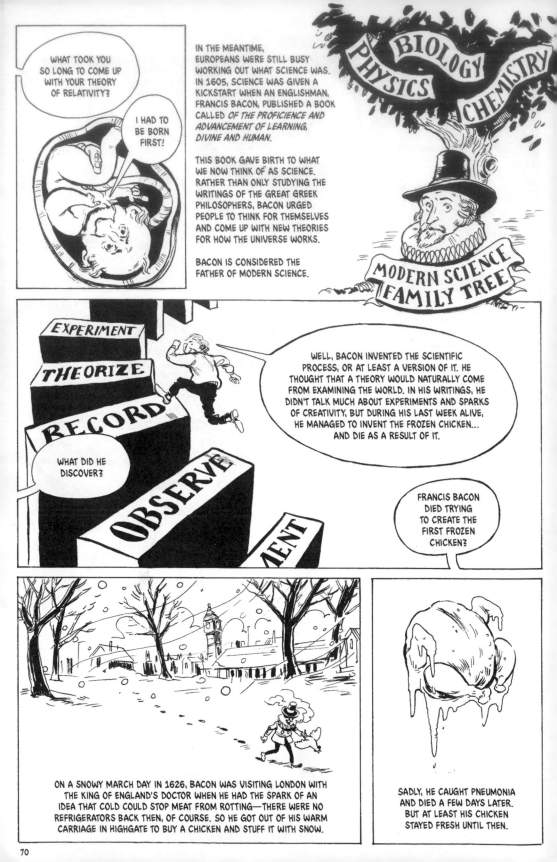

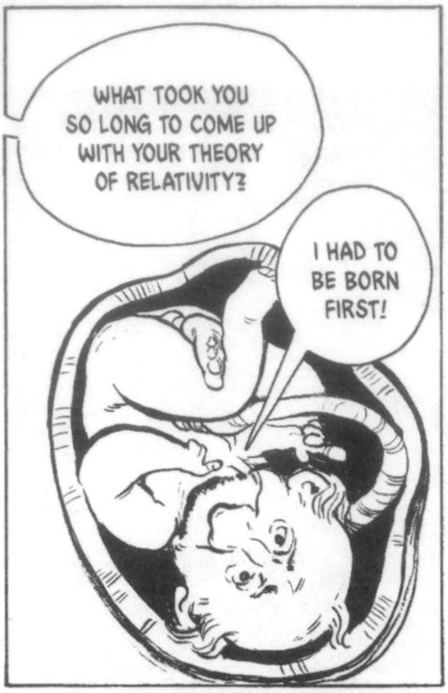
WHAT TOOK YOU SO LONG TO COME UP WITH YOUR THEORY OF RELATIVITY?
I HAD TO BE BORN FIRST!

IN THE MEANTIME, EUROPEANS WERE STILL BUSY WORKING OUT WHAT SCIENCE WAS. IN 1605, SCIENCE WAS GIVEN A KICKSTART WHEN AN ENGLISHMAN, FRANCIS BACON, PUBLISHED A BOOK CALLED *OF THE PROFICIENCE AND ADVANCEMENT OF LEARNING, DIVINE AND HUMAN*.

THIS BOOK GAVE BIRTH TO WHAT WE NOW THINK OF AS SCIENCE. RATHER THAN ONLY STUDYING THE WRITINGS OF THE GREAT GREEK PHILOSOPHERS, BACON URGED PEOPLE TO THINK FOR THEMSELVES AND COME UP WITH NEW THEORIES FOR HOW THE UNIVERSE WORKS.

BACON IS CONSIDERED THE FATHER OF MODERN SCIENCE.

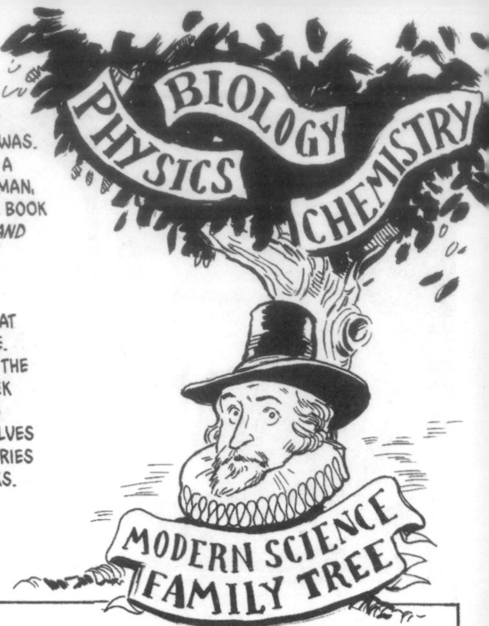
BIOLOGY
PHYSICS
CHEMISTRY
MODERN SCIENCE FAMILY TREE

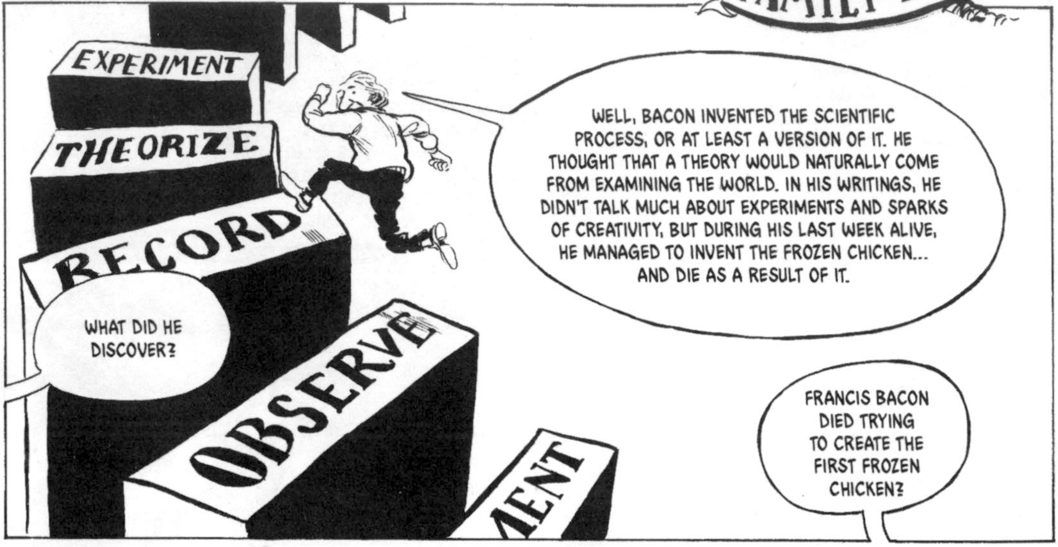
EXPERIMENT
THEORIZE
RECORD
OBSERVE
WELL, BACON INVENTED THE SCIENTIFIC PROCESS, OR AT LEAST A VERSION OF IT. HE THOUGHT THAT A THEORY WOULD NATURALLY COME FROM EXAMINING THE WORLD. IN HIS WRITINGS, HE DIDN'T TALK MUCH ABOUT EXPERIMENTS AND SPARKS OF CREATIVITY, BUT DURING HIS LAST WEEK ALIVE, HE MANAGED TO INVENT THE FROZEN CHICKEN... AND DIE AS A RESULT OF IT.
WHAT DID HE DISCOVER?
FRANCIS BACON DIED TRYING TO CREATE THE FIRST FROZEN CHICKEN?

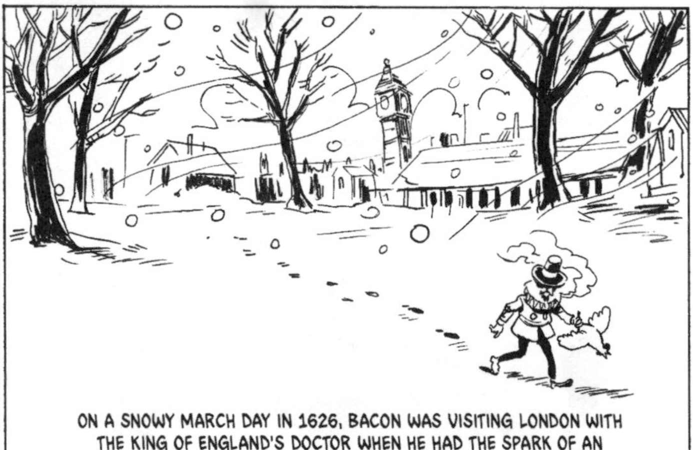

ON A SNOWY MARCH DAY IN 1626, BACON WAS VISITING LONDON WITH THE KING OF ENGLAND'S DOCTOR WHEN HE HAD THE SPARK OF AN IDEA THAT COLD COULD STOP MEAT FROM ROTTING—THERE WERE NO REFRIGERATORS BACK THEN, OF COURSE. SO HE GOT OUT OF HIS WARM CARRIAGE IN HIGHGATE TO BUY A CHICKEN AND STUFF IT WITH SNOW.

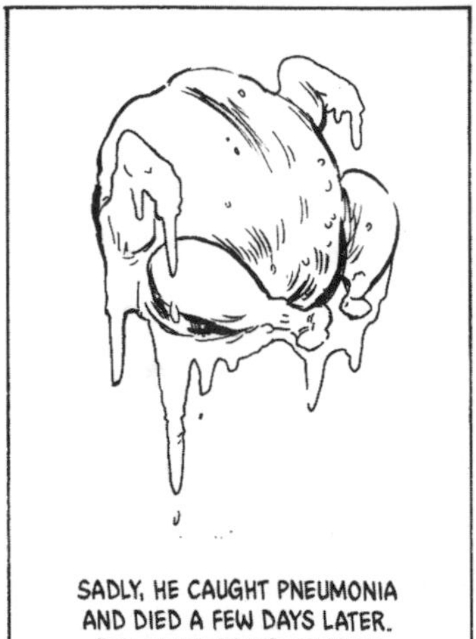

SADLY, HE CAUGHT PNEUMONIA AND DIED A FEW DAYS LATER. BUT AT LEAST HIS CHICKEN STAYED FRESH UNTIL THEN.

CHAPTER 13: A BRIEF HISTORY OF LIGHT

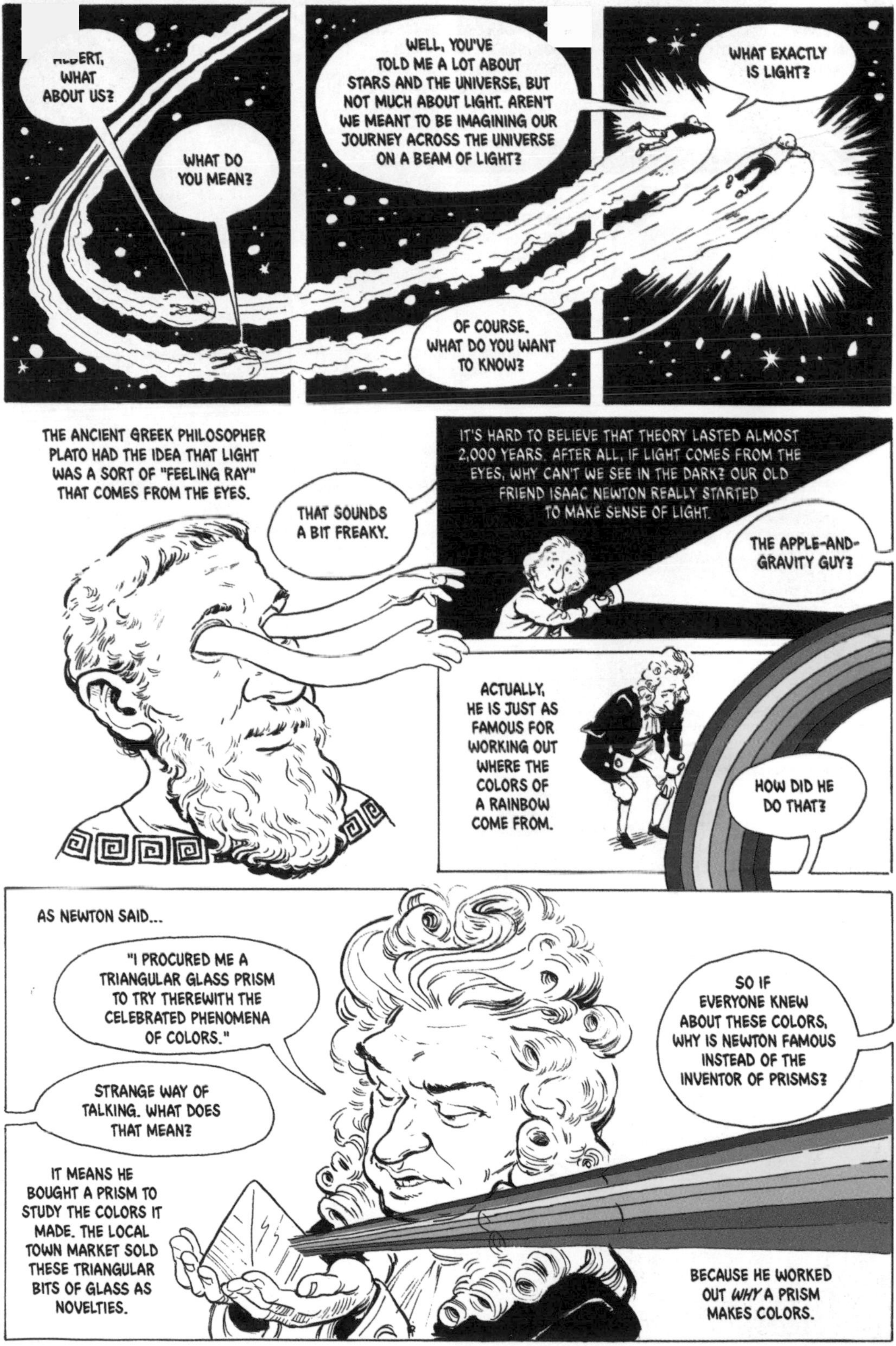
WHAT ABOUT US?
WHAT DO YOU MEAN?
WELL, YOU'VE TOLD ME A LOT ABOUT STARS AND THE UNIVERSE, BUT NOT MUCH ABOUT LIGHT. AREN'T WE MEANT TO BE IMAGINING OUR JOURNEY ACROSS THE UNIVERSE ON A BEAM OF LIGHT?
OF COURSE. WHAT DO YOU WANT TO KNOW?
WHAT EXACTLY IS LIGHT?
THE ANCIENT GREEK PHILOSOPHER PLATO HAD THE IDEA THAT LIGHT WAS A SORT OF "FEELING RAY" THAT COMES FROM THE EYES.
THAT SOUNDS A BIT FREAKY.
IT'S HARD TO BELIEVE THAT THEORY LASTED ALMOST 2,000 YEARS. AFTER ALL, IF LIGHT COMES FROM THE EYES, WHY CAN'T WE SEE IN THE DARK? OUR OLD FRIEND ISAAC NEWTON REALLY STARTED TO MAKE SENSE OF LIGHT.
THE APPLE-AND-GRAVITY GUY?
ACTUALLY, HE IS JUST AS FAMOUS FOR WORKING OUT WHERE THE COLORS OF A RAINBOW COME FROM.
HOW DID HE DO THAT?
AS NEWTON SAID...
"I PROCURED ME A TRIANGULAR GLASS PRISM TO TRY THEREWITH THE CELEBRATED PHENOMENA OF COLORS."
STRANGE WAY OF TALKING. WHAT DOES THAT MEAN?
IT MEANS HE BOUGHT A PRISM TO STUDY THE COLORS IT MADE. THE LOCAL TOWN MARKET SOLD THESE TRIANGULAR BITS OF GLASS AS NOVELTIES.
SO IF EVERYONE KNEW ABOUT THESE COLORS, WHY IS NEWTON FAMOUS INSTEAD OF THE INVENTOR OF PRISMS?
BECAUSE HE WORKED OUT WHY A PRISM MAKES COLORS.

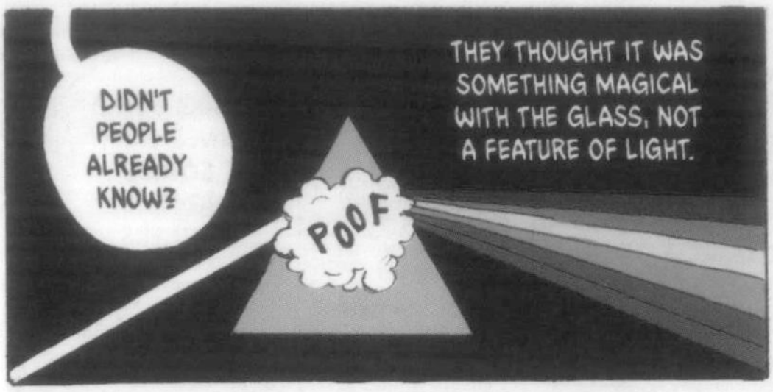
DIDN'T PEOPLE ALREADY KNOW?
THEY THOUGHT IT WAS SOMETHING MAGICAL WITH THE GLASS, NOT A FEATURE OF LIGHT.
POOF

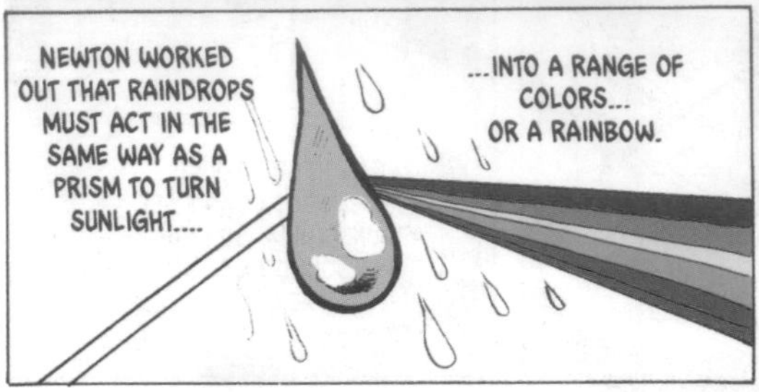
NEWTON WORKED OUT THAT RAINDROPS MUST ACT IN THE SAME WAY AS A PRISM TO TURN SUNLIGHT....
...INTO A RANGE OF COLORS... OR A RAINBOW.

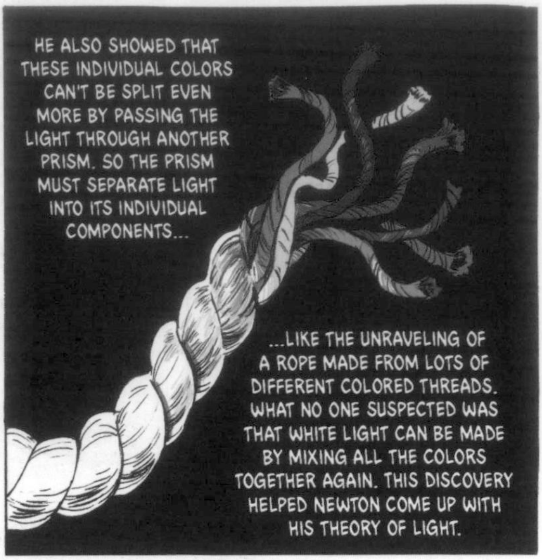
HE ALSO SHOWED THAT THESE INDIVIDUAL COLORS CAN'T BE SPLIT EVEN MORE BY PASSING THE LIGHT THROUGH ANOTHER PRISM. SO THE PRISM MUST SEPARATE LIGHT INTO ITS INDIVIDUAL COMPONENTS...
...LIKE THE UNRAVELING OF A ROPE MADE FROM LOTS OF DIFFERENT COLORED THREADS. WHAT NO ONE SUSPECTED WAS THAT WHITE LIGHT CAN BE MADE BY MIXING ALL THE COLORS TOGETHER AGAIN. THIS DISCOVERY HELPED NEWTON COME UP WITH HIS THEORY OF LIGHT.

NEWTON BELIEVED LIGHT WAS MADE UP OF LITTLE PARTICLES, OR CORPUSCLES, AS HE CALLED THEM, OF DIFFERENT TYPES, EACH ONE REPRESENTING A DIFFERENT COLOR ACROSS THE RAINBOW, OR SPECTRUM, AS SCIENTISTS CALL IT, FROM RED TO VIOLET. WHITE LIGHT IS AN EQUAL MIXTURE OF ALL THE DIFFERENT TYPES OF LIGHT PARTICLES. EVERY HUE IMAGINABLE CAN BE MADE BY MIXING THE DIFFERENT COLORED TYPES IN VARYING PROPORTIONS.

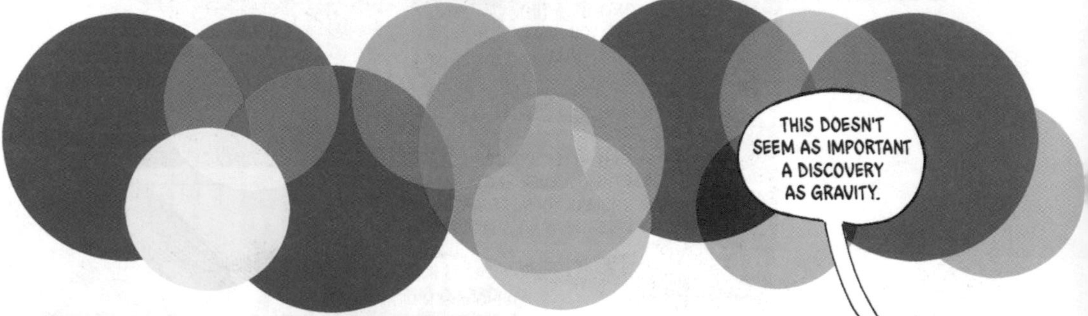
THIS DOESN'T SEEM AS IMPORTANT A DISCOVERY AS GRAVITY.

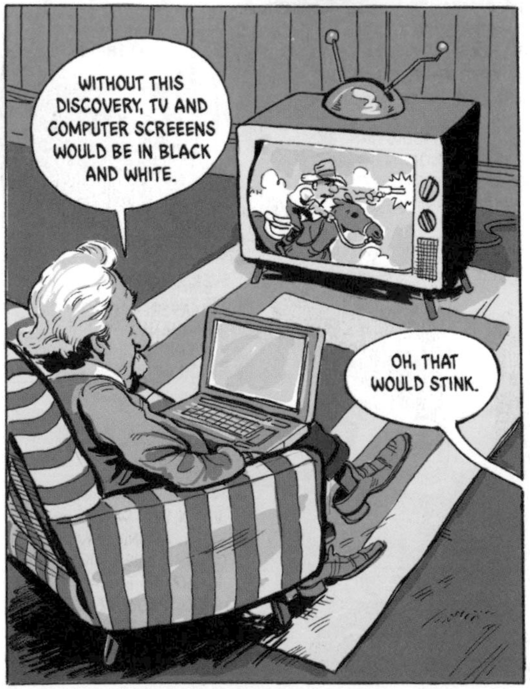
WITHOUT THIS DISCOVERY, TV AND COMPUTER SCREEENS WOULD BE IN BLACK AND WHITE.
OH, THAT WOULD STINK.

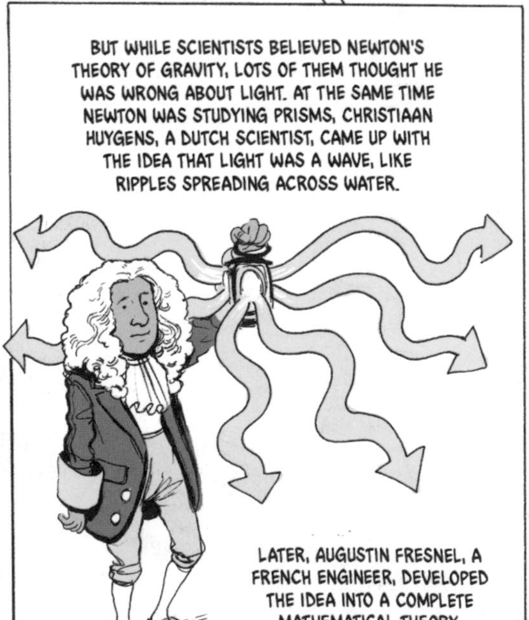
BUT WHILE SCIENTISTS BELIEVED NEWTON'S THEORY OF GRAVITY, LOTS OF THEM THOUGHT HE WAS WRONG ABOUT LIGHT. AT THE SAME TIME NEWTON WAS STUDYING PRISMS, CHRISTIAAN HUYGENS, A DUTCH SCIENTIST, CAME UP WITH THE IDEA THAT LIGHT WAS A WAVE, LIKE RIPPLES SPREADING ACROSS WATER.
LATER, AUGUSTIN FRESNEL, A FRENCH ENGINEER, DEVELOPED THE IDEA INTO A COMPLETE MATHEMATICAL THEORY.

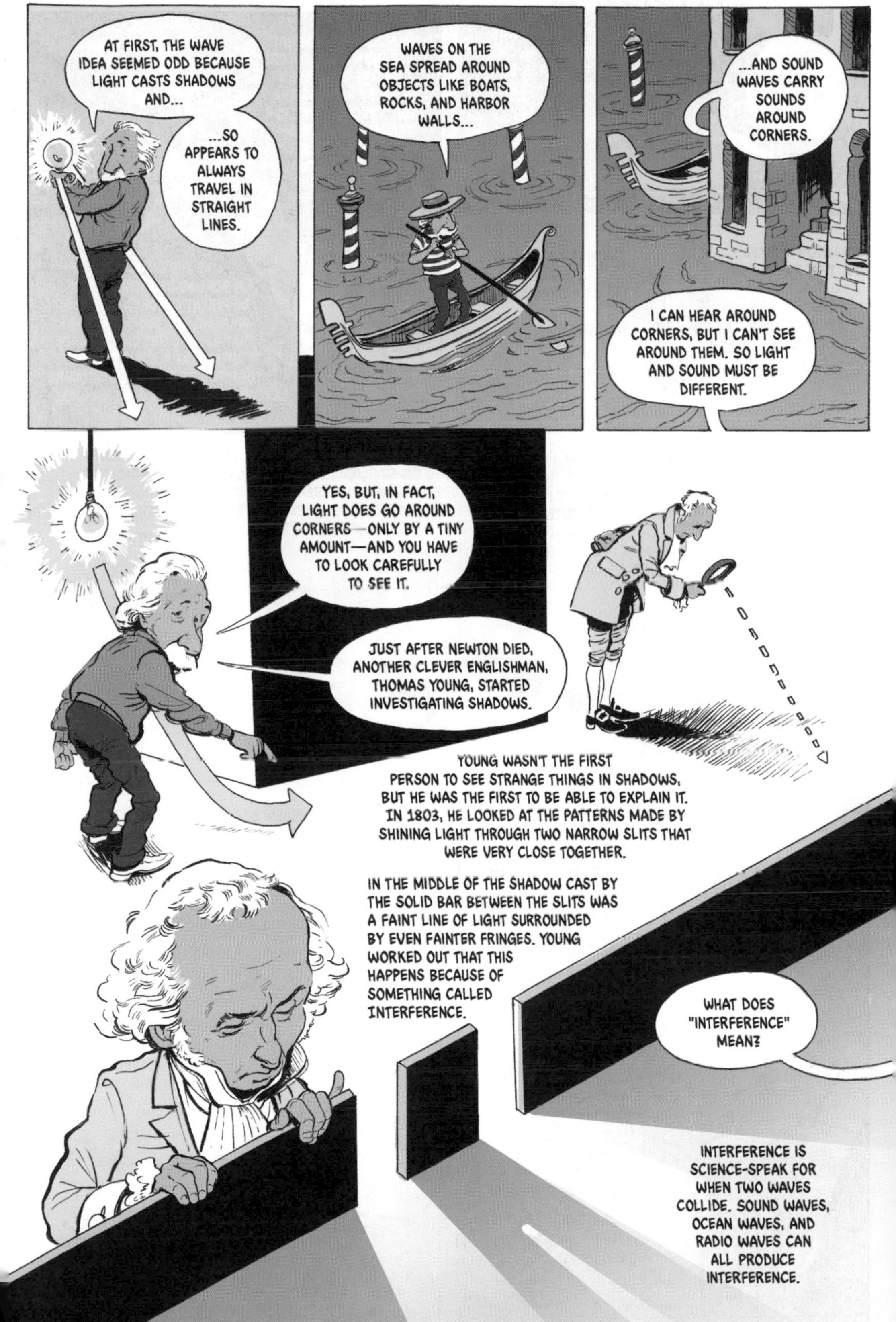
AT FIRST, THE WAVE IDEA SEEMED ODD BECAUSE LIGHT CASTS SHADOWS AND...
...SO APPEARS TO ALWAYS TRAVEL IN STRAIGHT LINES.
WAVES ON THE SEA SPREAD AROUND OBJECTS LIKE BOATS, ROCKS, AND HARBOR WALLS...
...AND SOUND WAVES CARRY SOUNDS AROUND CORNERS.
I CAN HEAR AROUND CORNERS, BUT I CAN'T SEE AROUND THEM. SO LIGHT AND SOUND MUST BE DIFFERENT.
YES, BUT, IN FACT, LIGHT DOES GO AROUND CORNERS—ONLY BY A TINY AMOUNT—AND YOU HAVE TO LOOK CAREFULLY TO SEE IT.
JUST AFTER NEWTON DIED, ANOTHER CLEVER ENGLISHMAN, THOMAS YOUNG, STARTED INVESTIGATING SHADOWS.
YOUNG WASN'T THE FIRST PERSON TO SEE STRANGE THINGS IN SHADOWS, BUT HE WAS THE FIRST TO BE ABLE TO EXPLAIN IT. IN 1803, HE LOOKED AT THE PATTERNS MADE BY SHINING LIGHT THROUGH TWO NARROW SLITS THAT WERE VERY CLOSE TOGETHER.
IN THE MIDDLE OF THE SHADOW CAST BY THE SOLID BAR BETWEEN THE SLITS WAS A FAINT LINE OF LIGHT SURROUNDED BY EVEN FAINTER FRINGES. YOUNG WORKED OUT THAT THIS HAPPENS BECAUSE OF SOMETHING CALLED INTERFERENCE.
WHAT DOES "INTERFERENCE" MEAN?
INTERFERENCE IS SCIENCE-SPEAK FOR WHEN TWO WAVES COLLIDE. SOUND WAVES, OCEAN WAVES, AND RADIO WAVES CAN ALL PRODUCE INTERFERENCE.

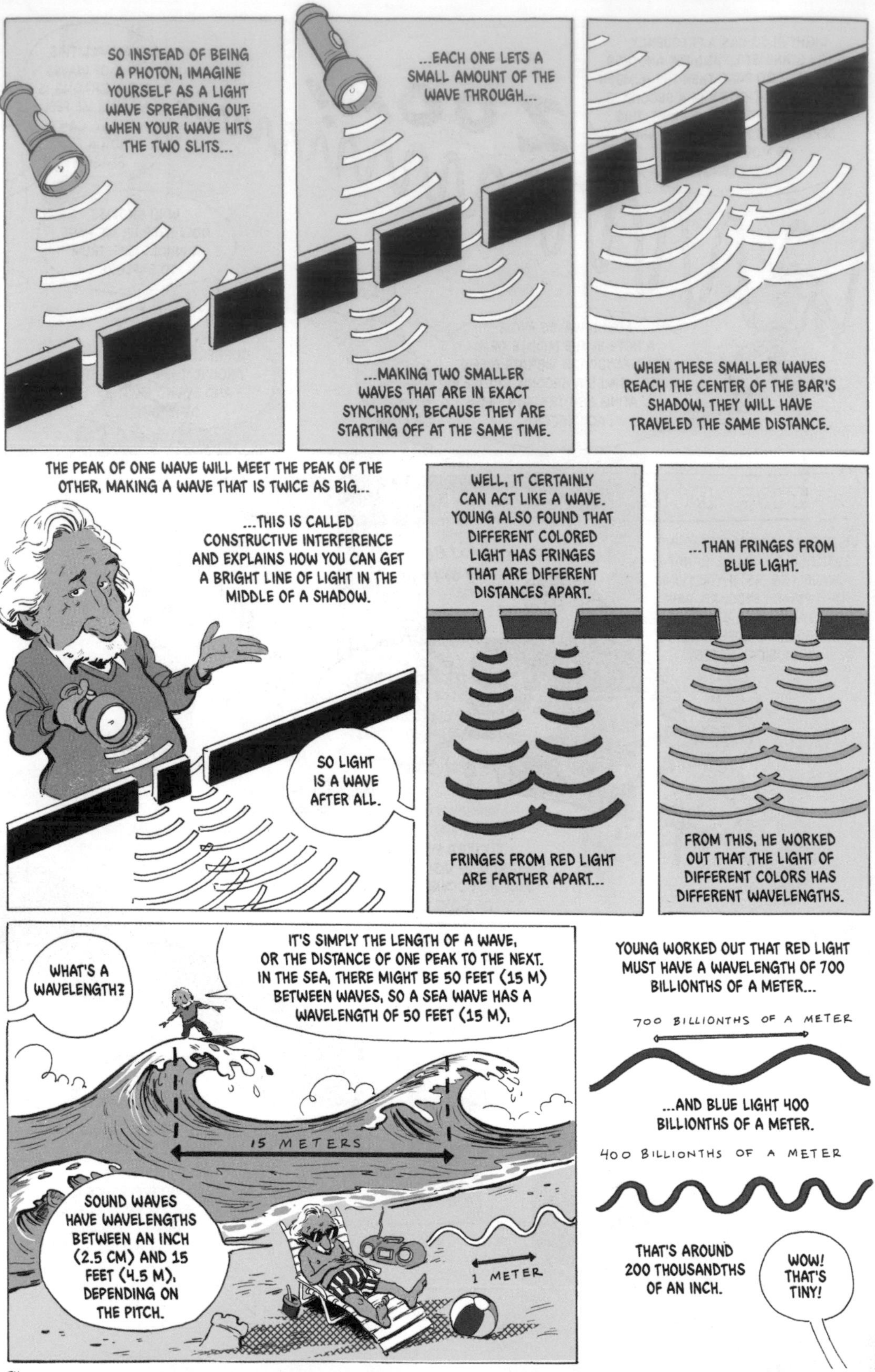
SO INSTEAD OF BEING A PHOTON, IMAGINE YOURSELF AS A LIGHT WAVE SPREADING OUT: WHEN YOUR WAVE HITS THE TWO SLITS...
...EACH ONE LETS A SMALL AMOUNT OF THE WAVE THROUGH...
...MAKING TWO SMALLER WAVES THAT ARE IN EXACT SYNCHRONY, BECAUSE THEY ARE STARTING OFF AT THE SAME TIME.
WHEN THESE SMALLER WAVES REACH THE CENTER OF THE BAR'S SHADOW, THEY WILL HAVE TRAVELED THE SAME DISTANCE.
THE PEAK OF ONE WAVE WILL MEET THE PEAK OF THE OTHER, MAKING A WAVE THAT IS TWICE AS BIG...
...THIS IS CALLED CONSTRUCTIVE INTERFERENCE AND EXPLAINS HOW YOU CAN GET A BRIGHT LINE OF LIGHT IN THE MIDDLE OF A SHADOW.
SO LIGHT IS A WAVE AFTER ALL.
WELL, IT CERTAINLY CAN ACT LIKE A WAVE. YOUNG ALSO FOUND THAT DIFFERENT COLORED LIGHT HAS FRINGES THAT ARE DIFFERENT DISTANCES APART.
FRINGES FROM RED LIGHT ARE FARTHER APART...
...THAN FRINGES FROM BLUE LIGHT.
FROM THIS, HE WORKED OUT THAT THE LIGHT OF DIFFERENT COLORS HAS DIFFERENT WAVELENGTHS.
WHAT'S A WAVELENGTH?
IT'S SIMPLY THE LENGTH OF A WAVE, OR THE DISTANCE OF ONE PEAK TO THE NEXT. IN THE SEA, THERE MIGHT BE 50 FEET (15 M) BETWEEN WAVES, SO A SEA WAVE HAS A WAVELENGTH OF 50 FEET (15 M).
15 METERS
SOUND WAVES HAVE WAVELENGTHS BETWEEN AN INCH (2.5 CM) AND 15 FEET (4.5 M), DEPENDING ON THE PITCH.
1 METER
YOUNG WORKED OUT THAT RED LIGHT MUST HAVE A WAVELENGTH OF 700 BILLIONTHS OF A METER...
700 BILLIONTHS OF A METER
...AND BLUE LIGHT 400 BILLIONTHS OF A METER.
400 BILLIONTHS OF A METER
THAT'S AROUND 200 THOUSANDTHS OF AN INCH.
WOW! THAT'S TINY!

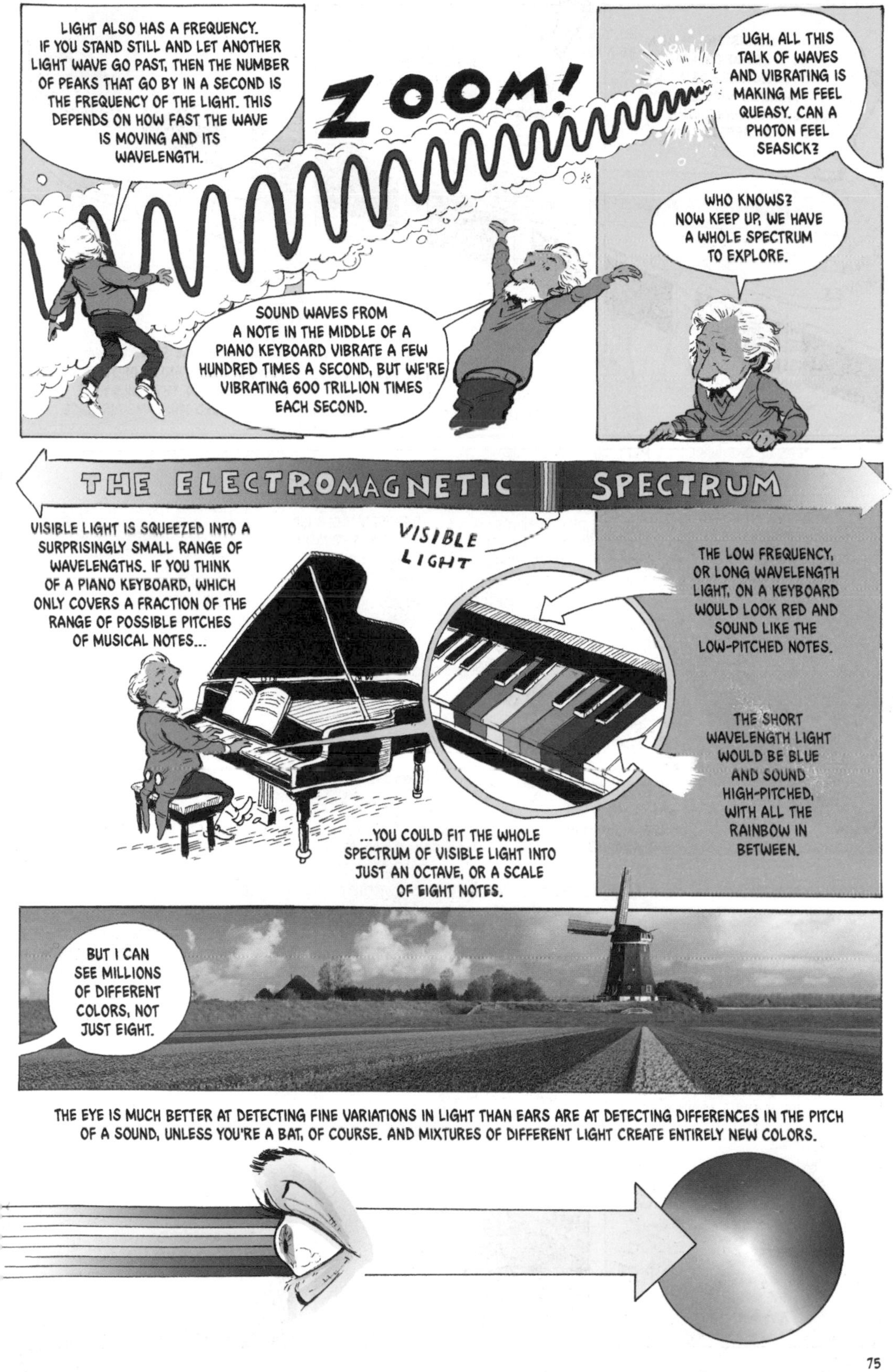
LIGHT ALSO HAS A FREQUENCY. IF YOU STAND STILL AND LET ANOTHER LIGHT WAVE GO PAST, THEN THE NUMBER OF PEAKS THAT GO BY IN A SECOND IS THE FREQUENCY OF THE LIGHT. THIS DEPENDS ON HOW FAST THE WAVE IS MOVING AND ITS WAVELENGTH.
ZOOM!
SOUND WAVES FROM A NOTE IN THE MIDDLE OF A PIANO KEYBOARD VIBRATE A FEW HUNDRED TIMES A SECOND, BUT WE'RE VIBRATING 600 TRILLION TIMES EACH SECOND.
UGH, ALL THIS TALK OF WAVES AND VIBRATING IS MAKING ME FEEL QUEASY. CAN A PHOTON FEEL SEASICK?
WHO KNOWS? NOW KEEP UP, WE HAVE A WHOLE SPECTRUM TO EXPLORE.
THE ELECTROMAGNETIC SPECTRUM
VISIBLE LIGHT
VISIBLE LIGHT IS SQUEEZED INTO A SURPRISINGLY SMALL RANGE OF WAVELENGTHS. IF YOU THINK OF A PIANO KEYBOARD, WHICH ONLY COVERS A FRACTION OF THE RANGE OF POSSIBLE PITCHES OF MUSICAL NOTES...
...YOU COULD FIT THE WHOLE SPECTRUM OF VISIBLE LIGHT INTO JUST AN OCTAVE, OR A SCALE OF EIGHT NOTES.
THE LOW FREQUENCY, OR LONG WAVELENGTH LIGHT, ON A KEYBOARD WOULD LOOK RED AND SOUND LIKE THE LOW-PITCHED NOTES.
THE SHORT WAVELENGTH LIGHT WOULD BE BLUE AND SOUND HIGH-PITCHED, WITH ALL THE RAINBOW IN BETWEEN.
BUT I CAN SEE MILLIONS OF DIFFERENT COLORS, NOT JUST EIGHT.
THE EYE IS MUCH BETTER AT DETECTING FINE VARIATIONS IN LIGHT THAN EARS ARE AT DETECTING DIFFERENCES IN THE PITCH OF A SOUND, UNLESS YOU'RE A BAT, OF COURSE. AND MIXTURES OF DIFFERENT LIGHT CREATE ENTIRELY NEW COLORS.

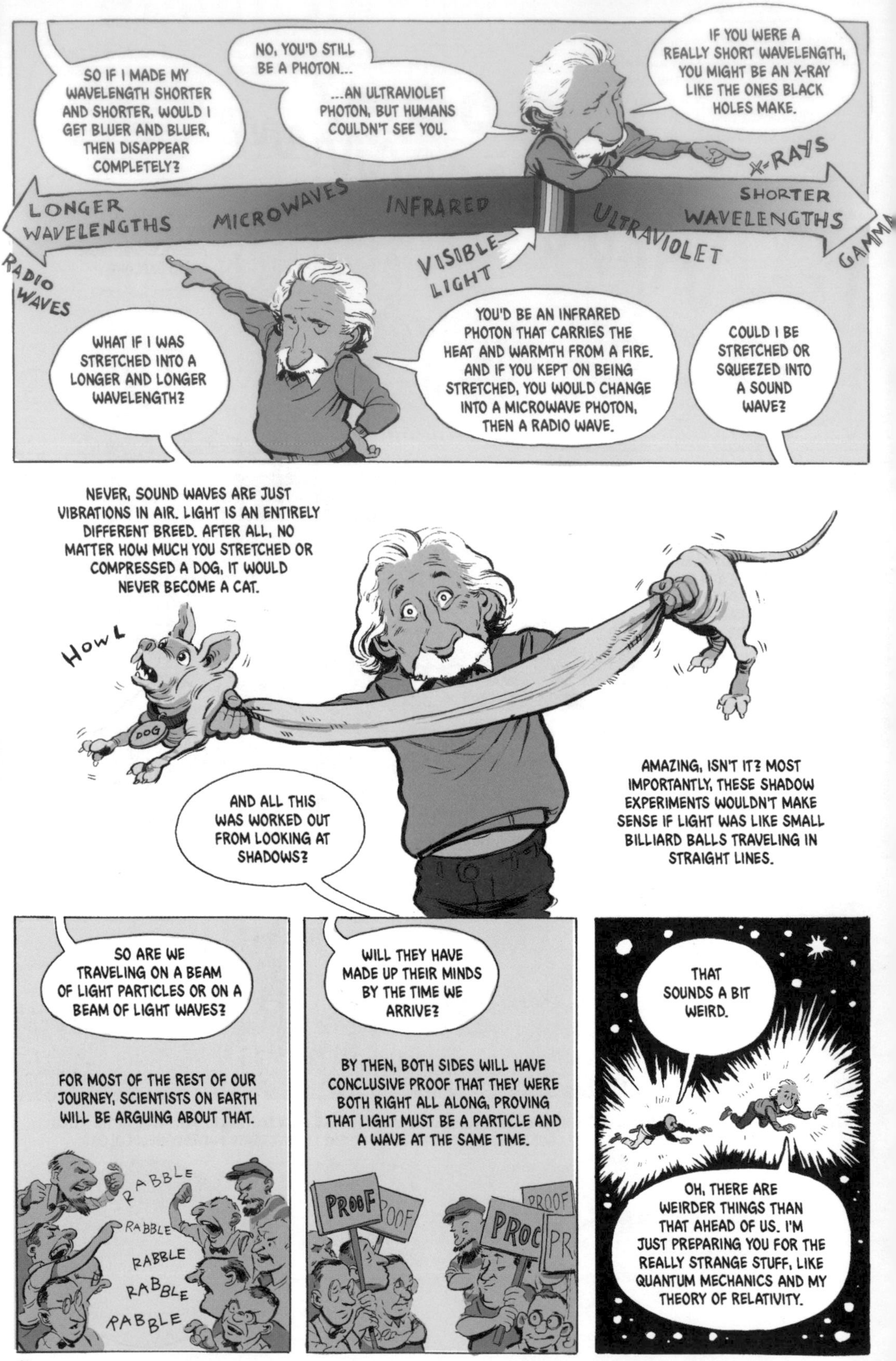
SO IF I MADE MY WAVELENGTH SHORTER AND SHORTER, WOULD I GET BLUER AND BLUER, THEN DISAPPEAR COMPLETELY?
NO, YOU'D STILL BE A PHOTON...
...AN ULTRAVIOLET PHOTON, BUT HUMANS COULDN'T SEE YOU.
IF YOU WERE A REALLY SHORT WAVELENGTH, YOU MIGHT BE AN X-RAY LIKE THE ONES BLACK HOLES MAKE.
X-RAYS
LONGER WAVELENGTHS
MICROWAVES
INFRARED
ULTRAVIOLET
SHORTER WAVELENGTHS
GAMMA
RADIO WAVES
VISIBLE LIGHT
WHAT IF I WAS STRETCHED INTO A LONGER AND LONGER WAVELENGTH?
YOU'D BE AN INFRARED PHOTON THAT CARRIES THE HEAT AND WARMTH FROM A FIRE. AND IF YOU KEPT ON BEING STRETCHED, YOU WOULD CHANGE INTO A MICROWAVE PHOTON, THEN A RADIO WAVE.
COULD I BE STRETCHED OR SQUEEZED INTO A SOUND WAVE?
NEVER, SOUND WAVES ARE JUST VIBRATIONS IN AIR. LIGHT IS AN ENTIRELY DIFFERENT BREED. AFTER ALL, NO MATTER HOW MUCH YOU STRETCHED OR COMPRESSED A DOG, IT WOULD NEVER BECOME A CAT.
HOWL
DOG
AND ALL THIS WAS WORKED OUT FROM LOOKING AT SHADOWS?
AMAZING, ISN'T IT? MOST IMPORTANTLY, THESE SHADOW EXPERIMENTS WOULDN'T MAKE SENSE IF LIGHT WAS LIKE SMALL BILLIARD BALLS TRAVELING IN STRAIGHT LINES.
SO ARE WE TRAVELING ON A BEAM OF LIGHT PARTICLES OR ON A BEAM OF LIGHT WAVES?
FOR MOST OF THE REST OF OUR JOURNEY, SCIENTISTS ON EARTH WILL BE ARGUING ABOUT THAT.
RABBLE
RABBLE
RABBLE
RABBLE
RABBLE
WILL THEY HAVE MADE UP THEIR MINDS BY THE TIME WE ARRIVE?
BY THEN, BOTH SIDES WILL HAVE CONCLUSIVE PROOF THAT THEY WERE BOTH RIGHT ALL ALONG, PROVING THAT LIGHT MUST BE A PARTICLE AND A WAVE AT THE SAME TIME.
PROOF
PROOF
THAT SOUNDS A BIT WEIRD.
OH, THERE ARE WEIRDER THINGS THAN THAT AHEAD OF US. I'M JUST PREPARING YOU FOR THE REALLY STRANGE STUFF, LIKE QUANTUM MECHANICS AND MY THEORY OF RELATIVITY.

CHAPTER 14: WHY YOU SHOULD NEVER TRUST A QUANTUM MECHANIC

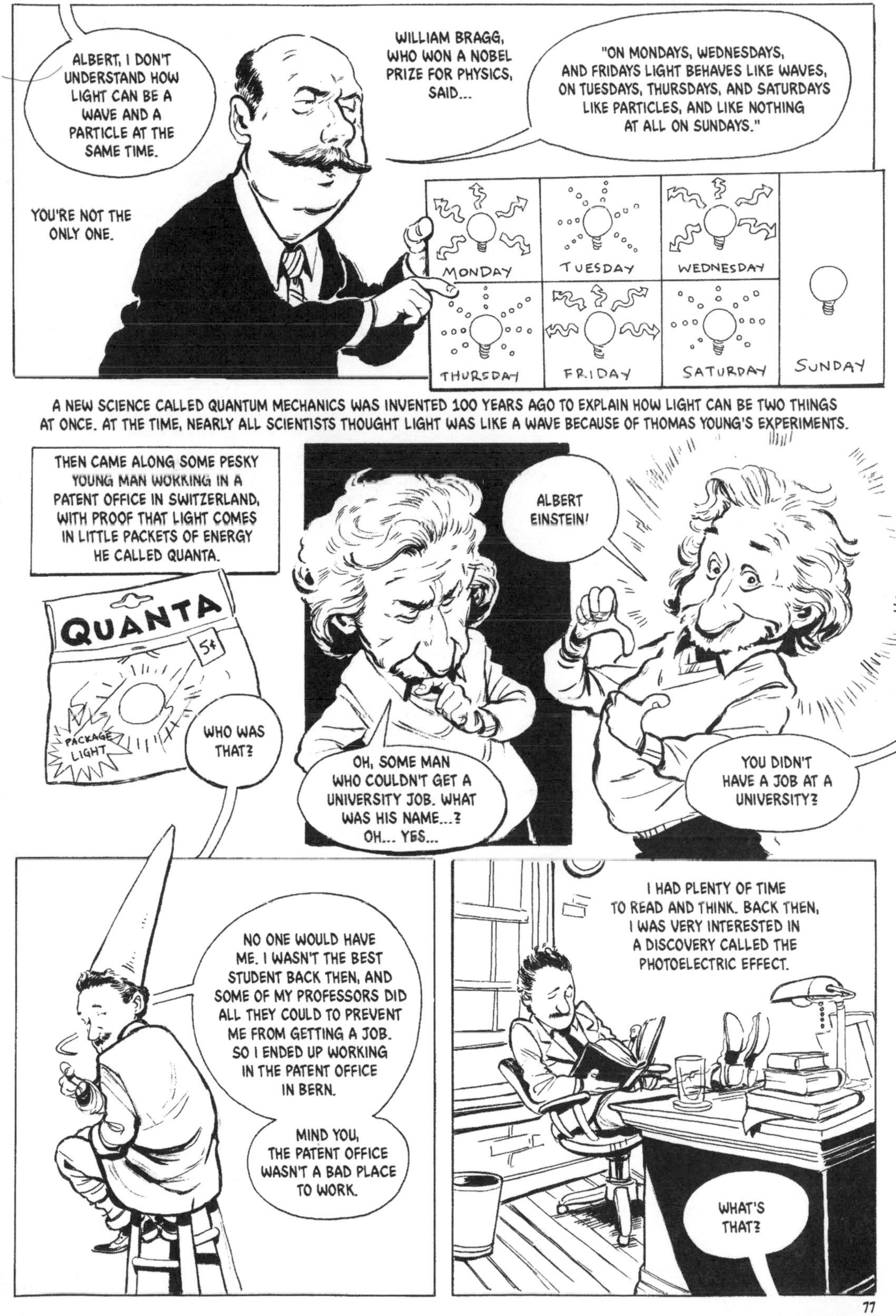
ALBERT, I DON'T UNDERSTAND HOW LIGHT CAN BE A WAVE AND A PARTICLE AT THE SAME TIME.
YOU'RE NOT THE ONLY ONE.
WILLIAM BRAGG, WHO WON A NOBEL PRIZE FOR PHYSICS, SAID...
"ON MONDAYS, WEDNESDAYS, AND FRIDAYS LIGHT BEHAVES LIKE WAVES, ON TUESDAYS, THURSDAYS, AND SATURDAYS LIKE PARTICLES, AND LIKE NOTHING AT ALL ON SUNDAYS."
MONDAY
TUESDAY
WEDNESDAY
THURSDAY
FRIDAY
SATURDAY
SUNDAY
A NEW SCIENCE CALLED QUANTUM MECHANICS WAS INVENTED 100 YEARS AGO TO EXPLAIN HOW LIGHT CAN BE TWO THINGS AT ONCE. AT THE TIME, NEARLY ALL SCIENTISTS THOUGHT LIGHT WAS LIKE A WAVE BECAUSE OF THOMAS YOUNG'S EXPERIMENTS.
THEN CAME ALONG SOME PESKY YOUNG MAN WORKING IN A PATENT OFFICE IN SWITZERLAND, WITH PROOF THAT LIGHT COMES IN LITTLE PACKETS OF ENERGY HE CALLED QUANTA.
QUANTA
5¢
PACKAGE LIGHT
WHO WAS THAT?
OH, SOME MAN WHO COULDN'T GET A UNIVERSITY JOB. WHAT WAS HIS NAME...? OH... YES...
ALBERT EINSTEIN!
YOU DIDN'T HAVE A JOB AT A UNIVERSITY?
NO ONE WOULD HAVE ME. I WASN'T THE BEST STUDENT BACK THEN, AND SOME OF MY PROFESSORS DID ALL THEY COULD TO PREVENT ME FROM GETTING A JOB. SO I ENDED UP WORKING IN THE PATENT OFFICE IN BERN.
MIND YOU, THE PATENT OFFICE WASN'T A BAD PLACE TO WORK.
I HAD PLENTY OF TIME TO READ AND THINK. BACK THEN, I WAS VERY INTERESTED IN A DISCOVERY CALLED THE PHOTOELECTRIC EFFECT.
WHAT'S THAT?

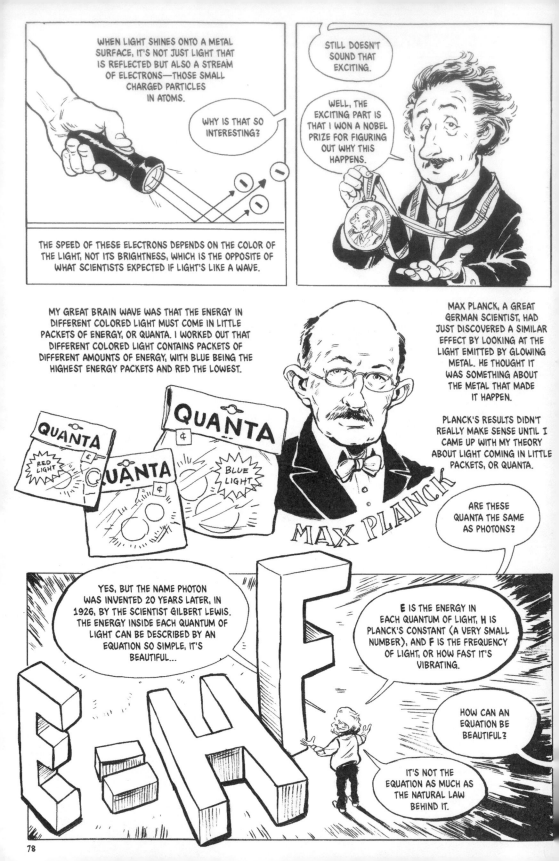

WHEN LIGHT SHINES ONTO A METAL SURFACE, IT'S NOT JUST LIGHT THAT IS REFLECTED BUT ALSO A STREAM OF ELECTRONS—THOSE SMALL CHARGED PARTICLES IN ATOMS.

WHY IS THAT SO INTERESTING?

STILL DOESN'T SOUND THAT EXCITING.

WELL, THE EXCITING PART IS THAT I WON A NOBEL PRIZE FOR FIGURING OUT WHY THIS HAPPENS.

THE SPEED OF THESE ELECTRONS DEPENDS ON THE COLOR OF THE LIGHT, NOT ITS BRIGHTNESS, WHICH IS THE OPPOSITE OF WHAT SCIENTISTS EXPECTED IF LIGHT'S LIKE A WAVE.

MY GREAT BRAIN WAVE WAS THAT THE ENERGY IN DIFFERENT COLORED LIGHT MUST COME IN LITTLE PACKETS OF ENERGY, OR QUANTA. I WORKED OUT THAT DIFFERENT COLORED LIGHT CONTAINS PACKETS OF DIFFERENT AMOUNTS OF ENERGY, WITH BLUE BEING THE HIGHEST ENERGY PACKETS AND RED THE LOWEST.

QUANTA
RED LIGHT
¢

QUANTA

QUANTA
¢

QUANTA
BLUE LIGHT

MAX PLANCK, A GREAT GERMAN SCIENTIST, HAD JUST DISCOVERED A SIMILAR EFFECT BY LOOKING AT THE LIGHT EMITTED BY GLOWING METAL. HE THOUGHT IT WAS SOMETHING ABOUT THE METAL THAT MADE IT HAPPEN.

PLANCK'S RESULTS DIDN'T REALLY MAKE SENSE UNTIL I CAME UP WITH MY THEORY ABOUT LIGHT COMING IN LITTLE PACKETS, OR QUANTA.

MAX PLANCK

ARE THESE QUANTA THE SAME AS PHOTONS?

YES, BUT THE NAME PHOTON WAS INVENTED 20 YEARS LATER, IN 1926, BY THE SCIENTIST GILBERT LEWIS. THE ENERGY INSIDE EACH QUANTUM OF LIGHT CAN BE DESCRIBED BY AN EQUATION SO SIMPLE, IT'S BEAUTIFUL...

$$E = H F$$

E IS THE ENERGY IN EACH QUANTUM OF LIGHT, H IS PLANCK'S CONSTANT (A VERY SMALL NUMBER), AND F IS THE FREQUENCY OF LIGHT, OR HOW FAST IT'S VIBRATING.

HOW CAN AN EQUATION BE BEAUTIFUL?

IT'S NOT THE EQUATION AS MUCH AS THE NATURAL LAW BEHIND IT.

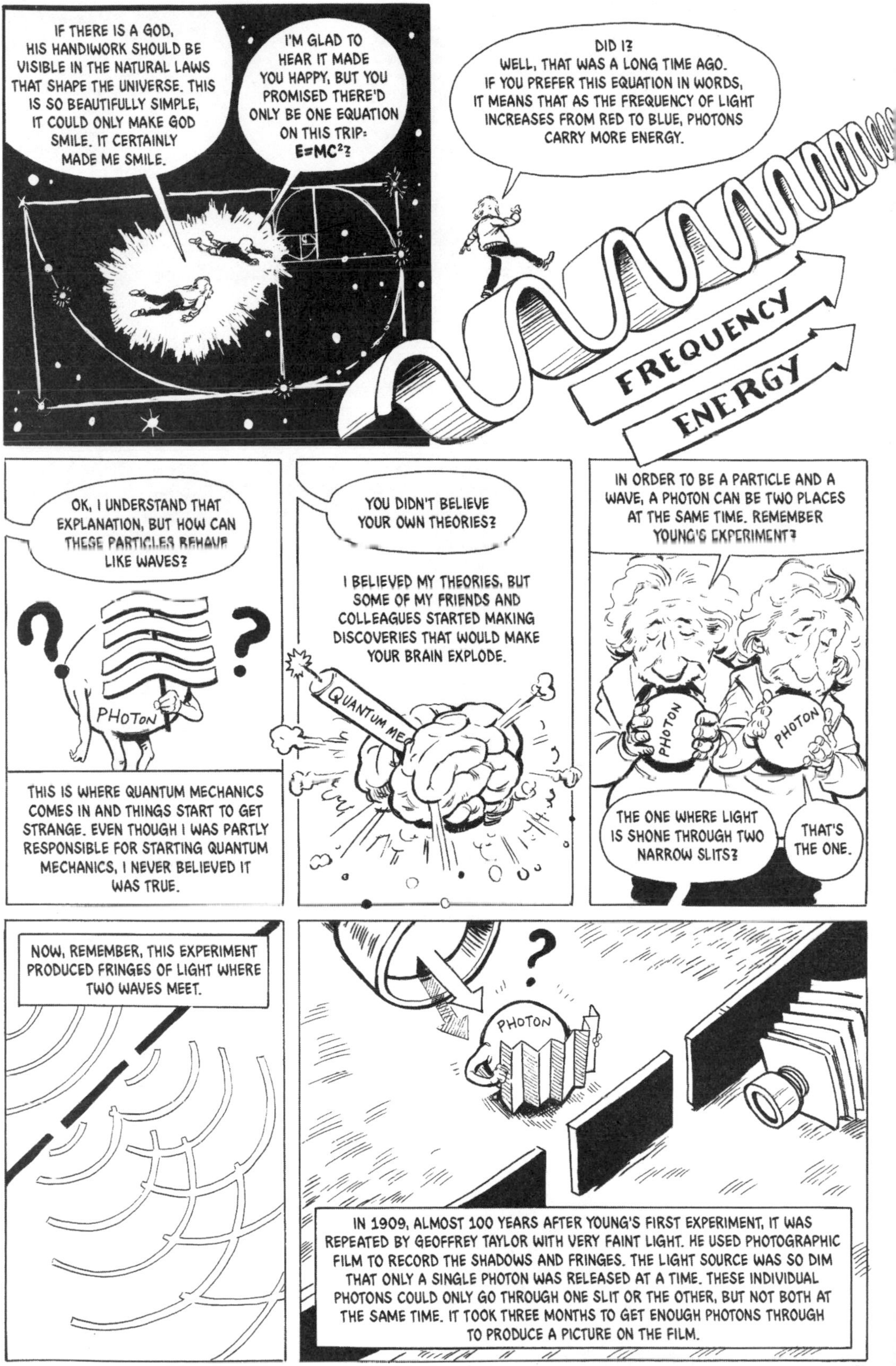
IF THERE IS A GOD, HIS HANDIWORK SHOULD BE VISIBLE IN THE NATURAL LAWS THAT SHAPE THE UNIVERSE. THIS IS SO BEAUTIFULLY SIMPLE, IT COULD ONLY MAKE GOD SMILE. IT CERTAINLY MADE ME SMILE.
I'M GLAD TO HEAR IT MADE YOU HAPPY, BUT YOU PROMISED THERE'D ONLY BE ONE EQUATION ON THIS TRIP: E=MC²?
DID I? WELL, THAT WAS A LONG TIME AGO. IF YOU PREFER THIS EQUATION IN WORDS, IT MEANS THAT AS THE FREQUENCY OF LIGHT INCREASES FROM RED TO BLUE, PHOTONS CARRY MORE ENERGY.
FREQUENCY
ENERGY
OK, I UNDERSTAND THAT EXPLANATION, BUT HOW CAN THESE PARTICLES BEHAVE LIKE WAVES?
PHOTON
THIS IS WHERE QUANTUM MECHANICS COMES IN AND THINGS START TO GET STRANGE. EVEN THOUGH I WAS PARTLY RESPONSIBLE FOR STARTING QUANTUM MECHANICS, I NEVER BELIEVED IT WAS TRUE.
YOU DIDN'T BELIEVE YOUR OWN THEORIES?
I BELIEVED MY THEORIES, BUT SOME OF MY FRIENDS AND COLLEAGUES STARTED MAKING DISCOVERIES THAT WOULD MAKE YOUR BRAIN EXPLODE.
QUANTUM ME
IN ORDER TO BE A PARTICLE AND A WAVE, A PHOTON CAN BE TWO PLACES AT THE SAME TIME. REMEMBER YOUNG'S EXPERIMENT?
PHOTON
PHOTON
THE ONE WHERE LIGHT IS SHONE THROUGH TWO NARROW SLITS?
THAT'S THE ONE.
NOW, REMEMBER, THIS EXPERIMENT PRODUCED FRINGES OF LIGHT WHERE TWO WAVES MEET.
PHOTON
IN 1909, ALMOST 100 YEARS AFTER YOUNG'S FIRST EXPERIMENT, IT WAS REPEATED BY GEOFFREY TAYLOR WITH VERY FAINT LIGHT. HE USED PHOTOGRAPHIC FILM TO RECORD THE SHADOWS AND FRINGES. THE LIGHT SOURCE WAS SO DIM THAT ONLY A SINGLE PHOTON WAS RELEASED AT A TIME. THESE INDIVIDUAL PHOTONS COULD ONLY GO THROUGH ONE SLIT OR THE OTHER, BUT NOT BOTH AT THE SAME TIME. IT TOOK THREE MONTHS TO GET ENOUGH PHOTONS THROUGH TO PRODUCE A PICTURE ON THE FILM.

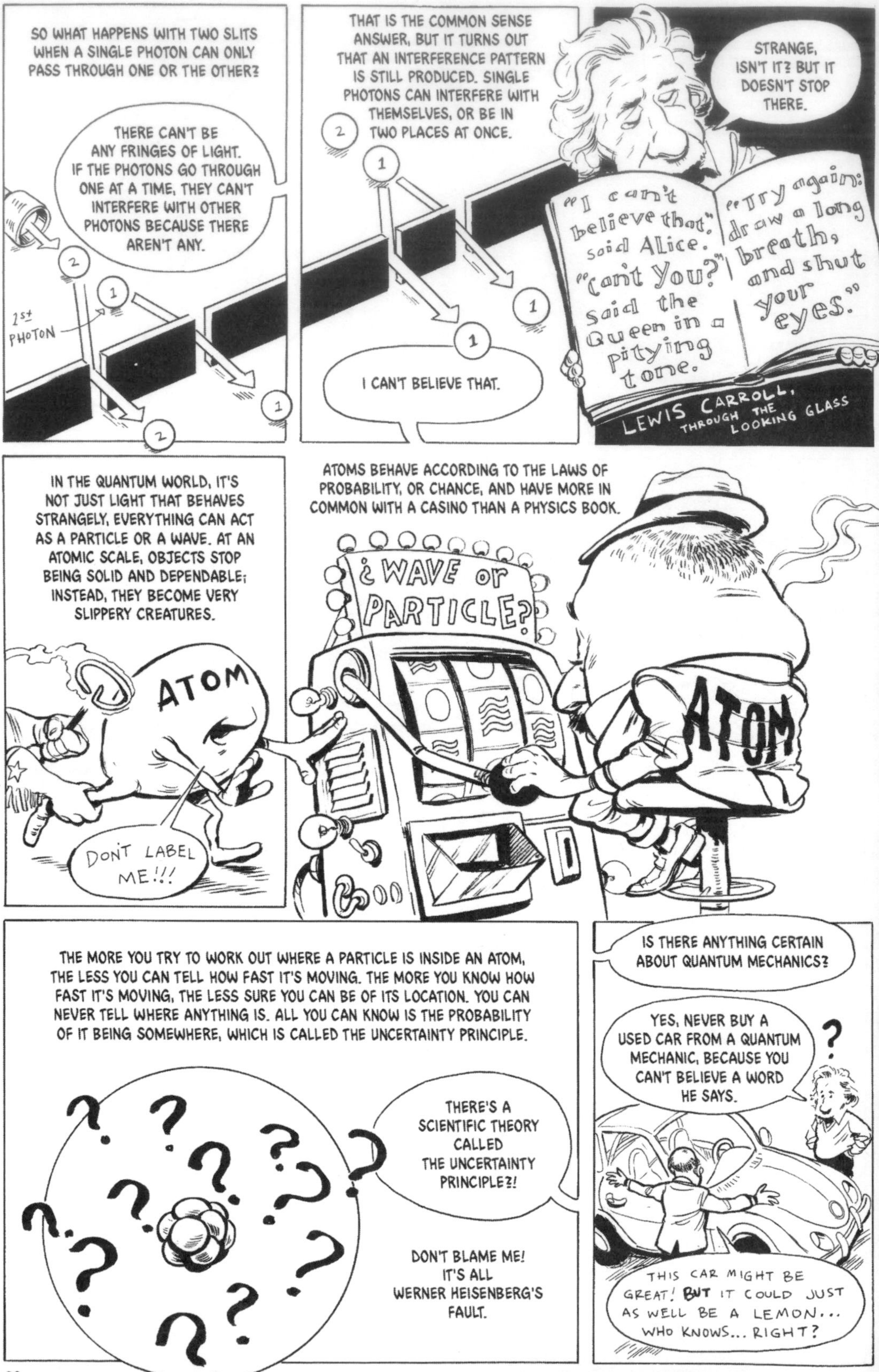
SO WHAT HAPPENS WITH TWO SLITS WHEN A SINGLE PHOTON CAN ONLY PASS THROUGH ONE OR THE OTHER?
THERE CAN'T BE ANY FRINGES OF LIGHT. IF THE PHOTONS GO THROUGH ONE AT A TIME, THEY CAN'T INTERFERE WITH OTHER PHOTONS BECAUSE THERE AREN'T ANY.
1ST PHOTON
THAT IS THE COMMON SENSE ANSWER, BUT IT TURNS OUT THAT AN INTERFERENCE PATTERN IS STILL PRODUCED. SINGLE PHOTONS CAN INTERFERE WITH THEMSELVES, OR BE IN TWO PLACES AT ONCE.
I CAN'T BELIEVE THAT.
STRANGE, ISN'T IT? BUT IT DOESN'T STOP THERE.
"I can't believe that," said Alice. "Can't you?" said the Queen in a pitying tone.
"Try again: draw a long breath, and shut your eyes."
LEWIS CARROLL, THROUGH THE LOOKING GLASS
IN THE QUANTUM WORLD, IT'S NOT JUST LIGHT THAT BEHAVES STRANGELY, EVERYTHING CAN ACT AS A PARTICLE OR A WAVE. AT AN ATOMIC SCALE, OBJECTS STOP BEING SOLID AND DEPENDABLE; INSTEAD, THEY BECOME VERY SLIPPERY CREATURES.
ATOM
DON'T LABEL ME!!!
ATOMS BEHAVE ACCORDING TO THE LAWS OF PROBABILITY, OR CHANCE, AND HAVE MORE IN COMMON WITH A CASINO THAN A PHYSICS BOOK.
¿WAVE or PARTICLE?
ATOM
THE MORE YOU TRY TO WORK OUT WHERE A PARTICLE IS INSIDE AN ATOM, THE LESS YOU CAN TELL HOW FAST IT'S MOVING. THE MORE YOU KNOW HOW FAST IT'S MOVING, THE LESS SURE YOU CAN BE OF ITS LOCATION. YOU CAN NEVER TELL WHERE ANYTHING IS. ALL YOU CAN KNOW IS THE PROBABILITY OF IT BEING SOMEWHERE, WHICH IS CALLED THE UNCERTAINTY PRINCIPLE.
THERE'S A SCIENTIFIC THEORY CALLED THE UNCERTAINTY PRINCIPLE?!
DON'T BLAME ME! IT'S ALL WERNER HEISENBERG'S FAULT.
IS THERE ANYTHING CERTAIN ABOUT QUANTUM MECHANICS?
YES, NEVER BUY A USED CAR FROM A QUANTUM MECHANIC, BECAUSE YOU CAN'T BELIEVE A WORD HE SAYS.
THIS CAR MIGHT BE GREAT! BUT IT COULD JUST AS WELL BE A LEMON... WHO KNOWS... RIGHT?

CHAPTER 15:
QUANTUM MECHANICS FOR CAT LOVERS: ISAAC NEWTON STRIKES BACK

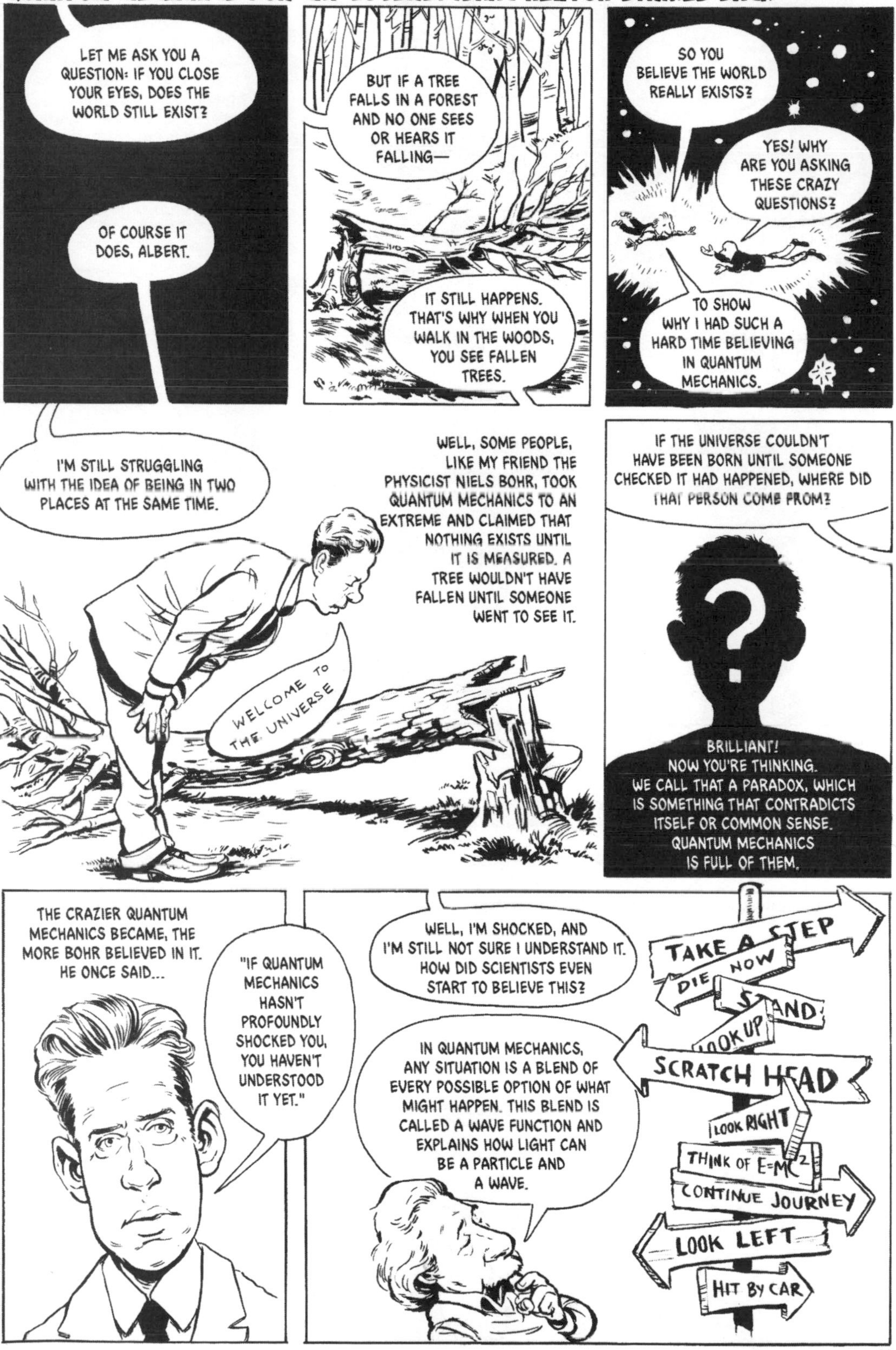

LET ME ASK YOU A QUESTION: IF YOU CLOSE YOUR EYES, DOES THE WORLD STILL EXIST?
OF COURSE IT DOES, ALBERT.
BUT IF A TREE FALLS IN A FOREST AND NO ONE SEES OR HEARS IT FALLING—
IT STILL HAPPENS. THAT'S WHY WHEN YOU WALK IN THE WOODS, YOU SEE FALLEN TREES.
SO YOU BELIEVE THE WORLD REALLY EXISTS?
YES! WHY ARE YOU ASKING THESE CRAZY QUESTIONS?
TO SHOW WHY I HAD SUCH A HARD TIME BELIEVING IN QUANTUM MECHANICS.
I'M STILL STRUGGLING WITH THE IDEA OF BEING IN TWO PLACES AT THE SAME TIME.
WELL, SOME PEOPLE, LIKE MY FRIEND THE PHYSICIST NIELS BOHR, TOOK QUANTUM MECHANICS TO AN EXTREME AND CLAIMED THAT NOTHING EXISTS UNTIL IT IS MEASURED. A TREE WOULDN'T HAVE FALLEN UNTIL SOMEONE WENT TO SEE IT.
WELCOME TO THE UNIVERSE
IF THE UNIVERSE COULDN'T HAVE BEEN BORN UNTIL SOMEONE CHECKED IT HAD HAPPENED, WHERE DID THAT PERSON COME FROM?
?
BRILLIANT! NOW YOU'RE THINKING. WE CALL THAT A PARADOX, WHICH IS SOMETHING THAT CONTRADICTS ITSELF OR COMMON SENSE. QUANTUM MECHANICS IS FULL OF THEM.
THE CRAZIER QUANTUM MECHANICS BECAME, THE MORE BOHR BELIEVED IN IT. HE ONCE SAID...
"IF QUANTUM MECHANICS HASN'T PROFOUNDLY SHOCKED YOU, YOU HAVEN'T UNDERSTOOD IT YET."
WELL, I'M SHOCKED, AND I'M STILL NOT SURE I UNDERSTAND IT. HOW DID SCIENTISTS EVEN START TO BELIEVE THIS?
IN QUANTUM MECHANICS, ANY SITUATION IS A BLEND OF EVERY POSSIBLE OPTION OF WHAT MIGHT HAPPEN. THIS BLEND IS CALLED A WAVE FUNCTION AND EXPLAINS HOW LIGHT CAN BE A PARTICLE AND A WAVE.
TAKE A STEP
DIE NOW
STAND
LOOK UP
SCRATCH HEAD
LOOK RIGHT
THINK OF E=MC2
CONTINUE JOURNEY
LOOK LEFT
HIT BY CAR

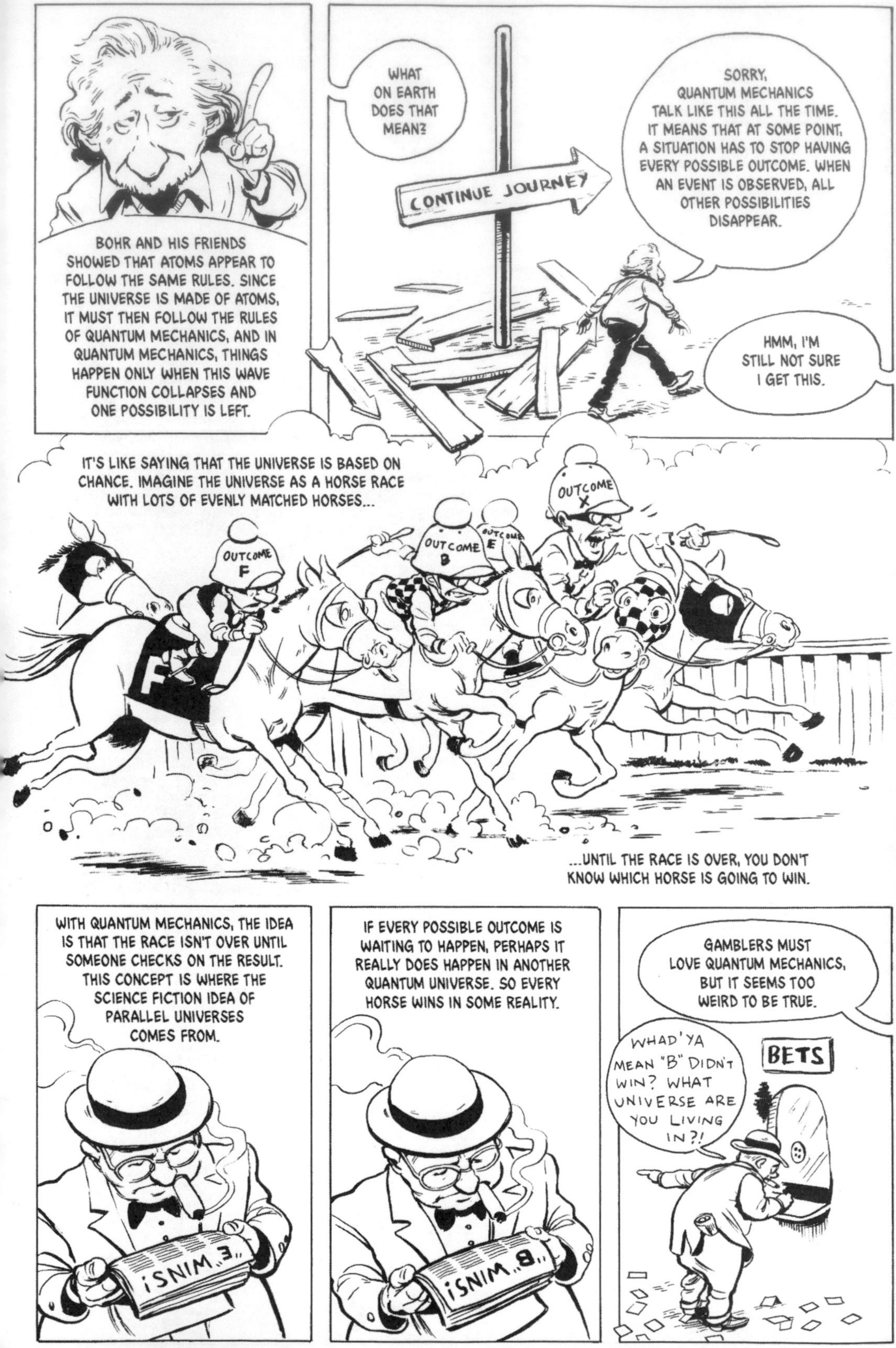

BOHR AND HIS FRIENDS SHOWED THAT ATOMS APPEAR TO FOLLOW THE SAME RULES. SINCE THE UNIVERSE IS MADE OF ATOMS, IT MUST THEN FOLLOW THE RULES OF QUANTUM MECHANICS, AND IN QUANTUM MECHANICS, THINGS HAPPEN ONLY WHEN THIS WAVE FUNCTION COLLAPSES AND ONE POSSIBILITY IS LEFT.
WHAT ON EARTH DOES THAT MEAN?
CONTINUE JOURNEY
SORRY, QUANTUM MECHANICS TALK LIKE THIS ALL THE TIME. IT MEANS THAT AT SOME POINT, A SITUATION HAS TO STOP HAVING EVERY POSSIBLE OUTCOME. WHEN AN EVENT IS OBSERVED, ALL OTHER POSSIBILITIES DISAPPEAR.
HMM, I'M STILL NOT SURE I GET THIS.
IT'S LIKE SAYING THAT THE UNIVERSE IS BASED ON CHANCE. IMAGINE THE UNIVERSE AS A HORSE RACE WITH LOTS OF EVENLY MATCHED HORSES...
OUTCOME F
OUTCOME B
OUTCOME E
OUTCOME X
F
...UNTIL THE RACE IS OVER, YOU DON'T KNOW WHICH HORSE IS GOING TO WIN.
WITH QUANTUM MECHANICS, THE IDEA IS THAT THE RACE ISN'T OVER UNTIL SOMEONE CHECKS ON THE RESULT. THIS CONCEPT IS WHERE THE SCIENCE FICTION IDEA OF PARALLEL UNIVERSES COMES FROM.
IF EVERY POSSIBLE OUTCOME IS WAITING TO HAPPEN, PERHAPS IT REALLY DOES HAPPEN IN ANOTHER QUANTUM UNIVERSE. SO EVERY HORSE WINS IN SOME REALITY.
GAMBLERS MUST LOVE QUANTUM MECHANICS, BUT IT SEEMS TOO WEIRD TO BE TRUE.
WHAD'YA MEAN "B" DIDN'T WIN? WHAT UNIVERSE ARE YOU LIVING IN?!
BETS

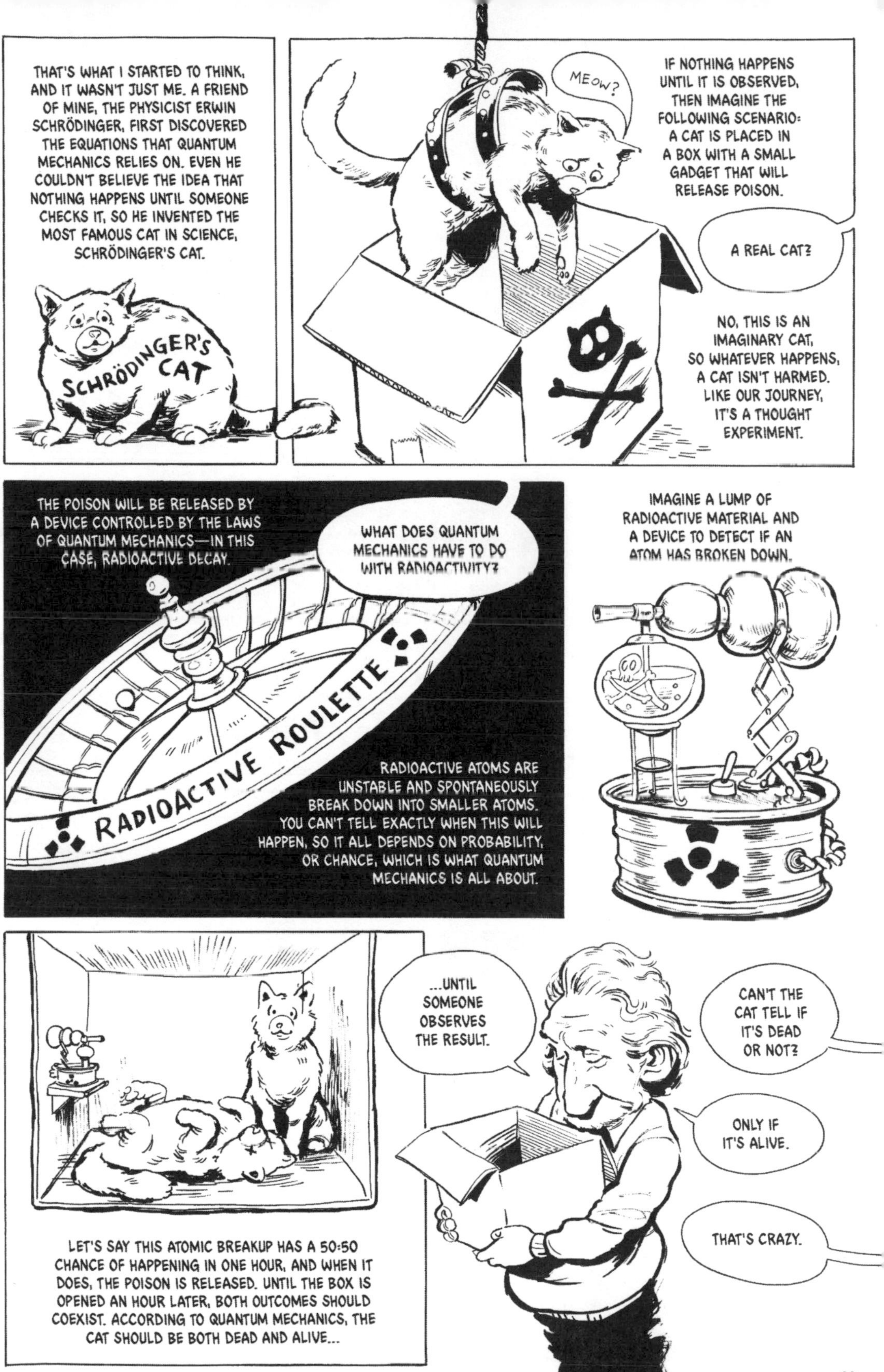
THAT'S WHAT I STARTED TO THINK, AND IT WASN'T JUST ME. A FRIEND OF MINE, THE PHYSICIST ERWIN SCHRÖDINGER, FIRST DISCOVERED THE EQUATIONS THAT QUANTUM MECHANICS RELIES ON. EVEN HE COULDN'T BELIEVE THE IDEA THAT NOTHING HAPPENS UNTIL SOMEONE CHECKS IT, SO HE INVENTED THE MOST FAMOUS CAT IN SCIENCE, SCHRÖDINGER'S CAT.
SCHRÖDINGER'S CAT
MEOW?
IF NOTHING HAPPENS UNTIL IT IS OBSERVED, THEN IMAGINE THE FOLLOWING SCENARIO: A CAT IS PLACED IN A BOX WITH A SMALL GADGET THAT WILL RELEASE POISON.
A REAL CAT?
NO, THIS IS AN IMAGINARY CAT, SO WHATEVER HAPPENS, A CAT ISN'T HARMED. LIKE OUR JOURNEY, IT'S A THOUGHT EXPERIMENT.
THE POISON WILL BE RELEASED BY A DEVICE CONTROLLED BY THE LAWS OF QUANTUM MECHANICS—IN THIS CASE, RADIOACTIVE DECAY.
WHAT DOES QUANTUM MECHANICS HAVE TO DO WITH RADIOACTIVITY?
RADIOACTIVE ROULETTE
RADIOACTIVE ATOMS ARE UNSTABLE AND SPONTANEOUSLY BREAK DOWN INTO SMALLER ATOMS. YOU CAN'T TELL EXACTLY WHEN THIS WILL HAPPEN, SO IT ALL DEPENDS ON PROBABILITY, OR CHANCE, WHICH IS WHAT QUANTUM MECHANICS IS ALL ABOUT.
IMAGINE A LUMP OF RADIOACTIVE MATERIAL AND A DEVICE TO DETECT IF AN ATOM HAS BROKEN DOWN.
LET'S SAY THIS ATOMIC BREAKUP HAS A 50:50 CHANCE OF HAPPENING IN ONE HOUR, AND WHEN IT DOES, THE POISON IS RELEASED. UNTIL THE BOX IS OPENED AN HOUR LATER, BOTH OUTCOMES SHOULD COEXIST. ACCORDING TO QUANTUM MECHANICS, THE CAT SHOULD BE BOTH DEAD AND ALIVE...
...UNTIL SOMEONE OBSERVES THE RESULT.
CAN'T THE CAT TELL IF IT'S DEAD OR NOT?
ONLY IF IT'S ALIVE.
THAT'S CRAZY.

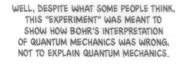

WELL, DESPITE WHAT SOME PEOPLE THINK, THIS "EXPERIMENT" WAS MEANT TO SHOW HOW BOHR'S INTERPRETATION OF QUANTUM MECHANICS WAS WRONG, NOT TO EXPLAIN QUANTUM MECHANICS.

I THINK QUANTUM MECHANICS MAKES MORE SENSE IF, RATHER THAN NEEDING A PERSON TO OBSERVE AN EVENT, IT REQUIRES A PARTICLE OR AN ATOM TO INTERACT WITH SOMETHING ELSE. THAT'S WHEN AN EVENT HAS TO COMMIT TO BEING ONE THING OR THE OTHER. SO FOR OUR CAT, THE RADIOACTIVE ATOM CAN BE BOTH INTACT AND BROKEN-DOWN AT THE SAME TIME, UNTIL THE RADIOACTIVITY DETECTOR IN THE BOX DETECTS IT. THIS WAY THE UNIVERSE CAN FOLLOW QUANTUM MECHANICS AT AN ATOMIC SCALE WITHOUT NEEDING A CAT THAT IS DEAD AND ALIVE AT THE SAME TIME.

THIS CRAZINESS OF A CAT THAT IS BOTH DEAD AND ALIVE ONLY APPLIES IF YOU STICK TO THE IDEA THAT NOTHING ACTUALLY HAPPENS UNTIL IT IS MEASURED BY A PERSON.

ATOM

ATOM

INTACT ATOM

BROKEN-DOWN ATOM

I THOUGHT YOU DIDN'T BELIEVE IN QUANTUM MECHANICS?

WELL, I DIDN'T BELIEVE THE EXTREME VERSION, BUT, PERHAPS, I'VE MELLOWED A BIT. ALL THIS CAT REALLY TELLS US ABOUT QUANTUM MECHANICS IS THAT TRYING TO USE IT TO EXPLAIN EVERYDAY LIFE DOESN'T WORK.

UNDERSTANDING ATOMS DOESN'T HELP YOU UNDERSTAND CATS...

...BUT, THEN AGAIN, UNDERSTANDING CATS DOESN'T HELP YOU UNDERSTAND ATOMS.

AT THE END OF THE DAY, WITHOUT QUANTUM MECHANICS, THERE ARE THINGS ABOUT LIGHT AND ATOMS WE JUST CAN'T EXPLAIN.

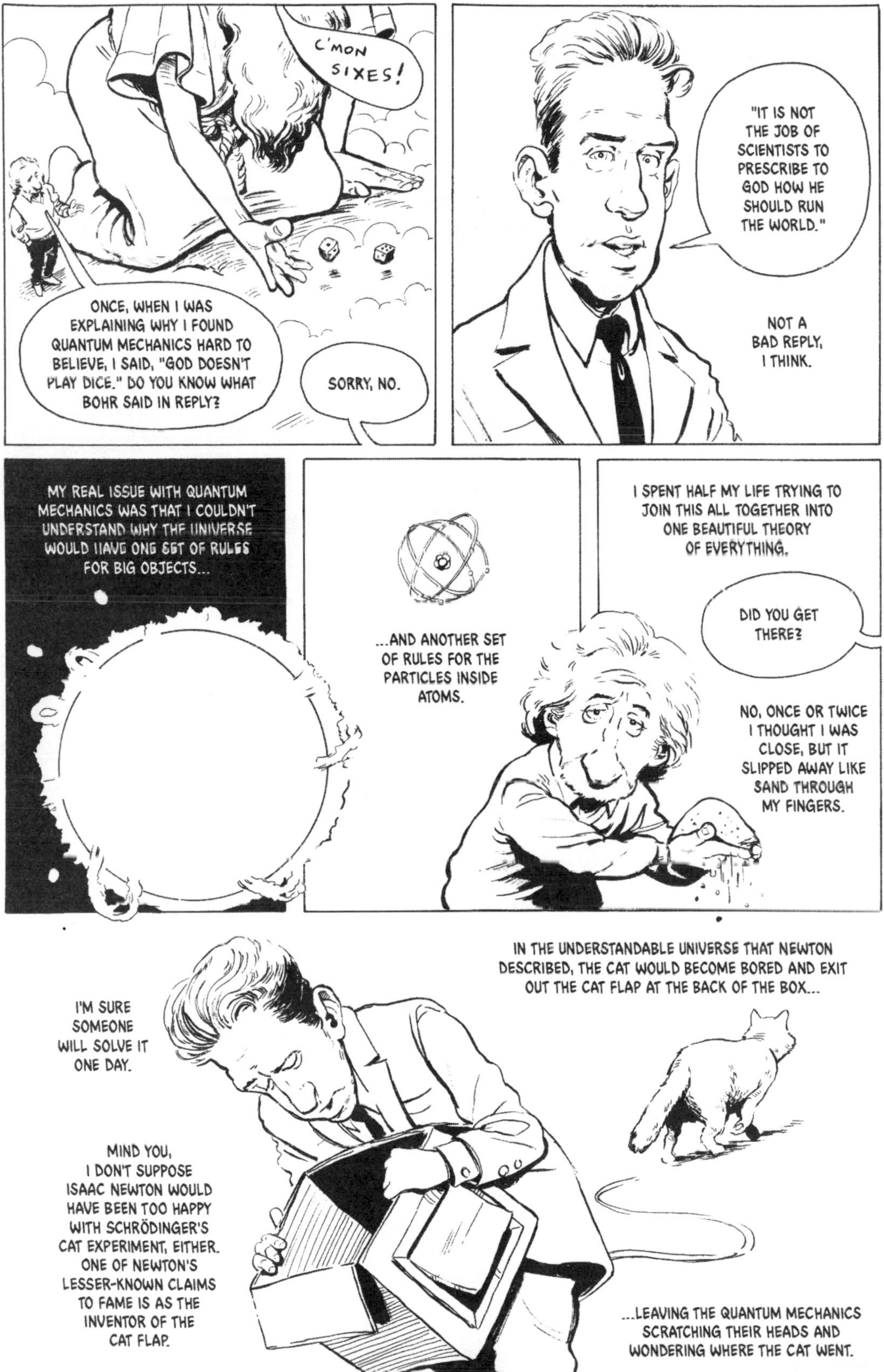
C'MON SIXES!
ONCE, WHEN I WAS EXPLAINING WHY I FOUND QUANTUM MECHANICS HARD TO BELIEVE, I SAID, "GOD DOESN'T PLAY DICE." DO YOU KNOW WHAT BOHR SAID IN REPLY?
SORRY, NO.
"IT IS NOT THE JOB OF SCIENTISTS TO PRESCRIBE TO GOD HOW HE SHOULD RUN THE WORLD."
NOT A BAD REPLY, I THINK.
MY REAL ISSUE WITH QUANTUM MECHANICS WAS THAT I COULDN'T UNDERSTAND WHY THE UNIVERSE WOULD HAVE ONE SET OF RULES FOR BIG OBJECTS...
...AND ANOTHER SET OF RULES FOR THE PARTICLES INSIDE ATOMS.
I SPENT HALF MY LIFE TRYING TO JOIN THIS ALL TOGETHER INTO ONE BEAUTIFUL THEORY OF EVERYTHING.
DID YOU GET THERE?
NO, ONCE OR TWICE I THOUGHT I WAS CLOSE, BUT IT SLIPPED AWAY LIKE SAND THROUGH MY FINGERS.
I'M SURE SOMEONE WILL SOLVE IT ONE DAY.
IN THE UNDERSTANDABLE UNIVERSE THAT NEWTON DESCRIBED, THE CAT WOULD BECOME BORED AND EXIT OUT THE CAT FLAP AT THE BACK OF THE BOX...
MIND YOU, I DON'T SUPPOSE ISAAC NEWTON WOULD HAVE BEEN TOO HAPPY WITH SCHRÖDINGER'S CAT EXPERIMENT, EITHER. ONE OF NEWTON'S LESSER-KNOWN CLAIMS TO FAME IS AS THE INVENTOR OF THE CAT FLAP.
...LEAVING THE QUANTUM MECHANICS SCRATCHING THEIR HEADS AND WONDERING WHERE THE CAT WENT.

CHAPTER 16: CLOCKING THE SPEED OF THE FASTEST THING IN THE UNIVERSE

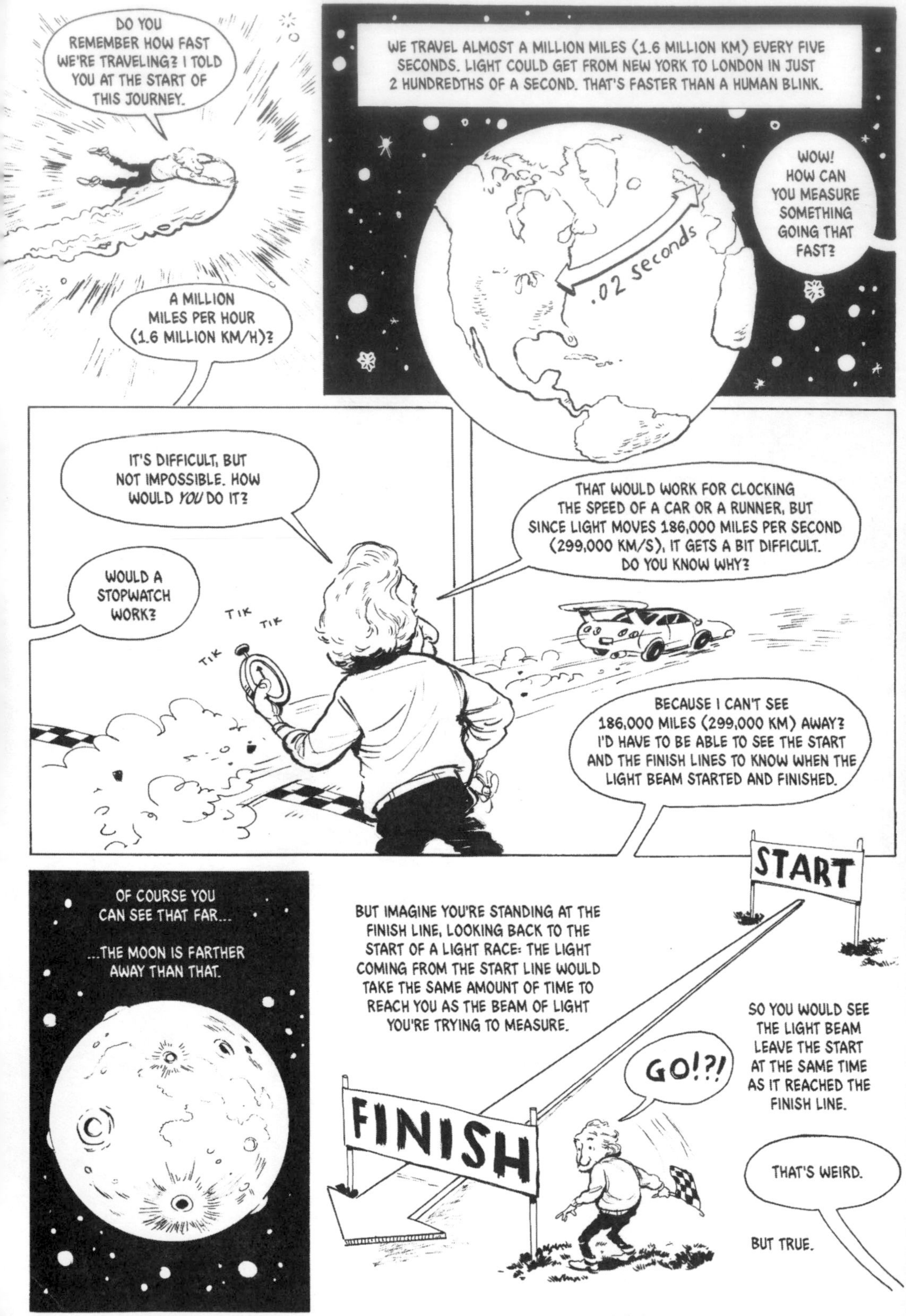
DO YOU REMEMBER HOW FAST WE'RE TRAVELING? I TOLD YOU AT THE START OF THIS JOURNEY.
A MILLION MILES PER HOUR (1.6 MILLION KM/H)?
WE TRAVEL ALMOST A MILLION MILES (1.6 MILLION KM) EVERY FIVE SECONDS. LIGHT COULD GET FROM NEW YORK TO LONDON IN JUST 2 HUNDREDTHS OF A SECOND. THAT'S FASTER THAN A HUMAN BLINK.
.02 seconds
WOW! HOW CAN YOU MEASURE SOMETHING GOING THAT FAST?
IT'S DIFFICULT, BUT NOT IMPOSSIBLE. HOW WOULD *YOU* DO IT?
WOULD A STOPWATCH WORK?
TIK TIK TIK
THAT WOULD WORK FOR CLOCKING THE SPEED OF A CAR OR A RUNNER, BUT SINCE LIGHT MOVES 186,000 MILES PER SECOND (299,000 KM/S), IT GETS A BIT DIFFICULT. DO YOU KNOW WHY?
BECAUSE I CAN'T SEE 186,000 MILES (299,000 KM) AWAY? I'D HAVE TO BE ABLE TO SEE THE START AND THE FINISH LINES TO KNOW WHEN THE LIGHT BEAM STARTED AND FINISHED.
OF COURSE YOU CAN SEE THAT FAR...
...THE MOON IS FARTHER AWAY THAN THAT.
BUT IMAGINE YOU'RE STANDING AT THE FINISH LINE, LOOKING BACK TO THE START OF A LIGHT RACE: THE LIGHT COMING FROM THE START LINE WOULD TAKE THE SAME AMOUNT OF TIME TO REACH YOU AS THE BEAM OF LIGHT YOU'RE TRYING TO MEASURE.
START
FINISH
GO!?!
SO YOU WOULD SEE THE LIGHT BEAM LEAVE THE START AT THE SAME TIME AS IT REACHED THE FINISH LINE.
THAT'S WEIRD.
BUT TRUE.

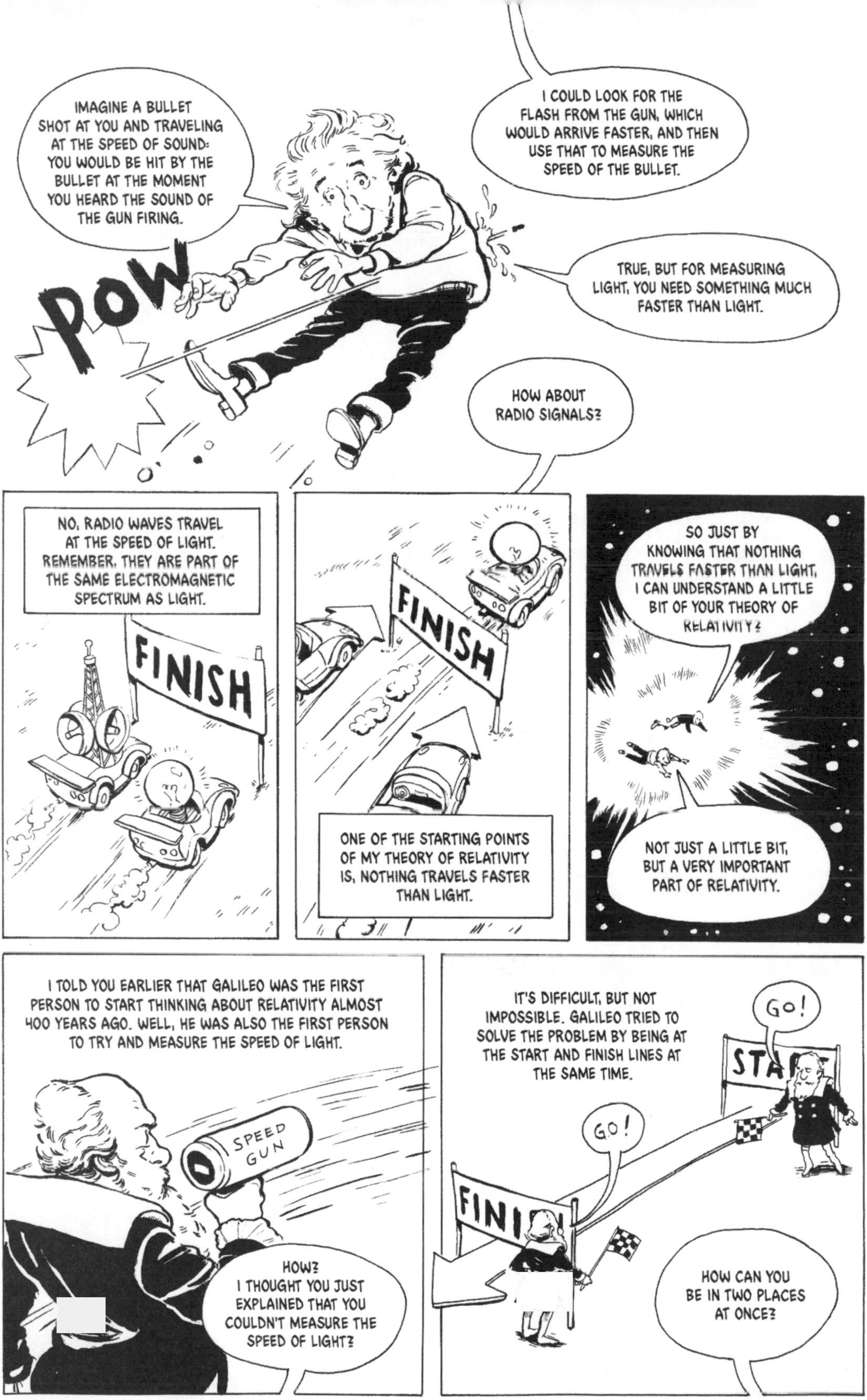
IMAGINE A BULLET SHOT AT YOU AND TRAVELING AT THE SPEED OF SOUND: YOU WOULD BE HIT BY THE BULLET AT THE MOMENT YOU HEARD THE SOUND OF THE GUN FIRING.
POW
I COULD LOOK FOR THE FLASH FROM THE GUN, WHICH WOULD ARRIVE FASTER, AND THEN USE THAT TO MEASURE THE SPEED OF THE BULLET.
TRUE, BUT FOR MEASURING LIGHT, YOU NEED SOMETHING MUCH FASTER THAN LIGHT.
HOW ABOUT RADIO SIGNALS?
NO, RADIO WAVES TRAVEL AT THE SPEED OF LIGHT. REMEMBER, THEY ARE PART OF THE SAME ELECTROMAGNETIC SPECTRUM AS LIGHT.
FINISH
FINISH
ONE OF THE STARTING POINTS OF MY THEORY OF RELATIVITY IS, NOTHING TRAVELS FASTER THAN LIGHT.
SO JUST BY KNOWING THAT NOTHING TRAVELS FASTER THAN LIGHT, I CAN UNDERSTAND A LITTLE BIT OF YOUR THEORY OF RELATIVITY?
NOT JUST A LITTLE BIT, BUT A VERY IMPORTANT PART OF RELATIVITY.
I TOLD YOU EARLIER THAT GALILEO WAS THE FIRST PERSON TO START THINKING ABOUT RELATIVITY ALMOST 400 YEARS AGO. WELL, HE WAS ALSO THE FIRST PERSON TO TRY AND MEASURE THE SPEED OF LIGHT.
SPEED GUN
HOW? I THOUGHT YOU JUST EXPLAINED THAT YOU COULDN'T MEASURE THE SPEED OF LIGHT?
IT'S DIFFICULT, BUT NOT IMPOSSIBLE. GALILEO TRIED TO SOLVE THE PROBLEM BY BEING AT THE START AND FINISH LINES AT THE SAME TIME.
GO!
GO!
HOW CAN YOU BE IN TWO PLACES AT ONCE?

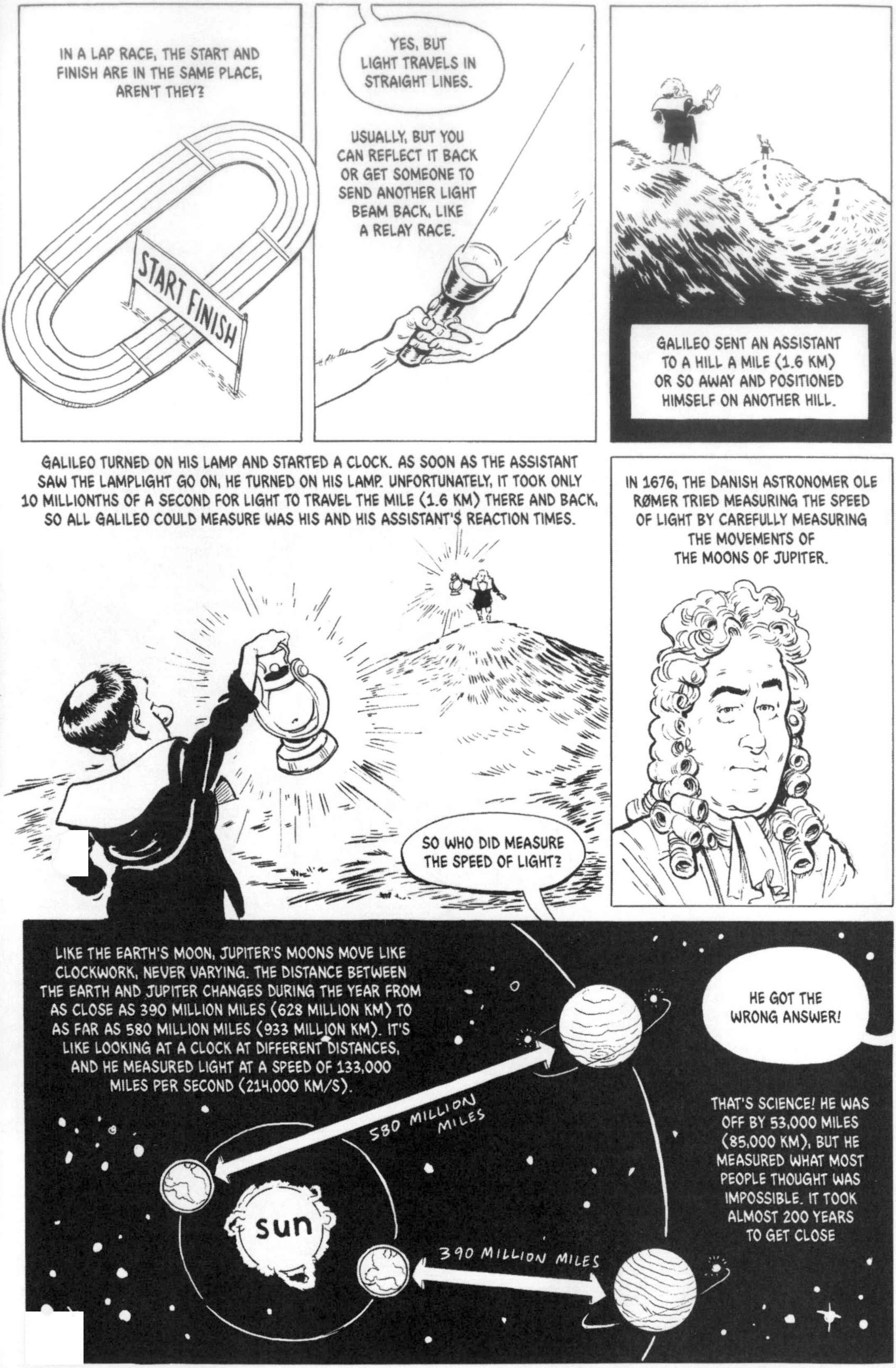
IN A LAP RACE, THE START AND FINISH ARE IN THE SAME PLACE, AREN'T THEY?
START FINISH
YES, BUT LIGHT TRAVELS IN STRAIGHT LINES.
USUALLY, BUT YOU CAN REFLECT IT BACK OR GET SOMEONE TO SEND ANOTHER LIGHT BEAM BACK, LIKE A RELAY RACE.
GALILEO SENT AN ASSISTANT TO A HILL A MILE (1.6 KM) OR SO AWAY AND POSITIONED HIMSELF ON ANOTHER HILL.
GALILEO TURNED ON HIS LAMP AND STARTED A CLOCK. AS SOON AS THE ASSISTANT SAW THE LAMPLIGHT GO ON, HE TURNED ON HIS LAMP. UNFORTUNATELY, IT TOOK ONLY 10 MILLIONTHS OF A SECOND FOR LIGHT TO TRAVEL THE MILE (1.6 KM) THERE AND BACK, SO ALL GALILEO COULD MEASURE WAS HIS AND HIS ASSISTANT'S REACTION TIMES.
SO WHO DID MEASURE THE SPEED OF LIGHT?
IN 1676, THE DANISH ASTRONOMER OLE RØMER TRIED MEASURING THE SPEED OF LIGHT BY CAREFULLY MEASURING THE MOVEMENTS OF THE MOONS OF JUPITER.
LIKE THE EARTH'S MOON, JUPITER'S MOONS MOVE LIKE CLOCKWORK, NEVER VARYING. THE DISTANCE BETWEEN THE EARTH AND JUPITER CHANGES DURING THE YEAR FROM AS CLOSE AS 390 MILLION MILES (628 MILLION KM) TO AS FAR AS 580 MILLION MILES (933 MILLION KM). IT'S LIKE LOOKING AT A CLOCK AT DIFFERENT DISTANCES, AND HE MEASURED LIGHT AT A SPEED OF 133,000 MILES PER SECOND (214,000 KM/S).
580 MILLION MILES
SUN
390 MILLION MILES
HE GOT THE WRONG ANSWER!
THAT'S SCIENCE! HE WAS OFF BY 53,000 MILES (85,000 KM), BUT HE MEASURED WHAT MOST PEOPLE THOUGHT WAS IMPOSSIBLE. IT TOOK ALMOST 200 YEARS TO GET CLOSE

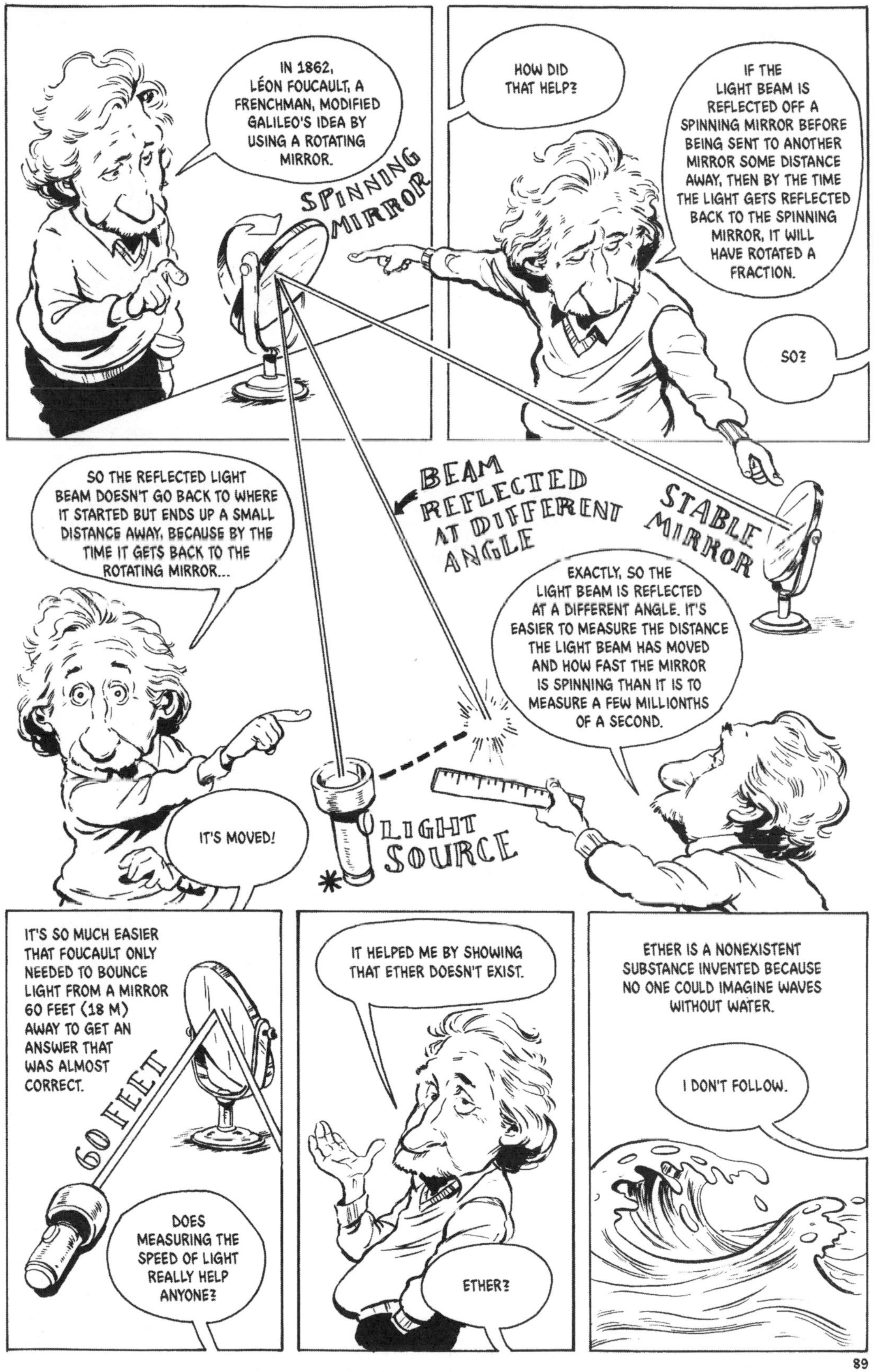
IN 1862, LÉON FOUCAULT, A FRENCHMAN, MODIFIED GALILEO'S IDEA BY USING A ROTATING MIRROR.
SPINNING MIRROR
HOW DID THAT HELP?
IF THE LIGHT BEAM IS REFLECTED OFF A SPINNING MIRROR BEFORE BEING SENT TO ANOTHER MIRROR SOME DISTANCE AWAY, THEN BY THE TIME THE LIGHT GETS REFLECTED BACK TO THE SPINNING MIRROR, IT WILL HAVE ROTATED A FRACTION.
SO?
SO THE REFLECTED LIGHT BEAM DOESN'T GO BACK TO WHERE IT STARTED BUT ENDS UP A SMALL DISTANCE AWAY, BECAUSE BY THE TIME IT GETS BACK TO THE ROTATING MIRROR...
BEAM REFLECTED AT DIFFERENT ANGLE
STABLE MIRROR
EXACTLY, SO THE LIGHT BEAM IS REFLECTED AT A DIFFERENT ANGLE. IT'S EASIER TO MEASURE THE DISTANCE THE LIGHT BEAM HAS MOVED AND HOW FAST THE MIRROR IS SPINNING THAN IT IS TO MEASURE A FEW MILLIONTHS OF A SECOND.
IT'S MOVED!
LIGHT SOURCE
IT'S SO MUCH EASIER THAT FOUCAULT ONLY NEEDED TO BOUNCE LIGHT FROM A MIRROR 60 FEET (18 M) AWAY TO GET AN ANSWER THAT WAS ALMOST CORRECT.
60 FEET
DOES MEASURING THE SPEED OF LIGHT REALLY HELP ANYONE?
IT HELPED ME BY SHOWING THAT ETHER DOESN'T EXIST.
ETHER?
ETHER IS A NONEXISTENT SUBSTANCE INVENTED BECAUSE NO ONE COULD IMAGINE WAVES WITHOUT WATER.
I DON'T FOLLOW.

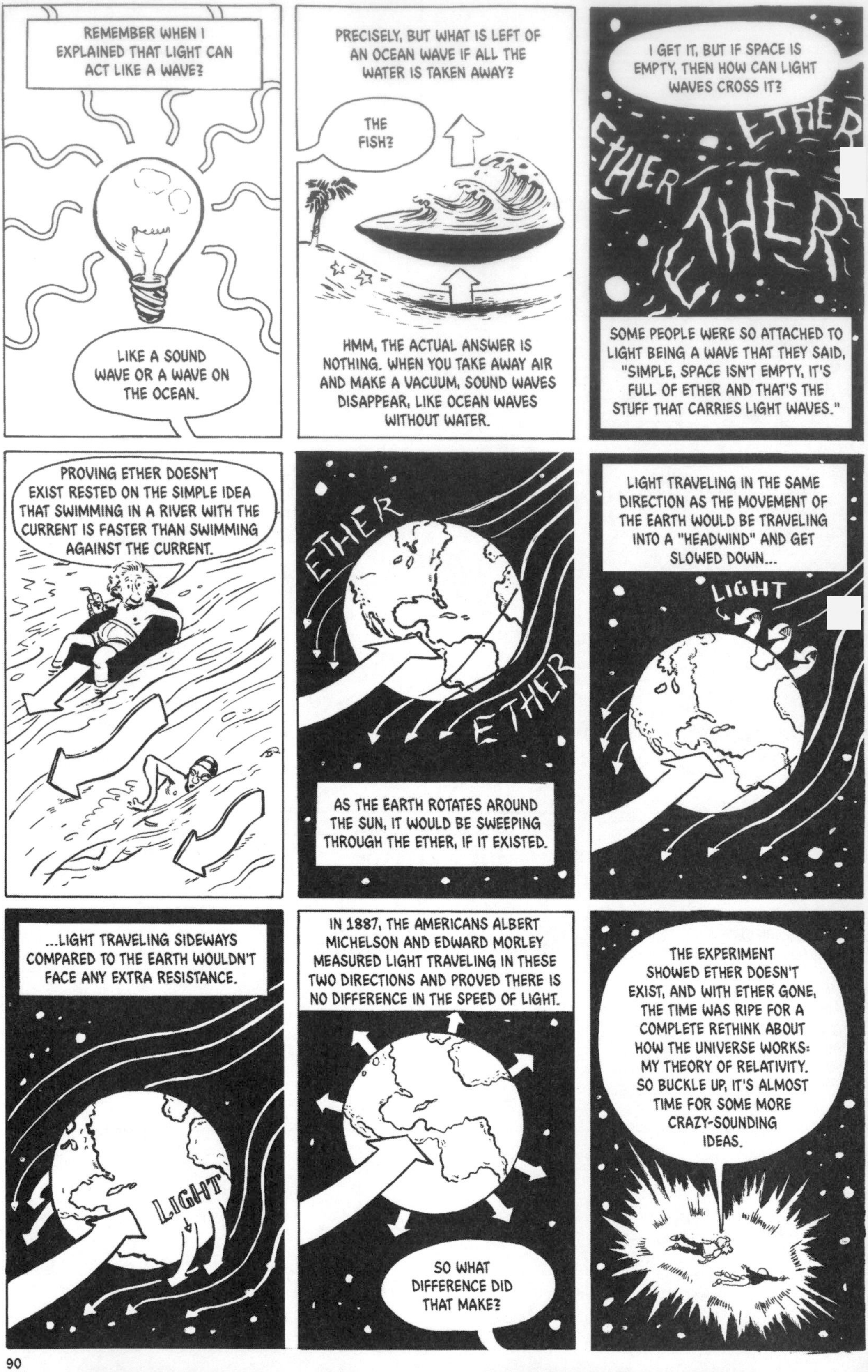

REMEMBER WHEN I EXPLAINED THAT LIGHT CAN ACT LIKE A WAVE?
LIKE A SOUND WAVE OR A WAVE ON THE OCEAN.
PRECISELY, BUT WHAT IS LEFT OF AN OCEAN WAVE IF ALL THE WATER IS TAKEN AWAY?
THE FISH?
HMM, THE ACTUAL ANSWER IS NOTHING. WHEN YOU TAKE AWAY AIR AND MAKE A VACUUM, SOUND WAVES DISAPPEAR, LIKE OCEAN WAVES WITHOUT WATER.
I GET IT, BUT IF SPACE IS EMPTY, THEN HOW CAN LIGHT WAVES CROSS IT?
ETHER
ETHER
ETHER
SOME PEOPLE WERE SO ATTACHED TO LIGHT BEING A WAVE THAT THEY SAID, "SIMPLE, SPACE ISN'T EMPTY, IT'S FULL OF ETHER AND THAT'S THE STUFF THAT CARRIES LIGHT WAVES."
PROVING ETHER DOESN'T EXIST RESTED ON THE SIMPLE IDEA THAT SWIMMING IN A RIVER WITH THE CURRENT IS FASTER THAN SWIMMING AGAINST THE CURRENT.
ETHER
ETHER
AS THE EARTH ROTATES AROUND THE SUN, IT WOULD BE SWEEPING THROUGH THE ETHER, IF IT EXISTED.
LIGHT TRAVELING IN THE SAME DIRECTION AS THE MOVEMENT OF THE EARTH WOULD BE TRAVELING INTO A "HEADWIND" AND GET SLOWED DOWN...
LIGHT
...LIGHT TRAVELING SIDEWAYS COMPARED TO THE EARTH WOULDN'T FACE ANY EXTRA RESISTANCE.
LIGHT
IN 1887, THE AMERICANS ALBERT MICHELSON AND EDWARD MORLEY MEASURED LIGHT TRAVELING IN THESE TWO DIRECTIONS AND PROVED THERE IS NO DIFFERENCE IN THE SPEED OF LIGHT.
SO WHAT DIFFERENCE DID THAT MAKE?
THE EXPERIMENT SHOWED ETHER DOESN'T EXIST, AND WITH ETHER GONE, THE TIME WAS RIPE FOR A COMPLETE RETHINK ABOUT HOW THE UNIVERSE WORKS: MY THEORY OF RELATIVITY. SO BUCKLE UP, IT'S ALMOST TIME FOR SOME MORE CRAZY-SOUNDING IDEAS.

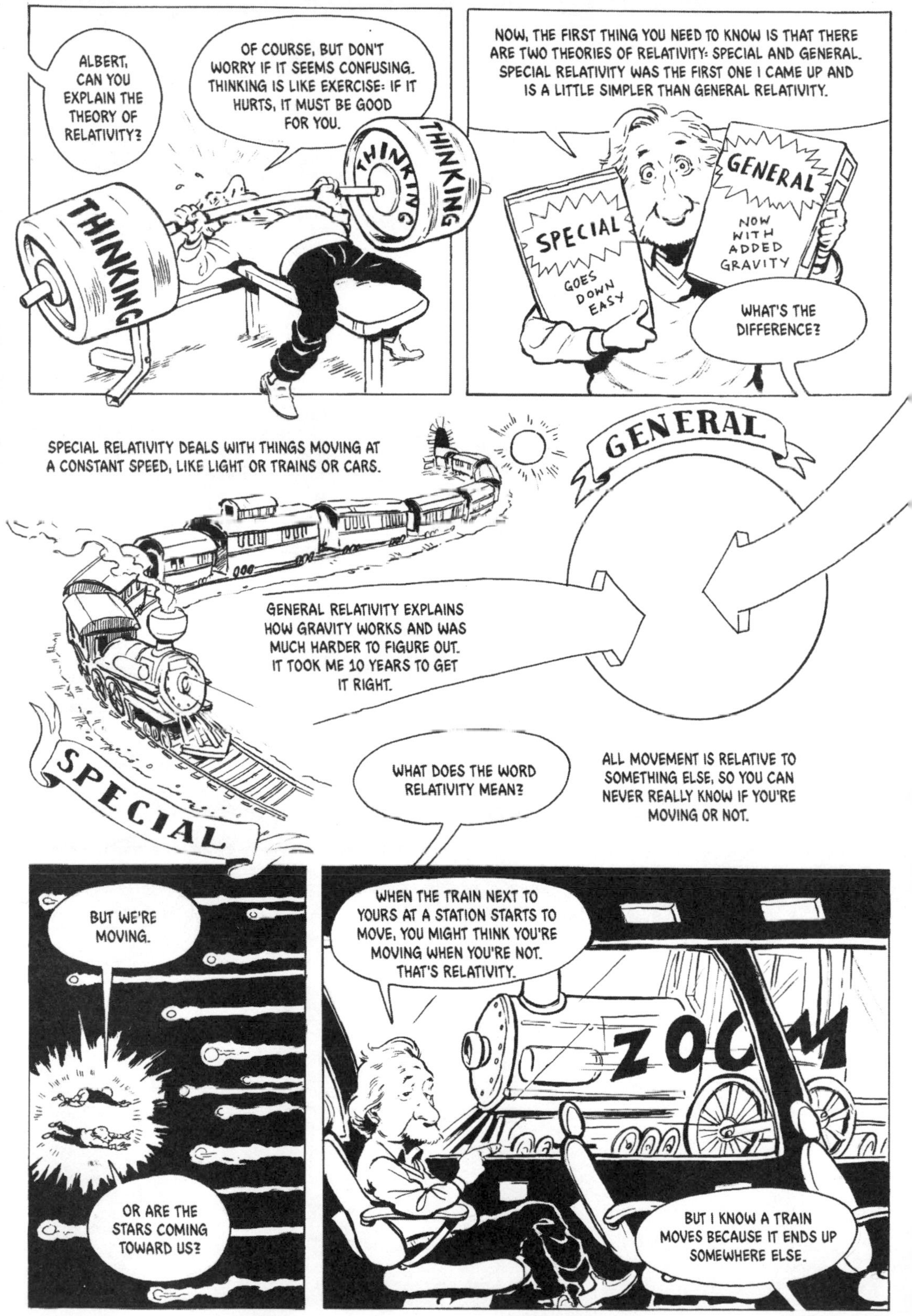
ALBERT, CAN YOU EXPLAIN THE THEORY OF RELATIVITY?
OF COURSE, BUT DON'T WORRY IF IT SEEMS CONFUSING. THINKING IS LIKE EXERCISE: IF IT HURTS, IT MUST BE GOOD FOR YOU.
THINKING
THINKING
THINKING
NOW, THE FIRST THING YOU NEED TO KNOW IS THAT THERE ARE TWO THEORIES OF RELATIVITY: SPECIAL AND GENERAL. SPECIAL RELATIVITY WAS THE FIRST ONE I CAME UP AND IS A LITTLE SIMPLER THAN GENERAL RELATIVITY.
SPECIAL
GOES DOWN EASY
GENERAL
NOW WITH ADDED GRAVITY
WHAT'S THE DIFFERENCE?
SPECIAL RELATIVITY DEALS WITH THINGS MOVING AT A CONSTANT SPEED, LIKE LIGHT OR TRAINS OR CARS.
GENERAL
GENERAL RELATIVITY EXPLAINS HOW GRAVITY WORKS AND WAS MUCH HARDER TO FIGURE OUT. IT TOOK ME 10 YEARS TO GET IT RIGHT.
SPECIAL
WHAT DOES THE WORD RELATIVITY MEAN?
ALL MOVEMENT IS RELATIVE TO SOMETHING ELSE, SO YOU CAN NEVER REALLY KNOW IF YOU'RE MOVING OR NOT.
BUT WE'RE MOVING.
OR ARE THE STARS COMING TOWARD US?
WHEN THE TRAIN NEXT TO YOURS AT A STATION STARTS TO MOVE, YOU MIGHT THINK YOU'RE MOVING WHEN YOU'RE NOT. THAT'S RELATIVITY.
ZOOM
BUT I KNOW A TRAIN MOVES BECAUSE IT ENDS UP SOMEWHERE ELSE.

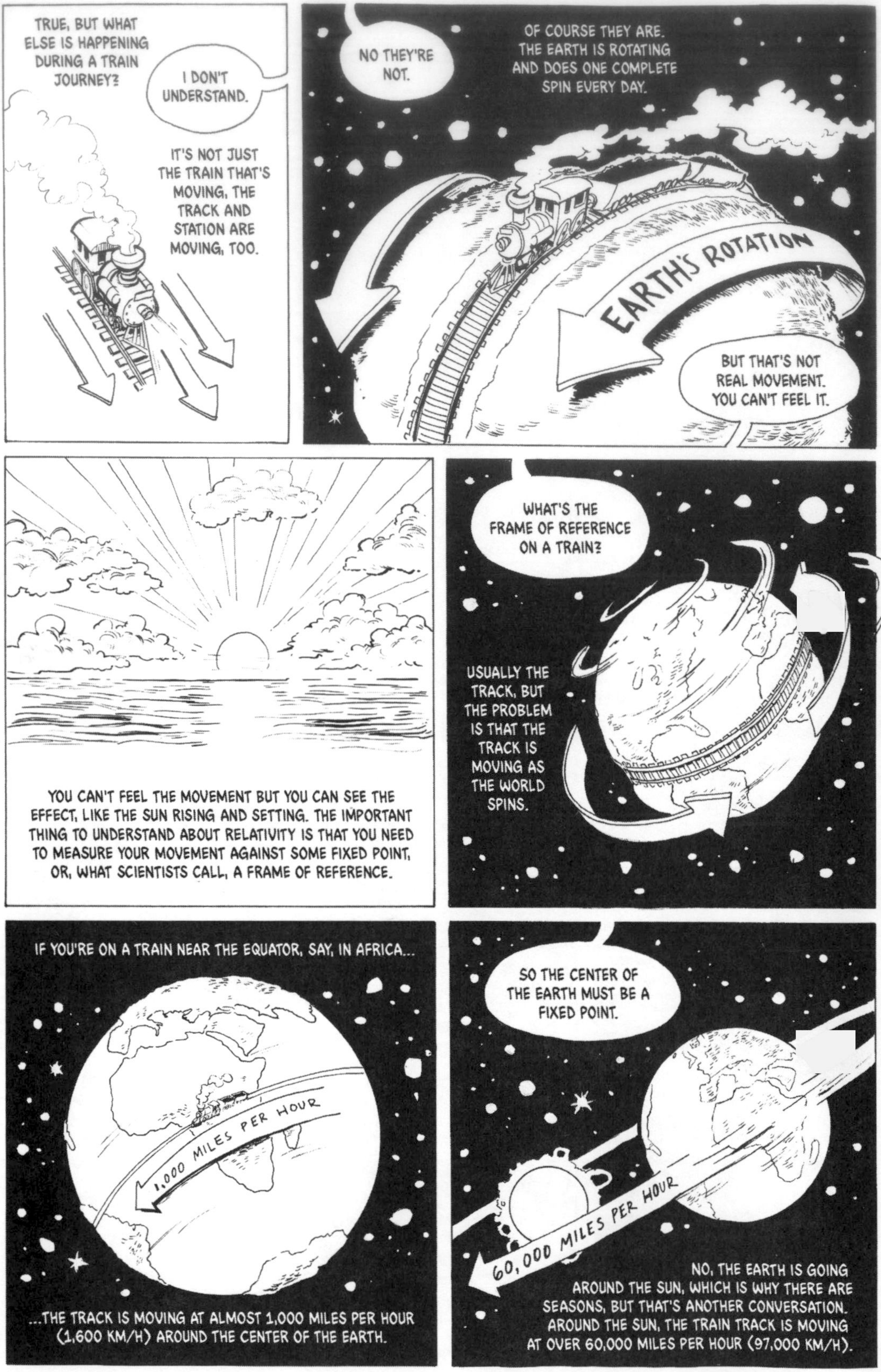
TRUE, BUT WHAT ELSE IS HAPPENING DURING A TRAIN JOURNEY?
I DON'T UNDERSTAND.
IT'S NOT JUST THE TRAIN THAT'S MOVING, THE TRACK AND STATION ARE MOVING, TOO.
NO THEY'RE NOT.
OF COURSE THEY ARE. THE EARTH IS ROTATING AND DOES ONE COMPLETE SPIN EVERY DAY.
EARTH'S ROTATION
BUT THAT'S NOT REAL MOVEMENT. YOU CAN'T FEEL IT.
YOU CAN'T FEEL THE MOVEMENT BUT YOU CAN SEE THE EFFECT, LIKE THE SUN RISING AND SETTING. THE IMPORTANT THING TO UNDERSTAND ABOUT RELATIVITY IS THAT YOU NEED TO MEASURE YOUR MOVEMENT AGAINST SOME FIXED POINT, OR, WHAT SCIENTISTS CALL, A FRAME OF REFERENCE.
WHAT'S THE FRAME OF REFERENCE ON A TRAIN?
USUALLY THE TRACK, BUT THE PROBLEM IS THAT THE TRACK IS MOVING AS THE WORLD SPINS.
IF YOU'RE ON A TRAIN NEAR THE EQUATOR, SAY, IN AFRICA...
1,000 MILES PER HOUR
...THE TRACK IS MOVING AT ALMOST 1,000 MILES PER HOUR (1,600 KM/H) AROUND THE CENTER OF THE EARTH.
SO THE CENTER OF THE EARTH MUST BE A FIXED POINT.
60,000 MILES PER HOUR
NO, THE EARTH IS GOING AROUND THE SUN, WHICH IS WHY THERE ARE SEASONS, BUT THAT'S ANOTHER CONVERSATION. AROUND THE SUN, THE TRAIN TRACK IS MOVING AT OVER 60,000 MILES PER HOUR (97,000 KM/H).

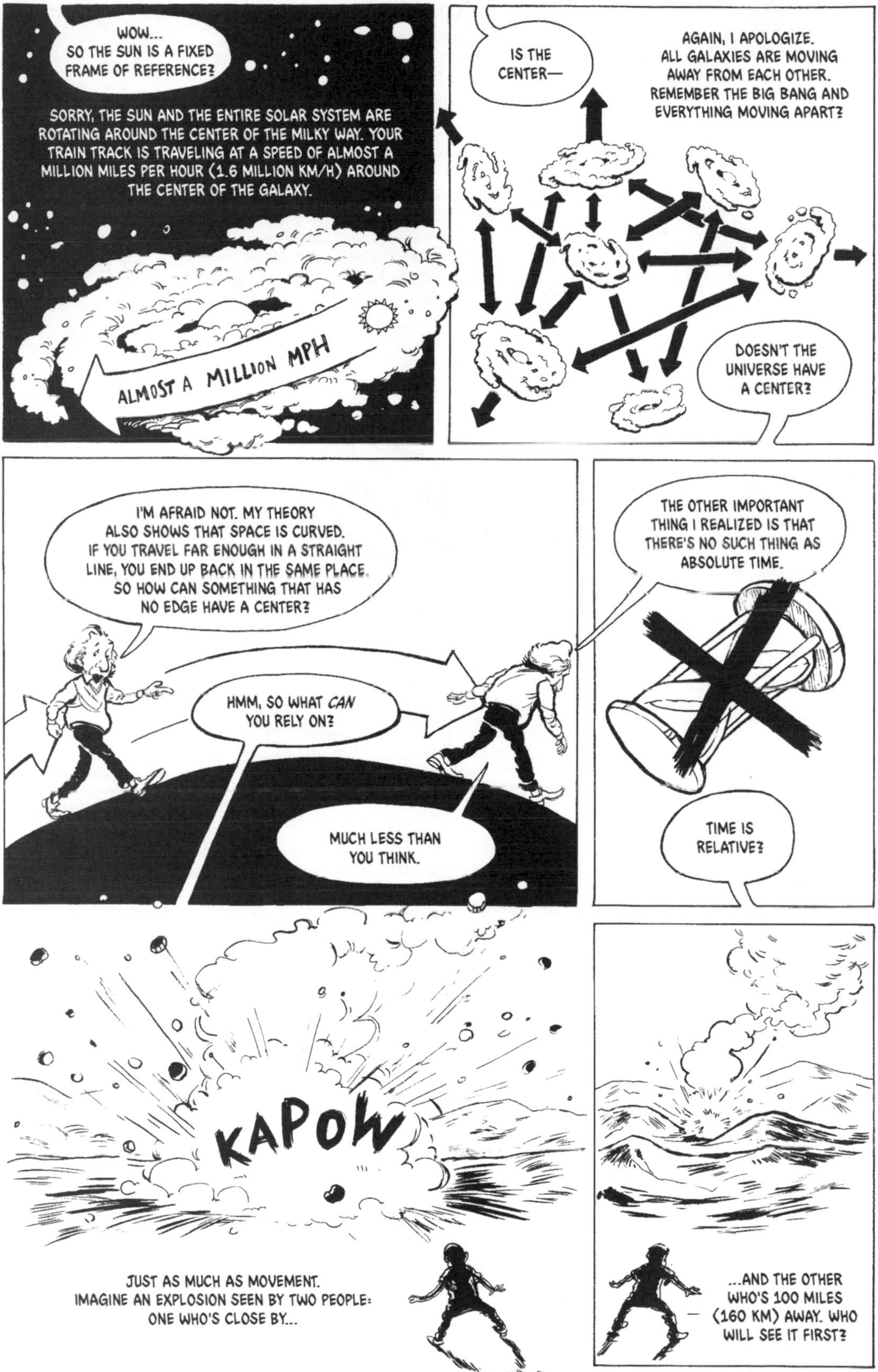
WOW... SO THE SUN IS A FIXED FRAME OF REFERENCE?
SORRY, THE SUN AND THE ENTIRE SOLAR SYSTEM ARE ROTATING AROUND THE CENTER OF THE MILKY WAY. YOUR TRAIN TRACK IS TRAVELING AT A SPEED OF ALMOST A MILLION MILES PER HOUR (1.6 MILLION KM/H) AROUND THE CENTER OF THE GALAXY.
ALMOST A MILLION MPH
IS THE CENTER—
AGAIN, I APOLOGIZE. ALL GALAXIES ARE MOVING AWAY FROM EACH OTHER. REMEMBER THE BIG BANG AND EVERYTHING MOVING APART?
DOESN'T THE UNIVERSE HAVE A CENTER?
I'M AFRAID NOT. MY THEORY ALSO SHOWS THAT SPACE IS CURVED. IF YOU TRAVEL FAR ENOUGH IN A STRAIGHT LINE, YOU END UP BACK IN THE SAME PLACE. SO HOW CAN SOMETHING THAT HAS NO EDGE HAVE A CENTER?
HMM, SO WHAT *CAN* YOU RELY ON?
MUCH LESS THAN YOU THINK.
THE OTHER IMPORTANT THING I REALIZED IS THAT THERE'S NO SUCH THING AS ABSOLUTE TIME.
TIME IS RELATIVE?
KAPOW
JUST AS MUCH AS MOVEMENT. IMAGINE AN EXPLOSION SEEN BY TWO PEOPLE: ONE WHO'S CLOSE BY...
...AND THE OTHER WHO'S 100 MILES (160 KM) AWAY. WHO WILL SEE IT FIRST?

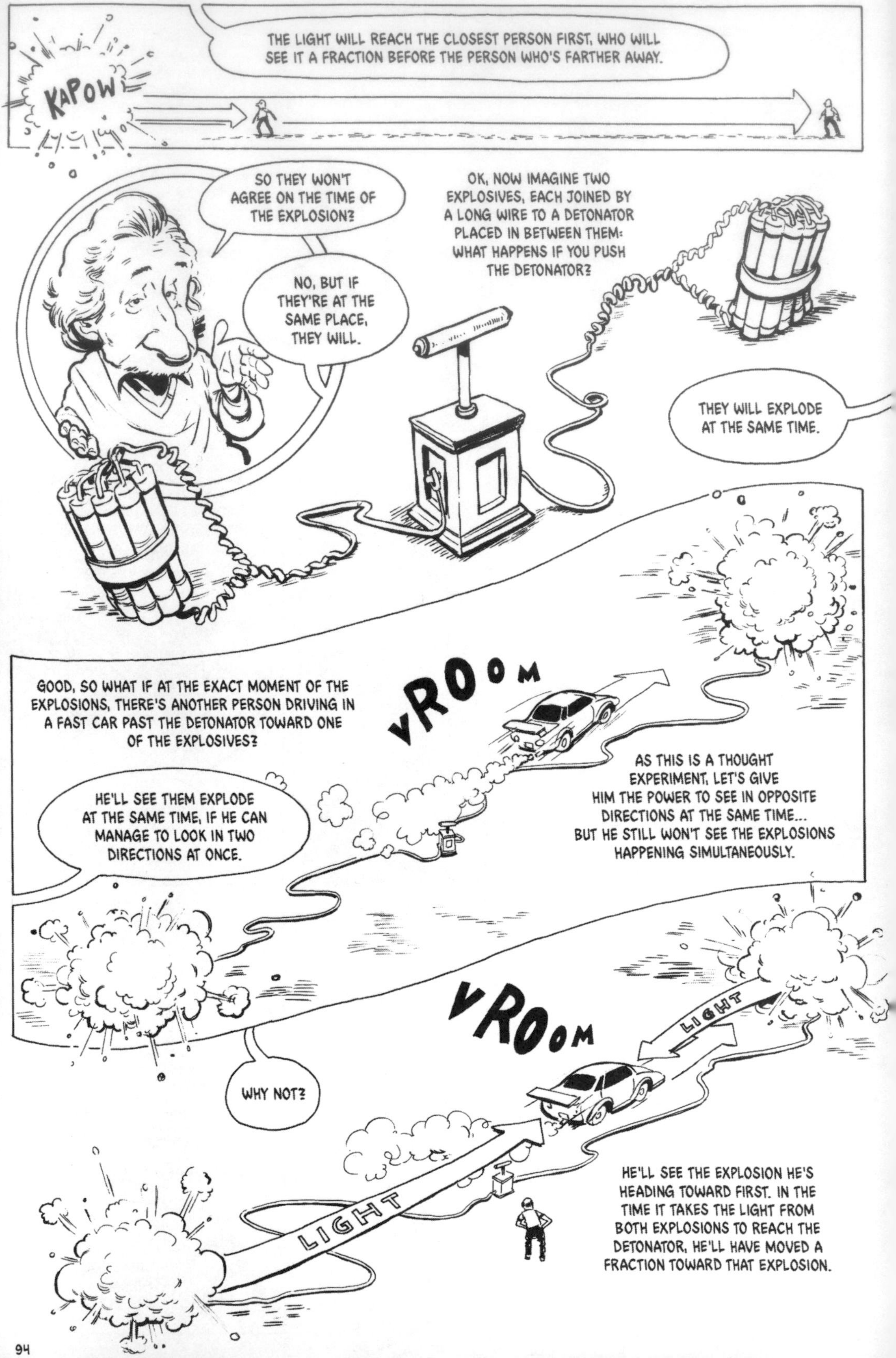
KAPOW
THE LIGHT WILL REACH THE CLOSEST PERSON FIRST, WHO WILL SEE IT A FRACTION BEFORE THE PERSON WHO'S FARTHER AWAY.
SO THEY WON'T AGREE ON THE TIME OF THE EXPLOSION?
NO, BUT IF THEY'RE AT THE SAME PLACE, THEY WILL.
OK, NOW IMAGINE TWO EXPLOSIVES, EACH JOINED BY A LONG WIRE TO A DETONATOR PLACED IN BETWEEN THEM: WHAT HAPPENS IF YOU PUSH THE DETONATOR?
THEY WILL EXPLODE AT THE SAME TIME.
GOOD, SO WHAT IF AT THE EXACT MOMENT OF THE EXPLOSIONS, THERE'S ANOTHER PERSON DRIVING IN A FAST CAR PAST THE DETONATOR TOWARD ONE OF THE EXPLOSIVES?
VROOM
HE'LL SEE THEM EXPLODE AT THE SAME TIME, IF HE CAN MANAGE TO LOOK IN TWO DIRECTIONS AT ONCE.
AS THIS IS A THOUGHT EXPERIMENT, LET'S GIVE HIM THE POWER TO SEE IN OPPOSITE DIRECTIONS AT THE SAME TIME... BUT HE STILL WON'T SEE THE EXPLOSIONS HAPPENING SIMULTANEOUSLY.
VROOM
LIGHT
WHY NOT?
LIGHT
HE'LL SEE THE EXPLOSION HE'S HEADING TOWARD FIRST. IN THE TIME IT TAKES THE LIGHT FROM BOTH EXPLOSIONS TO REACH THE DETONATOR, HE'LL HAVE MOVED A FRACTION TOWARD THAT EXPLOSION.

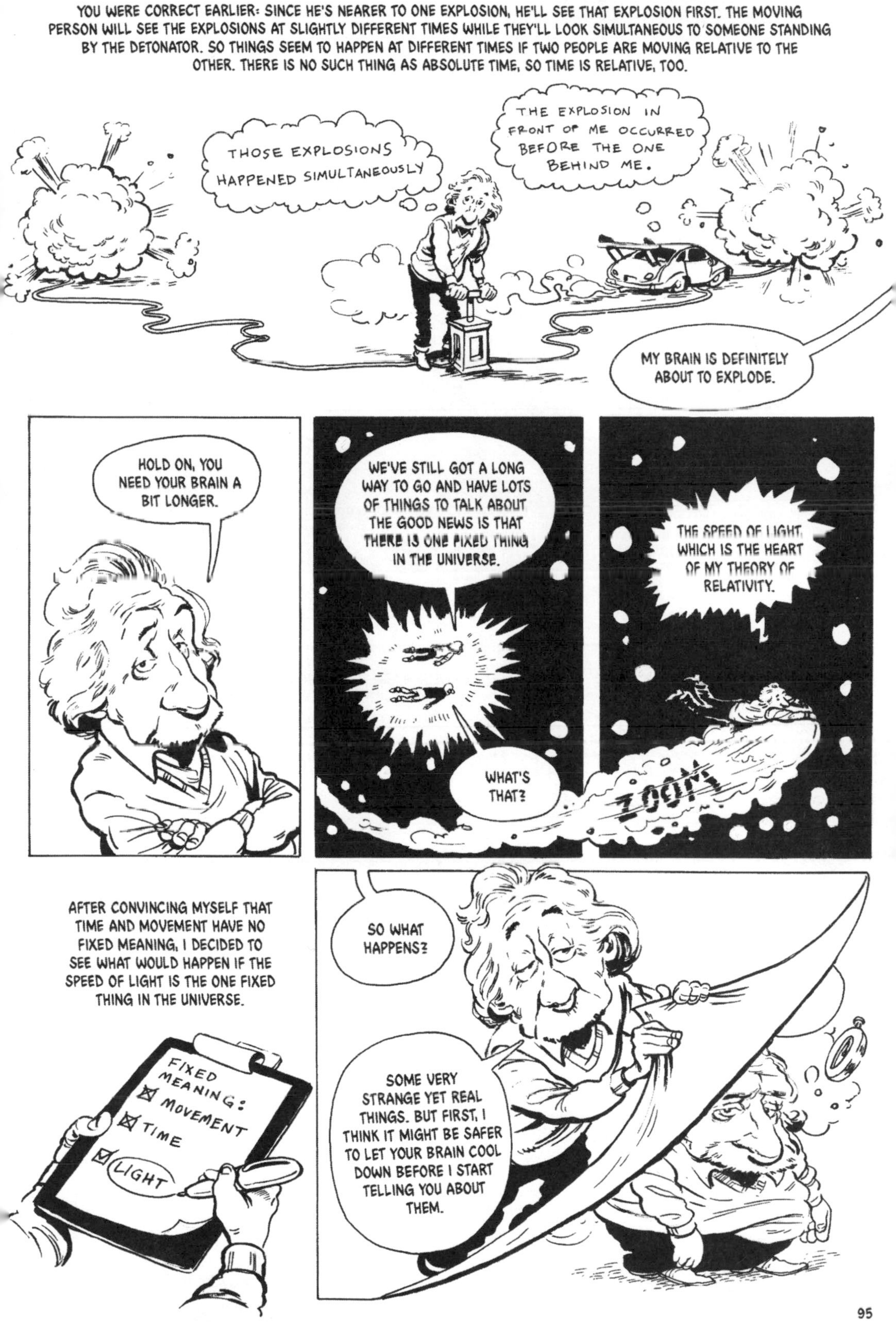
YOU WERE CORRECT EARLIER: SINCE HE'S NEARER TO ONE EXPLOSION, HE'LL SEE THAT EXPLOSION FIRST. THE MOVING PERSON WILL SEE THE EXPLOSIONS AT SLIGHTLY DIFFERENT TIMES WHILE THEY'LL LOOK SIMULTANEOUS TO SOMEONE STANDING BY THE DETONATOR. SO THINGS SEEM TO HAPPEN AT DIFFERENT TIMES IF TWO PEOPLE ARE MOVING RELATIVE TO THE OTHER. THERE IS NO SUCH THING AS ABSOLUTE TIME, SO TIME IS RELATIVE, TOO.
THOSE EXPLOSIONS HAPPENED SIMULTANEOUSLY
THE EXPLOSION IN FRONT OF ME OCCURRED BEFORE THE ONE BEHIND ME.
MY BRAIN IS DEFINITELY ABOUT TO EXPLODE.
HOLD ON, YOU NEED YOUR BRAIN A BIT LONGER.
WE'VE STILL GOT A LONG WAY TO GO AND HAVE LOTS OF THINGS TO TALK ABOUT THE GOOD NEWS IS THAT THERE IS ONE FIXED THING IN THE UNIVERSE.
WHAT'S THAT?
THE SPEED OF LIGHT, WHICH IS THE HEART OF MY THEORY OF RELATIVITY.
ZOOM
AFTER CONVINCING MYSELF THAT TIME AND MOVEMENT HAVE NO FIXED MEANING, I DECIDED TO SEE WHAT WOULD HAPPEN IF THE SPEED OF LIGHT IS THE ONE FIXED THING IN THE UNIVERSE.
FIXED MEANING:
MOVEMENT
TIME
LIGHT
SO WHAT HAPPENS?
SOME VERY STRANGE YET REAL THINGS. BUT FIRST, I THINK IT MIGHT BE SAFER TO LET YOUR BRAIN COOL DOWN BEFORE I START TELLING YOU ABOUT THEM.

CHAPTER 18: RELATIVELY WEIRD: HOW TO GROW YOUNGER, SHORTER, AND HEAVIER AT THE SAME TIME

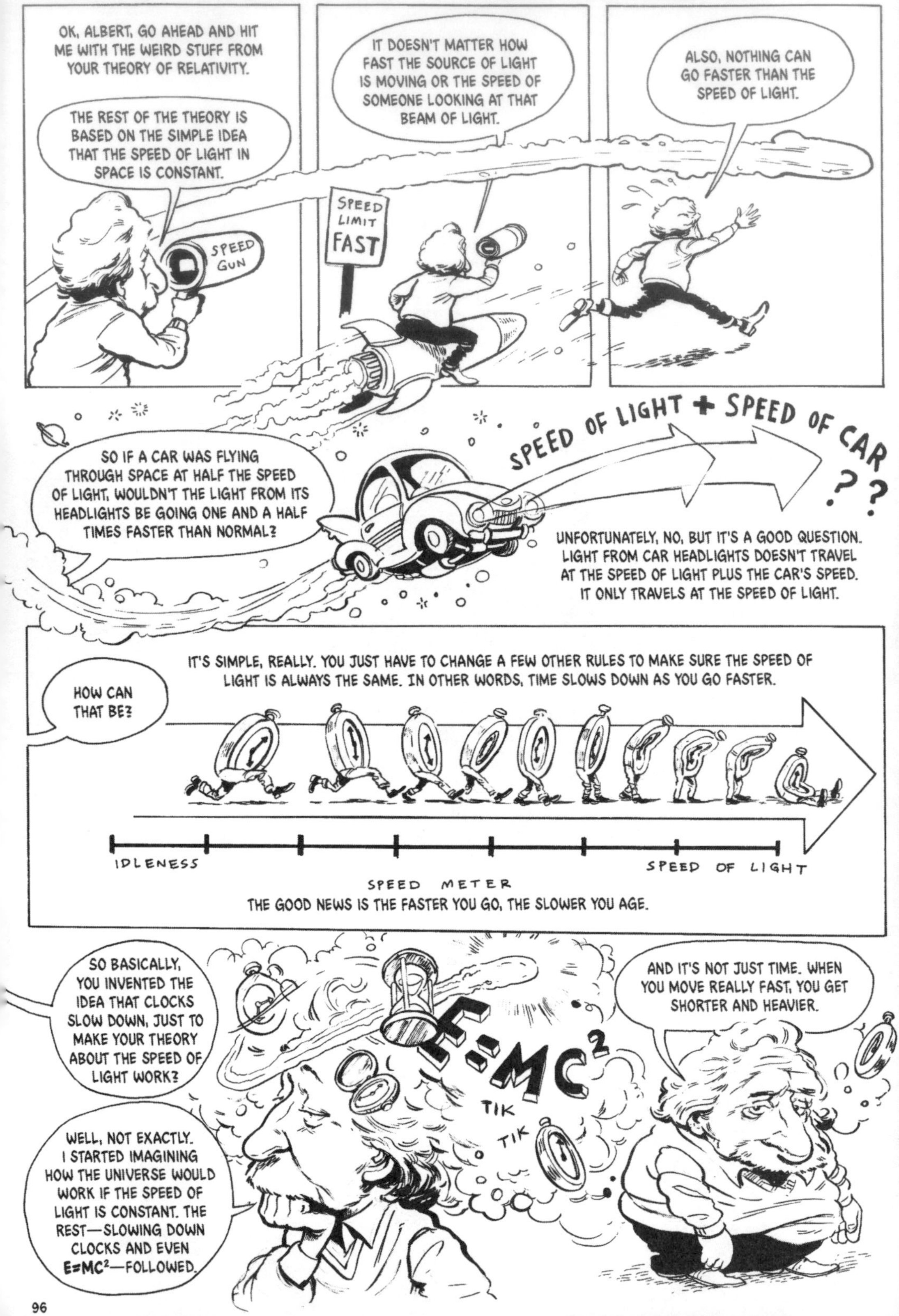

OK, ALBERT, GO AHEAD AND HIT ME WITH THE WEIRD STUFF FROM YOUR THEORY OF RELATIVITY.
THE REST OF THE THEORY IS BASED ON THE SIMPLE IDEA THAT THE SPEED OF LIGHT IN SPACE IS CONSTANT.
SPEED GUN
IT DOESN'T MATTER HOW FAST THE SOURCE OF LIGHT IS MOVING OR THE SPEED OF SOMEONE LOOKING AT THAT BEAM OF LIGHT.
SPEED LIMIT FAST
ALSO, NOTHING CAN GO FASTER THAN THE SPEED OF LIGHT.
SO IF A CAR WAS FLYING THROUGH SPACE AT HALF THE SPEED OF LIGHT, WOULDN'T THE LIGHT FROM ITS HEADLIGHTS BE GOING ONE AND A HALF TIMES FASTER THAN NORMAL?
SPEED OF LIGHT + SPEED OF CAR ??
UNFORTUNATELY, NO, BUT IT'S A GOOD QUESTION. LIGHT FROM CAR HEADLIGHTS DOESN'T TRAVEL AT THE SPEED OF LIGHT PLUS THE CAR'S SPEED. IT ONLY TRAVELS AT THE SPEED OF LIGHT.
HOW CAN THAT BE?
IT'S SIMPLE, REALLY. YOU JUST HAVE TO CHANGE A FEW OTHER RULES TO MAKE SURE THE SPEED OF LIGHT IS ALWAYS THE SAME. IN OTHER WORDS, TIME SLOWS DOWN AS YOU GO FASTER.
IDLENESS
SPEED OF LIGHT
SPEED METER
THE GOOD NEWS IS THE FASTER YOU GO, THE SLOWER YOU AGE.
SO BASICALLY, YOU INVENTED THE IDEA THAT CLOCKS SLOW DOWN, JUST TO MAKE YOUR THEORY ABOUT THE SPEED OF LIGHT WORK?
E=MC²
TIK
TIK
AND IT'S NOT JUST TIME. WHEN YOU MOVE REALLY FAST, YOU GET SHORTER AND HEAVIER.
WELL, NOT EXACTLY. I STARTED IMAGINING HOW THE UNIVERSE WOULD WORK IF THE SPEED OF LIGHT IS CONSTANT. THE REST—SLOWING DOWN CLOCKS AND EVEN E=MC²—FOLLOWED.

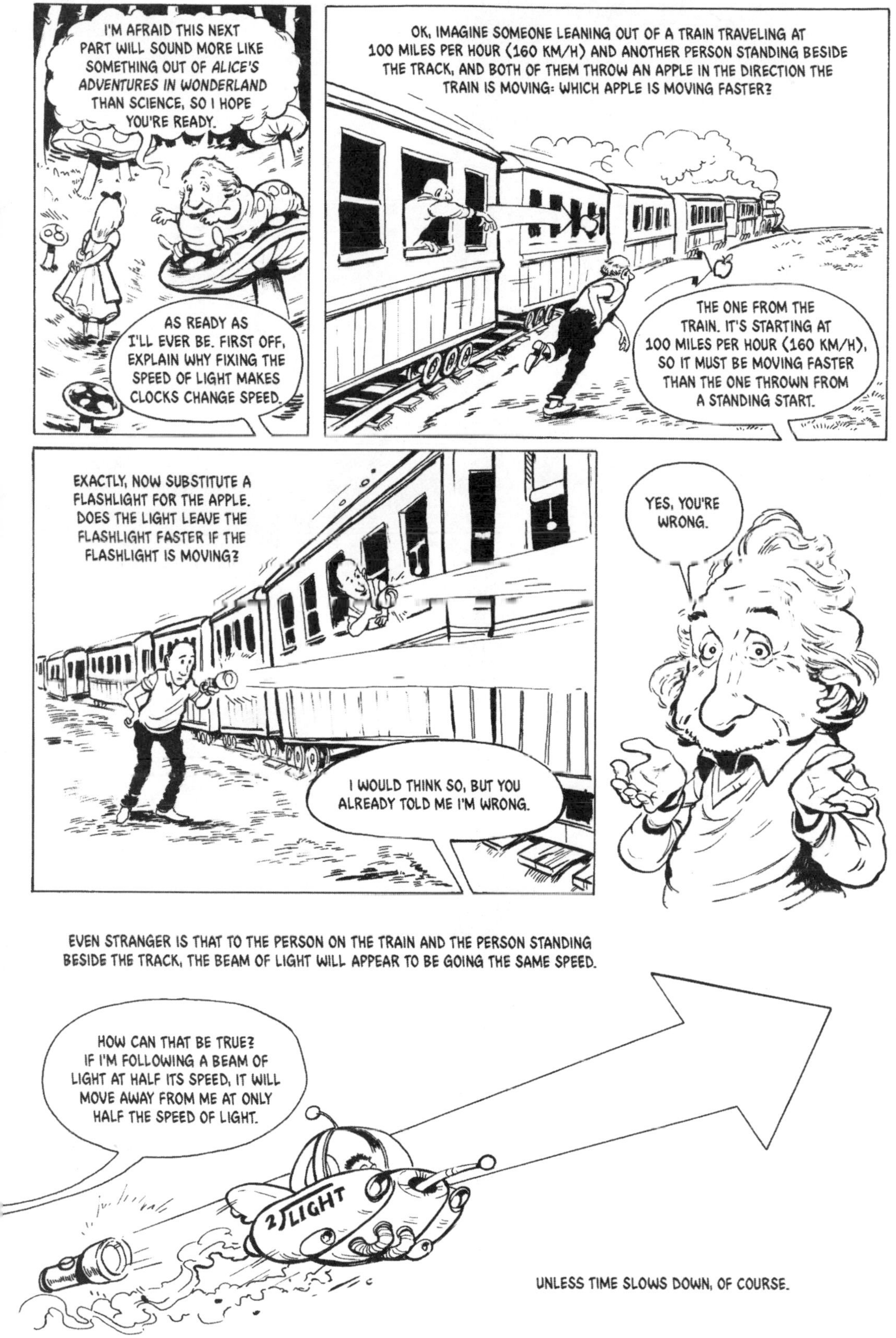
I'M AFRAID THIS NEXT PART WILL SOUND MORE LIKE SOMETHING OUT OF *ALICE'S ADVENTURES IN WONDERLAND* THAN SCIENCE, SO I HOPE YOU'RE READY.
AS READY AS I'LL EVER BE. FIRST OFF, EXPLAIN WHY FIXING THE SPEED OF LIGHT MAKES CLOCKS CHANGE SPEED.
OK, IMAGINE SOMEONE LEANING OUT OF A TRAIN TRAVELING AT 100 MILES PER HOUR (160 KM/H) AND ANOTHER PERSON STANDING BESIDE THE TRACK, AND BOTH OF THEM THROW AN APPLE IN THE DIRECTION THE TRAIN IS MOVING: WHICH APPLE IS MOVING FASTER?
THE ONE FROM THE TRAIN. IT'S STARTING AT 100 MILES PER HOUR (160 KM/H), SO IT MUST BE MOVING FASTER THAN THE ONE THROWN FROM A STANDING START.
EXACTLY, NOW SUBSTITUTE A FLASHLIGHT FOR THE APPLE. DOES THE LIGHT LEAVE THE FLASHLIGHT FASTER IF THE FLASHLIGHT IS MOVING?
I WOULD THINK SO, BUT YOU ALREADY TOLD ME I'M WRONG.
YES, YOU'RE WRONG.
EVEN STRANGER IS THAT TO THE PERSON ON THE TRAIN AND THE PERSON STANDING BESIDE THE TRACK, THE BEAM OF LIGHT WILL APPEAR TO BE GOING THE SAME SPEED.
HOW CAN THAT BE TRUE? IF I'M FOLLOWING A BEAM OF LIGHT AT HALF ITS SPEED, IT WILL MOVE AWAY FROM ME AT ONLY HALF THE SPEED OF LIGHT.
1/2 LIGHT
UNLESS TIME SLOWS DOWN, OF COURSE.

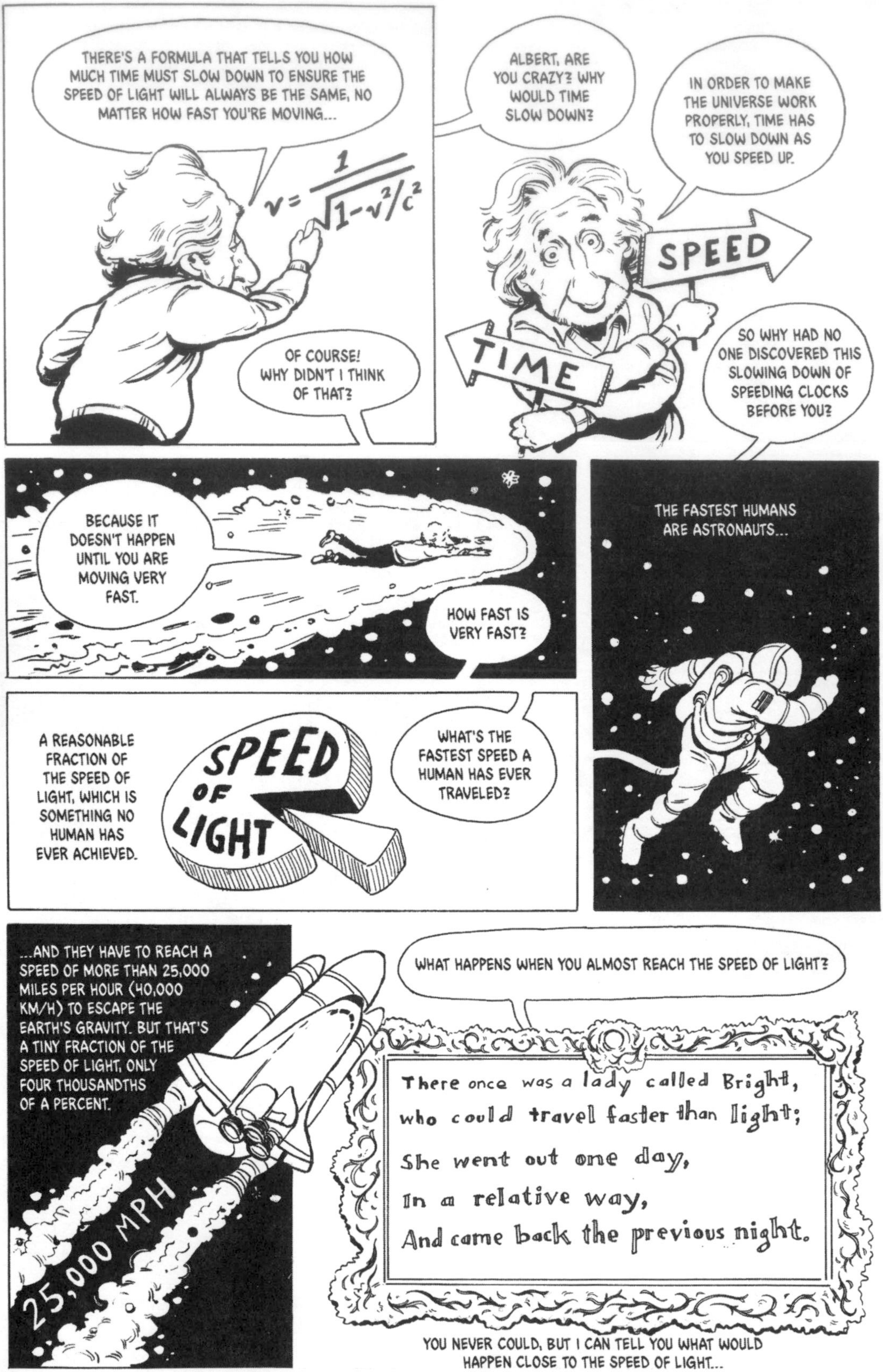
THERE'S A FORMULA THAT TELLS YOU HOW MUCH TIME MUST SLOW DOWN TO ENSURE THE SPEED OF LIGHT WILL ALWAYS BE THE SAME, NO MATTER HOW FAST YOU'RE MOVING...
$v = \frac{1}{\sqrt{1 - v^2/c^2}}$
OF COURSE! WHY DIDN'T I THINK OF THAT?
ALBERT, ARE YOU CRAZY? WHY WOULD TIME SLOW DOWN?
IN ORDER TO MAKE THE UNIVERSE WORK PROPERLY, TIME HAS TO SLOW DOWN AS YOU SPEED UP.
SPEED
TIME
SO WHY HAD NO ONE DISCOVERED THIS SLOWING DOWN OF SPEEDING CLOCKS BEFORE YOU?
BECAUSE IT DOESN'T HAPPEN UNTIL YOU ARE MOVING VERY FAST.
HOW FAST IS VERY FAST?
A REASONABLE FRACTION OF THE SPEED OF LIGHT, WHICH IS SOMETHING NO HUMAN HAS EVER ACHIEVED.
SPEED OF LIGHT
WHAT'S THE FASTEST SPEED A HUMAN HAS EVER TRAVELED?
THE FASTEST HUMANS ARE ASTRONAUTS...
...AND THEY HAVE TO REACH A SPEED OF MORE THAN 25,000 MILES PER HOUR (40,000 KM/H) TO ESCAPE THE EARTH'S GRAVITY. BUT THAT'S A TINY FRACTION OF THE SPEED OF LIGHT, ONLY FOUR THOUSANDTHS OF A PERCENT.
25,000 MPH
WHAT HAPPENS WHEN YOU ALMOST REACH THE SPEED OF LIGHT?
There once was a lady called Bright,
who could travel faster than light;
She went out one day,
In a relative way,
And came back the previous night.
YOU NEVER COULD, BUT I CAN TELL YOU WHAT WOULD HAPPEN CLOSE TO THE SPEED OF LIGHT...

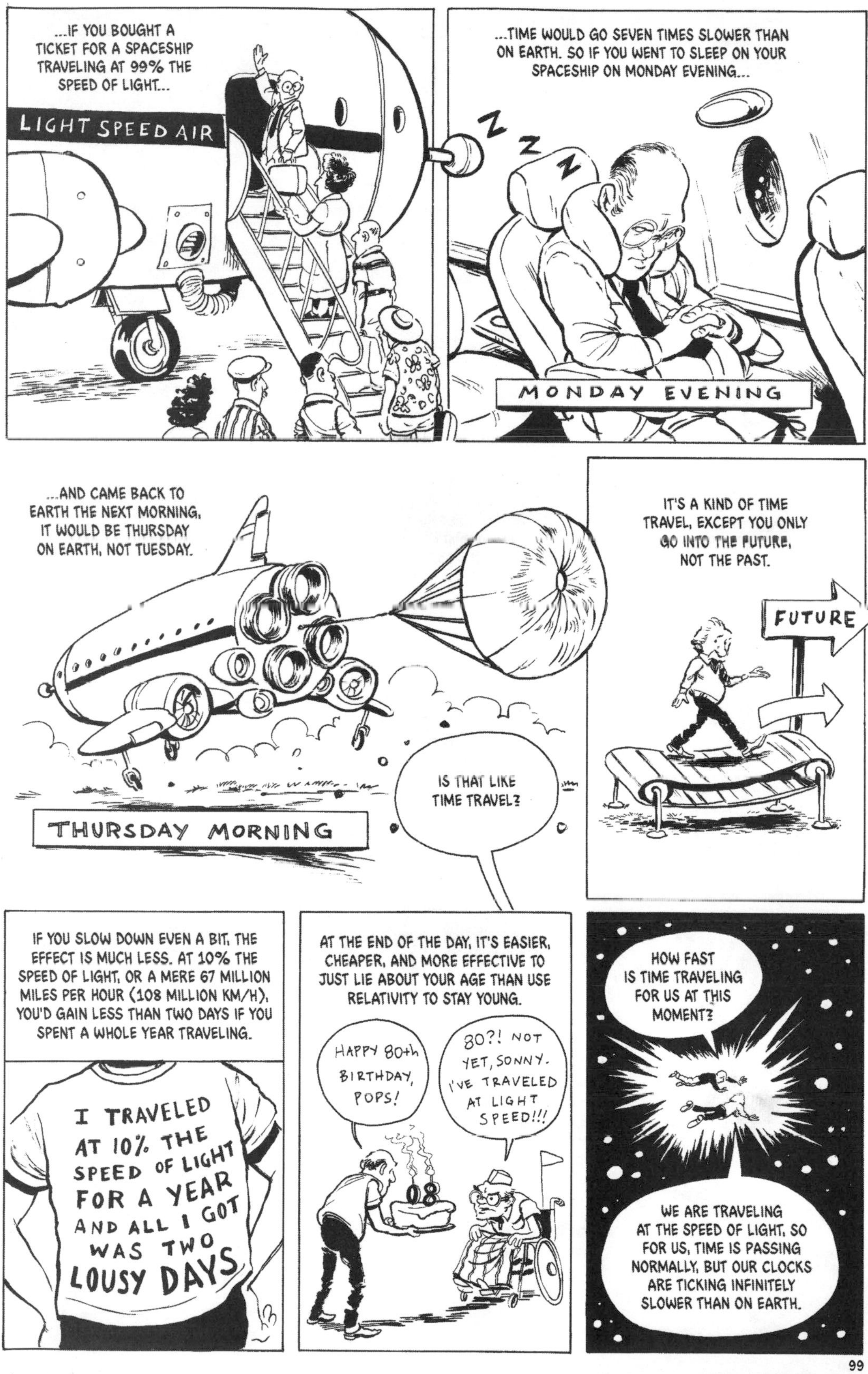
...IF YOU BOUGHT A TICKET FOR A SPACESHIP TRAVELING AT 99% THE SPEED OF LIGHT...
LIGHT SPEED AIR
...TIME WOULD GO SEVEN TIMES SLOWER THAN ON EARTH. SO IF YOU WENT TO SLEEP ON YOUR SPACESHIP ON MONDAY EVENING...
Z Z Z
MONDAY EVENING
...AND CAME BACK TO EARTH THE NEXT MORNING, IT WOULD BE THURSDAY ON EARTH, NOT TUESDAY.
THURSDAY MORNING
IS THAT LIKE TIME TRAVEL?
IT'S A KIND OF TIME TRAVEL, EXCEPT YOU ONLY GO INTO THE FUTURE, NOT THE PAST.
FUTURE
IF YOU SLOW DOWN EVEN A BIT, THE EFFECT IS MUCH LESS. AT 10% THE SPEED OF LIGHT, OR A MERE 67 MILLION MILES PER HOUR (108 MILLION KM/H), YOU'D GAIN LESS THAN TWO DAYS IF YOU SPENT A WHOLE YEAR TRAVELING.
I TRAVELED AT 10% THE SPEED OF LIGHT FOR A YEAR AND ALL I GOT WAS TWO LOUSY DAYS
AT THE END OF THE DAY, IT'S EASIER, CHEAPER, AND MORE EFFECTIVE TO JUST LIE ABOUT YOUR AGE THAN USE RELATIVITY TO STAY YOUNG.
HAPPY 80th BIRTHDAY, POPS!
80?! NOT YET, SONNY. I'VE TRAVELED AT LIGHT SPEED!!!
08
HOW FAST IS TIME TRAVELING FOR US AT THIS MOMENT?
WE ARE TRAVELING AT THE SPEED OF LIGHT, SO FOR US, TIME IS PASSING NORMALLY, BUT OUR CLOCKS ARE TICKING INFINITELY SLOWER THAN ON EARTH.

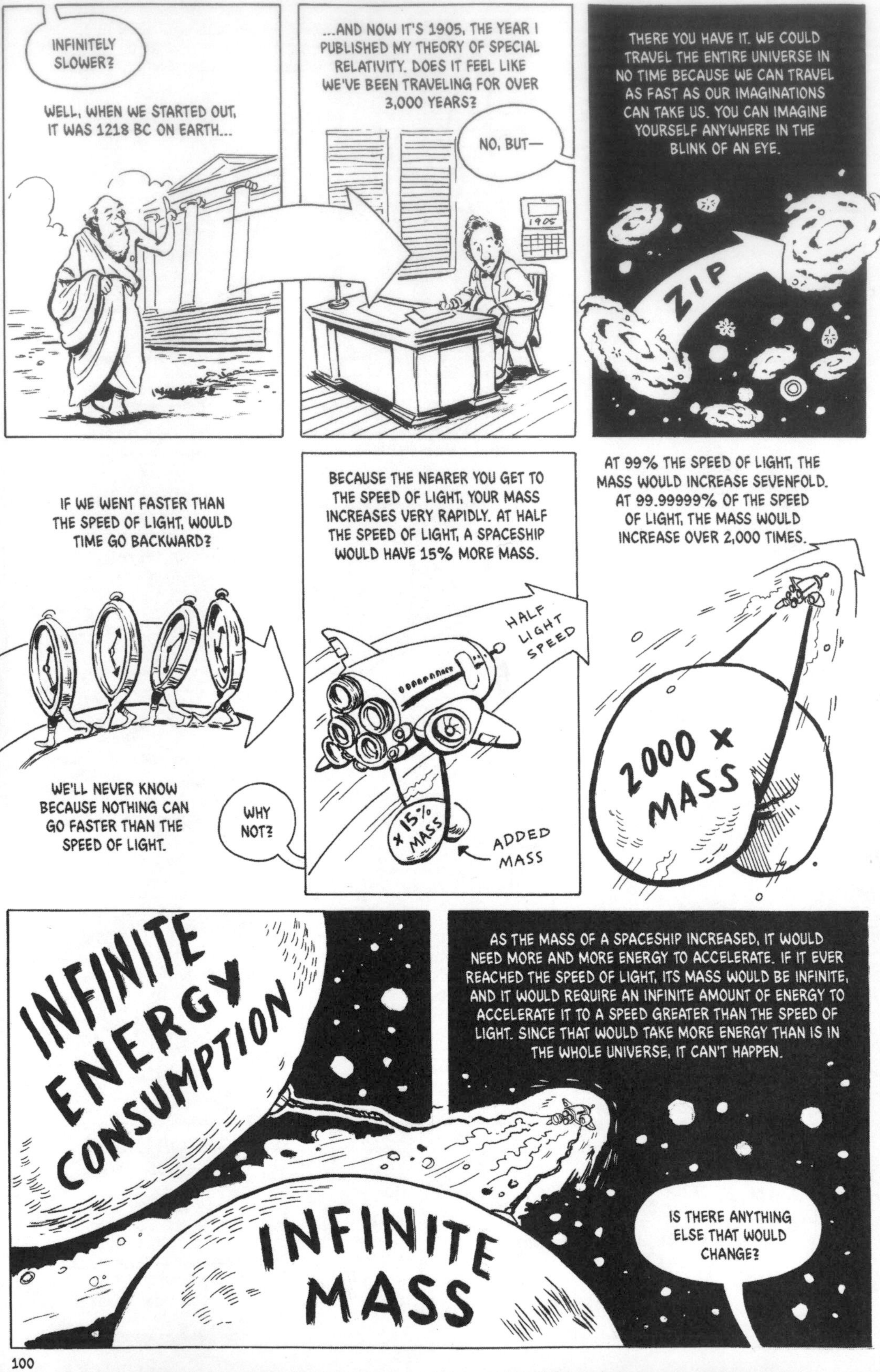

INFINITELY SLOWER?
WELL, WHEN WE STARTED OUT, IT WAS 1218 BC ON EARTH...
...AND NOW IT'S 1905, THE YEAR I PUBLISHED MY THEORY OF SPECIAL RELATIVITY. DOES IT FEEL LIKE WE'VE BEEN TRAVELING FOR OVER 3,000 YEARS?
NO, BUT—
1905
THERE YOU HAVE IT. WE COULD TRAVEL THE ENTIRE UNIVERSE IN NO TIME BECAUSE WE CAN TRAVEL AS FAST AS OUR IMAGINATIONS CAN TAKE US. YOU CAN IMAGINE YOURSELF ANYWHERE IN THE BLINK OF AN EYE.
ZIP
IF WE WENT FASTER THAN THE SPEED OF LIGHT, WOULD TIME GO BACKWARD?
WE'LL NEVER KNOW BECAUSE NOTHING CAN GO FASTER THAN THE SPEED OF LIGHT.
WHY NOT?
BECAUSE THE NEARER YOU GET TO THE SPEED OF LIGHT, YOUR MASS INCREASES VERY RAPIDLY. AT HALF THE SPEED OF LIGHT, A SPACESHIP WOULD HAVE 15% MORE MASS.
HALF LIGHT SPEED
x 15% MASS
ADDED MASS
AT 99% THE SPEED OF LIGHT, THE MASS WOULD INCREASE SEVENFOLD. AT 99.99999% OF THE SPEED OF LIGHT, THE MASS WOULD INCREASE OVER 2,000 TIMES.
2000 x MASS
INFINITE ENERGY CONSUMPTION
INFINITE MASS
AS THE MASS OF A SPACESHIP INCREASED, IT WOULD NEED MORE AND MORE ENERGY TO ACCELERATE. IF IT EVER REACHED THE SPEED OF LIGHT, ITS MASS WOULD BE INFINITE, AND IT WOULD REQUIRE AN INFINITE AMOUNT OF ENERGY TO ACCELERATE IT TO A SPEED GREATER THAN THE SPEED OF LIGHT. SINCE THAT WOULD TAKE MORE ENERGY THAN IS IN THE WHOLE UNIVERSE, IT CAN'T HAPPEN.
IS THERE ANYTHING ELSE THAT WOULD CHANGE?

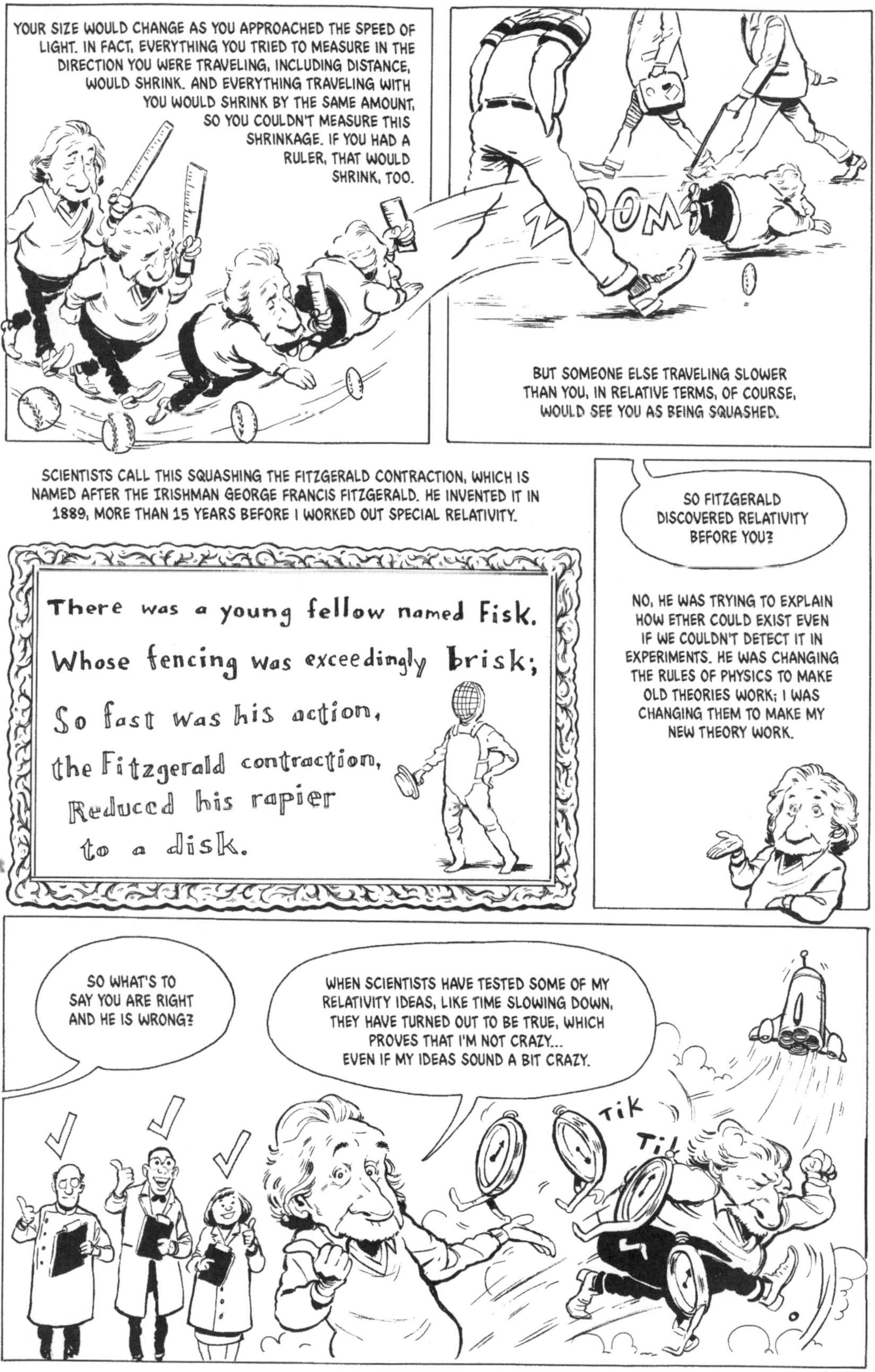

YOUR SIZE WOULD CHANGE AS YOU APPROACHED THE SPEED OF LIGHT. IN FACT, EVERYTHING YOU TRIED TO MEASURE IN THE DIRECTION YOU WERE TRAVELING, INCLUDING DISTANCE, WOULD SHRINK. AND EVERYTHING TRAVELING WITH YOU WOULD SHRINK BY THE SAME AMOUNT, SO YOU COULDN'T MEASURE THIS SHRINKAGE. IF YOU HAD A RULER, THAT WOULD SHRINK, TOO.
ZOOM
BUT SOMEONE ELSE TRAVELING SLOWER THAN YOU, IN RELATIVE TERMS, OF COURSE, WOULD SEE YOU AS BEING SQUASHED.
SCIENTISTS CALL THIS SQUASHING THE FITZGERALD CONTRACTION, WHICH IS NAMED AFTER THE IRISHMAN GEORGE FRANCIS FITZGERALD. HE INVENTED IT IN 1889, MORE THAN 15 YEARS BEFORE I WORKED OUT SPECIAL RELATIVITY.
There was a young fellow named Fisk.
Whose fencing was exceedingly brisk;
So fast was his action,
the Fitzgerald contraction,
Reduced his rapier
to a disk.
SO FITZGERALD DISCOVERED RELATIVITY BEFORE YOU?
NO, HE WAS TRYING TO EXPLAIN HOW ETHER COULD EXIST EVEN IF WE COULDN'T DETECT IT IN EXPERIMENTS. HE WAS CHANGING THE RULES OF PHYSICS TO MAKE OLD THEORIES WORK; I WAS CHANGING THEM TO MAKE MY NEW THEORY WORK.
SO WHAT'S TO SAY YOU ARE RIGHT AND HE IS WRONG?
WHEN SCIENTISTS HAVE TESTED SOME OF MY RELATIVITY IDEAS, LIKE TIME SLOWING DOWN, THEY HAVE TURNED OUT TO BE TRUE, WHICH PROVES THAT I'M NOT CRAZY... EVEN IF MY IDEAS SOUND A BIT CRAZY.
TIK

CHAPTER 19: YES, BUT IS IT TRUE? ALBERT EINSTEIN'S THEORIES UNDER THE MICROSCOPE

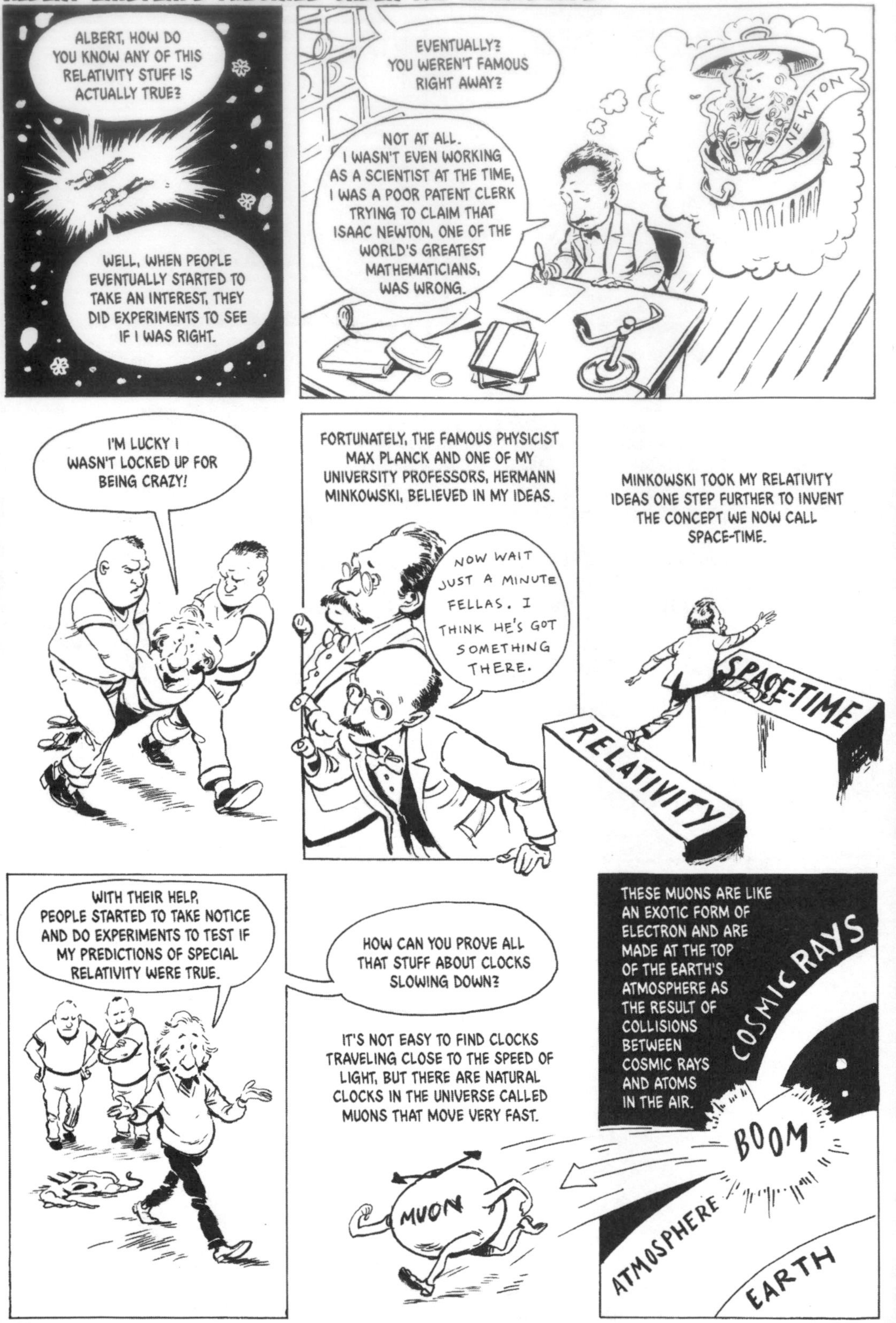
ALBERT, HOW DO YOU KNOW ANY OF THIS RELATIVITY STUFF IS ACTUALLY TRUE?
WELL, WHEN PEOPLE EVENTUALLY STARTED TO TAKE AN INTEREST, THEY DID EXPERIMENTS TO SEE IF I WAS RIGHT.
EVENTUALLY? YOU WEREN'T FAMOUS RIGHT AWAY?
NOT AT ALL. I WASN'T EVEN WORKING AS A SCIENTIST AT THE TIME, I WAS A POOR PATENT CLERK TRYING TO CLAIM THAT ISAAC NEWTON, ONE OF THE WORLD'S GREATEST MATHEMATICIANS, WAS WRONG.
NEWTON
I'M LUCKY I WASN'T LOCKED UP FOR BEING CRAZY!
FORTUNATELY, THE FAMOUS PHYSICIST MAX PLANCK AND ONE OF MY UNIVERSITY PROFESSORS, HERMANN MINKOWSKI, BELIEVED IN MY IDEAS.
NOW WAIT JUST A MINUTE FELLAS. I THINK HE'S GOT SOMETHING THERE.
MINKOWSKI TOOK MY RELATIVITY IDEAS ONE STEP FURTHER TO INVENT THE CONCEPT WE NOW CALL SPACE-TIME.
RELATIVITY
SPACE-TIME
WITH THEIR HELP, PEOPLE STARTED TO TAKE NOTICE AND DO EXPERIMENTS TO TEST IF MY PREDICTIONS OF SPECIAL RELATIVITY WERE TRUE.
HOW CAN YOU PROVE ALL THAT STUFF ABOUT CLOCKS SLOWING DOWN?
IT'S NOT EASY TO FIND CLOCKS TRAVELING CLOSE TO THE SPEED OF LIGHT, BUT THERE ARE NATURAL CLOCKS IN THE UNIVERSE CALLED MUONS THAT MOVE VERY FAST.
THESE MUONS ARE LIKE AN EXOTIC FORM OF ELECTRON AND ARE MADE AT THE TOP OF THE EARTH'S ATMOSPHERE AS THE RESULT OF COLLISIONS BETWEEN COSMIC RAYS AND ATOMS IN THE AIR.
COSMIC RAYS
BOOM
MUON
ATMOSPHERE
EARTH

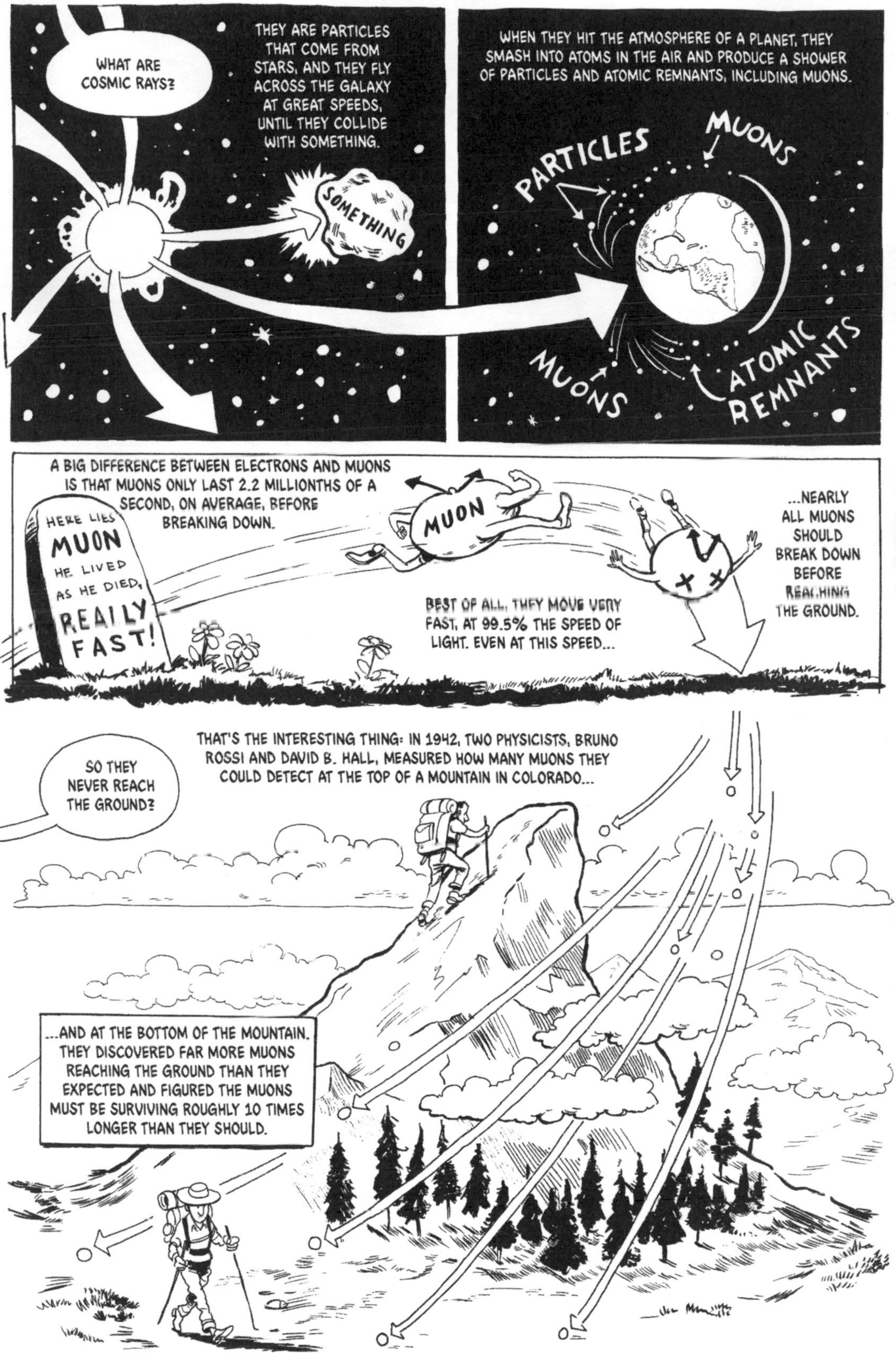

WHAT ARE COSMIC RAYS?
THEY ARE PARTICLES THAT COME FROM STARS, AND THEY FLY ACROSS THE GALAXY AT GREAT SPEEDS, UNTIL THEY COLLIDE WITH SOMETHING.
SOMETHING
WHEN THEY HIT THE ATMOSPHERE OF A PLANET, THEY SMASH INTO ATOMS IN THE AIR AND PRODUCE A SHOWER OF PARTICLES AND ATOMIC REMNANTS, INCLUDING MUONS.
PARTICLES
MUONS
MUONS
ATOMIC REMNANTS
A BIG DIFFERENCE BETWEEN ELECTRONS AND MUONS IS THAT MUONS ONLY LAST 2.2 MILLIONTHS OF A SECOND, ON AVERAGE, BEFORE BREAKING DOWN.
HERE LIES MUON HE LIVED AS HE DIED, REALLY FAST!
MUON
BEST OF ALL, THEY MOVE VERY FAST, AT 99.5% THE SPEED OF LIGHT. EVEN AT THIS SPEED...
...NEARLY ALL MUONS SHOULD BREAK DOWN BEFORE REACHING THE GROUND.
SO THEY NEVER REACH THE GROUND?
THAT'S THE INTERESTING THING: IN 1942, TWO PHYSICISTS, BRUNO ROSSI AND DAVID B. HALL, MEASURED HOW MANY MUONS THEY COULD DETECT AT THE TOP OF A MOUNTAIN IN COLORADO...
...AND AT THE BOTTOM OF THE MOUNTAIN. THEY DISCOVERED FAR MORE MUONS REACHING THE GROUND THAN THEY EXPECTED AND FIGURED THE MUONS MUST BE SURVIVING ROUGHLY 10 TIMES LONGER THAN THEY SHOULD.

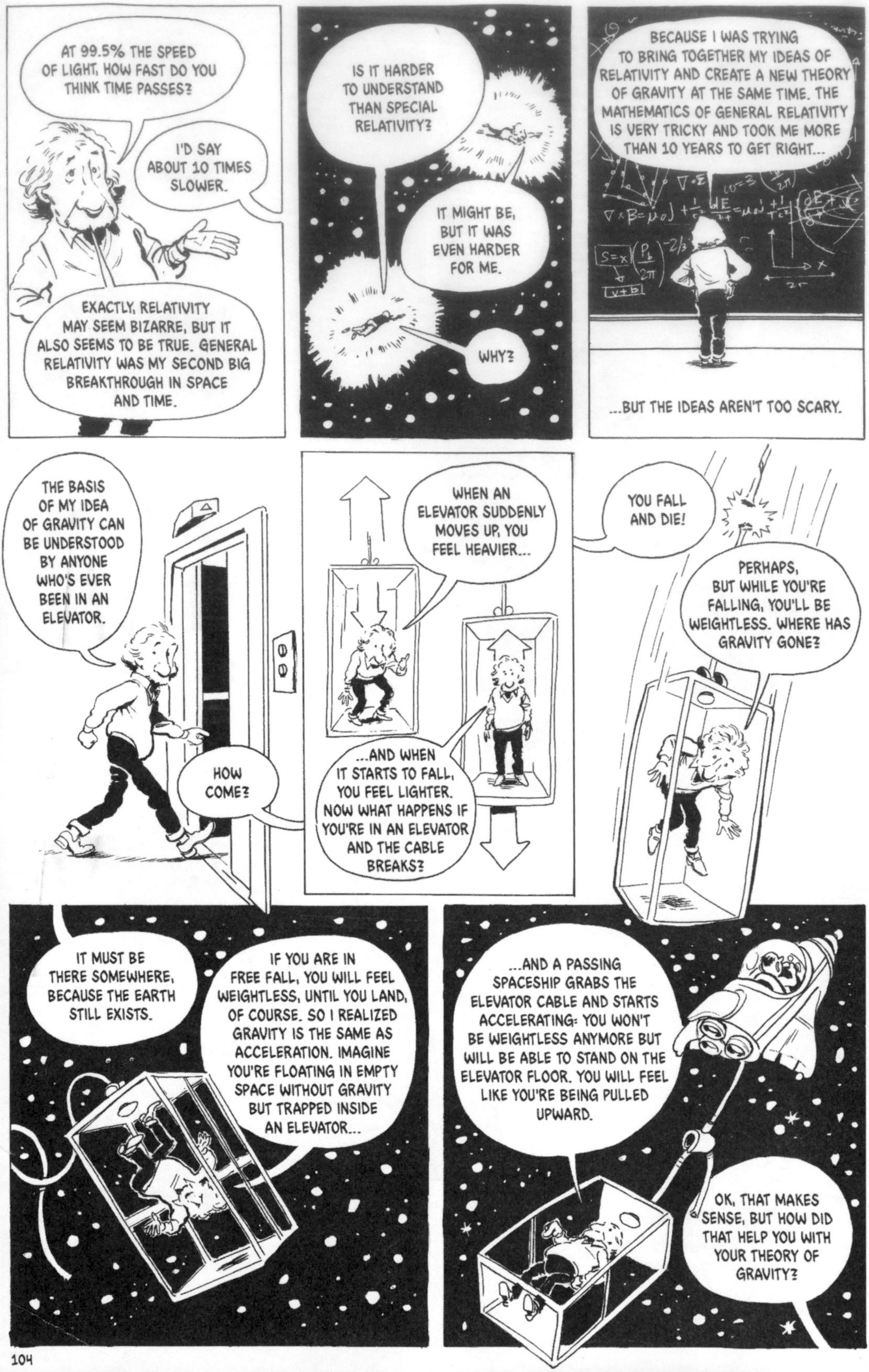
AT 99.5% THE SPEED OF LIGHT, HOW FAST DO YOU THINK TIME PASSES?
I'D SAY ABOUT 10 TIMES SLOWER.
EXACTLY. RELATIVITY MAY SEEM BIZARRE, BUT IT ALSO SEEMS TO BE TRUE. GENERAL RELATIVITY WAS MY SECOND BIG BREAKTHROUGH IN SPACE AND TIME.
IS IT HARDER TO UNDERSTAND THAN SPECIAL RELATIVITY?
IT MIGHT BE, BUT IT WAS EVEN HARDER FOR ME.
WHY?
BECAUSE I WAS TRYING TO BRING TOGETHER MY IDEAS OF RELATIVITY AND CREATE A NEW THEORY OF GRAVITY AT THE SAME TIME. THE MATHEMATICS OF GENERAL RELATIVITY IS VERY TRICKY AND TOOK ME MORE THAN 10 YEARS TO GET RIGHT...
...BUT THE IDEAS AREN'T TOO SCARY.
THE BASIS OF MY IDEA OF GRAVITY CAN BE UNDERSTOOD BY ANYONE WHO'S EVER BEEN IN AN ELEVATOR.
HOW COME?
WHEN AN ELEVATOR SUDDENLY MOVES UP, YOU FEEL HEAVIER...
...AND WHEN IT STARTS TO FALL, YOU FEEL LIGHTER. NOW WHAT HAPPENS IF YOU'RE IN AN ELEVATOR AND THE CABLE BREAKS?
YOU FALL AND DIE!
PERHAPS, BUT WHILE YOU'RE FALLING, YOU'LL BE WEIGHTLESS. WHERE HAS GRAVITY GONE?
IT MUST BE THERE SOMEWHERE, BECAUSE THE EARTH STILL EXISTS.
IF YOU ARE IN FREE FALL, YOU WILL FEEL WEIGHTLESS, UNTIL YOU LAND, OF COURSE. SO I REALIZED GRAVITY IS THE SAME AS ACCELERATION. IMAGINE YOU'RE FLOATING IN EMPTY SPACE WITHOUT GRAVITY BUT TRAPPED INSIDE AN ELEVATOR...
...AND A PASSING SPACESHIP GRABS THE ELEVATOR CABLE AND STARTS ACCELERATING: YOU WON'T BE WEIGHTLESS ANYMORE BUT WILL BE ABLE TO STAND ON THE ELEVATOR FLOOR. YOU WILL FEEL LIKE YOU'RE BEING PULLED UPWARD.
OK, THAT MAKES SENSE, BUT HOW DID THAT HELP YOU WITH YOUR THEORY OF GRAVITY?

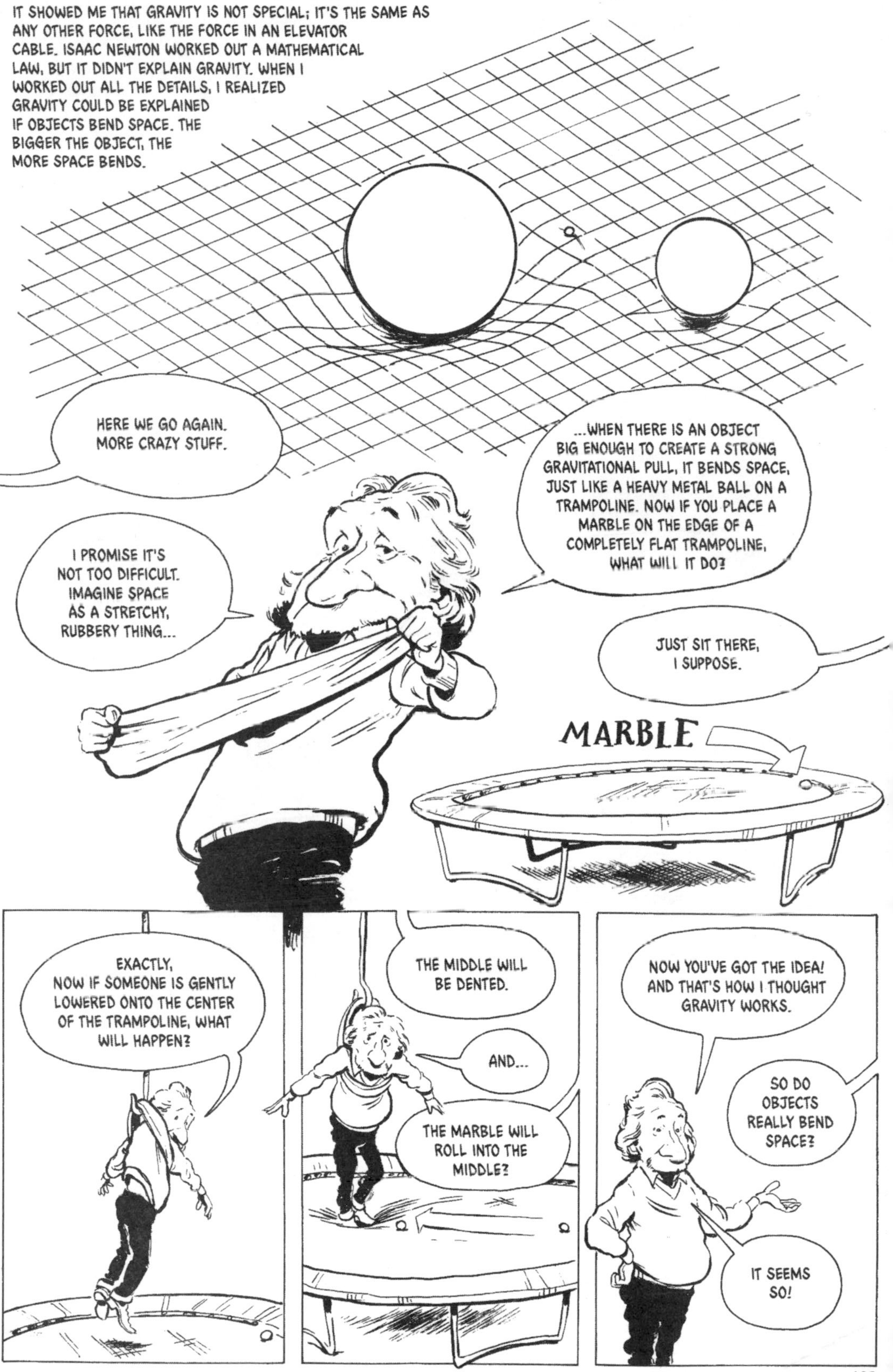
IT SHOWED ME THAT GRAVITY IS NOT SPECIAL; IT'S THE SAME AS ANY OTHER FORCE, LIKE THE FORCE IN AN ELEVATOR CABLE. ISAAC NEWTON WORKED OUT A MATHEMATICAL LAW, BUT IT DIDN'T EXPLAIN GRAVITY. WHEN I WORKED OUT ALL THE DETAILS, I REALIZED GRAVITY COULD BE EXPLAINED IF OBJECTS BEND SPACE. THE BIGGER THE OBJECT, THE MORE SPACE BENDS.
HERE WE GO AGAIN. MORE CRAZY STUFF.
I PROMISE IT'S NOT TOO DIFFICULT. IMAGINE SPACE AS A STRETCHY, RUBBERY THING...
...WHEN THERE IS AN OBJECT BIG ENOUGH TO CREATE A STRONG GRAVITATIONAL PULL, IT BENDS SPACE, JUST LIKE A HEAVY METAL BALL ON A TRAMPOLINE. NOW IF YOU PLACE A MARBLE ON THE EDGE OF A COMPLETELY FLAT TRAMPOLINE, WHAT WILL IT DO?
JUST SIT THERE, I SUPPOSE.
MARBLE
EXACTLY, NOW IF SOMEONE IS GENTLY LOWERED ONTO THE CENTER OF THE TRAMPOLINE, WHAT WILL HAPPEN?
THE MIDDLE WILL BE DENTED.
AND...
THE MARBLE WILL ROLL INTO THE MIDDLE?
NOW YOU'VE GOT THE IDEA! AND THAT'S HOW I THOUGHT GRAVITY WORKS.
SO DO OBJECTS REALLY BEND SPACE?
IT SEEMS SO!

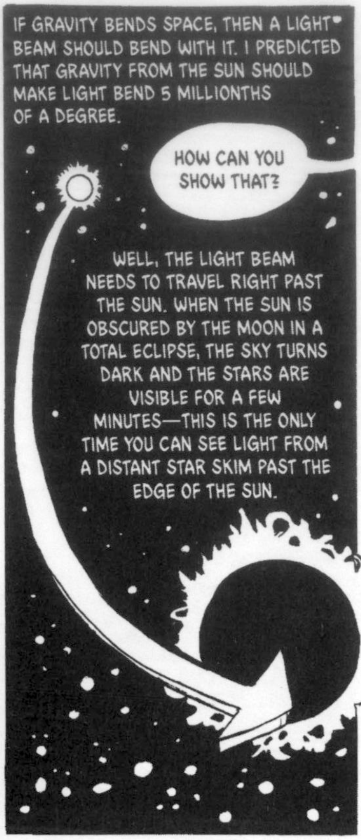
IF GRAVITY BENDS SPACE, THEN A LIGHT BEAM SHOULD BEND WITH IT. I PREDICTED THAT GRAVITY FROM THE SUN SHOULD MAKE LIGHT BEND 5 MILLIONTHS OF A DEGREE.
HOW CAN YOU SHOW THAT?
WELL, THE LIGHT BEAM NEEDS TO TRAVEL RIGHT PAST THE SUN. WHEN THE SUN IS OBSCURED BY THE MOON IN A TOTAL ECLIPSE, THE SKY TURNS DARK AND THE STARS ARE VISIBLE FOR A FEW MINUTES—THIS IS THE ONLY TIME YOU CAN SEE LIGHT FROM A DISTANT STAR SKIM PAST THE EDGE OF THE SUN.

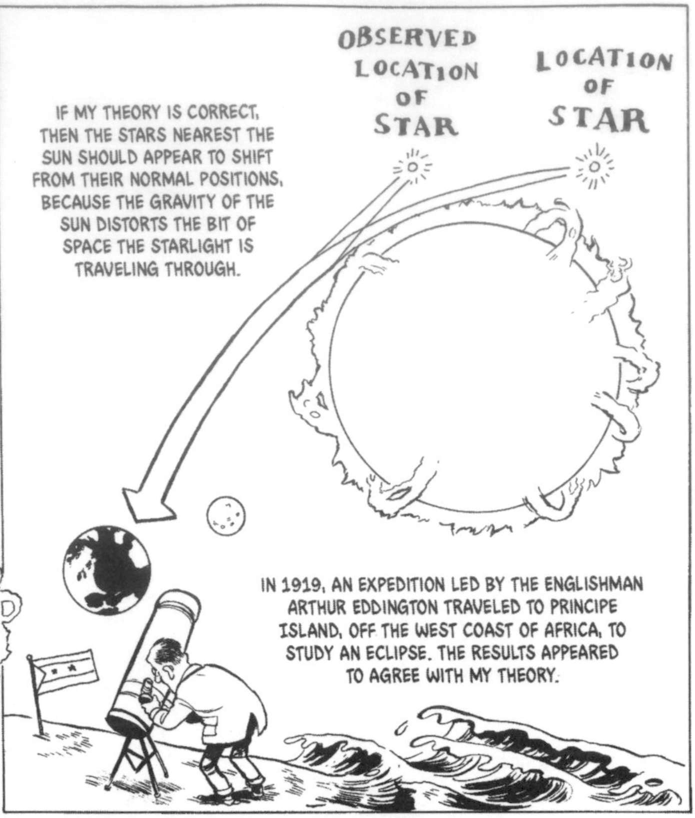
IF MY THEORY IS CORRECT, THEN THE STARS NEAREST THE SUN SHOULD APPEAR TO SHIFT FROM THEIR NORMAL POSITIONS, BECAUSE THE GRAVITY OF THE SUN DISTORTS THE BIT OF SPACE THE STARLIGHT IS TRAVELING THROUGH.
OBSERVED LOCATION OF STAR
LOCATION OF STAR
IN 1919, AN EXPEDITION LED BY THE ENGLISHMAN ARTHUR EDDINGTON TRAVELED TO PRINCIPE ISLAND, OFF THE WEST COAST OF AFRICA, TO STUDY AN ECLIPSE. THE RESULTS APPEARED TO AGREE WITH MY THEORY.

IN GENERAL RELATIVITY, I DISCOVERED IT'S NOT JUST SPEED THAT SLOWS DOWN TIME, GRAVITY CAN, TOO. BUT TO GET BIG EFFECTS, YOU NEED A LOT OF GRAVITY. REMEMBER BLACK HOLES?

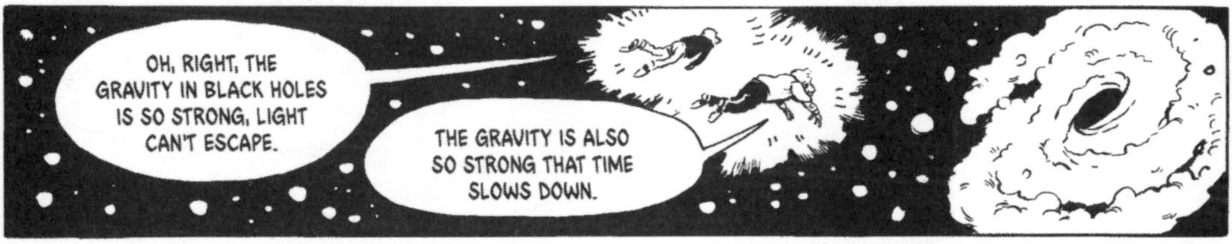
OH, RIGHT, THE GRAVITY IN BLACK HOLES IS SO STRONG, LIGHT CAN'T ESCAPE.
THE GRAVITY IS ALSO SO STRONG THAT TIME SLOWS DOWN.

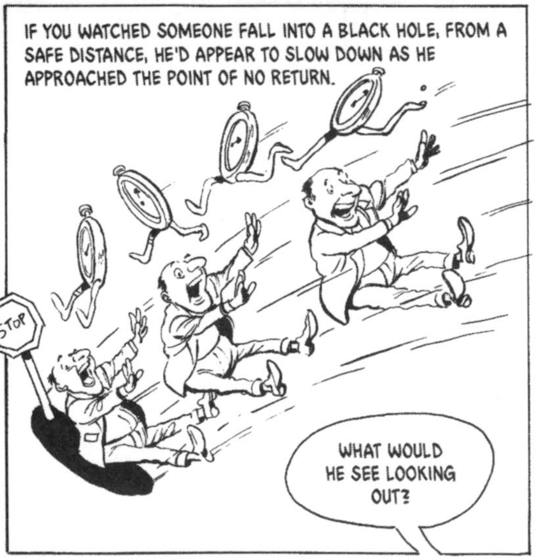
IF YOU WATCHED SOMEONE FALL INTO A BLACK HOLE, FROM A SAFE DISTANCE, HE'D APPEAR TO SLOW DOWN AS HE APPROACHED THE POINT OF NO RETURN.
STOP
WHAT WOULD HE SEE LOOKING OUT?

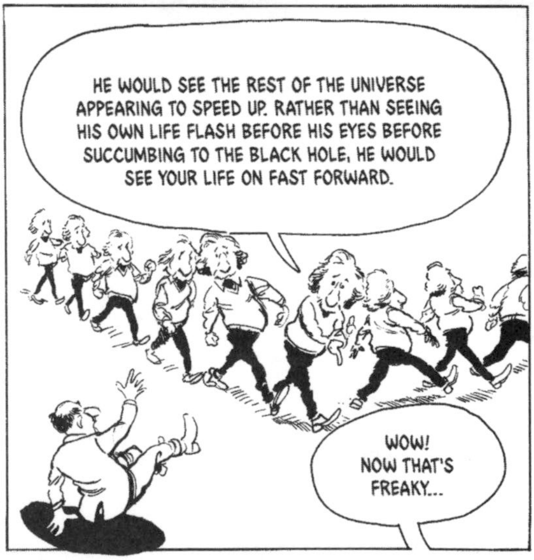
HE WOULD SEE THE REST OF THE UNIVERSE APPEARING TO SPEED UP. RATHER THAN SEEING HIS OWN LIFE FLASH BEFORE HIS EYES BEFORE SUCCUMBING TO THE BLACK HOLE, HE WOULD SEE YOUR LIFE ON FAST FORWARD.
WOW! NOW THAT'S FREAKY...

CHAPTER 20:
LIFE IN THE GALACTIC SUBURBS AND WHY NEWS TAKES SO LONG TO GET THERE

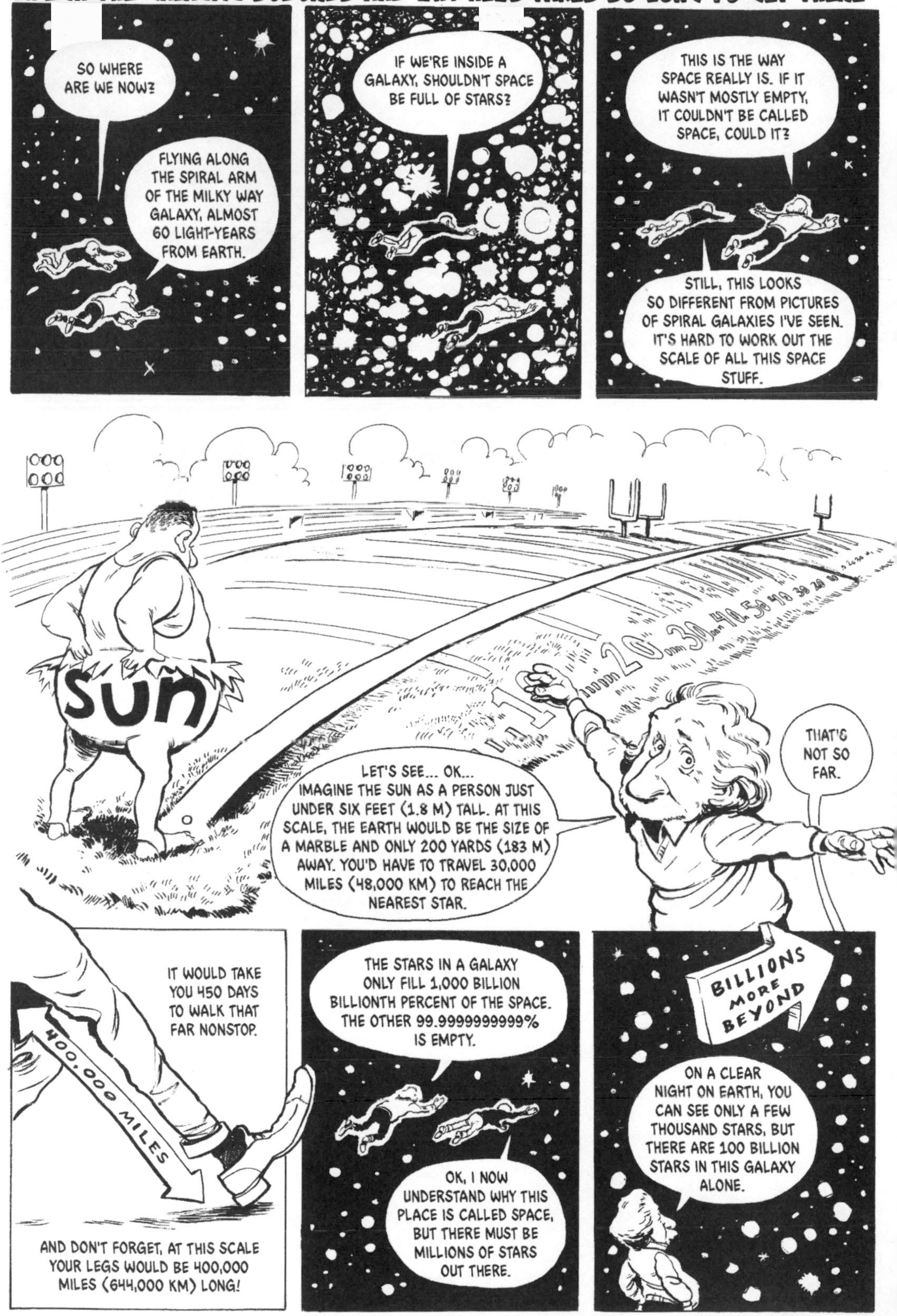
SO WHERE ARE WE NOW?
FLYING ALONG THE SPIRAL ARM OF THE MILKY WAY GALAXY, ALMOST 60 LIGHT-YEARS FROM EARTH.
IF WE'RE INSIDE A GALAXY, SHOULDN'T SPACE BE FULL OF STARS?
THIS IS THE WAY SPACE REALLY IS. IF IT WASN'T MOSTLY EMPTY, IT COULDN'T BE CALLED SPACE, COULD IT?
STILL, THIS LOOKS SO DIFFERENT FROM PICTURES OF SPIRAL GALAXIES I'VE SEEN. IT'S HARD TO WORK OUT THE SCALE OF ALL THIS SPACE STUFF.
SUN
LET'S SEE... OK... IMAGINE THE SUN AS A PERSON JUST UNDER SIX FEET (1.8 M) TALL. AT THIS SCALE, THE EARTH WOULD BE THE SIZE OF A MARBLE AND ONLY 200 YARDS (183 M) AWAY. YOU'D HAVE TO TRAVEL 30,000 MILES (48,000 KM) TO REACH THE NEAREST STAR.
THAT'S NOT SO FAR.
IT WOULD TAKE YOU 450 DAYS TO WALK THAT FAR NONSTOP.
400,000 MILES
AND DON'T FORGET, AT THIS SCALE YOUR LEGS WOULD BE 400,000 MILES (644,000 KM) LONG!
THE STARS IN A GALAXY ONLY FILL 1,000 BILLION BILLIONTH PERCENT OF THE SPACE. THE OTHER 99.9999999999% IS EMPTY.
OK, I NOW UNDERSTAND WHY THIS PLACE IS CALLED SPACE, BUT THERE MUST BE MILLIONS OF STARS OUT THERE.
BILLIONS MORE BEYOND
ON A CLEAR NIGHT ON EARTH, YOU CAN SEE ONLY A FEW THOUSAND STARS, BUT THERE ARE 100 BILLION STARS IN THIS GALAXY ALONE.

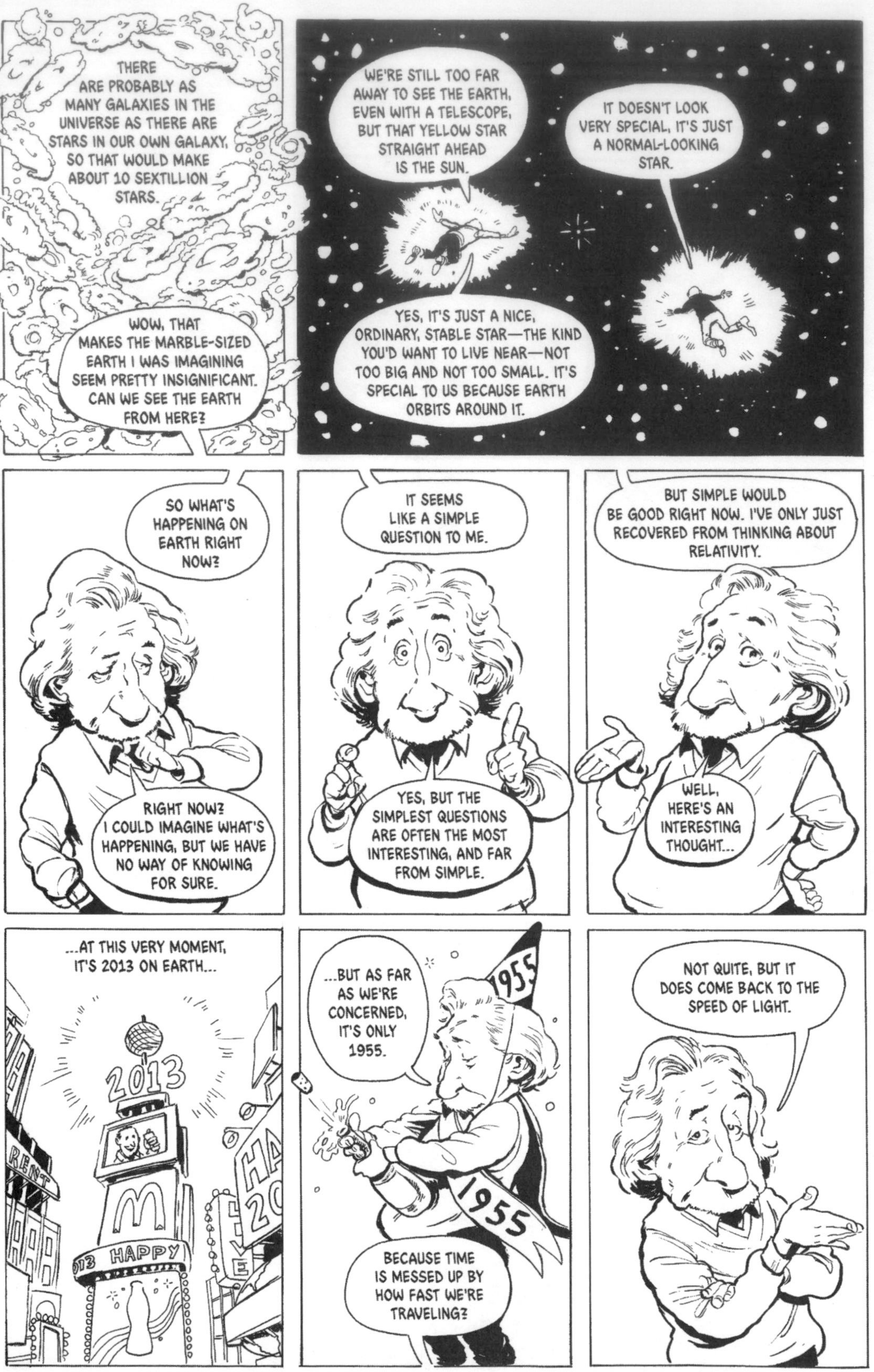
THERE ARE PROBABLY AS MANY GALAXIES IN THE UNIVERSE AS THERE ARE STARS IN OUR OWN GALAXY, SO THAT WOULD MAKE ABOUT 10 SEXTILLION STARS.
WOW, THAT MAKES THE MARBLE-SIZED EARTH I WAS IMAGINING SEEM PRETTY INSIGNIFICANT. CAN WE SEE THE EARTH FROM HERE?
WE'RE STILL TOO FAR AWAY TO SEE THE EARTH, EVEN WITH A TELESCOPE, BUT THAT YELLOW STAR STRAIGHT AHEAD IS THE SUN.
IT DOESN'T LOOK VERY SPECIAL, IT'S JUST A NORMAL-LOOKING STAR.
YES, IT'S JUST A NICE, ORDINARY, STABLE STAR—THE KIND YOU'D WANT TO LIVE NEAR—NOT TOO BIG AND NOT TOO SMALL. IT'S SPECIAL TO US BECAUSE EARTH ORBITS AROUND IT.
SO WHAT'S HAPPENING ON EARTH RIGHT NOW?
RIGHT NOW? I COULD IMAGINE WHAT'S HAPPENING, BUT WE HAVE NO WAY OF KNOWING FOR SURE.
IT SEEMS LIKE A SIMPLE QUESTION TO ME.
YES, BUT THE SIMPLEST QUESTIONS ARE OFTEN THE MOST INTERESTING, AND FAR FROM SIMPLE.
BUT SIMPLE WOULD BE GOOD RIGHT NOW. I'VE ONLY JUST RECOVERED FROM THINKING ABOUT RELATIVITY.
WELL, HERE'S AN INTERESTING THOUGHT...
...AT THIS VERY MOMENT, IT'S 2013 ON EARTH...
2013
RENT
HAPPY
...BUT AS FAR AS WE'RE CONCERNED, IT'S ONLY 1955.
1955
1955
BECAUSE TIME IS MESSED UP BY HOW FAST WE'RE TRAVELING?
NOT QUITE, BUT IT DOES COME BACK TO THE SPEED OF LIGHT.

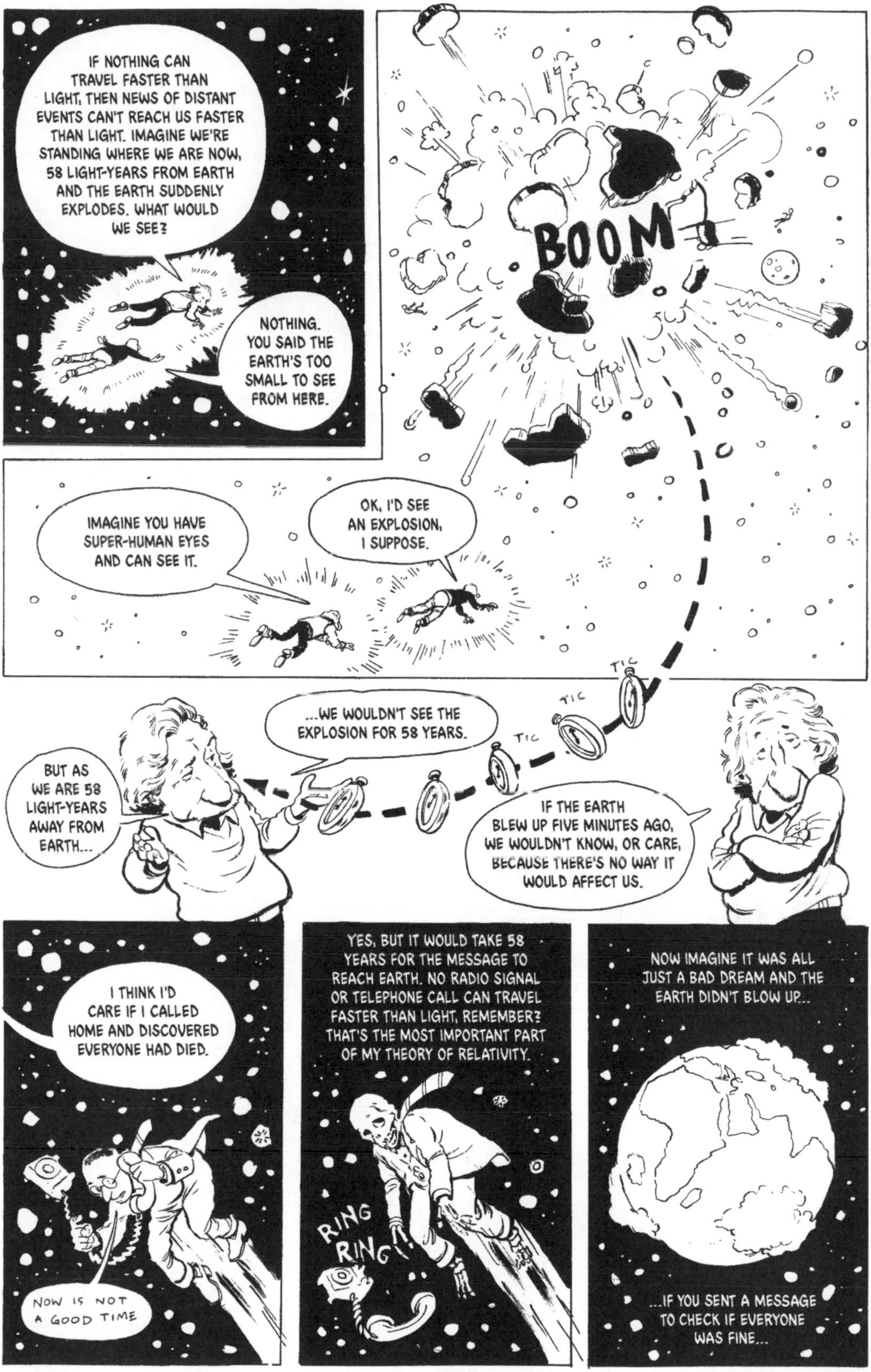
IF NOTHING CAN TRAVEL FASTER THAN LIGHT, THEN NEWS OF DISTANT EVENTS CAN'T REACH US FASTER THAN LIGHT. IMAGINE WE'RE STANDING WHERE WE ARE NOW, 58 LIGHT-YEARS FROM EARTH AND THE EARTH SUDDENLY EXPLODES. WHAT WOULD WE SEE?
NOTHING. YOU SAID THE EARTH'S TOO SMALL TO SEE FROM HERE.
BOOM
IMAGINE YOU HAVE SUPER-HUMAN EYES AND CAN SEE IT.
OK, I'D SEE AN EXPLOSION, I SUPPOSE.
TIC
TIC
TIC
BUT AS WE ARE 58 LIGHT-YEARS AWAY FROM EARTH...
...WE WOULDN'T SEE THE EXPLOSION FOR 58 YEARS.
IF THE EARTH BLEW UP FIVE MINUTES AGO, WE WOULDN'T KNOW, OR CARE, BECAUSE THERE'S NO WAY IT WOULD AFFECT US.
I THINK I'D CARE IF I CALLED HOME AND DISCOVERED EVERYONE HAD DIED.
NOW IS NOT A GOOD TIME
YES, BUT IT WOULD TAKE 58 YEARS FOR THE MESSAGE TO REACH EARTH. NO RADIO SIGNAL OR TELEPHONE CALL CAN TRAVEL FASTER THAN LIGHT, REMEMBER? THAT'S THE MOST IMPORTANT PART OF MY THEORY OF RELATIVITY.
RING RING
NOW IMAGINE IT WAS ALL JUST A BAD DREAM AND THE EARTH DIDN'T BLOW UP...
...IF YOU SENT A MESSAGE TO CHECK IF EVERYONE WAS FINE...

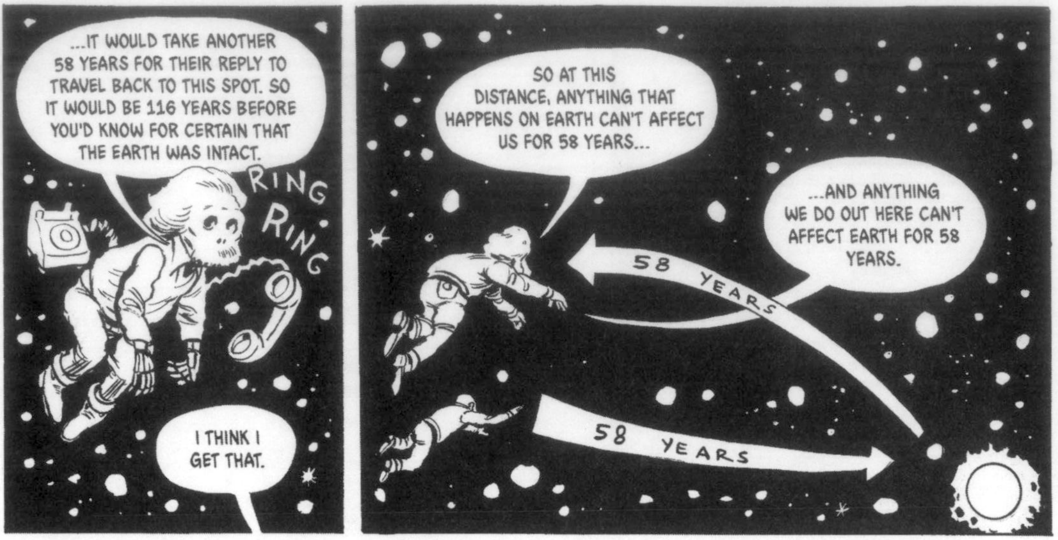
...IT WOULD TAKE ANOTHER 58 YEARS FOR THEIR REPLY TO TRAVEL BACK TO THIS SPOT. SO IT WOULD BE 116 YEARS BEFORE YOU'D KNOW FOR CERTAIN THAT THE EARTH WAS INTACT.
RING RING
I THINK I GET THAT.
SO AT THIS DISTANCE, ANYTHING THAT HAPPENS ON EARTH CAN'T AFFECT US FOR 58 YEARS...
...AND ANYTHING WE DO OUT HERE CAN'T AFFECT EARTH FOR 58 YEARS.
58 YEARS
58 YEARS

WHEN ANYTHING BIG HAPPENS, THE POSSIBLE CONSEQUENCES START SPREADING LIKE RIPPLES THROUGH SPACE AT THE SPEED OF LIGHT. UNTIL LIGHT, OR ANY SIGNAL, FROM THAT EVENT REACHES A CERTAIN POINT IN TIME AND SPACE, IT'S AS IF IT NEVER HAPPENED.

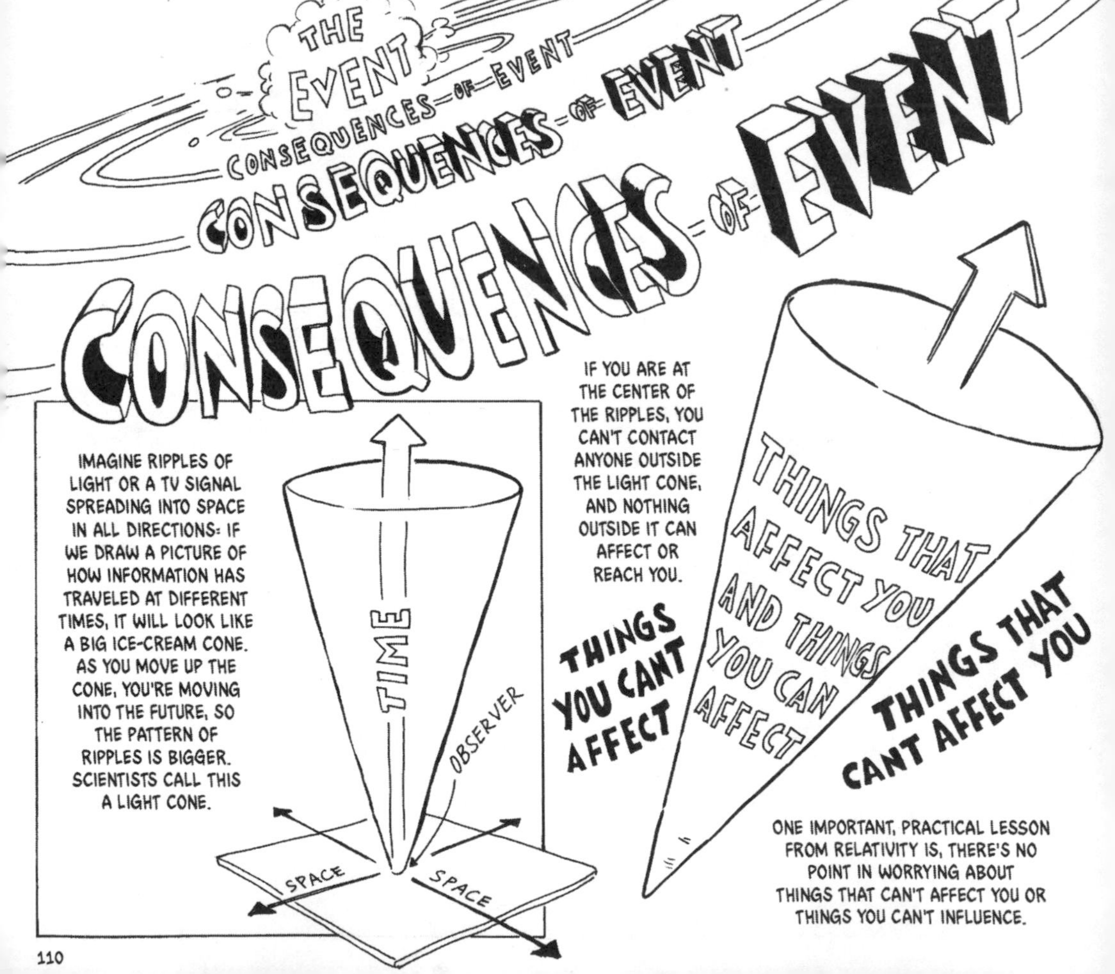
THE EVENT
CONSEQUENCES OF EVENT
CONSEQUENCES OF EVENT
CONSEQUENCES OF EVENT
CONSEQUENCES OF EVENT
CONSEQUENCES
IMAGINE RIPPLES OF LIGHT OR A TV SIGNAL SPREADING INTO SPACE IN ALL DIRECTIONS: IF WE DRAW A PICTURE OF HOW INFORMATION HAS TRAVELED AT DIFFERENT TIMES, IT WILL LOOK LIKE A BIG ICE-CREAM CONE. AS YOU MOVE UP THE CONE, YOU'RE MOVING INTO THE FUTURE, SO THE PATTERN OF RIPPLES IS BIGGER. SCIENTISTS CALL THIS A LIGHT CONE.
TIME
OBSERVER
SPACE
SPACE
THINGS YOU CANT AFFECT
THINGS THAT AFFECT YOU AND THINGS YOU CAN AFFECT
THINGS THAT CANT AFFECT YOU

IF YOU ARE AT THE CENTER OF THE RIPPLES, YOU CAN'T CONTACT ANYONE OUTSIDE THE LIGHT CONE, AND NOTHING OUTSIDE IT CAN AFFECT OR REACH YOU.

ONE IMPORTANT, PRACTICAL LESSON FROM RELATIVITY IS, THERE'S NO POINT IN WORRYING ABOUT THINGS THAT CAN'T AFFECT YOU OR THINGS YOU CAN'T INFLUENCE.

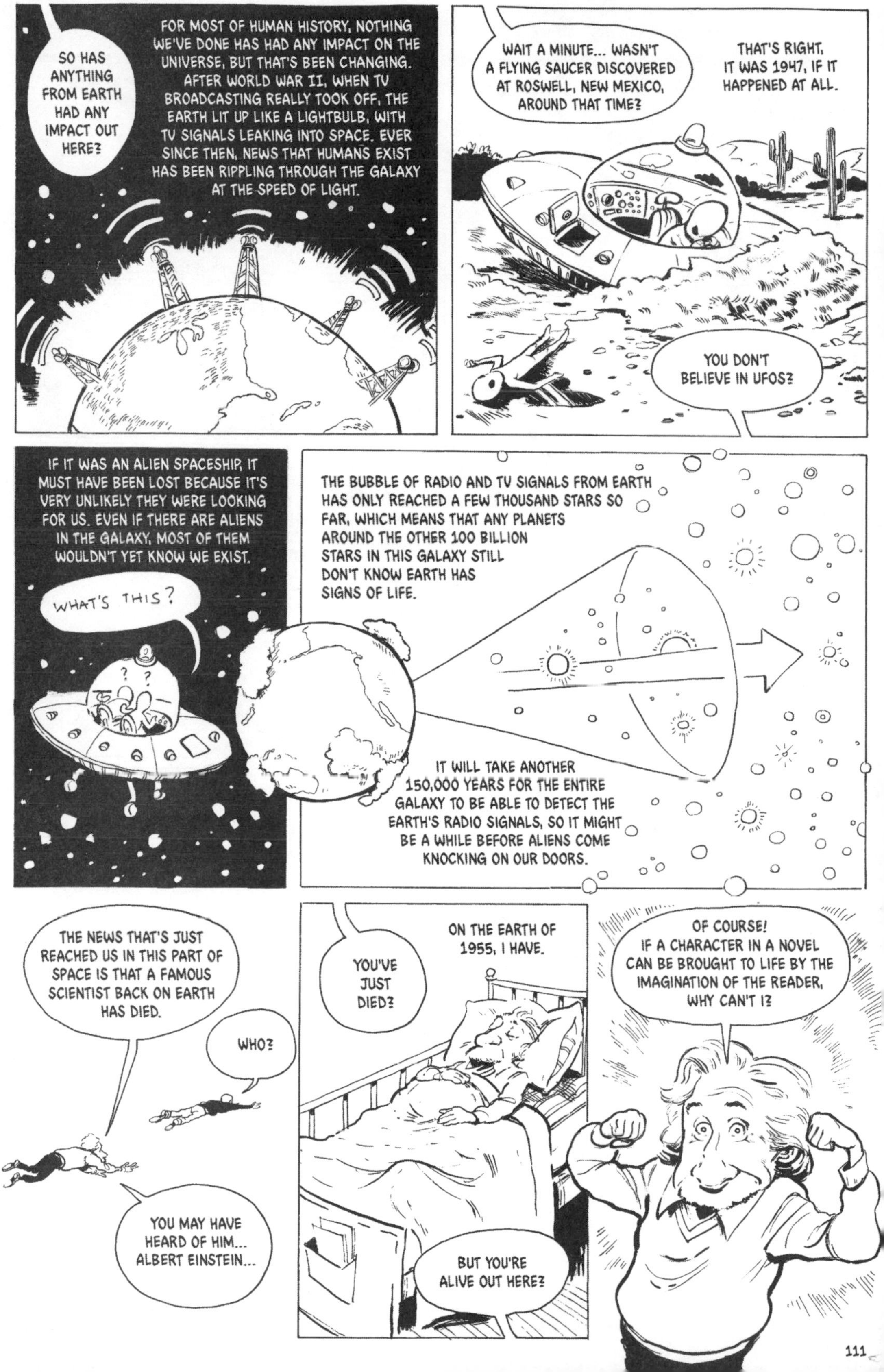
SO HAS ANYTHING FROM EARTH HAD ANY IMPACT OUT HERE?
FOR MOST OF HUMAN HISTORY, NOTHING WE'VE DONE HAS HAD ANY IMPACT ON THE UNIVERSE, BUT THAT'S BEEN CHANGING. AFTER WORLD WAR II, WHEN TV BROADCASTING REALLY TOOK OFF, THE EARTH LIT UP LIKE A LIGHTBULB, WITH TV SIGNALS LEAKING INTO SPACE. EVER SINCE THEN, NEWS THAT HUMANS EXIST HAS BEEN RIPPLING THROUGH THE GALAXY AT THE SPEED OF LIGHT.
WAIT A MINUTE... WASN'T A FLYING SAUCER DISCOVERED AT ROSWELL, NEW MEXICO, AROUND THAT TIME?
THAT'S RIGHT, IT WAS 1947, IF IT HAPPENED AT ALL.
YOU DON'T BELIEVE IN UFOS?
IF IT WAS AN ALIEN SPACESHIP, IT MUST HAVE BEEN LOST BECAUSE IT'S VERY UNLIKELY THEY WERE LOOKING FOR US. EVEN IF THERE ARE ALIENS IN THE GALAXY, MOST OF THEM WOULDN'T YET KNOW WE EXIST.
WHAT'S THIS?
THE BUBBLE OF RADIO AND TV SIGNALS FROM EARTH HAS ONLY REACHED A FEW THOUSAND STARS SO FAR, WHICH MEANS THAT ANY PLANETS AROUND THE OTHER 100 BILLION STARS IN THIS GALAXY STILL DON'T KNOW EARTH HAS SIGNS OF LIFE.
IT WILL TAKE ANOTHER 150,000 YEARS FOR THE ENTIRE GALAXY TO BE ABLE TO DETECT THE EARTH'S RADIO SIGNALS, SO IT MIGHT BE A WHILE BEFORE ALIENS COME KNOCKING ON OUR DOORS.
THE NEWS THAT'S JUST REACHED US IN THIS PART OF SPACE IS THAT A FAMOUS SCIENTIST BACK ON EARTH HAS DIED.
WHO?
YOU MAY HAVE HEARD OF HIM... ALBERT EINSTEIN...
YOU'VE JUST DIED?
ON THE EARTH OF 1955, I HAVE.
BUT YOU'RE ALIVE OUT HERE?
OF COURSE! IF A CHARACTER IN A NOVEL CAN BE BROUGHT TO LIFE BY THE IMAGINATION OF THE READER, WHY CAN'T I?

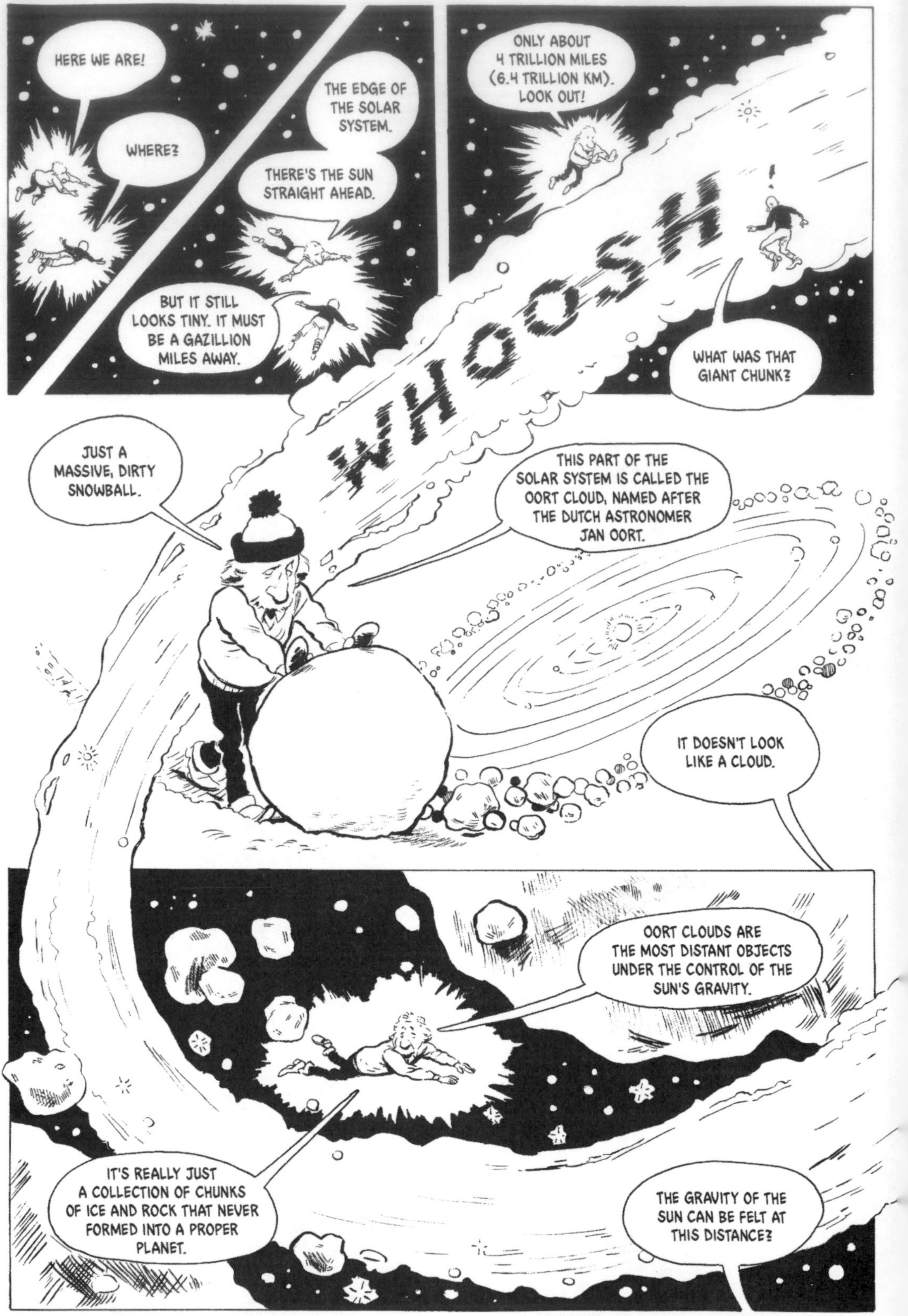
HERE WE ARE!
WHERE?
THE EDGE OF THE SOLAR SYSTEM.
THERE'S THE SUN STRAIGHT AHEAD.
BUT IT STILL LOOKS TINY. IT MUST BE A GAZILLION MILES AWAY.
ONLY ABOUT 4 TRILLION MILES (6.4 TRILLION KM). LOOK OUT!
WHOOOSH!
WHAT WAS THAT GIANT CHUNK?
JUST A MASSIVE, DIRTY SNOWBALL.
THIS PART OF THE SOLAR SYSTEM IS CALLED THE OORT CLOUD, NAMED AFTER THE DUTCH ASTRONOMER JAN OORT.
IT DOESN'T LOOK LIKE A CLOUD.
OORT CLOUDS ARE THE MOST DISTANT OBJECTS UNDER THE CONTROL OF THE SUN'S GRAVITY.
IT'S REALLY JUST A COLLECTION OF CHUNKS OF ICE AND ROCK THAT NEVER FORMED INTO A PROPER PLANET.
THE GRAVITY OF THE SUN CAN BE FELT AT THIS DISTANCE?

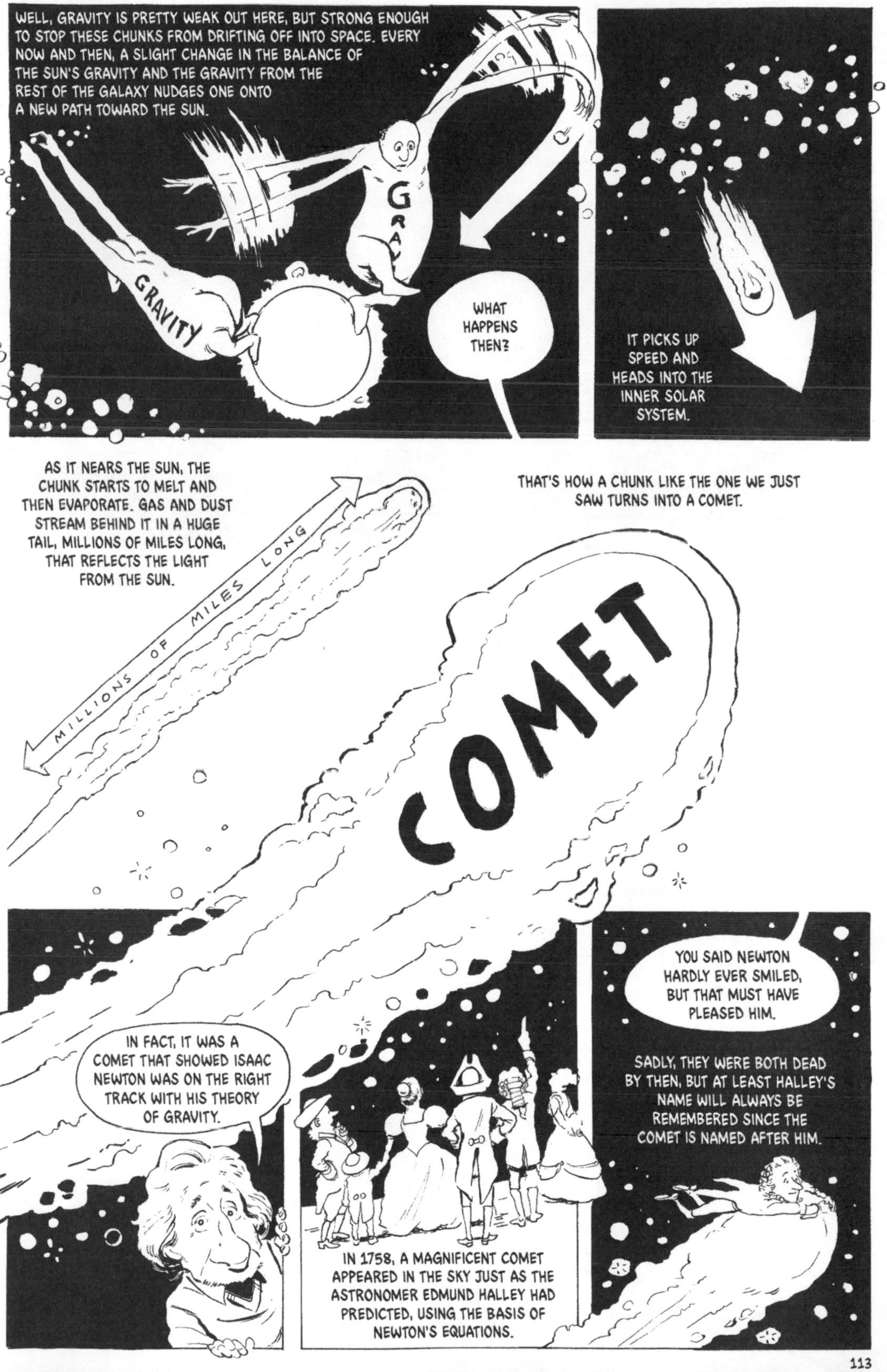
WELL, GRAVITY IS PRETTY WEAK OUT HERE, BUT STRONG ENOUGH TO STOP THESE CHUNKS FROM DRIFTING OFF INTO SPACE. EVERY NOW AND THEN, A SLIGHT CHANGE IN THE BALANCE OF THE SUN'S GRAVITY AND THE GRAVITY FROM THE REST OF THE GALAXY NUDGES ONE ONTO A NEW PATH TOWARD THE SUN.
GRAVITY
GRAV
WHAT HAPPENS THEN?
IT PICKS UP SPEED AND HEADS INTO THE INNER SOLAR SYSTEM.
AS IT NEARS THE SUN, THE CHUNK STARTS TO MELT AND THEN EVAPORATE. GAS AND DUST STREAM BEHIND IT IN A HUGE TAIL, MILLIONS OF MILES LONG, THAT REFLECTS THE LIGHT FROM THE SUN.
MILLIONS OF MILES LONG
THAT'S HOW A CHUNK LIKE THE ONE WE JUST SAW TURNS INTO A COMET.
COMET
IN FACT, IT WAS A COMET THAT SHOWED ISAAC NEWTON WAS ON THE RIGHT TRACK WITH HIS THEORY OF GRAVITY.
IN 1758, A MAGNIFICENT COMET APPEARED IN THE SKY JUST AS THE ASTRONOMER EDMUND HALLEY HAD PREDICTED, USING THE BASIS OF NEWTON'S EQUATIONS.
YOU SAID NEWTON HARDLY EVER SMILED, BUT THAT MUST HAVE PLEASED HIM.
SADLY, THEY WERE BOTH DEAD BY THEN, BUT AT LEAST HALLEY'S NAME WILL ALWAYS BE REMEMBERED SINCE THE COMET IS NAMED AFTER HIM.

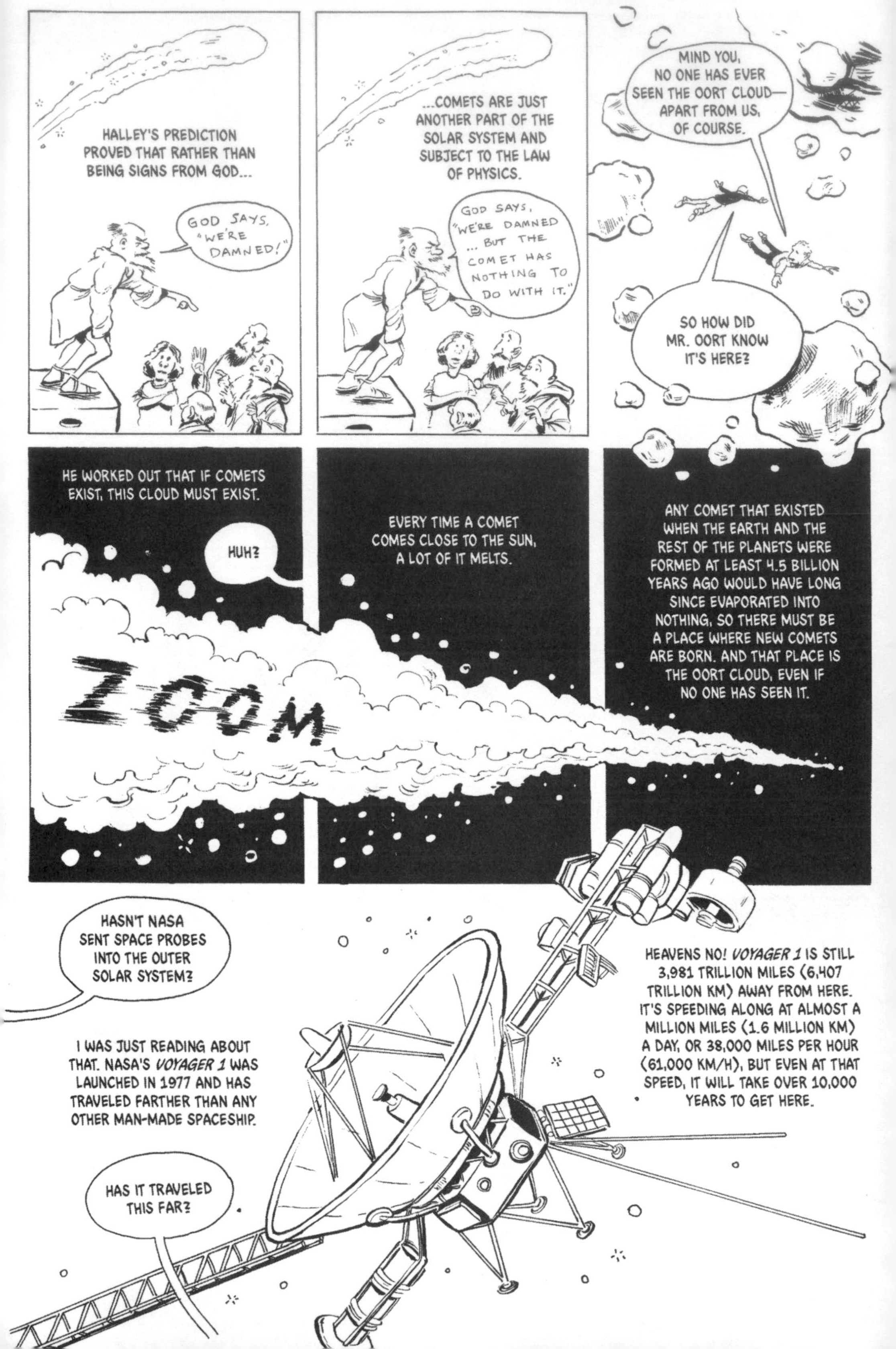
HALLEY'S PREDICTION PROVED THAT RATHER THAN BEING SIGNS FROM GOD...
GOD SAYS, "WE'RE DAMNED!"
...COMETS ARE JUST ANOTHER PART OF THE SOLAR SYSTEM AND SUBJECT TO THE LAW OF PHYSICS.
GOD SAYS, "WE'RE DAMNED ... BUT THE COMET HAS NOTHING TO DO WITH IT."
MIND YOU, NO ONE HAS EVER SEEN THE OORT CLOUD— APART FROM US, OF COURSE.
SO HOW DID MR. OORT KNOW IT'S HERE?
HE WORKED OUT THAT IF COMETS EXIST, THIS CLOUD MUST EXIST.
HUH?
ZOOM
EVERY TIME A COMET COMES CLOSE TO THE SUN, A LOT OF IT MELTS.
ANY COMET THAT EXISTED WHEN THE EARTH AND THE REST OF THE PLANETS WERE FORMED AT LEAST 4.5 BILLION YEARS AGO WOULD HAVE LONG SINCE EVAPORATED INTO NOTHING, SO THERE MUST BE A PLACE WHERE NEW COMETS ARE BORN. AND THAT PLACE IS THE OORT CLOUD, EVEN IF NO ONE HAS SEEN IT.
HASN'T NASA SENT SPACE PROBES INTO THE OUTER SOLAR SYSTEM?
I WAS JUST READING ABOUT THAT. NASA'S *VOYAGER 1* WAS LAUNCHED IN 1977 AND HAS TRAVELED FARTHER THAN ANY OTHER MAN-MADE SPACESHIP.
HAS IT TRAVELED THIS FAR?
HEAVENS NO! *VOYAGER 1* IS STILL 3,981 TRILLION MILES (6,407 TRILLION KM) AWAY FROM HERE. IT'S SPEEDING ALONG AT ALMOST A MILLION MILES (1.6 MILLION KM) A DAY, OR 38,000 MILES PER HOUR (61,000 KM/H), BUT EVEN AT THAT SPEED, IT WILL TAKE OVER 10,000 YEARS TO GET HERE.

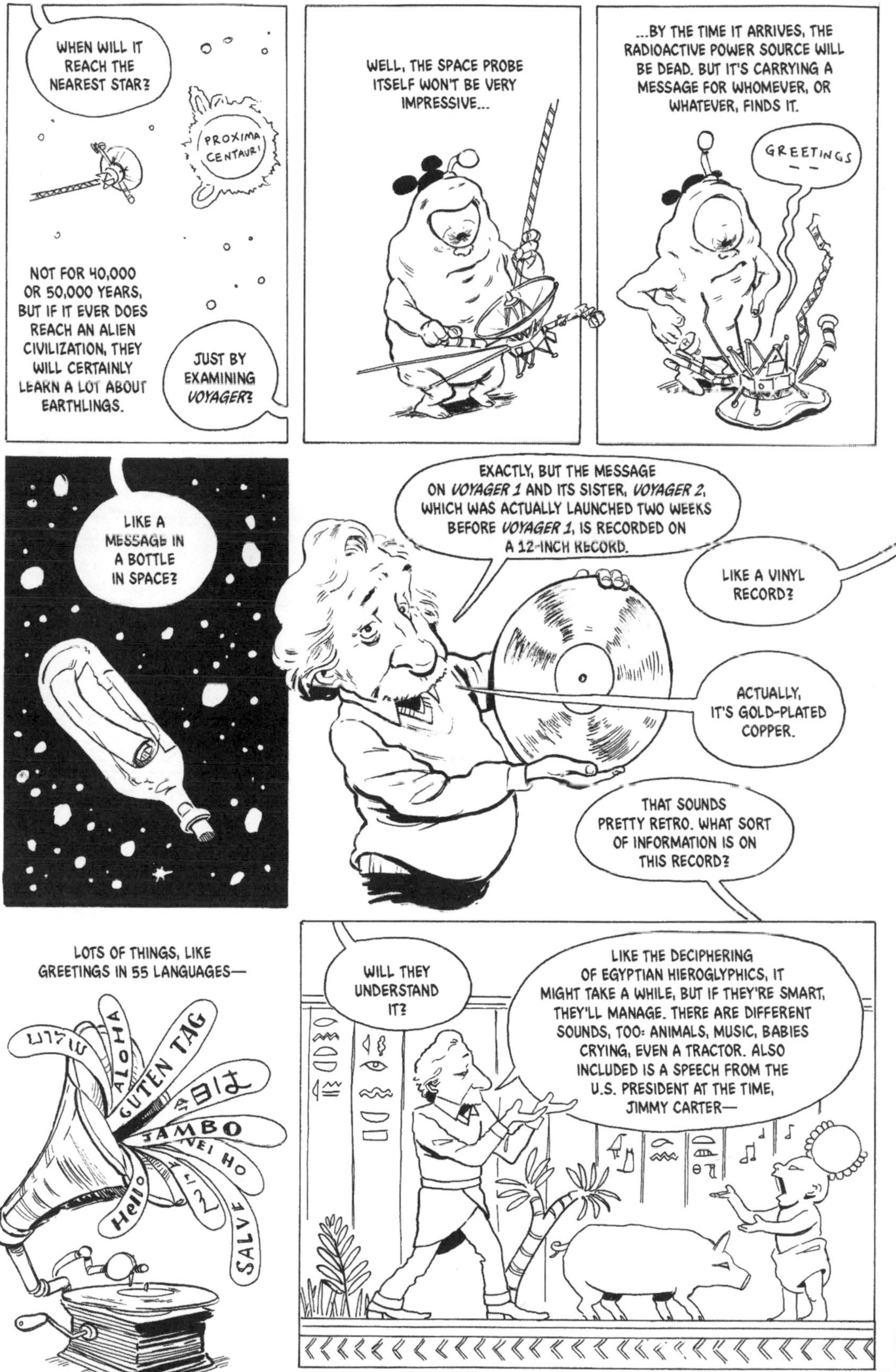
WHEN WILL IT REACH THE NEAREST STAR?
PROXIMA CENTAURI
NOT FOR 40,000 OR 50,000 YEARS, BUT IF IT EVER DOES REACH AN ALIEN CIVILIZATION, THEY WILL CERTAINLY LEARN A LOT ABOUT EARTHLINGS.
JUST BY EXAMINING *VOYAGER*?
WELL, THE SPACE PROBE ITSELF WON'T BE VERY IMPRESSIVE...
...BY THE TIME IT ARRIVES, THE RADIOACTIVE POWER SOURCE WILL BE DEAD. BUT IT'S CARRYING A MESSAGE FOR WHOMEVER, OR WHATEVER, FINDS IT.
GREETINGS --
LIKE A MESSAGE IN A BOTTLE IN SPACE?
EXACTLY, BUT THE MESSAGE ON *VOYAGER 1* AND ITS SISTER, *VOYAGER 2*, WHICH WAS ACTUALLY LAUNCHED TWO WEEKS BEFORE *VOYAGER 1*, IS RECORDED ON A 12-INCH RECORD.
LIKE A VINYL RECORD?
ACTUALLY, IT'S GOLD-PLATED COPPER.
THAT SOUNDS PRETTY RETRO. WHAT SORT OF INFORMATION IS ON THIS RECORD?
LOTS OF THINGS, LIKE GREETINGS IN 55 LANGUAGES—
ALOHA
GUTEN TAG
今日は
JAMBO
NEI HO
Hello
SALVE
WILL THEY UNDERSTAND IT?
LIKE THE DECIPHERING OF EGYPTIAN HIEROGLYPHICS, IT MIGHT TAKE A WHILE, BUT IF THEY'RE SMART, THEY'LL MANAGE. THERE ARE DIFFERENT SOUNDS, TOO: ANIMALS, MUSIC, BABIES CRYING, EVEN A TRACTOR. ALSO INCLUDED IS A SPEECH FROM THE U.S. PRESIDENT AT THE TIME, JIMMY CARTER—

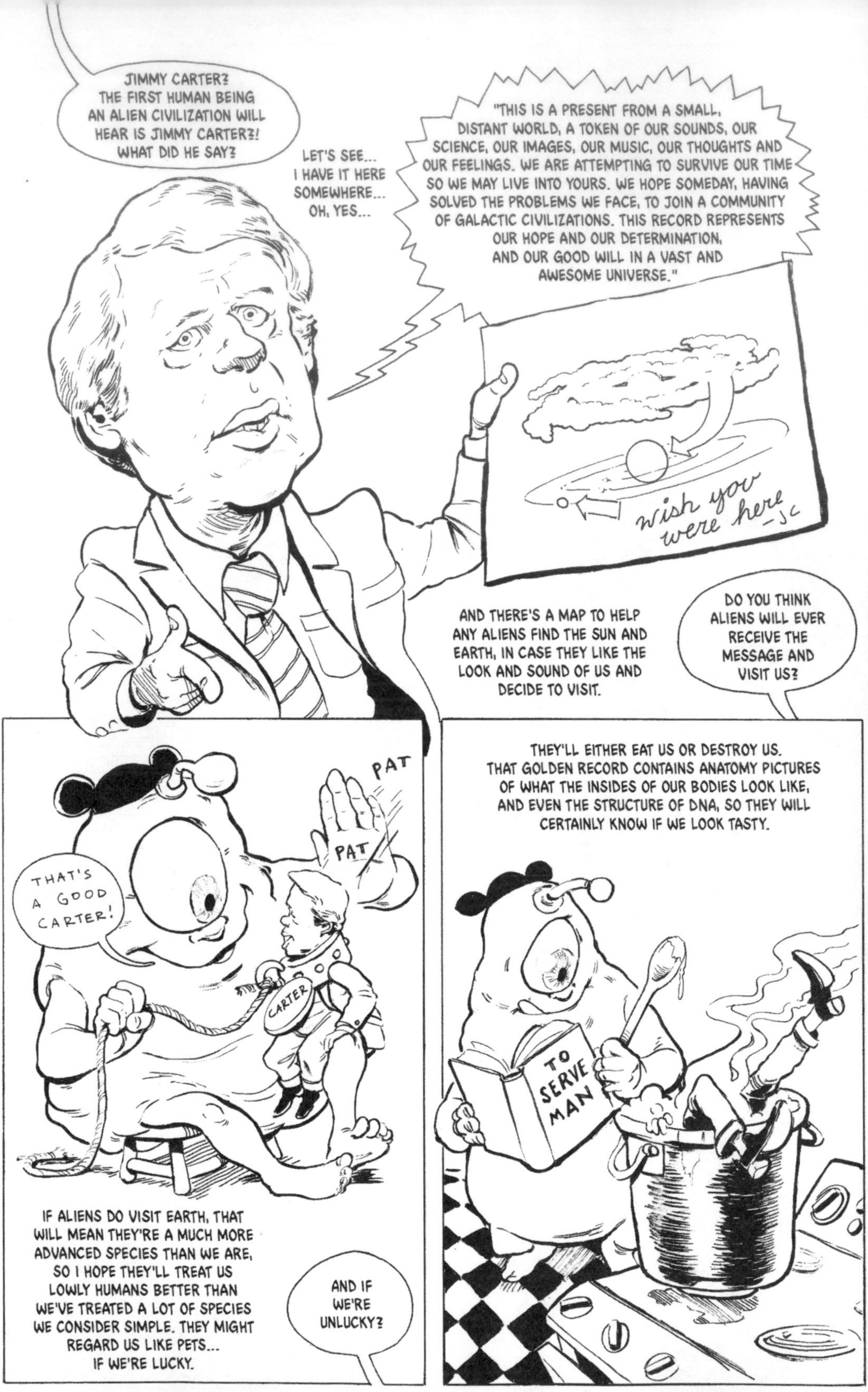
JIMMY CARTER? THE FIRST HUMAN BEING AN ALIEN CIVILIZATION WILL HEAR IS JIMMY CARTER?! WHAT DID HE SAY?
LET'S SEE... I HAVE IT HERE SOMEWHERE... OH, YES...
"THIS IS A PRESENT FROM A SMALL, DISTANT WORLD, A TOKEN OF OUR SOUNDS, OUR SCIENCE, OUR IMAGES, OUR MUSIC, OUR THOUGHTS AND OUR FEELINGS. WE ARE ATTEMPTING TO SURVIVE OUR TIME SO WE MAY LIVE INTO YOURS. WE HOPE SOMEDAY, HAVING SOLVED THE PROBLEMS WE FACE, TO JOIN A COMMUNITY OF GALACTIC CIVILIZATIONS. THIS RECORD REPRESENTS OUR HOPE AND OUR DETERMINATION, AND OUR GOOD WILL IN A VAST AND AWESOME UNIVERSE."
wish you were here -JC
AND THERE'S A MAP TO HELP ANY ALIENS FIND THE SUN AND EARTH, IN CASE THEY LIKE THE LOOK AND SOUND OF US AND DECIDE TO VISIT.
DO YOU THINK ALIENS WILL EVER RECEIVE THE MESSAGE AND VISIT US?
PAT
PAT
THAT'S A GOOD CARTER!
CARTER
IF ALIENS DO VISIT EARTH, THAT WILL MEAN THEY'RE A MUCH MORE ADVANCED SPECIES THAN WE ARE, SO I HOPE THEY'LL TREAT US LOWLY HUMANS BETTER THAN WE'VE TREATED A LOT OF SPECIES WE CONSIDER SIMPLE. THEY MIGHT REGARD US LIKE PETS... IF WE'RE LUCKY.
AND IF WE'RE UNLUCKY?
THEY'LL EITHER EAT US OR DESTROY US. THAT GOLDEN RECORD CONTAINS ANATOMY PICTURES OF WHAT THE INSIDES OF OUR BODIES LOOK LIKE, AND EVEN THE STRUCTURE OF DNA, SO THEY WILL CERTAINLY KNOW IF WE LOOK TASTY.
TO SERVE MAN

CHAPTER 22: TO BE OR NOT TO BE? IS THAT A PLANET I SEE BEFORE ME?

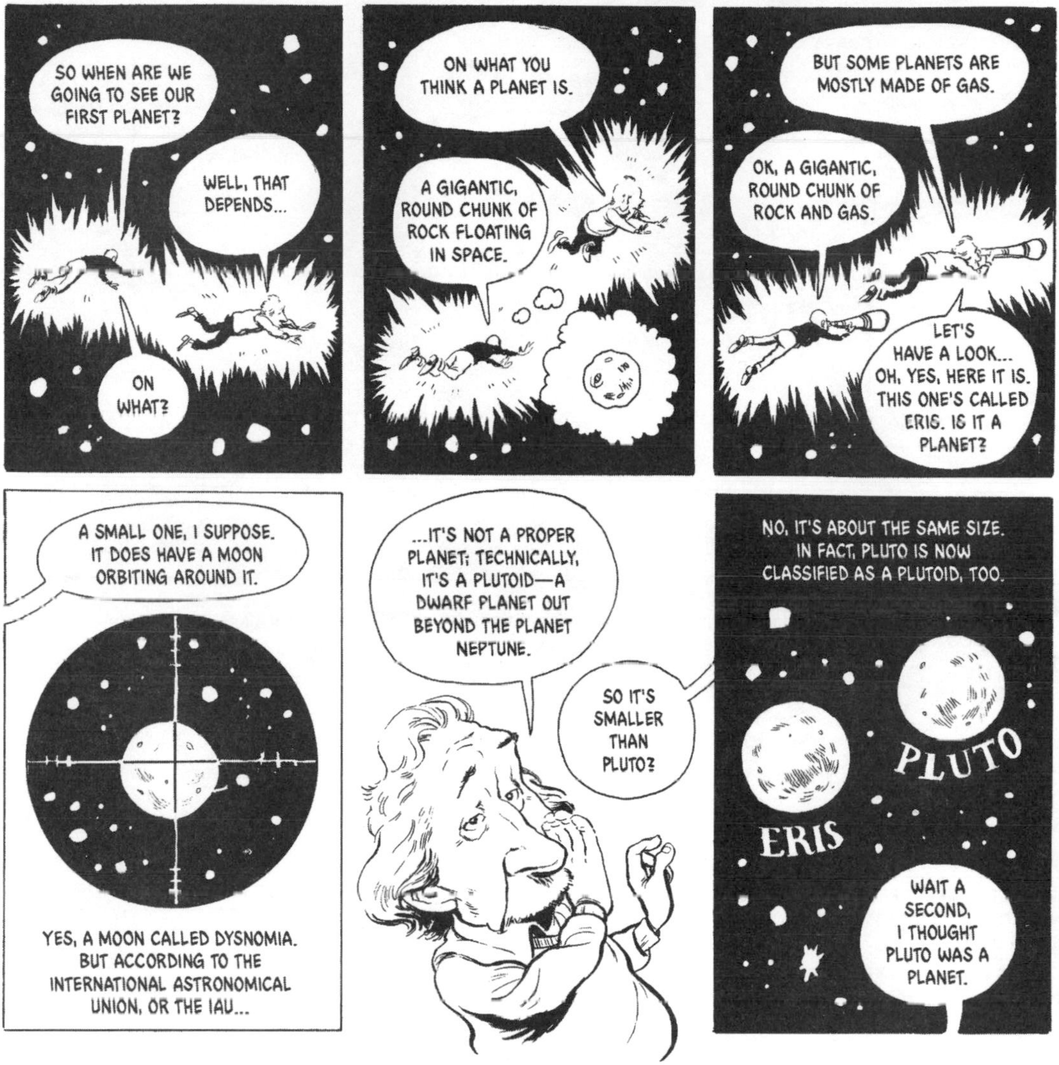
SO WHEN ARE WE GOING TO SEE OUR FIRST PLANET?
WELL, THAT DEPENDS...
ON WHAT?
ON WHAT YOU THINK A PLANET IS.
A GIGANTIC, ROUND CHUNK OF ROCK FLOATING IN SPACE.
BUT SOME PLANETS ARE MOSTLY MADE OF GAS.
OK, A GIGANTIC, ROUND CHUNK OF ROCK AND GAS.
LET'S HAVE A LOOK... OH, YES, HERE IT IS. THIS ONE'S CALLED ERIS. IS IT A PLANET?
A SMALL ONE, I SUPPOSE. IT DOES HAVE A MOON ORBITING AROUND IT.
YES, A MOON CALLED DYSNOMIA. BUT ACCORDING TO THE INTERNATIONAL ASTRONOMICAL UNION, OR THE IAU...
...IT'S NOT A PROPER PLANET; TECHNICALLY, IT'S A PLUTOID—A DWARF PLANET OUT BEYOND THE PLANET NEPTUNE.
SO IT'S SMALLER THAN PLUTO?
NO, IT'S ABOUT THE SAME SIZE. IN FACT, PLUTO IS NOW CLASSIFIED AS A PLUTOID, TOO.
PLUTO
ERIS
WAIT A SECOND, I THOUGHT PLUTO WAS A PLANET.

I REMEMBER HOW EXCITING IT WAS WHEN PLUTO WAS DISCOVERED BACK IN 1930, BUT IN 2006, AFTER ASTRONOMERS STARTED DISCOVERING SMALL PLANET-LIKE OBJECTS SUCH AS ERIS, THE IAU REVISED THE DEFINITION OF A PLANET...

THE IAU'S REVISED DEFINITION OF A PLANET:

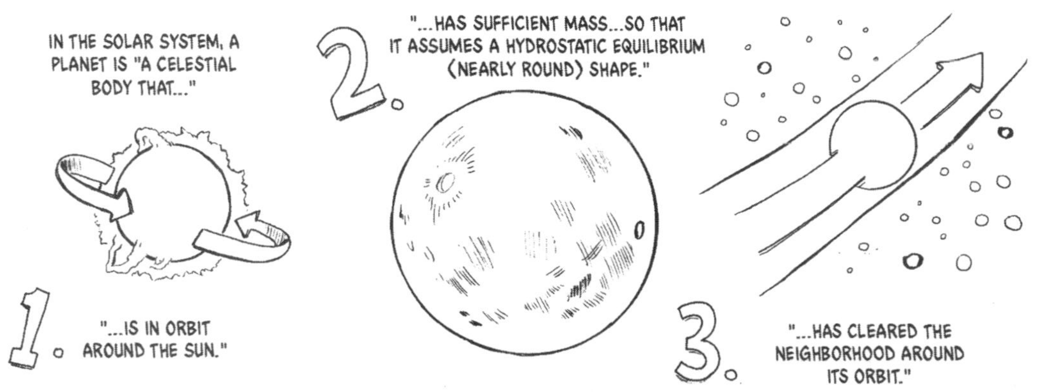
IN THE SOLAR SYSTEM, A PLANET IS "A CELESTIAL BODY THAT..."
1.
"...IS IN ORBIT AROUND THE SUN."
2.
"...HAS SUFFICIENT MASS...SO THAT IT ASSUMES A HYDROSTATIC EQUILIBRIUM (NEARLY ROUND) SHAPE."
3.
"...HAS CLEARED THE NEIGHBORHOOD AROUND ITS ORBIT."

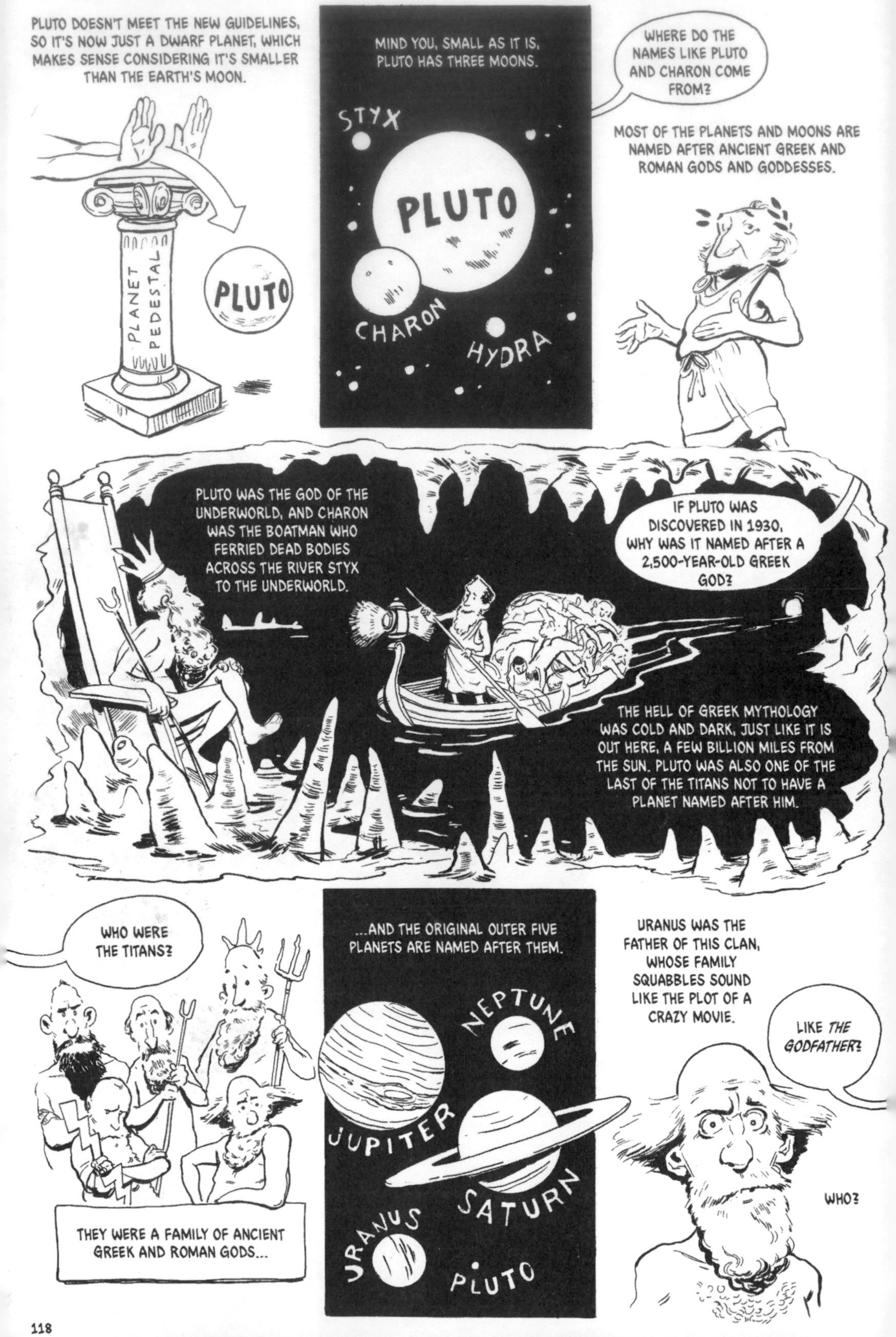

PLUTO DOESN'T MEET THE NEW GUIDELINES, SO IT'S NOW JUST A DWARF PLANET, WHICH MAKES SENSE CONSIDERING IT'S SMALLER THAN THE EARTH'S MOON.
PLANET PEDESTAL
PLUTO
MIND YOU, SMALL AS IT IS, PLUTO HAS THREE MOONS.
STYX
PLUTO
CHARON
HYDRA
WHERE DO THE NAMES LIKE PLUTO AND CHARON COME FROM?
MOST OF THE PLANETS AND MOONS ARE NAMED AFTER ANCIENT GREEK AND ROMAN GODS AND GODDESSES.
PLUTO WAS THE GOD OF THE UNDERWORLD, AND CHARON WAS THE BOATMAN WHO FERRIED DEAD BODIES ACROSS THE RIVER STYX TO THE UNDERWORLD.
IF PLUTO WAS DISCOVERED IN 1930, WHY WAS IT NAMED AFTER A 2,500-YEAR-OLD GREEK GOD?
THE HELL OF GREEK MYTHOLOGY WAS COLD AND DARK, JUST LIKE IT IS OUT HERE, A FEW BILLION MILES FROM THE SUN. PLUTO WAS ALSO ONE OF THE LAST OF THE TITANS NOT TO HAVE A PLANET NAMED AFTER HIM.
WHO WERE THE TITANS?
THEY WERE A FAMILY OF ANCIENT GREEK AND ROMAN GODS...
...AND THE ORIGINAL OUTER FIVE PLANETS ARE NAMED AFTER THEM.
NEPTUNE
JUPITER
SATURN
URANUS
PLUTO
URANUS WAS THE FATHER OF THIS CLAN, WHOSE FAMILY SQUABBLES SOUND LIKE THE PLOT OF A CRAZY MOVIE.
LIKE *THE GODFATHER*?
WHO?

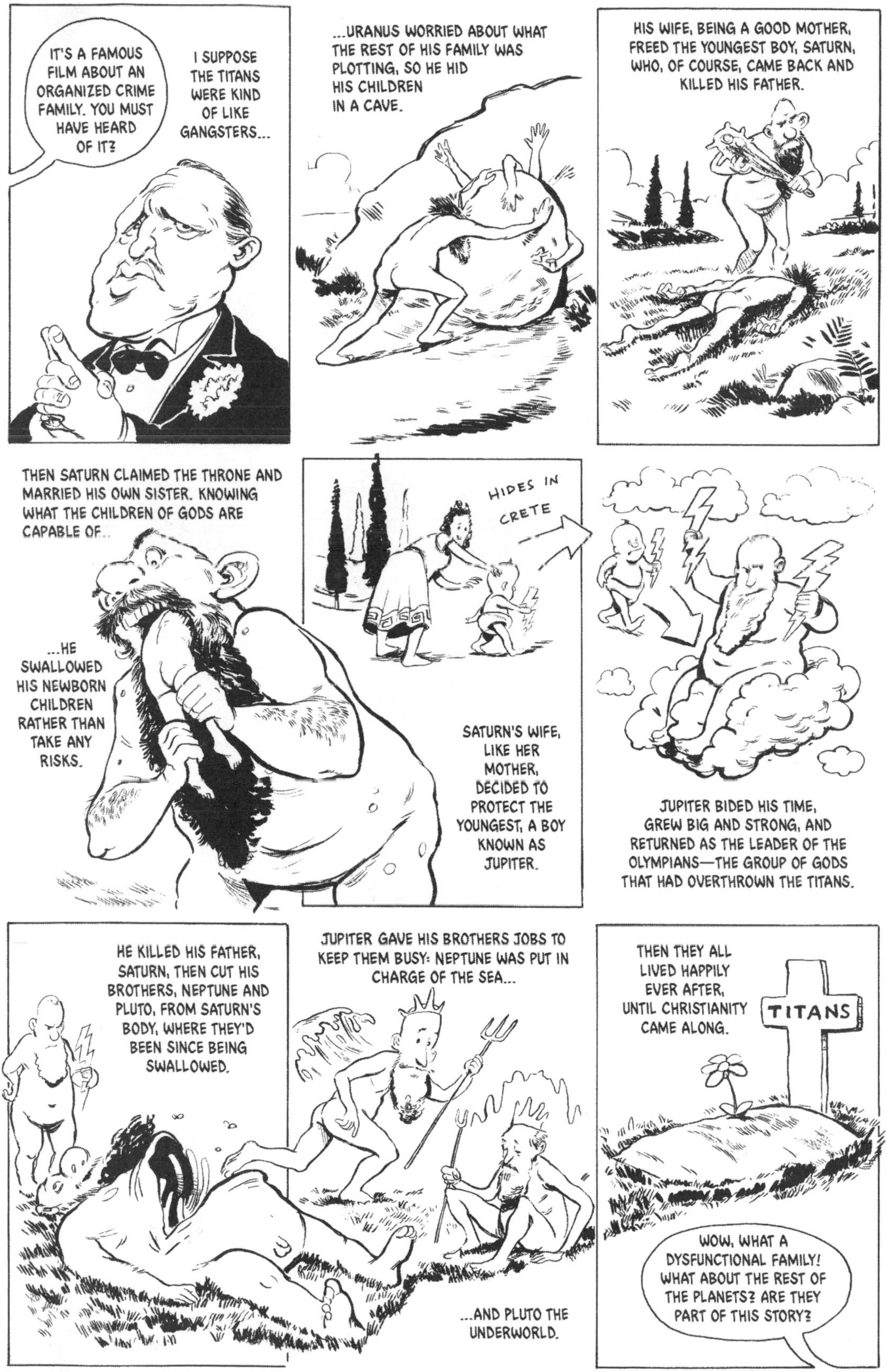
IT'S A FAMOUS FILM ABOUT AN ORGANIZED CRIME FAMILY. YOU MUST HAVE HEARD OF IT?
I SUPPOSE THE TITANS WERE KIND OF LIKE GANGSTERS...
...URANUS WORRIED ABOUT WHAT THE REST OF HIS FAMILY WAS PLOTTING, SO HE HID HIS CHILDREN IN A CAVE.
HIS WIFE, BEING A GOOD MOTHER, FREED THE YOUNGEST BOY, SATURN, WHO, OF COURSE, CAME BACK AND KILLED HIS FATHER.
THEN SATURN CLAIMED THE THRONE AND MARRIED HIS OWN SISTER. KNOWING WHAT THE CHILDREN OF GODS ARE CAPABLE OF...
...HE SWALLOWED HIS NEWBORN CHILDREN RATHER THAN TAKE ANY RISKS.
HIDES IN CRETE
SATURN'S WIFE, LIKE HER MOTHER, DECIDED TO PROTECT THE YOUNGEST, A BOY KNOWN AS JUPITER.
JUPITER BIDED HIS TIME, GREW BIG AND STRONG, AND RETURNED AS THE LEADER OF THE OLYMPIANS—THE GROUP OF GODS THAT HAD OVERTHROWN THE TITANS.
HE KILLED HIS FATHER, SATURN, THEN CUT HIS BROTHERS, NEPTUNE AND PLUTO, FROM SATURN'S BODY, WHERE THEY'D BEEN SINCE BEING SWALLOWED.
JUPITER GAVE HIS BROTHERS JOBS TO KEEP THEM BUSY: NEPTUNE WAS PUT IN CHARGE OF THE SEA...
...AND PLUTO THE UNDERWORLD.
THEN THEY ALL LIVED HAPPILY EVER AFTER, UNTIL CHRISTIANITY CAME ALONG.
TITANS
WOW, WHAT A DYSFUNCTIONAL FAMILY! WHAT ABOUT THE REST OF THE PLANETS? ARE THEY PART OF THIS STORY?

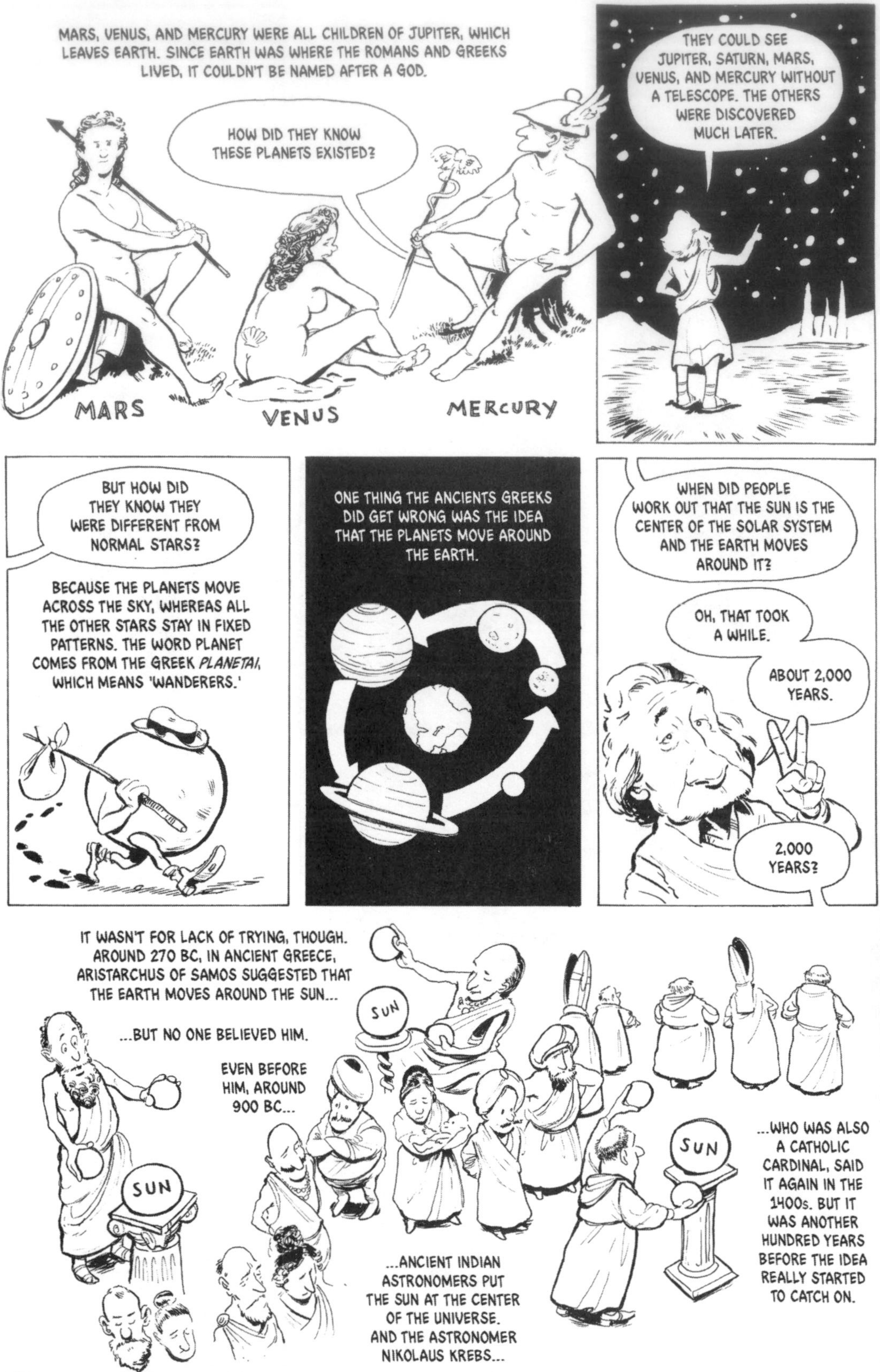
MARS, VENUS, AND MERCURY WERE ALL CHILDREN OF JUPITER, WHICH LEAVES EARTH. SINCE EARTH WAS WHERE THE ROMANS AND GREEKS LIVED, IT COULDN'T BE NAMED AFTER A GOD.
HOW DID THEY KNOW THESE PLANETS EXISTED?
MARS
VENUS
MERCURY
THEY COULD SEE JUPITER, SATURN, MARS, VENUS, AND MERCURY WITHOUT A TELESCOPE. THE OTHERS WERE DISCOVERED MUCH LATER.
BUT HOW DID THEY KNOW THEY WERE DIFFERENT FROM NORMAL STARS?
BECAUSE THE PLANETS MOVE ACROSS THE SKY, WHEREAS ALL THE OTHER STARS STAY IN FIXED PATTERNS. THE WORD PLANET COMES FROM THE GREEK *PLANETAI*, WHICH MEANS 'WANDERERS.'
ONE THING THE ANCIENTS GREEKS DID GET WRONG WAS THE IDEA THAT THE PLANETS MOVE AROUND THE EARTH.
WHEN DID PEOPLE WORK OUT THAT THE SUN IS THE CENTER OF THE SOLAR SYSTEM AND THE EARTH MOVES AROUND IT?
OH, THAT TOOK A WHILE.
ABOUT 2,000 YEARS.
2,000 YEARS?
IT WASN'T FOR LACK OF TRYING, THOUGH. AROUND 270 BC, IN ANCIENT GREECE, ARISTARCHUS OF SAMOS SUGGESTED THAT THE EARTH MOVES AROUND THE SUN...
SUN
...BUT NO ONE BELIEVED HIM.
EVEN BEFORE HIM, AROUND 900 BC...
SUN
...ANCIENT INDIAN ASTRONOMERS PUT THE SUN AT THE CENTER OF THE UNIVERSE. AND THE ASTRONOMER NIKOLAUS KREBS...
SUN
...WHO WAS ALSO A CATHOLIC CARDINAL, SAID IT AGAIN IN THE 1400s. BUT IT WAS ANOTHER HUNDRED YEARS BEFORE THE IDEA REALLY STARTED TO CATCH ON.

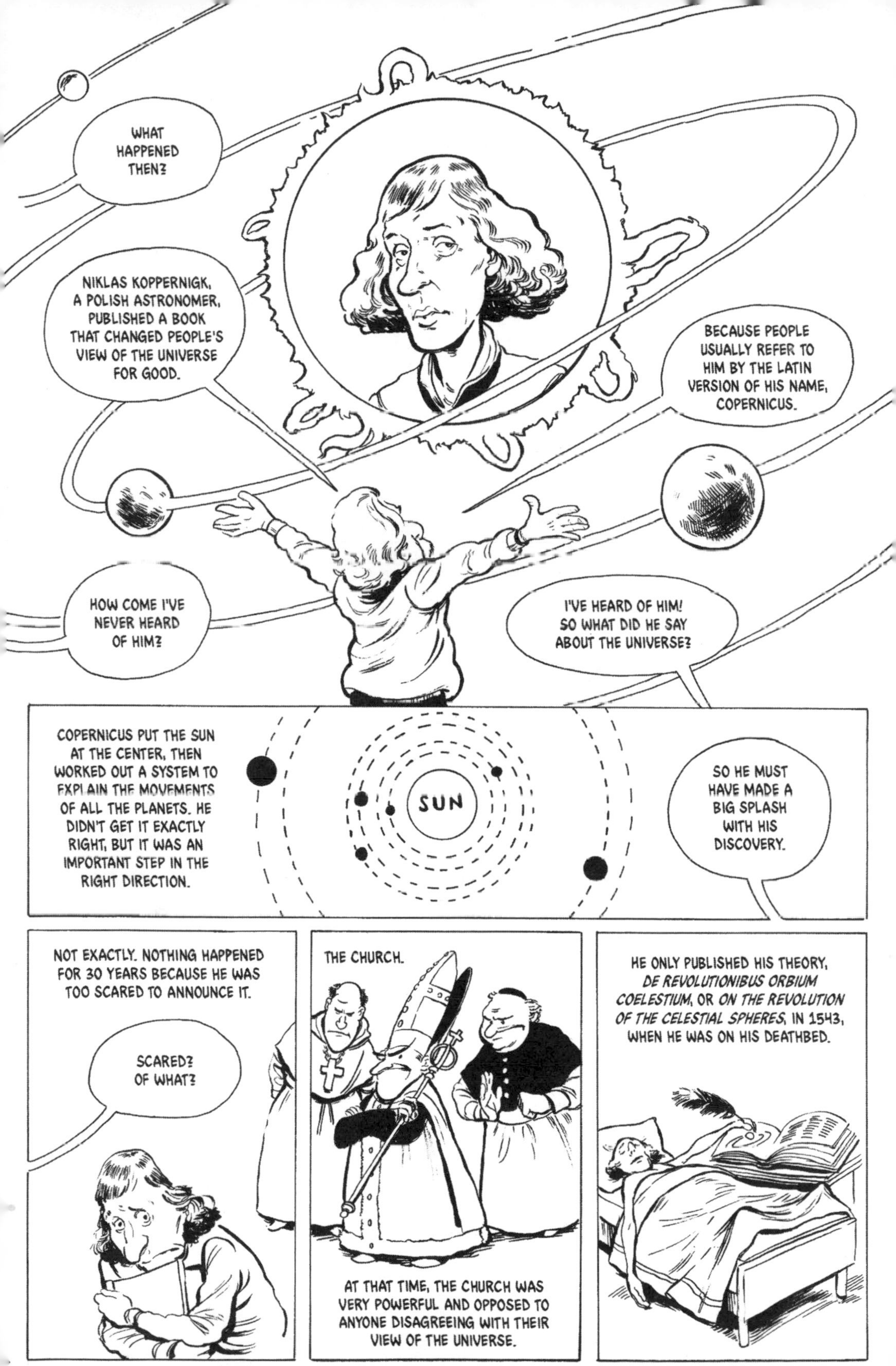
WHAT HAPPENED THEN?
NIKLAS KOPPERNIGK, A POLISH ASTRONOMER, PUBLISHED A BOOK THAT CHANGED PEOPLE'S VIEW OF THE UNIVERSE FOR GOOD.
BECAUSE PEOPLE USUALLY REFER TO HIM BY THE LATIN VERSION OF HIS NAME, COPERNICUS.
HOW COME I'VE NEVER HEARD OF HIM?
I'VE HEARD OF HIM! SO WHAT DID HE SAY ABOUT THE UNIVERSE?
COPERNICUS PUT THE SUN AT THE CENTER, THEN WORKED OUT A SYSTEM TO EXPLAIN THE MOVEMENTS OF ALL THE PLANETS. HE DIDN'T GET IT EXACTLY RIGHT, BUT IT WAS AN IMPORTANT STEP IN THE RIGHT DIRECTION.
SUN
SO HE MUST HAVE MADE A BIG SPLASH WITH HIS DISCOVERY.
NOT EXACTLY. NOTHING HAPPENED FOR 30 YEARS BECAUSE HE WAS TOO SCARED TO ANNOUNCE IT.
SCARED? OF WHAT?
THE CHURCH.
AT THAT TIME, THE CHURCH WAS VERY POWERFUL AND OPPOSED TO ANYONE DISAGREEING WITH THEIR VIEW OF THE UNIVERSE.
HE ONLY PUBLISHED HIS THEORY, *DE REVOLUTIONIBUS ORBIUM COELESTIUM*, OR *ON THE REVOLUTION OF THE CELESTIAL SPHERES*, IN 1543, WHEN HE WAS ON HIS DEATHBED.

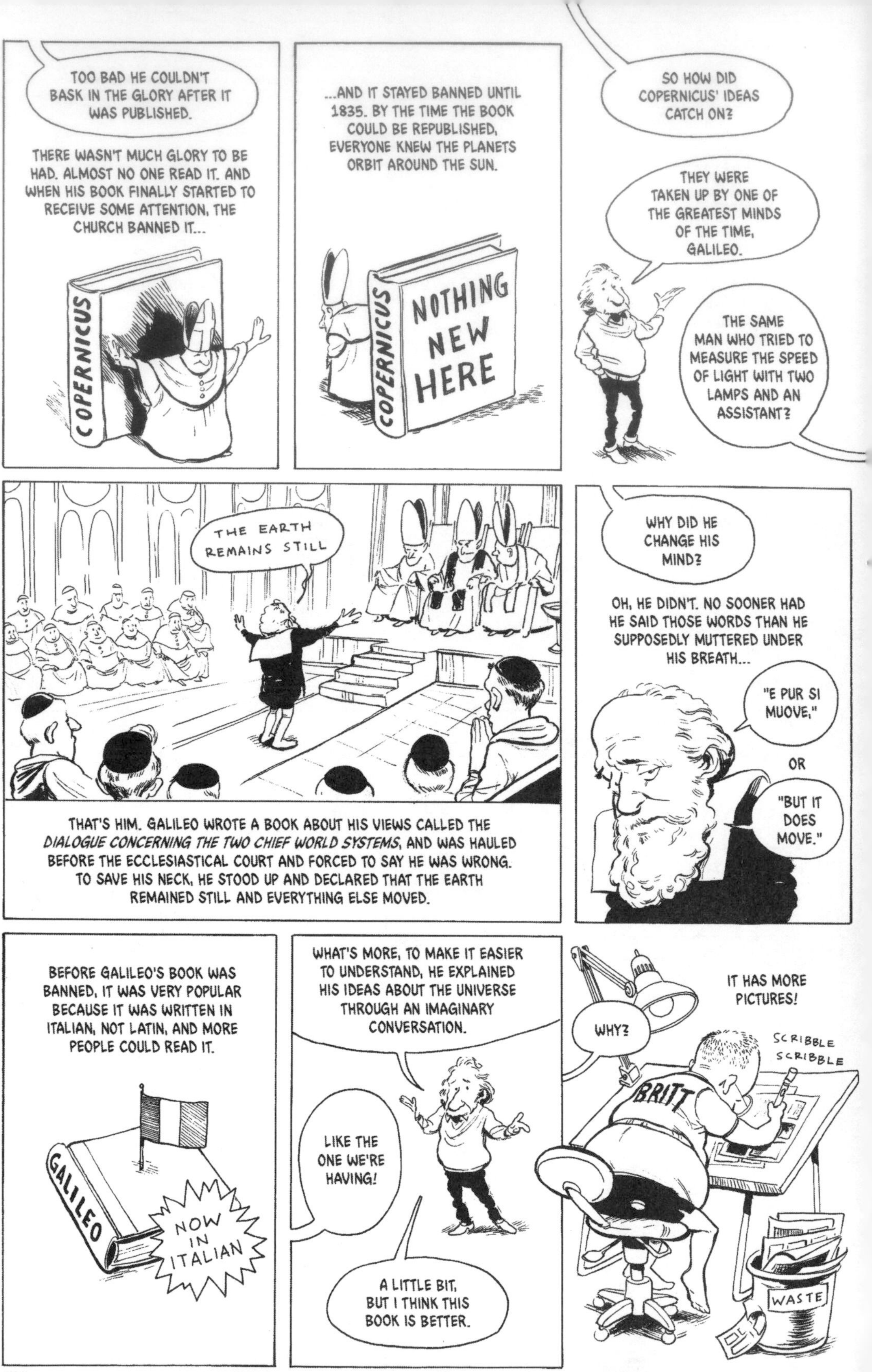
TOO BAD HE COULDN'T BASK IN THE GLORY AFTER IT WAS PUBLISHED.
THERE WASN'T MUCH GLORY TO BE HAD. ALMOST NO ONE READ IT. AND WHEN HIS BOOK FINALLY STARTED TO RECEIVE SOME ATTENTION, THE CHURCH BANNED IT...
COPERNICUS
...AND IT STAYED BANNED UNTIL 1835. BY THE TIME THE BOOK COULD BE REPUBLISHED, EVERYONE KNEW THE PLANETS ORBIT AROUND THE SUN.
COPERNICUS
NOTHING NEW HERE
SO HOW DID COPERNICUS' IDEAS CATCH ON?
THEY WERE TAKEN UP BY ONE OF THE GREATEST MINDS OF THE TIME, GALILEO.
THE SAME MAN WHO TRIED TO MEASURE THE SPEED OF LIGHT WITH TWO LAMPS AND AN ASSISTANT?
THE EARTH REMAINS STILL
THAT'S HIM. GALILEO WROTE A BOOK ABOUT HIS VIEWS CALLED THE *DIALOGUE CONCERNING THE TWO CHIEF WORLD SYSTEMS*, AND WAS HAULED BEFORE THE ECCLESIASTICAL COURT AND FORCED TO SAY HE WAS WRONG. TO SAVE HIS NECK, HE STOOD UP AND DECLARED THAT THE EARTH REMAINED STILL AND EVERYTHING ELSE MOVED.
WHY DID HE CHANGE HIS MIND?
OH, HE DIDN'T. NO SOONER HAD HE SAID THOSE WORDS THAN HE SUPPOSEDLY MUTTERED UNDER HIS BREATH...
"E PUR SI MUOVE,"
OR
"BUT IT DOES MOVE."
BEFORE GALILEO'S BOOK WAS BANNED, IT WAS VERY POPULAR BECAUSE IT WAS WRITTEN IN ITALIAN, NOT LATIN, AND MORE PEOPLE COULD READ IT.
GALILEO
NOW IN ITALIAN
WHAT'S MORE, TO MAKE IT EASIER TO UNDERSTAND, HE EXPLAINED HIS IDEAS ABOUT THE UNIVERSE THROUGH AN IMAGINARY CONVERSATION.
LIKE THE ONE WE'RE HAVING!
A LITTLE BIT, BUT I THINK THIS BOOK IS BETTER.
WHY?
IT HAS MORE PICTURES!
SCRIBBLE SCRIBBLE
BRITT
WASTE

CHAPTER 23:
IF I HAVE SEEN FURTHER: THE NOSE, THE MOOSE, AND THE TELESCOPE

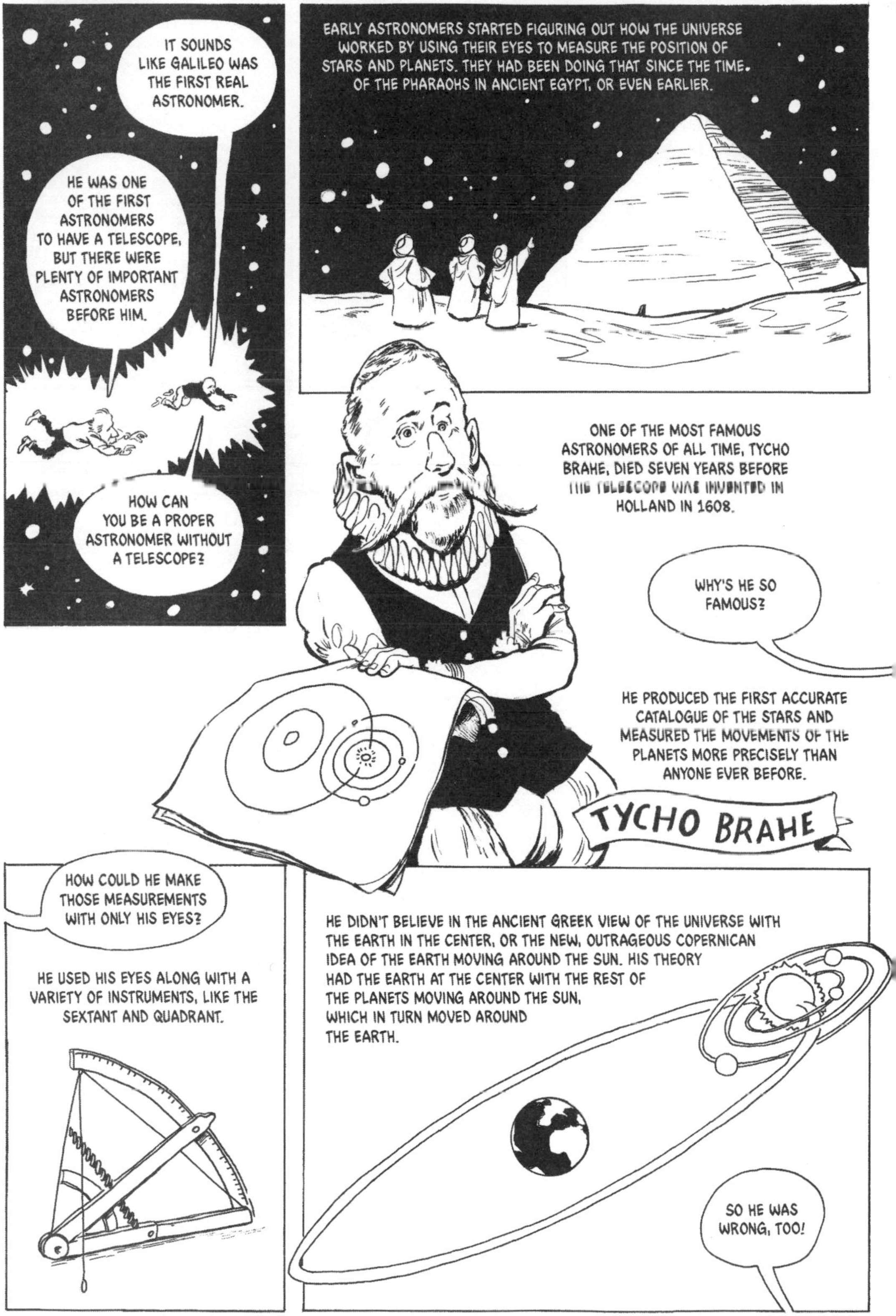
IT SOUNDS LIKE GALILEO WAS THE FIRST REAL ASTRONOMER.
HE WAS ONE OF THE FIRST ASTRONOMERS TO HAVE A TELESCOPE, BUT THERE WERE PLENTY OF IMPORTANT ASTRONOMERS BEFORE HIM.
HOW CAN YOU BE A PROPER ASTRONOMER WITHOUT A TELESCOPE?
EARLY ASTRONOMERS STARTED FIGURING OUT HOW THE UNIVERSE WORKED BY USING THEIR EYES TO MEASURE THE POSITION OF STARS AND PLANETS. THEY HAD BEEN DOING THAT SINCE THE TIME OF THE PHARAOHS IN ANCIENT EGYPT, OR EVEN EARLIER.
ONE OF THE MOST FAMOUS ASTRONOMERS OF ALL TIME, TYCHO BRAHE, DIED SEVEN YEARS BEFORE THE TELESCOPE WAS INVENTED IN HOLLAND IN 1608.
WHY'S HE SO FAMOUS?
HE PRODUCED THE FIRST ACCURATE CATALOGUE OF THE STARS AND MEASURED THE MOVEMENTS OF THE PLANETS MORE PRECISELY THAN ANYONE EVER BEFORE.
TYCHO BRAHE
HOW COULD HE MAKE THOSE MEASUREMENTS WITH ONLY HIS EYES?
HE USED HIS EYES ALONG WITH A VARIETY OF INSTRUMENTS, LIKE THE SEXTANT AND QUADRANT.
HE DIDN'T BELIEVE IN THE ANCIENT GREEK VIEW OF THE UNIVERSE WITH THE EARTH IN THE CENTER, OR THE NEW, OUTRAGEOUS COPERNICAN IDEA OF THE EARTH MOVING AROUND THE SUN. HIS THEORY HAD THE EARTH AT THE CENTER WITH THE REST OF THE PLANETS MOVING AROUND THE SUN, WHICH IN TURN MOVED AROUND THE EARTH.
SO HE WAS WRONG, TOO!

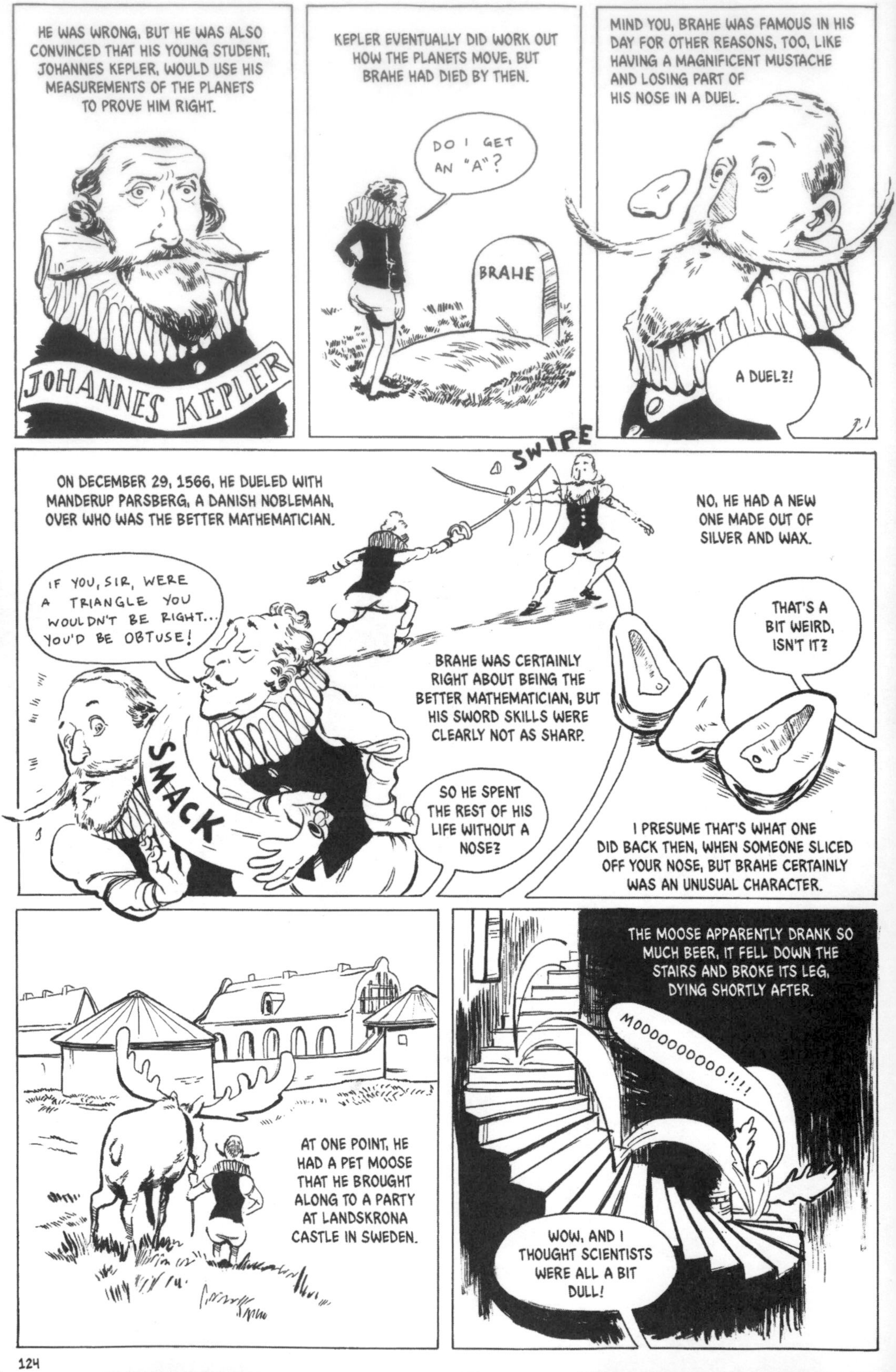

HE WAS WRONG, BUT HE WAS ALSO CONVINCED THAT HIS YOUNG STUDENT, JOHANNES KEPLER, WOULD USE HIS MEASUREMENTS OF THE PLANETS TO PROVE HIM RIGHT.
JOHANNES KEPLER
KEPLER EVENTUALLY DID WORK OUT HOW THE PLANETS MOVE, BUT BRAHE HAD DIED BY THEN.
DO I GET AN "A"?
BRAHE
MIND YOU, BRAHE WAS FAMOUS IN HIS DAY FOR OTHER REASONS, TOO, LIKE HAVING A MAGNIFICENT MUSTACHE AND LOSING PART OF HIS NOSE IN A DUEL.
A DUEL?!
ON DECEMBER 29, 1566, HE DUELED WITH MANDERUP PARSBERG, A DANISH NOBLEMAN, OVER WHO WAS THE BETTER MATHEMATICIAN.
SWIPE
IF YOU, SIR, WERE A TRIANGLE YOU WOULDN'T BE RIGHT... YOU'D BE OBTUSE!
SMACK
BRAHE WAS CERTAINLY RIGHT ABOUT BEING THE BETTER MATHEMATICIAN, BUT HIS SWORD SKILLS WERE CLEARLY NOT AS SHARP.
SO HE SPENT THE REST OF HIS LIFE WITHOUT A NOSE?
NO, HE HAD A NEW ONE MADE OUT OF SILVER AND WAX.
THAT'S A BIT WEIRD, ISN'T IT?
I PRESUME THAT'S WHAT ONE DID BACK THEN, WHEN SOMEONE SLICED OFF YOUR NOSE, BUT BRAHE CERTAINLY WAS AN UNUSUAL CHARACTER.
AT ONE POINT, HE HAD A PET MOOSE THAT HE BROUGHT ALONG TO A PARTY AT LANDSKRONA CASTLE IN SWEDEN.
THE MOOSE APPARENTLY DRANK SO MUCH BEER, IT FELL DOWN THE STAIRS AND BROKE ITS LEG, DYING SHORTLY AFTER.
MOOOOOOOOOO!!!!
WOW, AND I THOUGHT SCIENTISTS WERE ALL A BIT DULL!

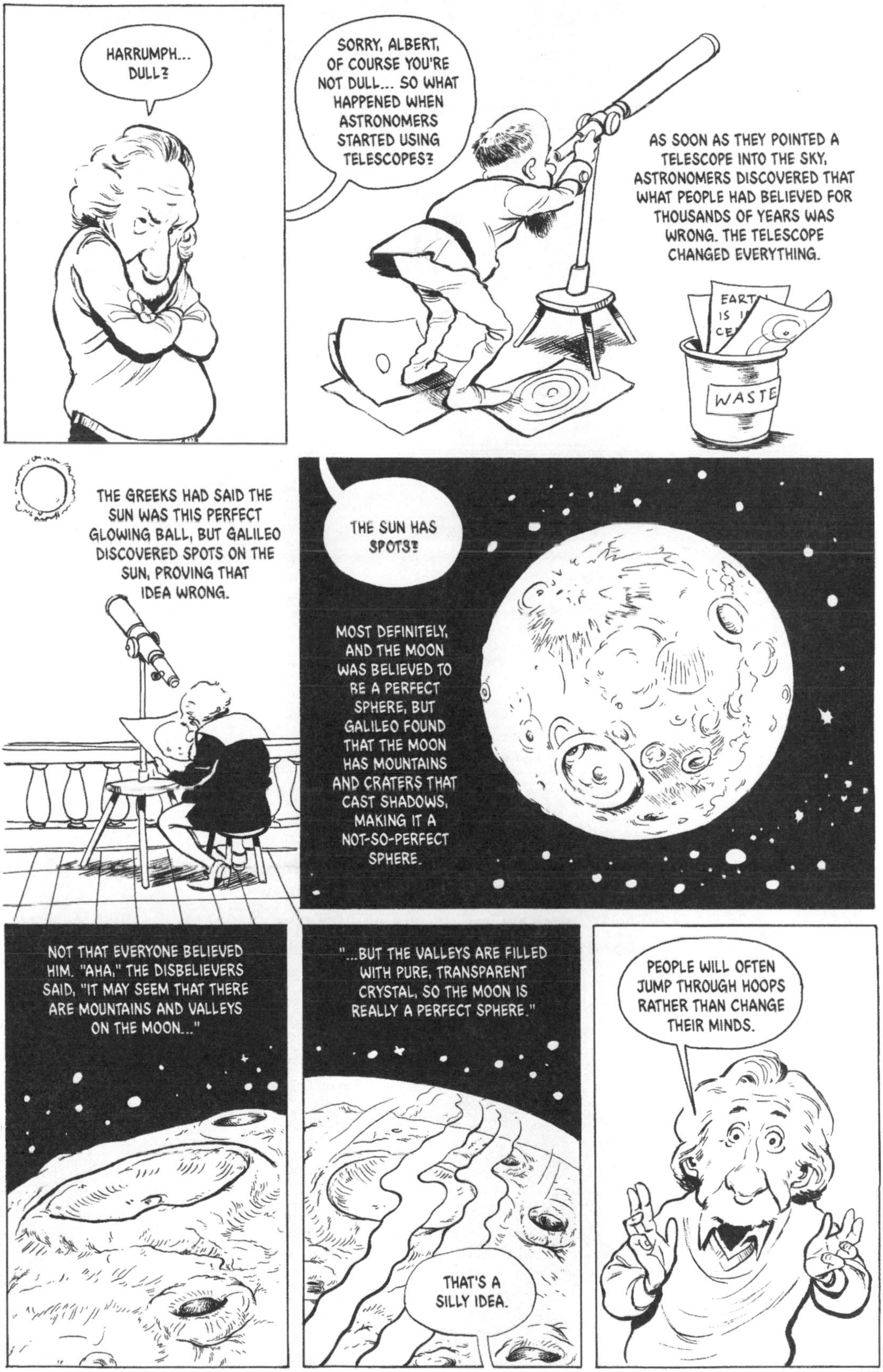
HARRUMPH... DULL?
SORRY, ALBERT, OF COURSE YOU'RE NOT DULL... SO WHAT HAPPENED WHEN ASTRONOMERS STARTED USING TELESCOPES?
AS SOON AS THEY POINTED A TELESCOPE INTO THE SKY, ASTRONOMERS DISCOVERED THAT WHAT PEOPLE HAD BELIEVED FOR THOUSANDS OF YEARS WAS WRONG. THE TELESCOPE CHANGED EVERYTHING.
WASTE
THE GREEKS HAD SAID THE SUN WAS THIS PERFECT GLOWING BALL, BUT GALILEO DISCOVERED SPOTS ON THE SUN, PROVING THAT IDEA WRONG.
THE SUN HAS SPOTS?
MOST DEFINITELY, AND THE MOON WAS BELIEVED TO BE A PERFECT SPHERE, BUT GALILEO FOUND THAT THE MOON HAS MOUNTAINS AND CRATERS THAT CAST SHADOWS, MAKING IT A NOT-SO-PERFECT SPHERE.
NOT THAT EVERYONE BELIEVED HIM. "AHA," THE DISBELIEVERS SAID, "IT MAY SEEM THAT THERE ARE MOUNTAINS AND VALLEYS ON THE MOON..."
"...BUT THE VALLEYS ARE FILLED WITH PURE, TRANSPARENT CRYSTAL, SO THE MOON IS REALLY A PERFECT SPHERE."
THAT'S A SILLY IDEA.
PEOPLE WILL OFTEN JUMP THROUGH HOOPS RATHER THAN CHANGE THEIR MINDS.

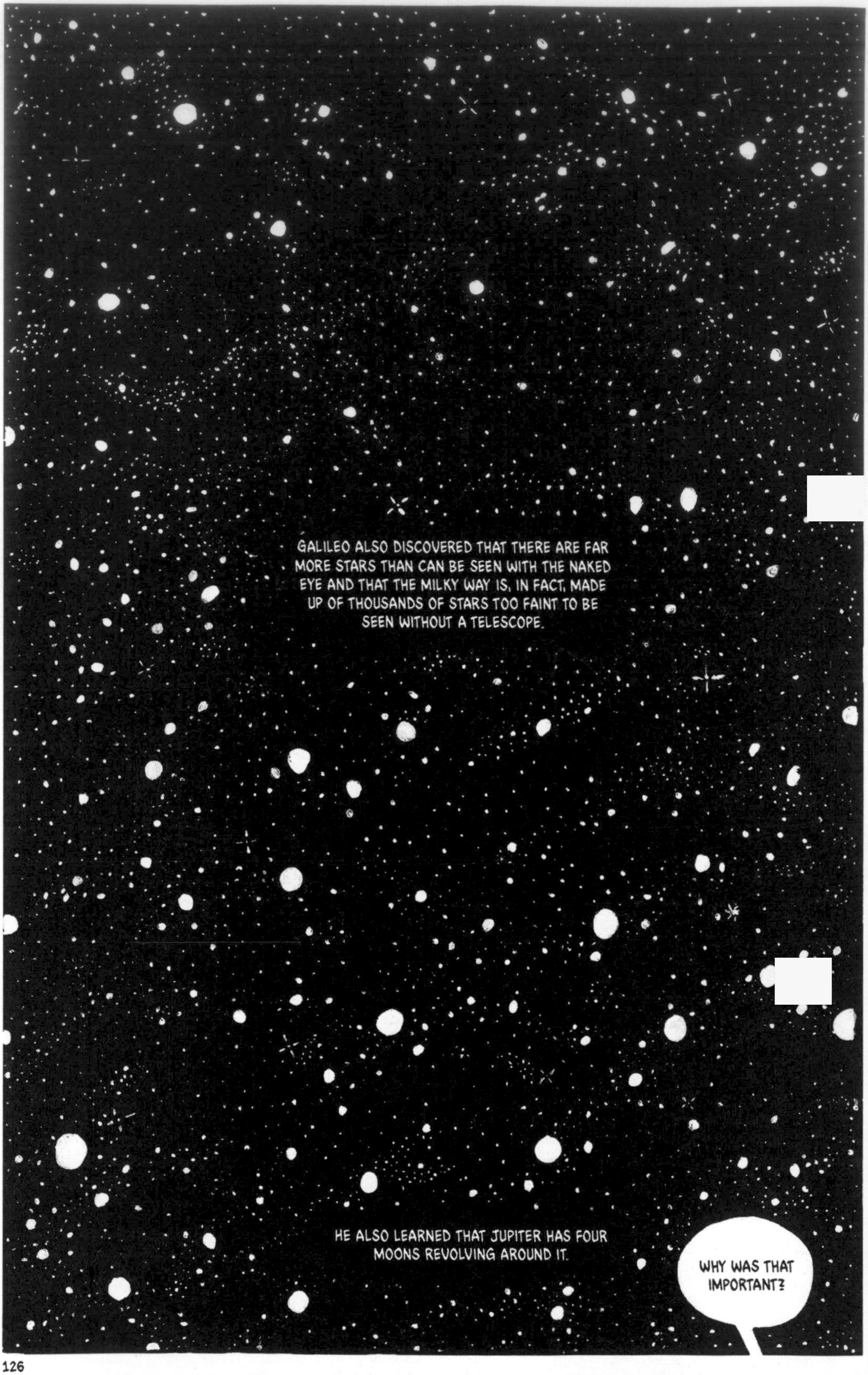
GALILEO ALSO DISCOVERED THAT THERE ARE FAR MORE STARS THAN CAN BE SEEN WITH THE NAKED EYE AND THAT THE MILKY WAY IS, IN FACT, MADE UP OF THOUSANDS OF STARS TOO FAINT TO BE SEEN WITHOUT A TELESCOPE.
HE ALSO LEARNED THAT JUPITER HAS FOUR MOONS REVOLVING AROUND IT.
WHY WAS THAT IMPORTANT?

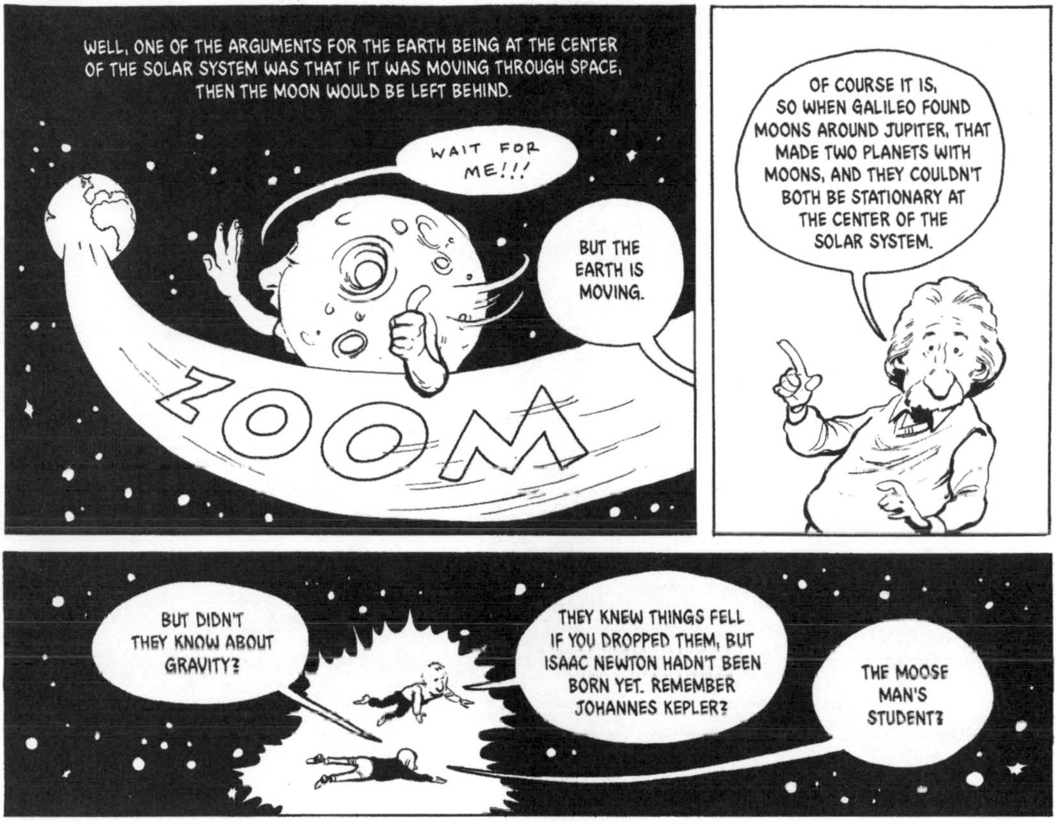
WELL, ONE OF THE ARGUMENTS FOR THE EARTH BEING AT THE CENTER OF THE SOLAR SYSTEM WAS THAT IF IT WAS MOVING THROUGH SPACE, THEN THE MOON WOULD BE LEFT BEHIND.
WAIT FOR ME!!!
BUT THE EARTH IS MOVING.
ZOOM
OF COURSE IT IS, SO WHEN GALILEO FOUND MOONS AROUND JUPITER, THAT MADE TWO PLANETS WITH MOONS, AND THEY COULDN'T BOTH BE STATIONARY AT THE CENTER OF THE SOLAR SYSTEM.
BUT DIDN'T THEY KNOW ABOUT GRAVITY?
THEY KNEW THINGS FELL IF YOU DROPPED THEM, BUT ISAAC NEWTON HADN'T BEEN BORN YET. REMEMBER JOHANNES KEPLER?
THE MOOSE MAN'S STUDENT?

THAT'S HIM. KEPLER WORKED OUT THREE SIMPLE RULES, HIS LAWS OF PLANETARY MOTION, TO EXPLAIN HOW THE PLANETS MOVE...

KEPLER'S LAWS OF PLANETARY MOTION:

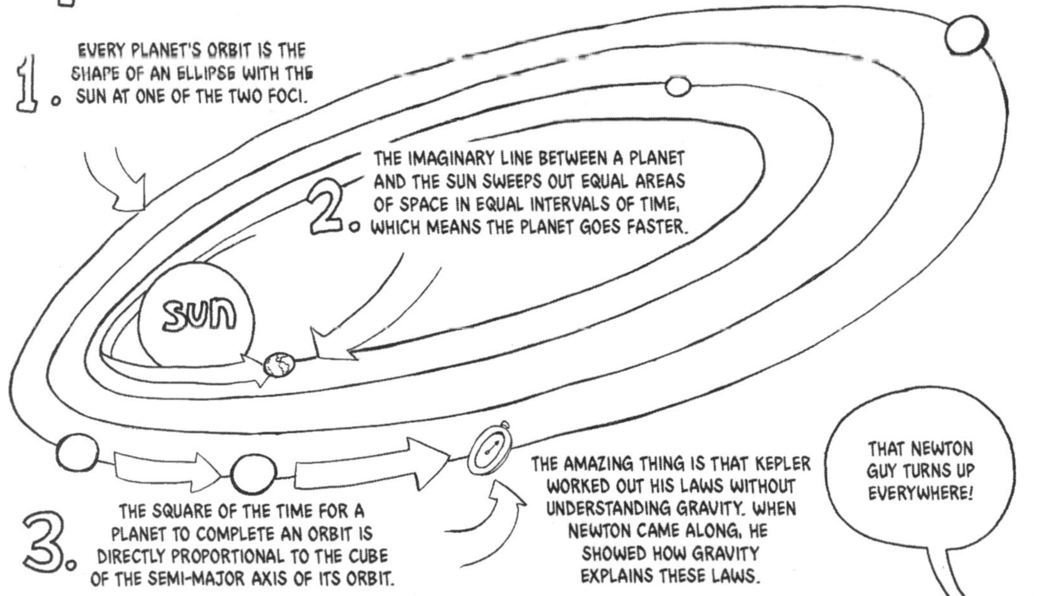
1. EVERY PLANET'S ORBIT IS THE SHAPE OF AN ELLIPSE WITH THE SUN AT ONE OF THE TWO FOCI.
2. THE IMAGINARY LINE BETWEEN A PLANET AND THE SUN SWEEPS OUT EQUAL AREAS OF SPACE IN EQUAL INTERVALS OF TIME, WHICH MEANS THE PLANET GOES FASTER.
SUN
3. THE SQUARE OF THE TIME FOR A PLANET TO COMPLETE AN ORBIT IS DIRECTLY PROPORTIONAL TO THE CUBE OF THE SEMI-MAJOR AXIS OF ITS ORBIT.
THE AMAZING THING IS THAT KEPLER WORKED OUT HIS LAWS WITHOUT UNDERSTANDING GRAVITY. WHEN NEWTON CAME ALONG, HE SHOWED HOW GRAVITY EXPLAINS THESE LAWS.
THAT NEWTON GUY TURNS UP EVERYWHERE!

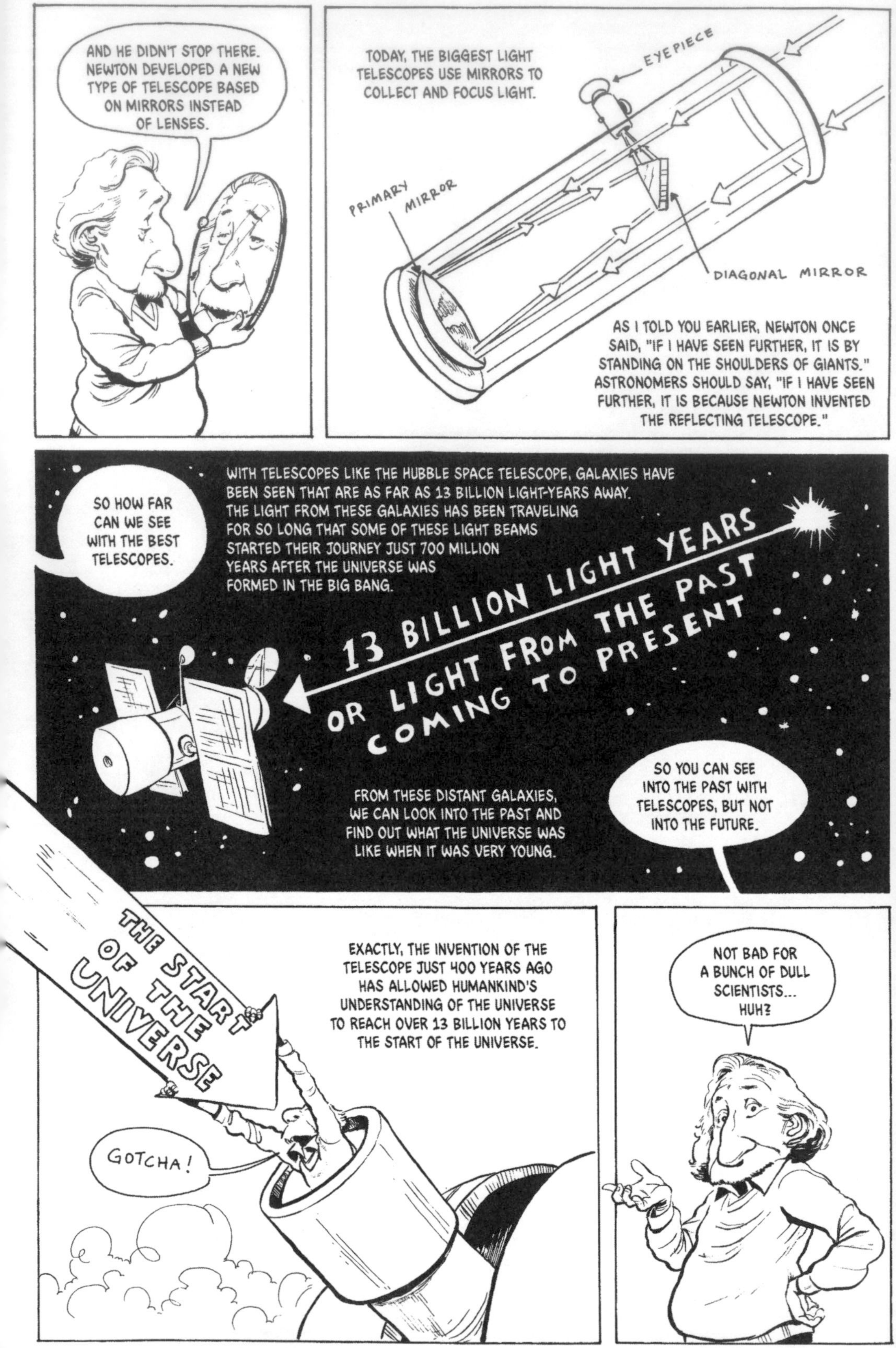
AND HE DIDN'T STOP THERE. NEWTON DEVELOPED A NEW TYPE OF TELESCOPE BASED ON MIRRORS INSTEAD OF LENSES.
TODAY, THE BIGGEST LIGHT TELESCOPES USE MIRRORS TO COLLECT AND FOCUS LIGHT.
EYEPIECE
PRIMARY MIRROR
DIAGONAL MIRROR
AS I TOLD YOU EARLIER, NEWTON ONCE SAID, "IF I HAVE SEEN FURTHER, IT IS BY STANDING ON THE SHOULDERS OF GIANTS." ASTRONOMERS SHOULD SAY, "IF I HAVE SEEN FURTHER, IT IS BECAUSE NEWTON INVENTED THE REFLECTING TELESCOPE."
SO HOW FAR CAN WE SEE WITH THE BEST TELESCOPES.
WITH TELESCOPES LIKE THE HUBBLE SPACE TELESCOPE, GALAXIES HAVE BEEN SEEN THAT ARE AS FAR AS 13 BILLION LIGHT-YEARS AWAY. THE LIGHT FROM THESE GALAXIES HAS BEEN TRAVELING FOR SO LONG THAT SOME OF THESE LIGHT BEAMS STARTED THEIR JOURNEY JUST 700 MILLION YEARS AFTER THE UNIVERSE WAS FORMED IN THE BIG BANG.
13 BILLION LIGHT YEARS
OR LIGHT FROM THE PAST COMING TO PRESENT
FROM THESE DISTANT GALAXIES, WE CAN LOOK INTO THE PAST AND FIND OUT WHAT THE UNIVERSE WAS LIKE WHEN IT WAS VERY YOUNG.
SO YOU CAN SEE INTO THE PAST WITH TELESCOPES, BUT NOT INTO THE FUTURE.
THE START OF THE UNIVERSE
EXACTLY, THE INVENTION OF THE TELESCOPE JUST 400 YEARS AGO HAS ALLOWED HUMANKIND'S UNDERSTANDING OF THE UNIVERSE TO REACH OVER 13 BILLION YEARS TO THE START OF THE UNIVERSE.
GOTCHA!
NOT BAD FOR A BUNCH OF DULL SCIENTISTS... HUH?

CHAPTER 24: ARE THERE PLANETS AROUND OTHER STARS?

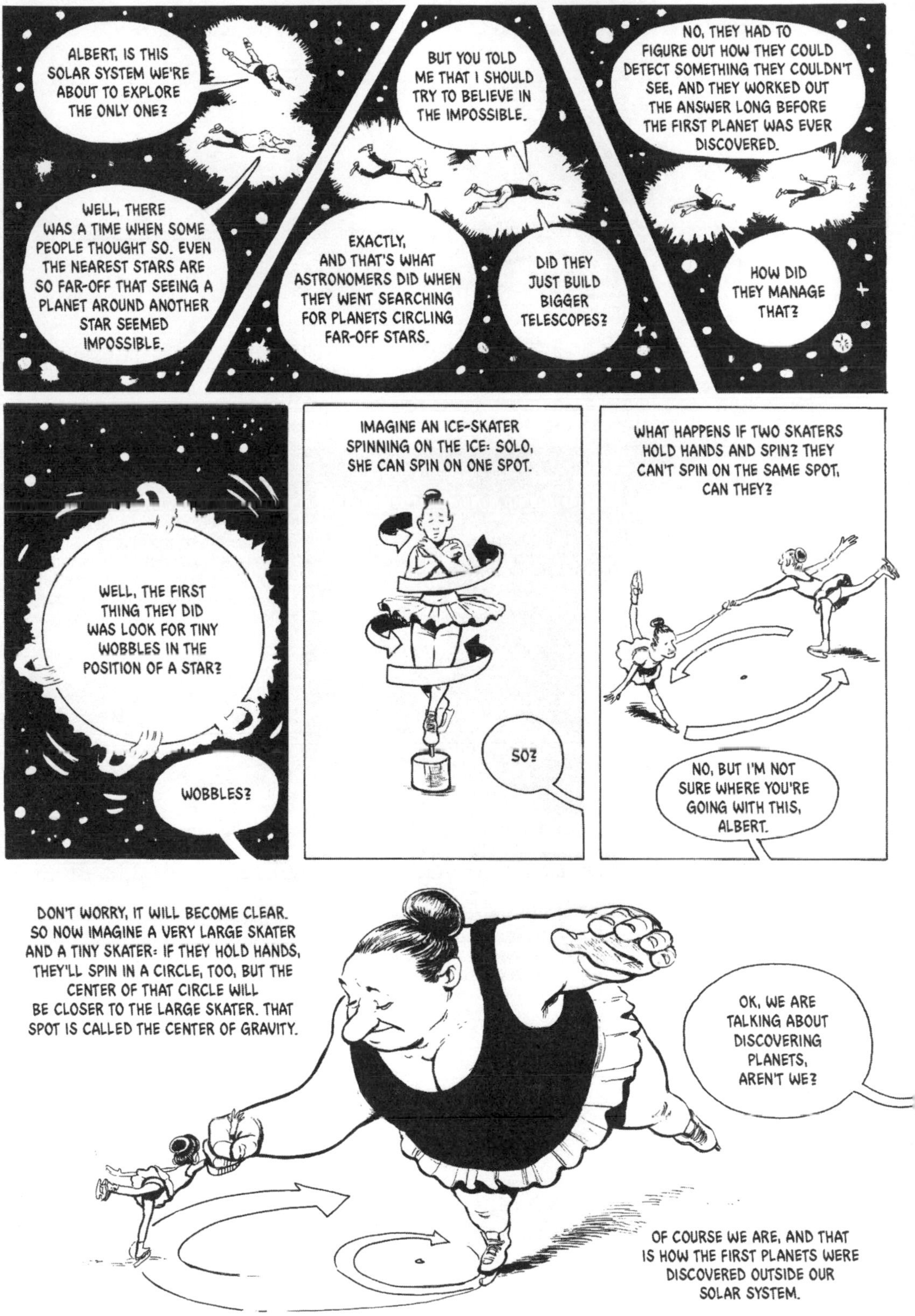
ALBERT, IS THIS SOLAR SYSTEM WE'RE ABOUT TO EXPLORE THE ONLY ONE?
WELL, THERE WAS A TIME WHEN SOME PEOPLE THOUGHT SO. EVEN THE NEAREST STARS ARE SO FAR-OFF THAT SEEING A PLANET AROUND ANOTHER STAR SEEMED IMPOSSIBLE.
BUT YOU TOLD ME THAT I SHOULD TRY TO BELIEVE IN THE IMPOSSIBLE.
EXACTLY, AND THAT'S WHAT ASTRONOMERS DID WHEN THEY WENT SEARCHING FOR PLANETS CIRCLING FAR-OFF STARS.
DID THEY JUST BUILD BIGGER TELESCOPES?
NO, THEY HAD TO FIGURE OUT HOW THEY COULD DETECT SOMETHING THEY COULDN'T SEE, AND THEY WORKED OUT THE ANSWER LONG BEFORE THE FIRST PLANET WAS EVER DISCOVERED.
HOW DID THEY MANAGE THAT?
WELL, THE FIRST THING THEY DID WAS LOOK FOR TINY WOBBLES IN THE POSITION OF A STAR?
WOBBLES?
IMAGINE AN ICE-SKATER SPINNING ON THE ICE: SOLO, SHE CAN SPIN ON ONE SPOT.
SO?
WHAT HAPPENS IF TWO SKATERS HOLD HANDS AND SPIN? THEY CAN'T SPIN ON THE SAME SPOT, CAN THEY?
NO, BUT I'M NOT SURE WHERE YOU'RE GOING WITH THIS, ALBERT.
DON'T WORRY, IT WILL BECOME CLEAR. SO NOW IMAGINE A VERY LARGE SKATER AND A TINY SKATER: IF THEY HOLD HANDS, THEY'LL SPIN IN A CIRCLE, TOO, BUT THE CENTER OF THAT CIRCLE WILL BE CLOSER TO THE LARGE SKATER. THAT SPOT IS CALLED THE CENTER OF GRAVITY.
OK, WE ARE TALKING ABOUT DISCOVERING PLANETS, AREN'T WE?
OF COURSE WE ARE, AND THAT IS HOW THE FIRST PLANETS WERE DISCOVERED OUTSIDE OUR SOLAR SYSTEM.

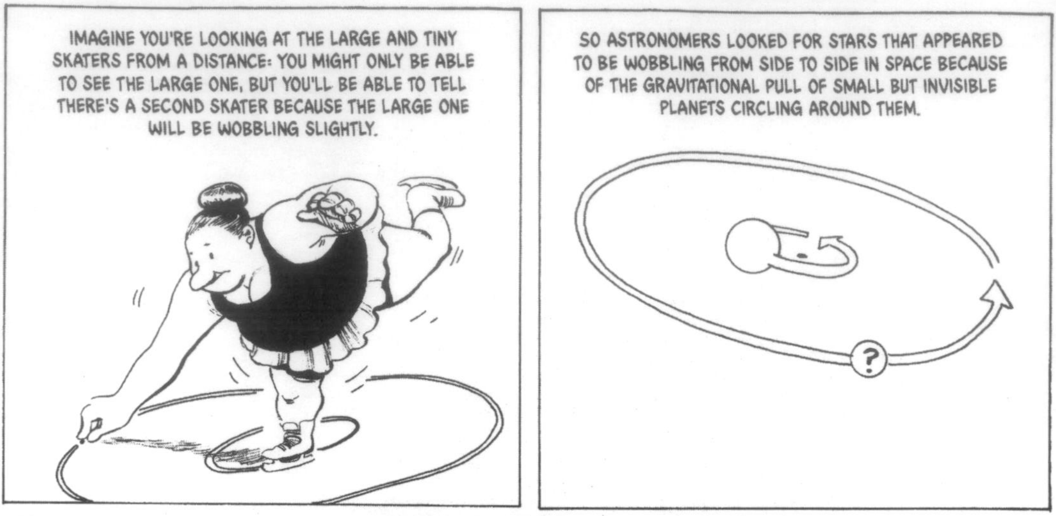
IMAGINE YOU'RE LOOKING AT THE LARGE AND TINY SKATERS FROM A DISTANCE: YOU MIGHT ONLY BE ABLE TO SEE THE LARGE ONE, BUT YOU'LL BE ABLE TO TELL THERE'S A SECOND SKATER BECAUSE THE LARGE ONE WILL BE WOBBLING SLIGHTLY.
SO ASTRONOMERS LOOKED FOR STARS THAT APPEARED TO BE WOBBLING FROM SIDE TO SIDE IN SPACE BECAUSE OF THE GRAVITATIONAL PULL OF SMALL BUT INVISIBLE PLANETS CIRCLING AROUND THEM.
?

SOME PEOPLE CLAIMED TO HAVE FOUND WOBBLING STARS AS LONG AGO AS 1855, BUT IT WASN'T UNTIL 1988 THAT THREE CANADIAN ASTRONOMERS, BRUCE CAMPBELL, G. A. H. WALKER, AND S. YANG, FOUND THE FIRST STAR WITH A DEFINITE PLANET. EVEN THEN IT TOOK UNTIL 2003 FOR SCIENTISTS TO CONFIRM IT REALLY WAS A PLANET.

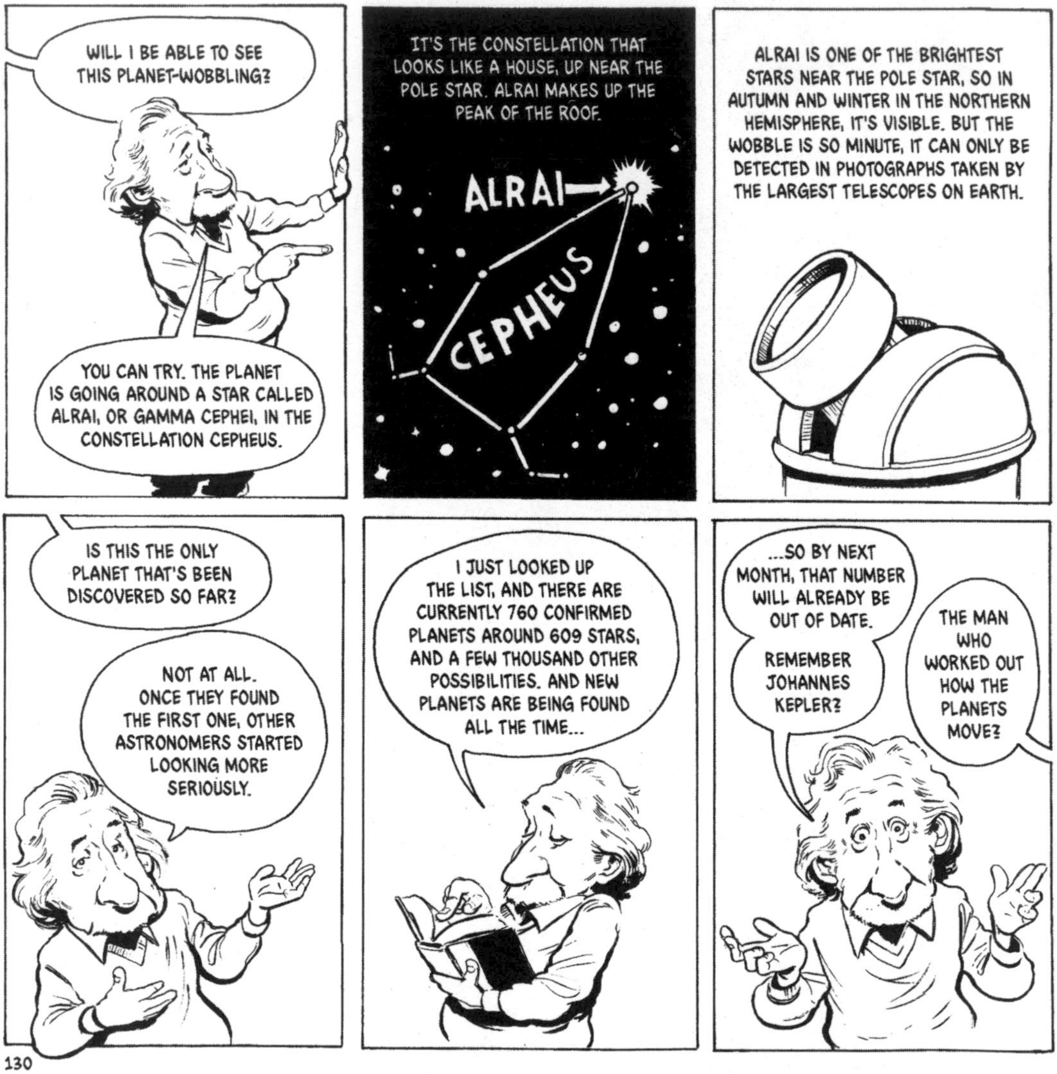
WILL I BE ABLE TO SEE THIS PLANET-WOBBLING?
YOU CAN TRY. THE PLANET IS GOING AROUND A STAR CALLED ALRAI, OR GAMMA CEPHEI, IN THE CONSTELLATION CEPHEUS.
IT'S THE CONSTELLATION THAT LOOKS LIKE A HOUSE, UP NEAR THE POLE STAR. ALRAI MAKES UP THE PEAK OF THE ROOF.
ALRAI
CEPHEUS
ALRAI IS ONE OF THE BRIGHTEST STARS NEAR THE POLE STAR, SO IN AUTUMN AND WINTER IN THE NORTHERN HEMISPHERE, IT'S VISIBLE. BUT THE WOBBLE IS SO MINUTE, IT CAN ONLY BE DETECTED IN PHOTOGRAPHS TAKEN BY THE LARGEST TELESCOPES ON EARTH.
IS THIS THE ONLY PLANET THAT'S BEEN DISCOVERED SO FAR?
NOT AT ALL. ONCE THEY FOUND THE FIRST ONE, OTHER ASTRONOMERS STARTED LOOKING MORE SERIOUSLY.
I JUST LOOKED UP THE LIST, AND THERE ARE CURRENTLY 760 CONFIRMED PLANETS AROUND 609 STARS, AND A FEW THOUSAND OTHER POSSIBILITIES. AND NEW PLANETS ARE BEING FOUND ALL THE TIME...
...SO BY NEXT MONTH, THAT NUMBER WILL ALREADY BE OUT OF DATE.
REMEMBER JOHANNES KEPLER?
THE MAN WHO WORKED OUT HOW THE PLANETS MOVE?

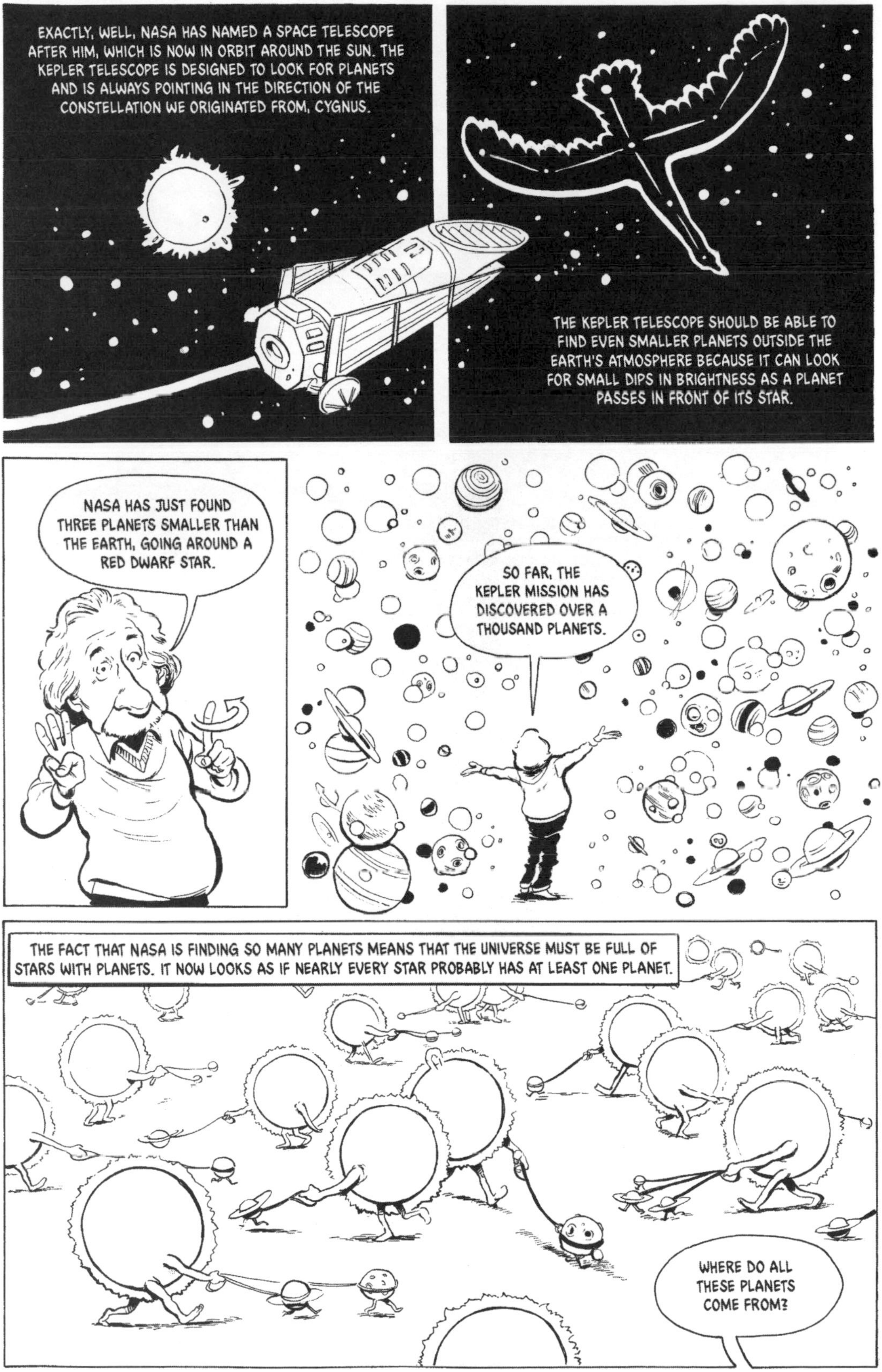
EXACTLY. WELL, NASA HAS NAMED A SPACE TELESCOPE AFTER HIM, WHICH IS NOW IN ORBIT AROUND THE SUN. THE KEPLER TELESCOPE IS DESIGNED TO LOOK FOR PLANETS AND IS ALWAYS POINTING IN THE DIRECTION OF THE CONSTELLATION WE ORIGINATED FROM, CYGNUS.
THE KEPLER TELESCOPE SHOULD BE ABLE TO FIND EVEN SMALLER PLANETS OUTSIDE THE EARTH'S ATMOSPHERE BECAUSE IT CAN LOOK FOR SMALL DIPS IN BRIGHTNESS AS A PLANET PASSES IN FRONT OF ITS STAR.
NASA HAS JUST FOUND THREE PLANETS SMALLER THAN THE EARTH, GOING AROUND A RED DWARF STAR.
SO FAR, THE KEPLER MISSION HAS DISCOVERED OVER A THOUSAND PLANETS.
THE FACT THAT NASA IS FINDING SO MANY PLANETS MEANS THAT THE UNIVERSE MUST BE FULL OF STARS WITH PLANETS. IT NOW LOOKS AS IF NEARLY EVERY STAR PROBABLY HAS AT LEAST ONE PLANET.
WHERE DO ALL THESE PLANETS COME FROM?

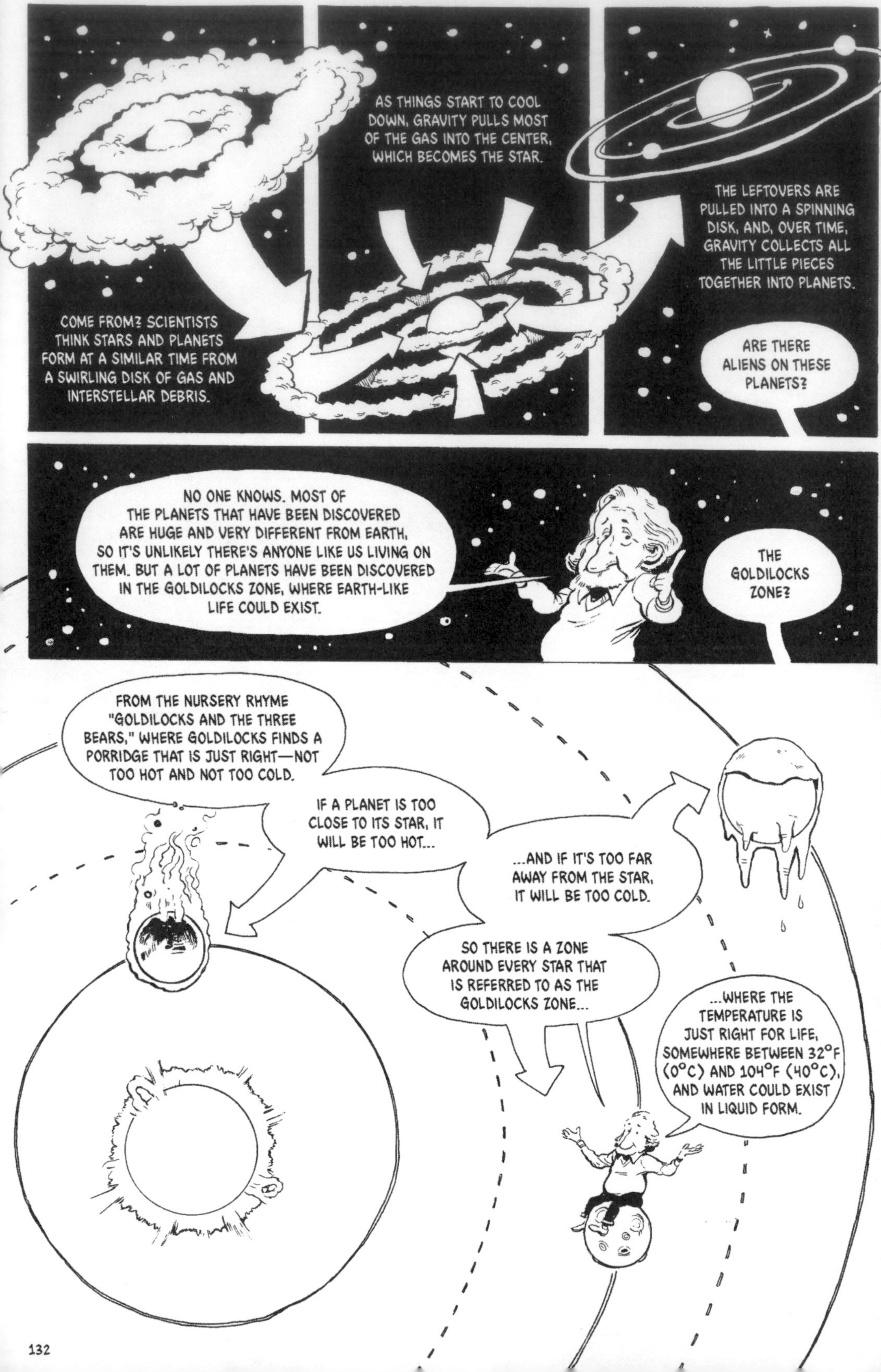
COME FROM? SCIENTISTS THINK STARS AND PLANETS FORM AT A SIMILAR TIME FROM A SWIRLING DISK OF GAS AND INTERSTELLAR DEBRIS.
AS THINGS START TO COOL DOWN, GRAVITY PULLS MOST OF THE GAS INTO THE CENTER, WHICH BECOMES THE STAR.
THE LEFTOVERS ARE PULLED INTO A SPINNING DISK, AND, OVER TIME, GRAVITY COLLECTS ALL THE LITTLE PIECES TOGETHER INTO PLANETS.
ARE THERE ALIENS ON THESE PLANETS?
NO ONE KNOWS. MOST OF THE PLANETS THAT HAVE BEEN DISCOVERED ARE HUGE AND VERY DIFFERENT FROM EARTH, SO IT'S UNLIKELY THERE'S ANYONE LIKE US LIVING ON THEM. BUT A LOT OF PLANETS HAVE BEEN DISCOVERED IN THE GOLDILOCKS ZONE, WHERE EARTH-LIKE LIFE COULD EXIST.
THE GOLDILOCKS ZONE?
FROM THE NURSERY RHYME "GOLDILOCKS AND THE THREE BEARS," WHERE GOLDILOCKS FINDS A PORRIDGE THAT IS JUST RIGHT—NOT TOO HOT AND NOT TOO COLD.
IF A PLANET IS TOO CLOSE TO ITS STAR, IT WILL BE TOO HOT...
...AND IF IT'S TOO FAR AWAY FROM THE STAR, IT WILL BE TOO COLD.
SO THERE IS A ZONE AROUND EVERY STAR THAT IS REFERRED TO AS THE GOLDILOCKS ZONE...
...WHERE THE TEMPERATURE IS JUST RIGHT FOR LIFE, SOMEWHERE BETWEEN 32°F (0°C) AND 104°F (40°C), AND WATER COULD EXIST IN LIQUID FORM.

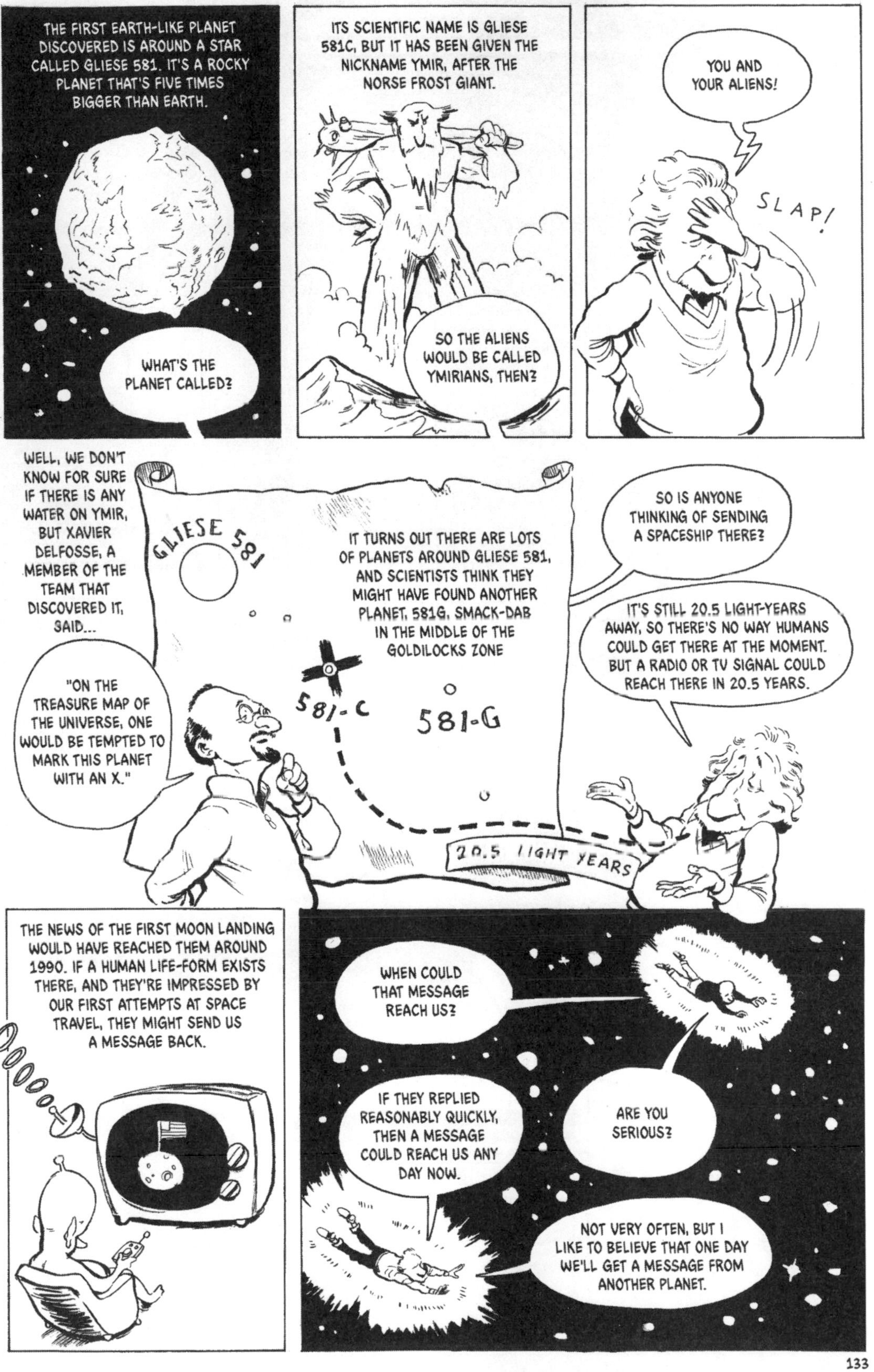
THE FIRST EARTH-LIKE PLANET DISCOVERED IS AROUND A STAR CALLED GLIESE 581. IT'S A ROCKY PLANET THAT'S FIVE TIMES BIGGER THAN EARTH.
WHAT'S THE PLANET CALLED?
ITS SCIENTIFIC NAME IS GLIESE 581C, BUT IT HAS BEEN GIVEN THE NICKNAME YMIR, AFTER THE NORSE FROST GIANT.
SO THE ALIENS WOULD BE CALLED YMIRIANS, THEN?
YOU AND YOUR ALIENS!
SLAP!
WELL, WE DON'T KNOW FOR SURE IF THERE IS ANY WATER ON YMIR, BUT XAVIER DELFOSSE, A MEMBER OF THE TEAM THAT DISCOVERED IT, SAID...
GLIESE 581
IT TURNS OUT THERE ARE LOTS OF PLANETS AROUND GLIESE 581, AND SCIENTISTS THINK THEY MIGHT HAVE FOUND ANOTHER PLANET, 581G, SMACK-DAB IN THE MIDDLE OF THE GOLDILOCKS ZONE
SO IS ANYONE THINKING OF SENDING A SPACESHIP THERE?
IT'S STILL 20.5 LIGHT-YEARS AWAY, SO THERE'S NO WAY HUMANS COULD GET THERE AT THE MOMENT. BUT A RADIO OR TV SIGNAL COULD REACH THERE IN 20.5 YEARS.
"ON THE TREASURE MAP OF THE UNIVERSE, ONE WOULD BE TEMPTED TO MARK THIS PLANET WITH AN X."
581-C
581-G
20.5 LIGHT YEARS
THE NEWS OF THE FIRST MOON LANDING WOULD HAVE REACHED THEM AROUND 1990. IF A HUMAN LIFE-FORM EXISTS THERE, AND THEY'RE IMPRESSED BY OUR FIRST ATTEMPTS AT SPACE TRAVEL, THEY MIGHT SEND US A MESSAGE BACK.
WHEN COULD THAT MESSAGE REACH US?
IF THEY REPLIED REASONABLY QUICKLY, THEN A MESSAGE COULD REACH US ANY DAY NOW.
ARE YOU SERIOUS?
NOT VERY OFTEN, BUT I LIKE TO BELIEVE THAT ONE DAY WE'LL GET A MESSAGE FROM ANOTHER PLANET.

CHAPTER 25: THE OUTER, BIGGER PLANETS OF THE SOLAR SYSTEM

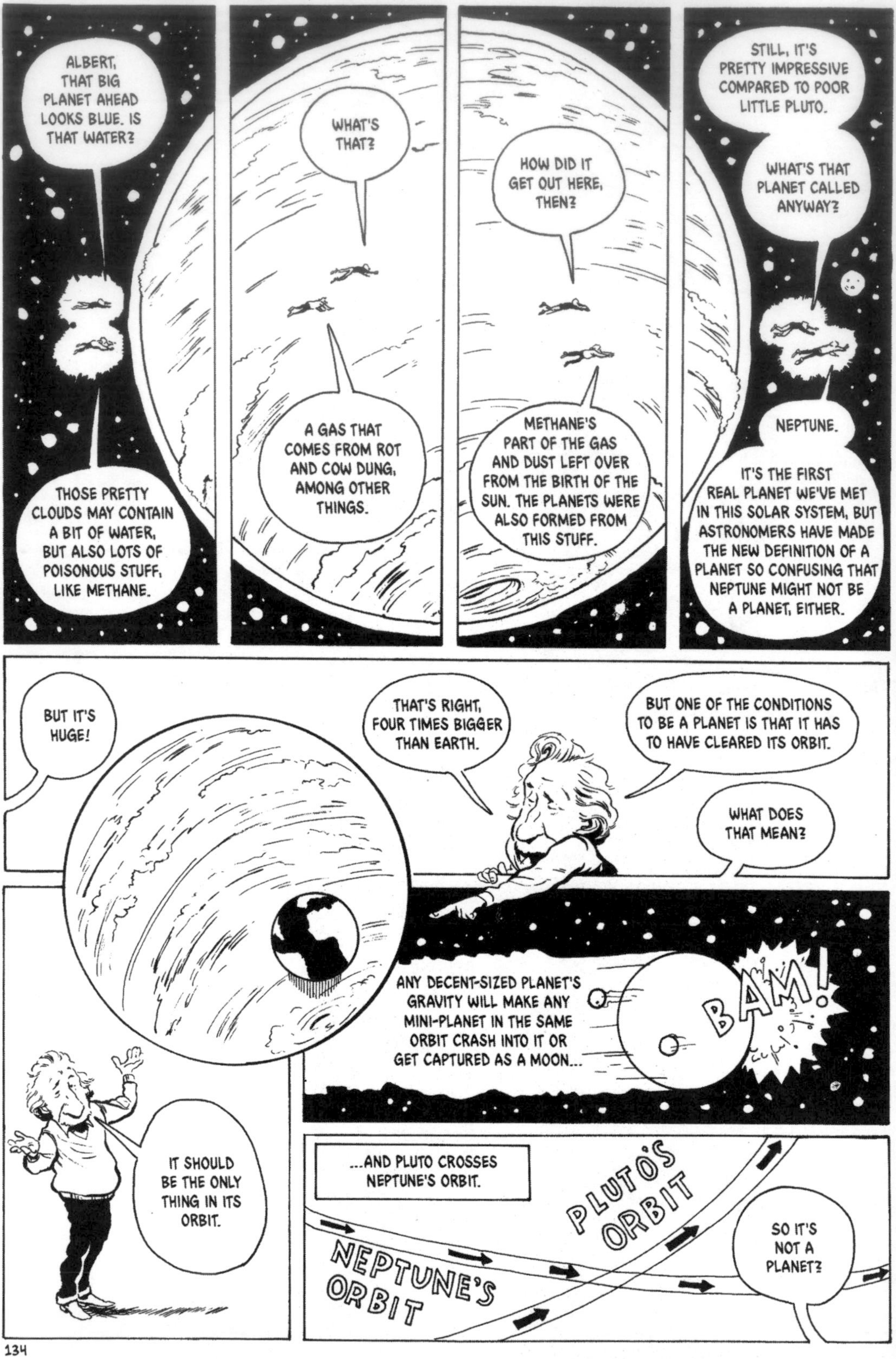
ALBERT, THAT BIG PLANET AHEAD LOOKS BLUE. IS THAT WATER?
THOSE PRETTY CLOUDS MAY CONTAIN A BIT OF WATER, BUT ALSO LOTS OF POISONOUS STUFF, LIKE METHANE.
WHAT'S THAT?
A GAS THAT COMES FROM ROT AND COW DUNG, AMONG OTHER THINGS.
HOW DID IT GET OUT HERE, THEN?
METHANE'S PART OF THE GAS AND DUST LEFT OVER FROM THE BIRTH OF THE SUN. THE PLANETS WERE ALSO FORMED FROM THIS STUFF.
STILL, IT'S PRETTY IMPRESSIVE COMPARED TO POOR LITTLE PLUTO.
WHAT'S THAT PLANET CALLED ANYWAY?
NEPTUNE.
IT'S THE FIRST REAL PLANET WE'VE MET IN THIS SOLAR SYSTEM, BUT ASTRONOMERS HAVE MADE THE NEW DEFINITION OF A PLANET SO CONFUSING THAT NEPTUNE MIGHT NOT BE A PLANET, EITHER.
BUT IT'S HUGE!
THAT'S RIGHT, FOUR TIMES BIGGER THAN EARTH.
BUT ONE OF THE CONDITIONS TO BE A PLANET IS THAT IT HAS TO HAVE CLEARED ITS ORBIT.
WHAT DOES THAT MEAN?
IT SHOULD BE THE ONLY THING IN ITS ORBIT.
ANY DECENT-SIZED PLANET'S GRAVITY WILL MAKE ANY MINI-PLANET IN THE SAME ORBIT CRASH INTO IT OR GET CAPTURED AS A MOON...
BAM!
...AND PLUTO CROSSES NEPTUNE'S ORBIT.
PLUTO'S ORBIT
NEPTUNE'S ORBIT
SO IT'S NOT A PLANET?

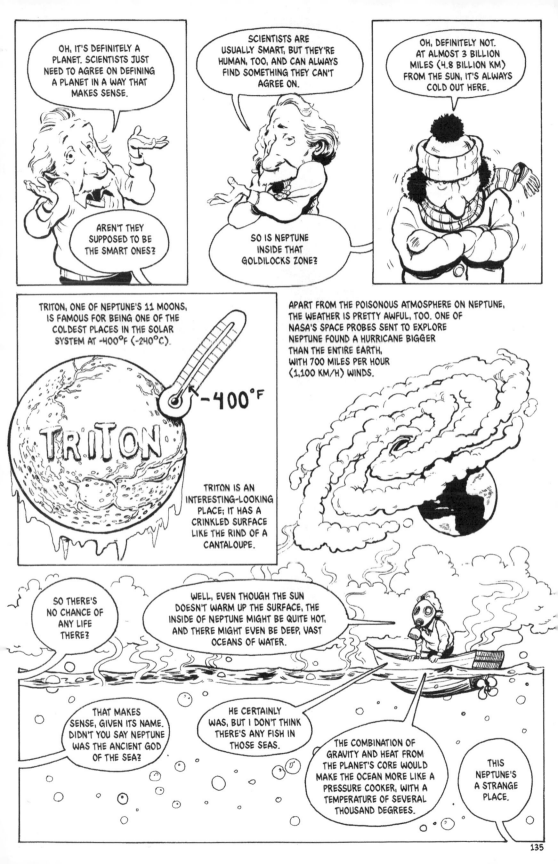

OH, IT'S DEFINITELY A PLANET. SCIENTISTS JUST NEED TO AGREE ON DEFINING A PLANET IN A WAY THAT MAKES SENSE.

AREN'T THEY SUPPOSED TO BE THE SMART ONES?

SCIENTISTS ARE USUALLY SMART, BUT THEY'RE HUMAN, TOO, AND CAN ALWAYS FIND SOMETHING THEY CAN'T AGREE ON.

SO IS NEPTUNE INSIDE THAT GOLDILOCKS ZONE?

OH, DEFINITELY NOT. AT ALMOST 3 BILLION MILES (4.8 BILLION KM) FROM THE SUN, IT'S ALWAYS COLD OUT HERE.

TRITON, ONE OF NEPTUNE'S 11 MOONS, IS FAMOUS FOR BEING ONE OF THE COLDEST PLACES IN THE SOLAR SYSTEM AT -400°F (-240°C).

-400°F

TRITON

TRITON IS AN INTERESTING-LOOKING PLACE; IT HAS A CRINKLED SURFACE LIKE THE RIND OF A CANTALOUPE.

APART FROM THE POISONOUS ATMOSPHERE ON NEPTUNE, THE WEATHER IS PRETTY AWFUL, TOO. ONE OF NASA'S SPACE PROBES SENT TO EXPLORE NEPTUNE FOUND A HURRICANE BIGGER THAN THE ENTIRE EARTH, WITH 700 MILES PER HOUR (1,100 KM/H) WINDS.

SO THERE'S NO CHANCE OF ANY LIFE THERE?

WELL, EVEN THOUGH THE SUN DOESN'T WARM UP THE SURFACE, THE INSIDE OF NEPTUNE MIGHT BE QUITE HOT, AND THERE MIGHT EVEN BE DEEP, VAST OCEANS OF WATER.

THAT MAKES SENSE, GIVEN ITS NAME. DIDN'T YOU SAY NEPTUNE WAS THE ANCIENT GOD OF THE SEA?

HE CERTAINLY WAS, BUT I DON'T THINK THERE'S ANY FISH IN THOSE SEAS.

THE COMBINATION OF GRAVITY AND HEAT FROM THE PLANET'S CORE WOULD MAKE THE OCEAN MORE LIKE A PRESSURE COOKER, WITH A TEMPERATURE OF SEVERAL THOUSAND DEGREES.

THIS NEPTUNE'S A STRANGE PLACE.

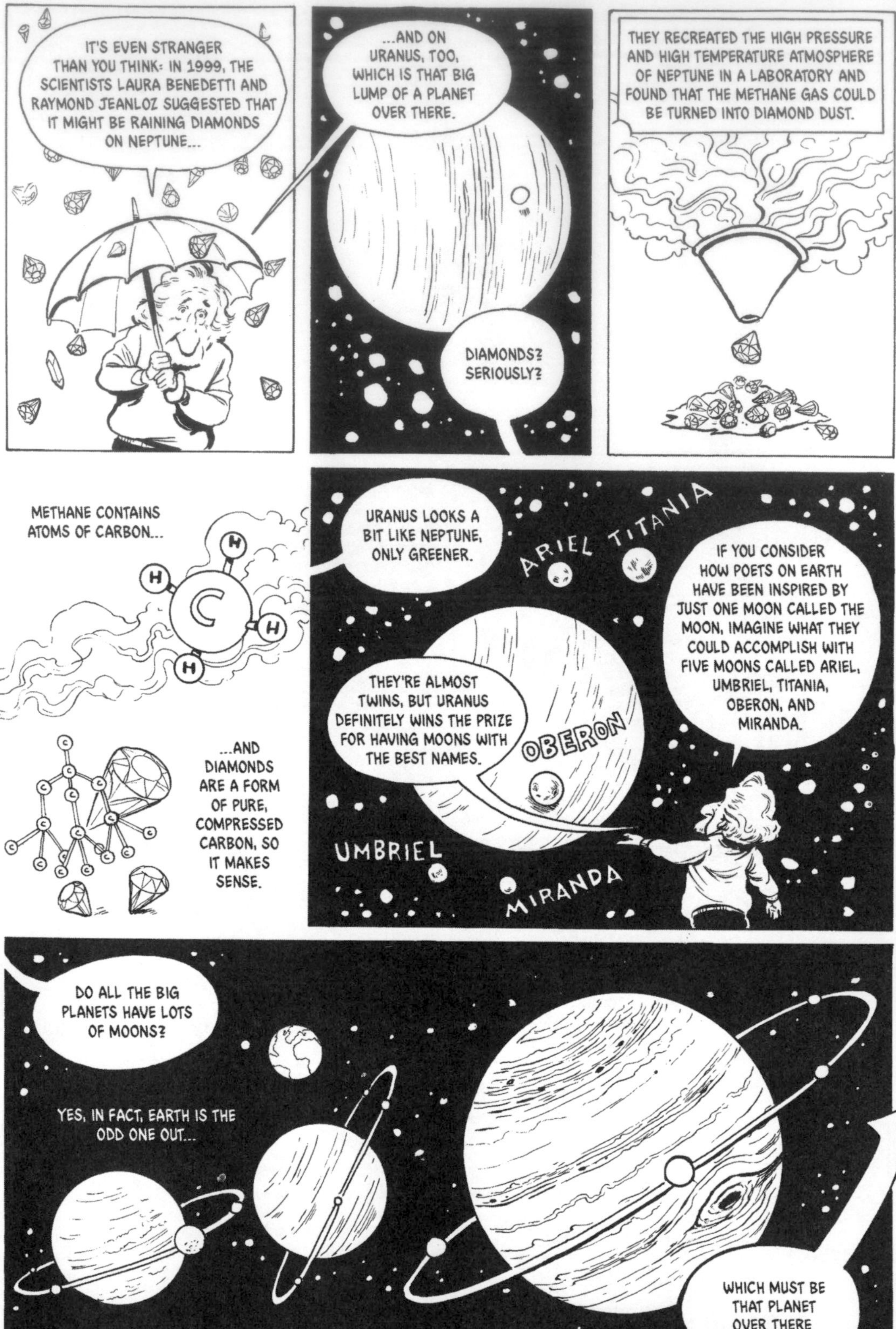
IT'S EVEN STRANGER THAN YOU THINK: IN 1999, THE SCIENTISTS LAURA BENEDETTI AND RAYMOND JEANLOZ SUGGESTED THAT IT MIGHT BE RAINING DIAMONDS ON NEPTUNE...
...AND ON URANUS, TOO, WHICH IS THAT BIG LUMP OF A PLANET OVER THERE.
DIAMONDS? SERIOUSLY?
THEY RECREATED THE HIGH PRESSURE AND HIGH TEMPERATURE ATMOSPHERE OF NEPTUNE IN A LABORATORY AND FOUND THAT THE METHANE GAS COULD BE TURNED INTO DIAMOND DUST.
METHANE CONTAINS ATOMS OF CARBON...
H
C
...AND DIAMONDS ARE A FORM OF PURE, COMPRESSED CARBON, SO IT MAKES SENSE.
URANUS LOOKS A BIT LIKE NEPTUNE, ONLY GREENER.
ARIEL
TITANIA
IF YOU CONSIDER HOW POETS ON EARTH HAVE BEEN INSPIRED BY JUST ONE MOON CALLED THE MOON, IMAGINE WHAT THEY COULD ACCOMPLISH WITH FIVE MOONS CALLED ARIEL, UMBRIEL, TITANIA, OBERON, AND MIRANDA.
THEY'RE ALMOST TWINS, BUT URANUS DEFINITELY WINS THE PRIZE FOR HAVING MOONS WITH THE BEST NAMES.
OBERON
UMBRIEL
MIRANDA
DO ALL THE BIG PLANETS HAVE LOTS OF MOONS?
YES, IN FACT, EARTH IS THE ODD ONE OUT...
WHICH MUST BE THAT PLANET OVER THERE.
...IT'S THE ONLY PLANET IN THE SOLAR SYSTEM WITH A SINGLE MOON.
THE BIG PLANETS ALL HAVE RINGS, TOO, BUT NONE COMPARE TO THE RINGS OF SATURN.

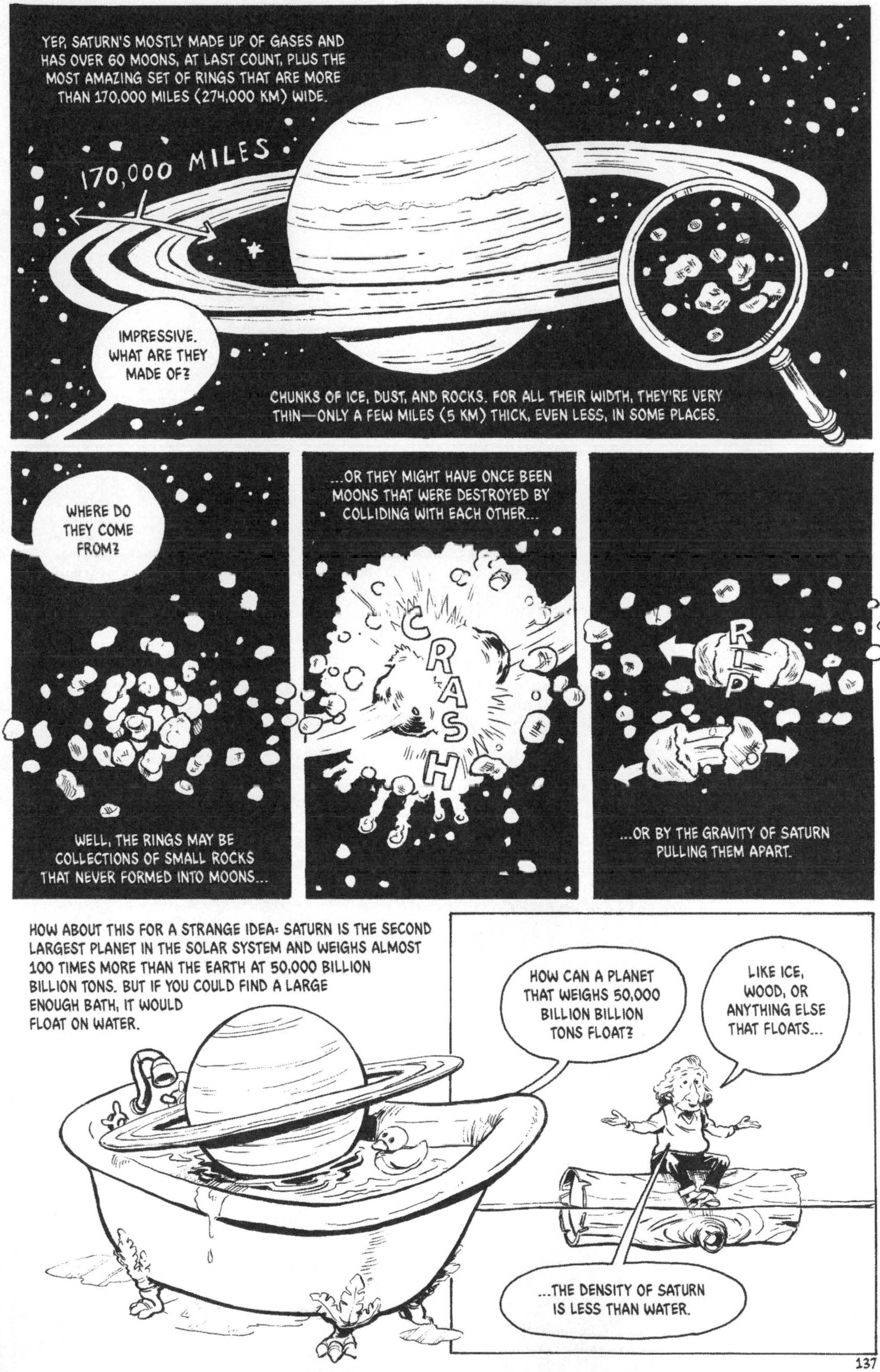

YEP, SATURN'S MOSTLY MADE UP OF GASES AND HAS OVER 60 MOONS, AT LAST COUNT, PLUS THE MOST AMAZING SET OF RINGS THAT ARE MORE THAN 170,000 MILES (274,000 KM) WIDE.
170,000 MILES
IMPRESSIVE. WHAT ARE THEY MADE OF?
CHUNKS OF ICE, DUST, AND ROCKS. FOR ALL THEIR WIDTH, THEY'RE VERY THIN—ONLY A FEW MILES (5 KM) THICK, EVEN LESS, IN SOME PLACES.
WHERE DO THEY COME FROM?
WELL, THE RINGS MAY BE COLLECTIONS OF SMALL ROCKS THAT NEVER FORMED INTO MOONS...
...OR THEY MIGHT HAVE ONCE BEEN MOONS THAT WERE DESTROYED BY COLLIDING WITH EACH OTHER...
CRASH
RIP
...OR BY THE GRAVITY OF SATURN PULLING THEM APART.
HOW ABOUT THIS FOR A STRANGE IDEA: SATURN IS THE SECOND LARGEST PLANET IN THE SOLAR SYSTEM AND WEIGHS ALMOST 100 TIMES MORE THAN THE EARTH AT 50,000 BILLION BILLION TONS. BUT IF YOU COULD FIND A LARGE ENOUGH BATH, IT WOULD FLOAT ON WATER.
HOW CAN A PLANET THAT WEIGHS 50,000 BILLION BILLION TONS FLOAT?
LIKE ICE, WOOD, OR ANYTHING ELSE THAT FLOATS...
...THE DENSITY OF SATURN IS LESS THAN WATER.

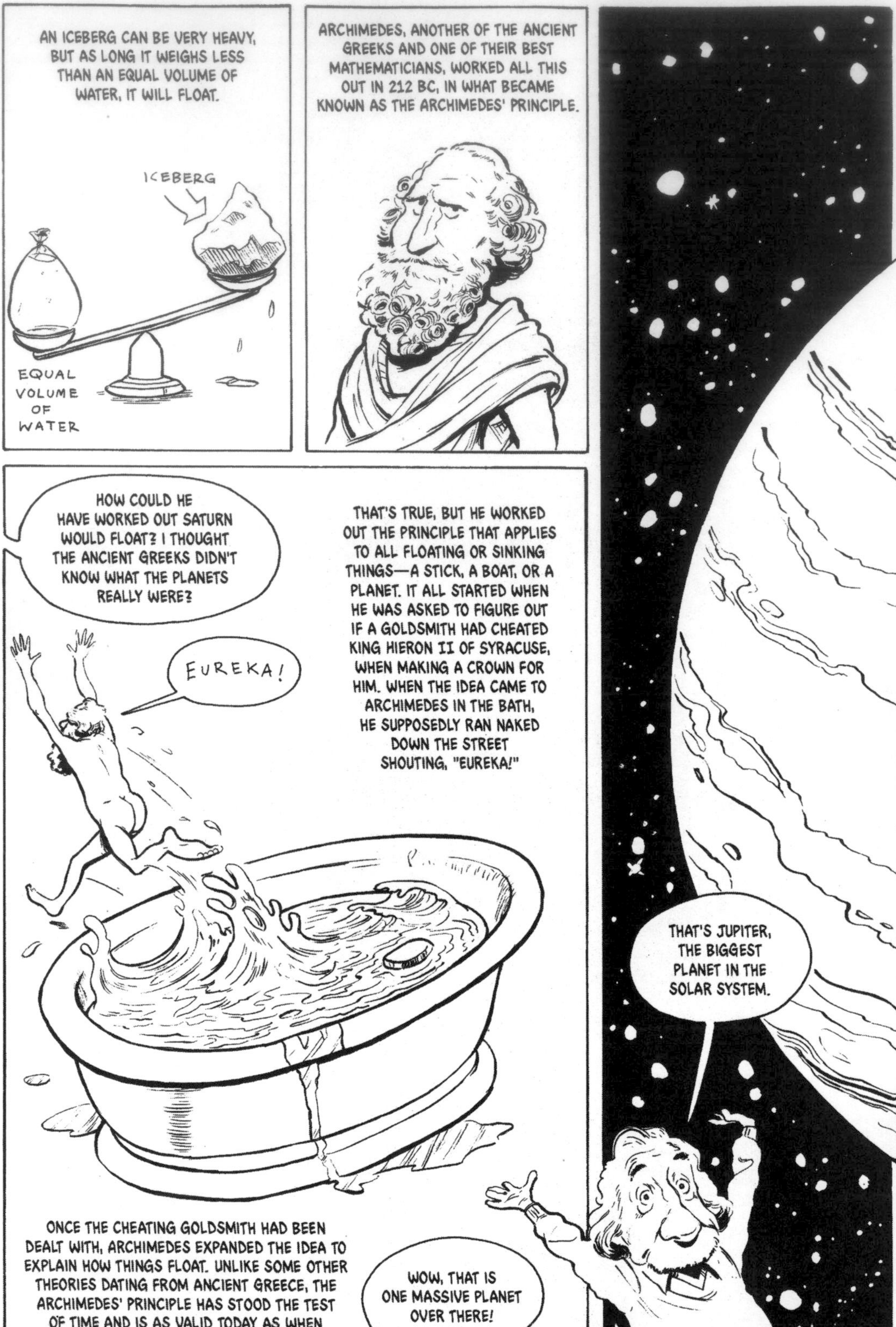
AN ICEBERG CAN BE VERY HEAVY, BUT AS LONG IT WEIGHS LESS THAN AN EQUAL VOLUME OF WATER, IT WILL FLOAT.
ICEBERG
EQUAL VOLUME OF WATER
ARCHIMEDES, ANOTHER OF THE ANCIENT GREEKS AND ONE OF THEIR BEST MATHEMATICIANS, WORKED ALL THIS OUT IN 212 BC, IN WHAT BECAME KNOWN AS THE ARCHIMEDES' PRINCIPLE.
HOW COULD HE HAVE WORKED OUT SATURN WOULD FLOAT? I THOUGHT THE ANCIENT GREEKS DIDN'T KNOW WHAT THE PLANETS REALLY WERE?
THAT'S TRUE, BUT HE WORKED OUT THE PRINCIPLE THAT APPLIES TO ALL FLOATING OR SINKING THINGS—A STICK, A BOAT, OR A PLANET. IT ALL STARTED WHEN HE WAS ASKED TO FIGURE OUT IF A GOLDSMITH HAD CHEATED KING HIERON II OF SYRACUSE, WHEN MAKING A CROWN FOR HIM. WHEN THE IDEA CAME TO ARCHIMEDES IN THE BATH, HE SUPPOSEDLY RAN NAKED DOWN THE STREET SHOUTING, "EUREKA!"
EUREKA!
ONCE THE CHEATING GOLDSMITH HAD BEEN DEALT WITH, ARCHIMEDES EXPANDED THE IDEA TO EXPLAIN HOW THINGS FLOAT. UNLIKE SOME OTHER THEORIES DATING FROM ANCIENT GREECE, THE ARCHIMEDES' PRINCIPLE HAS STOOD THE TEST OF TIME AND IS AS VALID TODAY AS WHEN ARCHIMEDES LEAPT OUT OF HIS BATH DRIPPING WITH WATER AND ENTHUSIASM.
WOW, THAT IS ONE MASSIVE PLANET OVER THERE!
THAT'S JUPITER, THE BIGGEST PLANET IN THE SOLAR SYSTEM.

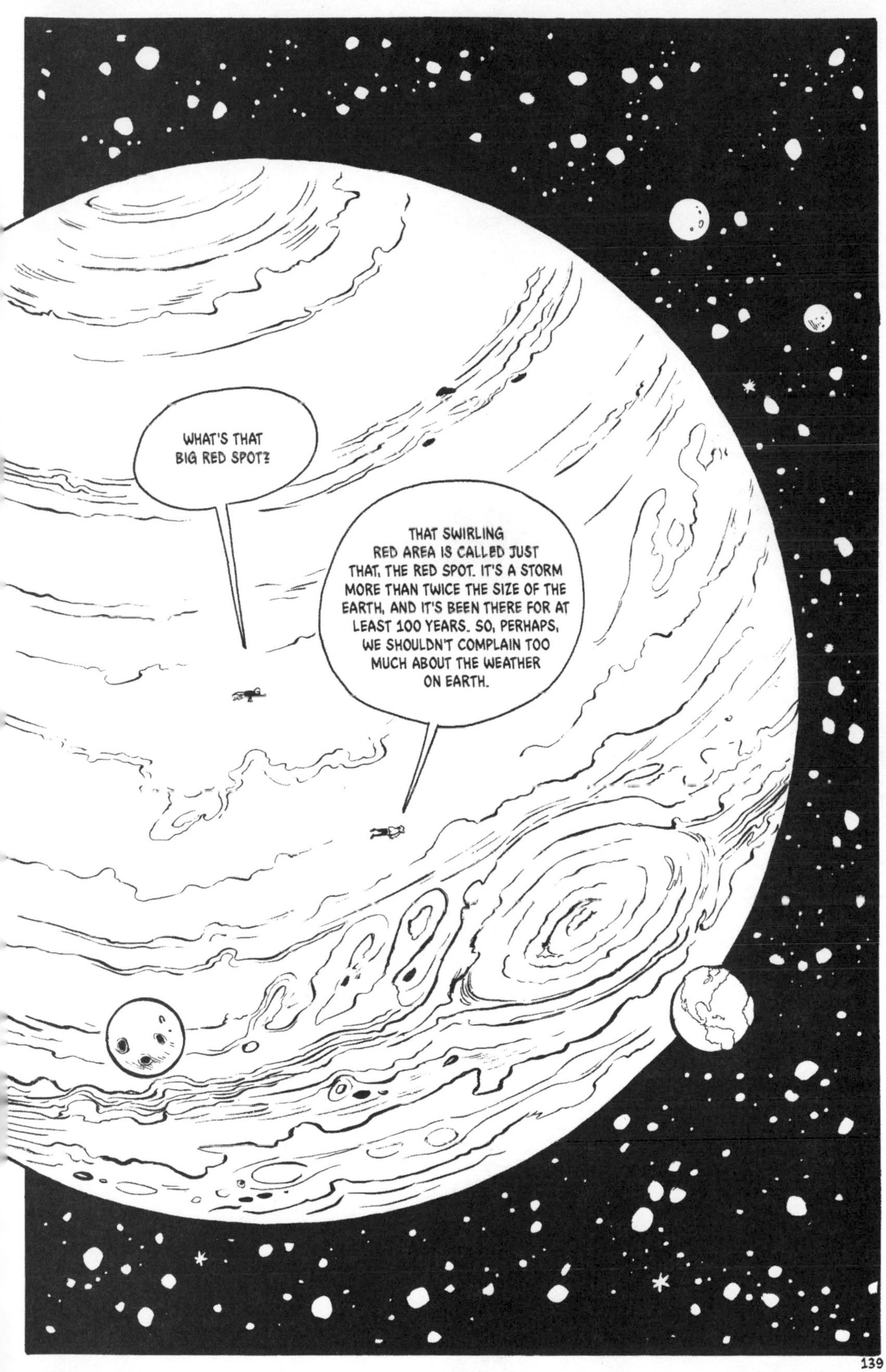
WHAT'S THAT BIG RED SPOT?
THAT SWIRLING RED AREA IS CALLED JUST THAT, THE RED SPOT. IT'S A STORM MORE THAN TWICE THE SIZE OF THE EARTH, AND IT'S BEEN THERE FOR AT LEAST 100 YEARS. SO, PERHAPS, WE SHOULDN'T COMPLAIN TOO MUCH ABOUT THE WEATHER ON EARTH.

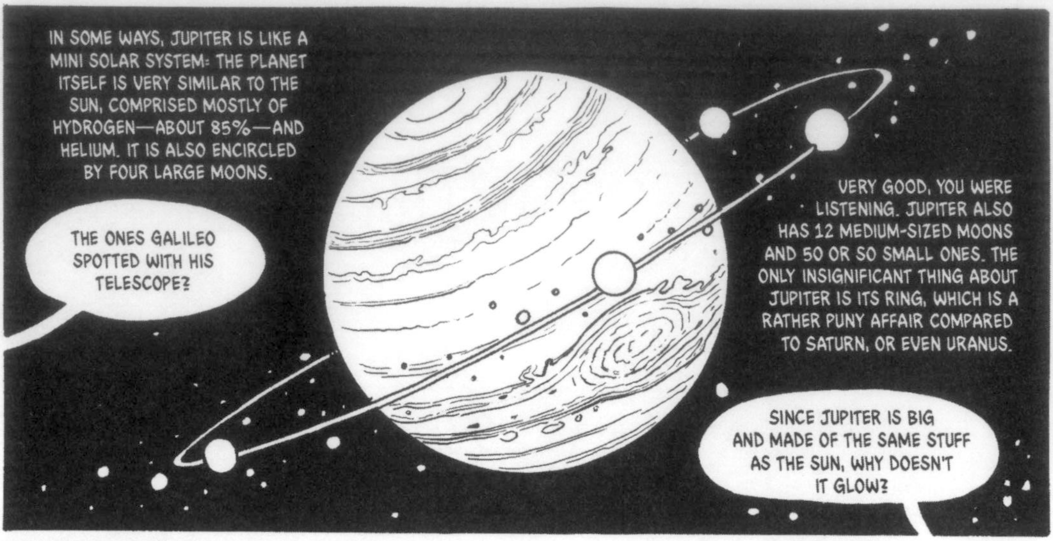
IN SOME WAYS, JUPITER IS LIKE A MINI SOLAR SYSTEM: THE PLANET ITSELF IS VERY SIMILAR TO THE SUN, COMPRISED MOSTLY OF HYDROGEN—ABOUT 85%—AND HELIUM. IT IS ALSO ENCIRCLED BY FOUR LARGE MOONS.
THE ONES GALILEO SPOTTED WITH HIS TELESCOPE?
VERY GOOD, YOU WERE LISTENING. JUPITER ALSO HAS 12 MEDIUM-SIZED MOONS AND 50 OR SO SMALL ONES. THE ONLY INSIGNIFICANT THING ABOUT JUPITER IS ITS RING, WHICH IS A RATHER PUNY AFFAIR COMPARED TO SATURN, OR EVEN URANUS.
SINCE JUPITER IS BIG AND MADE OF THE SAME STUFF AS THE SUN, WHY DOESN'T IT GLOW?

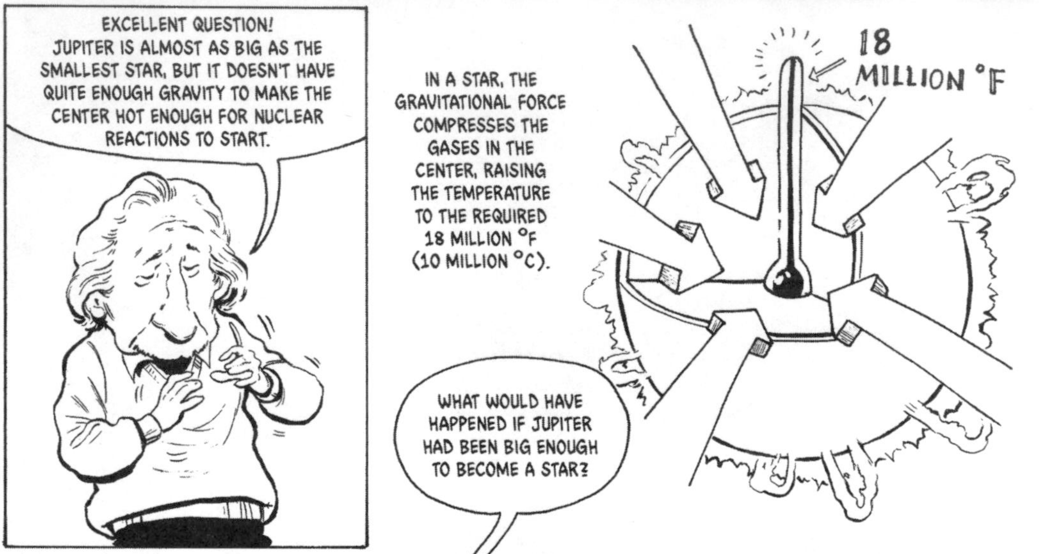
EXCELLENT QUESTION! JUPITER IS ALMOST AS BIG AS THE SMALLEST STAR, BUT IT DOESN'T HAVE QUITE ENOUGH GRAVITY TO MAKE THE CENTER HOT ENOUGH FOR NUCLEAR REACTIONS TO START.
IN A STAR, THE GRAVITATIONAL FORCE COMPRESSES THE GASES IN THE CENTER, RAISING THE TEMPERATURE TO THE REQUIRED 18 MILLION °F (10 MILLION °C).
18 MILLION °F
WHAT WOULD HAVE HAPPENED IF JUPITER HAD BEEN BIG ENOUGH TO BECOME A STAR?

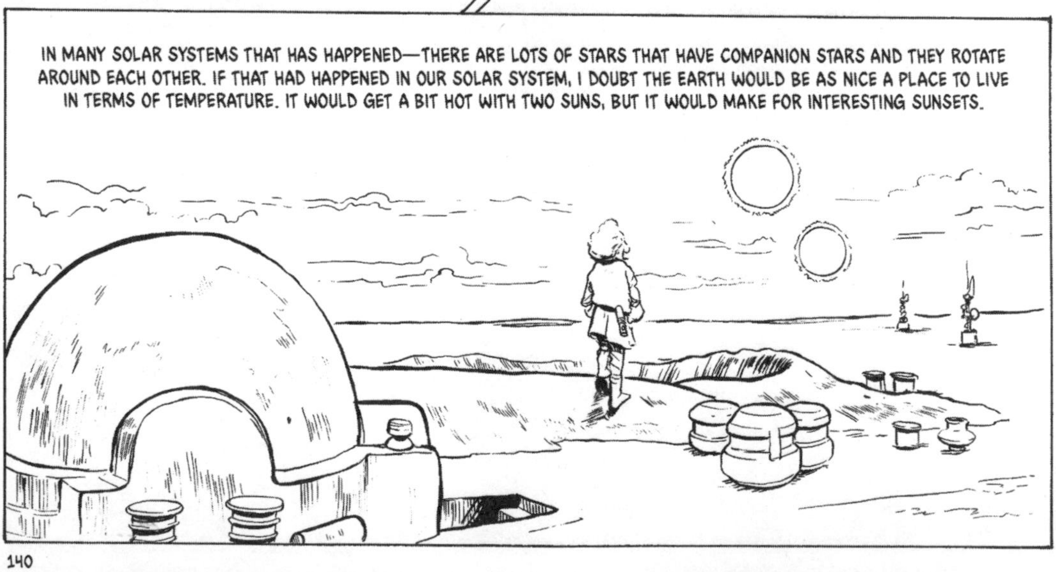
IN MANY SOLAR SYSTEMS THAT HAS HAPPENED—THERE ARE LOTS OF STARS THAT HAVE COMPANION STARS AND THEY ROTATE AROUND EACH OTHER. IF THAT HAD HAPPENED IN OUR SOLAR SYSTEM, I DOUBT THE EARTH WOULD BE AS NICE A PLACE TO LIVE IN TERMS OF TEMPERATURE. IT WOULD GET A BIT HOT WITH TWO SUNS, BUT IT WOULD MAKE FOR INTERESTING SUNSETS.

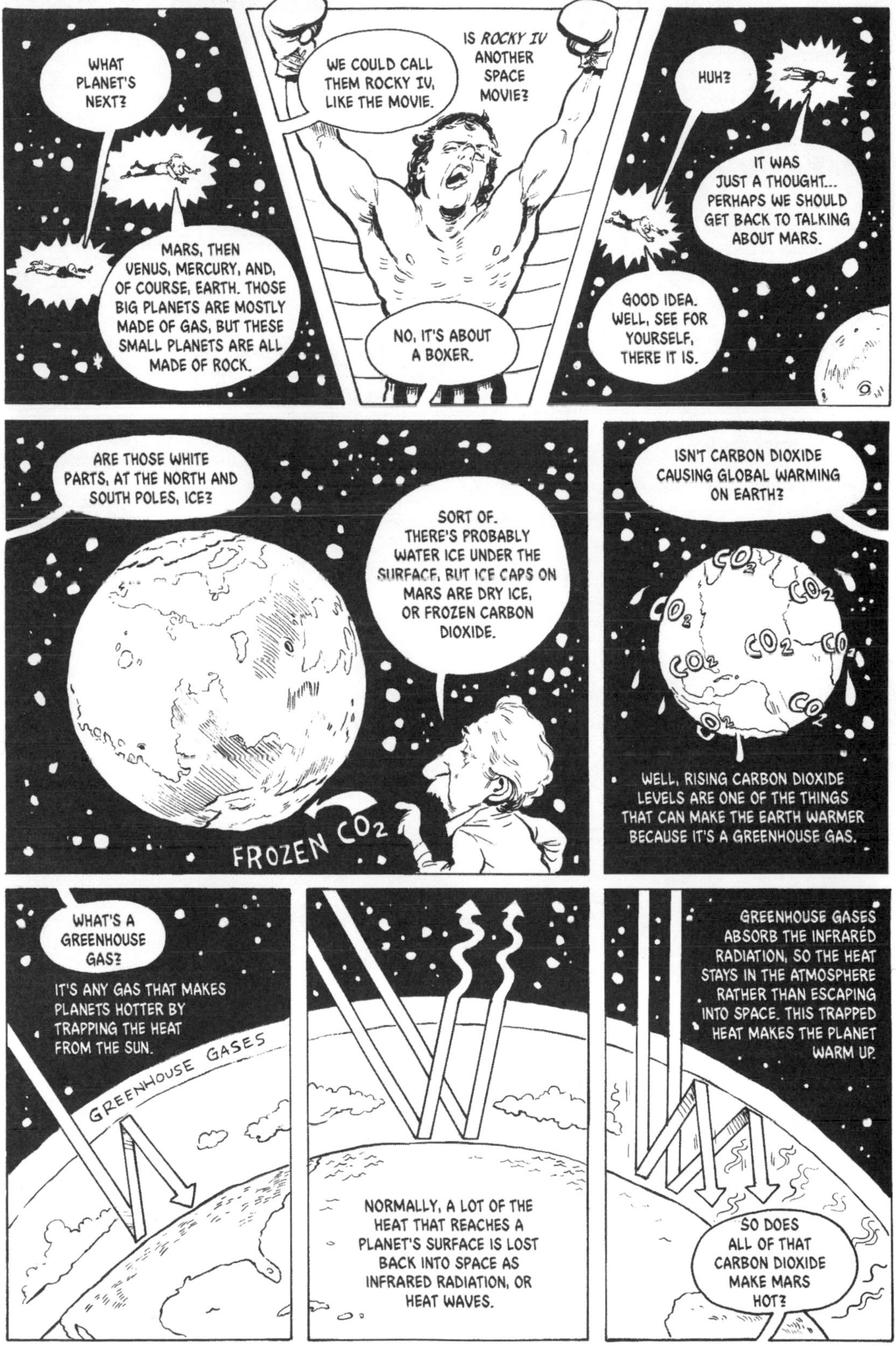
WHAT PLANET'S NEXT?
MARS, THEN VENUS, MERCURY, AND, OF COURSE, EARTH. THOSE BIG PLANETS ARE MOSTLY MADE OF GAS, BUT THESE SMALL PLANETS ARE ALL MADE OF ROCK.
WE COULD CALL THEM ROCKY IV, LIKE THE MOVIE.
IS *ROCKY IV* ANOTHER SPACE MOVIE?
NO, IT'S ABOUT A BOXER.
HUH?
IT WAS JUST A THOUGHT... PERHAPS WE SHOULD GET BACK TO TALKING ABOUT MARS.
GOOD IDEA. WELL, SEE FOR YOURSELF, THERE IT IS.
ARE THOSE WHITE PARTS, AT THE NORTH AND SOUTH POLES, ICE?
SORT OF. THERE'S PROBABLY WATER ICE UNDER THE SURFACE, BUT ICE CAPS ON MARS ARE DRY ICE, OR FROZEN CARBON DIOXIDE.
FROZEN CO_2
ISN'T CARBON DIOXIDE CAUSING GLOBAL WARMING ON EARTH?
CO_2
WELL, RISING CARBON DIOXIDE LEVELS ARE ONE OF THE THINGS THAT CAN MAKE THE EARTH WARMER BECAUSE IT'S A GREENHOUSE GAS.
WHAT'S A GREENHOUSE GAS?
IT'S ANY GAS THAT MAKES PLANETS HOTTER BY TRAPPING THE HEAT FROM THE SUN.
GREENHOUSE GASES
NORMALLY, A LOT OF THE HEAT THAT REACHES A PLANET'S SURFACE IS LOST BACK INTO SPACE AS INFRARED RADIATION, OR HEAT WAVES.
GREENHOUSE GASES ABSORB THE INFRARED RADIATION, SO THE HEAT STAYS IN THE ATMOSPHERE RATHER THAN ESCAPING INTO SPACE. THIS TRAPPED HEAT MAKES THE PLANET WARM UP.
SO DOES ALL OF THAT CARBON DIOXIDE MAKE MARS HOT?

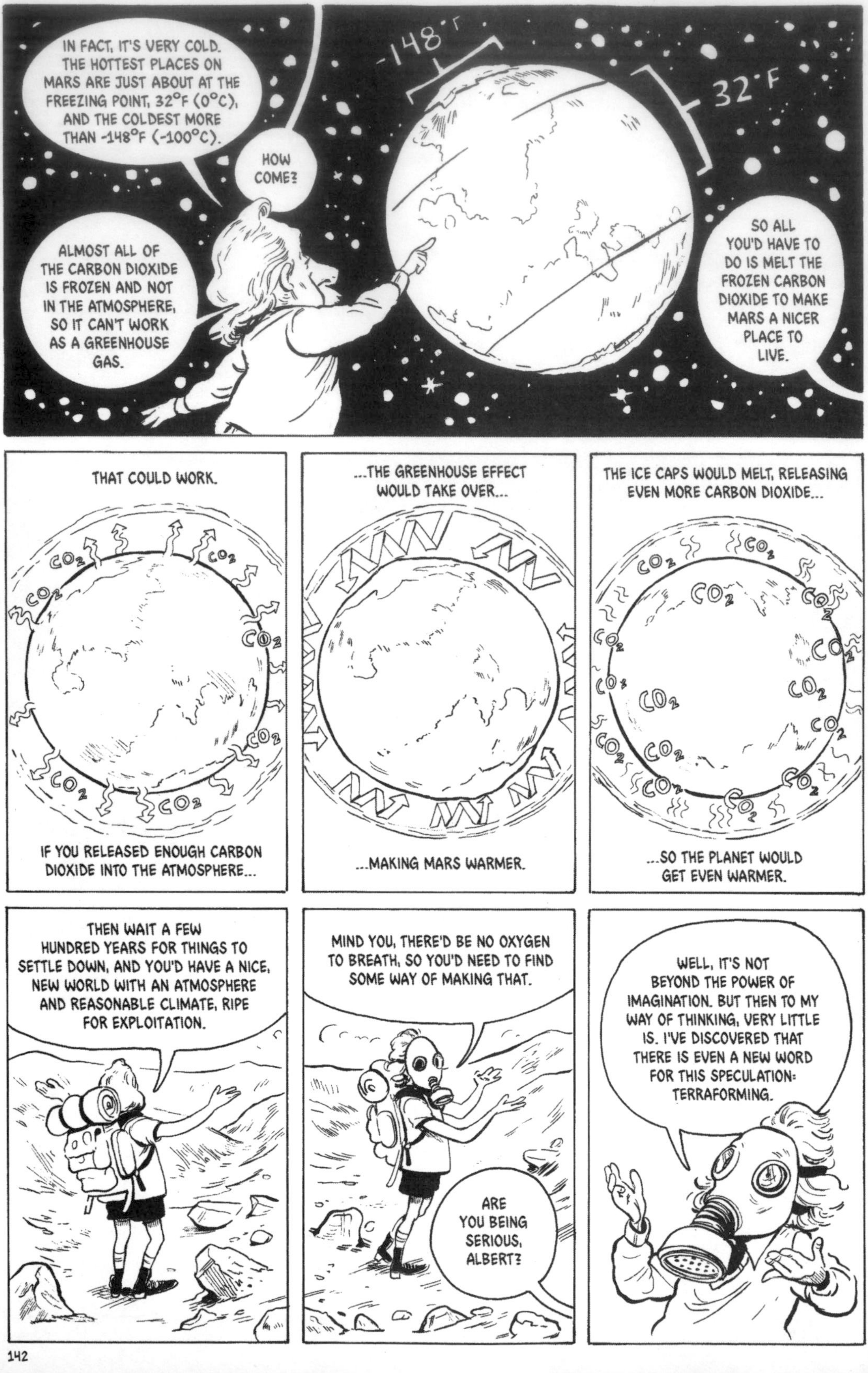
IN FACT, IT'S VERY COLD. THE HOTTEST PLACES ON MARS ARE JUST ABOUT AT THE FREEZING POINT, 32°F (0°C), AND THE COLDEST MORE THAN -148°F (-100°C).
HOW COME?
ALMOST ALL OF THE CARBON DIOXIDE IS FROZEN AND NOT IN THE ATMOSPHERE, SO IT CAN'T WORK AS A GREENHOUSE GAS.
-148°F
32°F
SO ALL YOU'D HAVE TO DO IS MELT THE FROZEN CARBON DIOXIDE TO MAKE MARS A NICER PLACE TO LIVE.
THAT COULD WORK.
CO_2
IF YOU RELEASED ENOUGH CARBON DIOXIDE INTO THE ATMOSPHERE...
...THE GREENHOUSE EFFECT WOULD TAKE OVER...
...MAKING MARS WARMER.
THE ICE CAPS WOULD MELT, RELEASING EVEN MORE CARBON DIOXIDE...
CO_2
...SO THE PLANET WOULD GET EVEN WARMER.
THEN WAIT A FEW HUNDRED YEARS FOR THINGS TO SETTLE DOWN, AND YOU'D HAVE A NICE, NEW WORLD WITH AN ATMOSPHERE AND REASONABLE CLIMATE, RIPE FOR EXPLOITATION.
MIND YOU, THERE'D BE NO OXYGEN TO BREATH, SO YOU'D NEED TO FIND SOME WAY OF MAKING THAT.
ARE YOU BEING SERIOUS, ALBERT?
WELL, IT'S NOT BEYOND THE POWER OF IMAGINATION. BUT THEN TO MY WAY OF THINKING, VERY LITTLE IS. I'VE DISCOVERED THAT THERE IS EVEN A NEW WORD FOR THIS SPECULATION: TERRAFORMING.

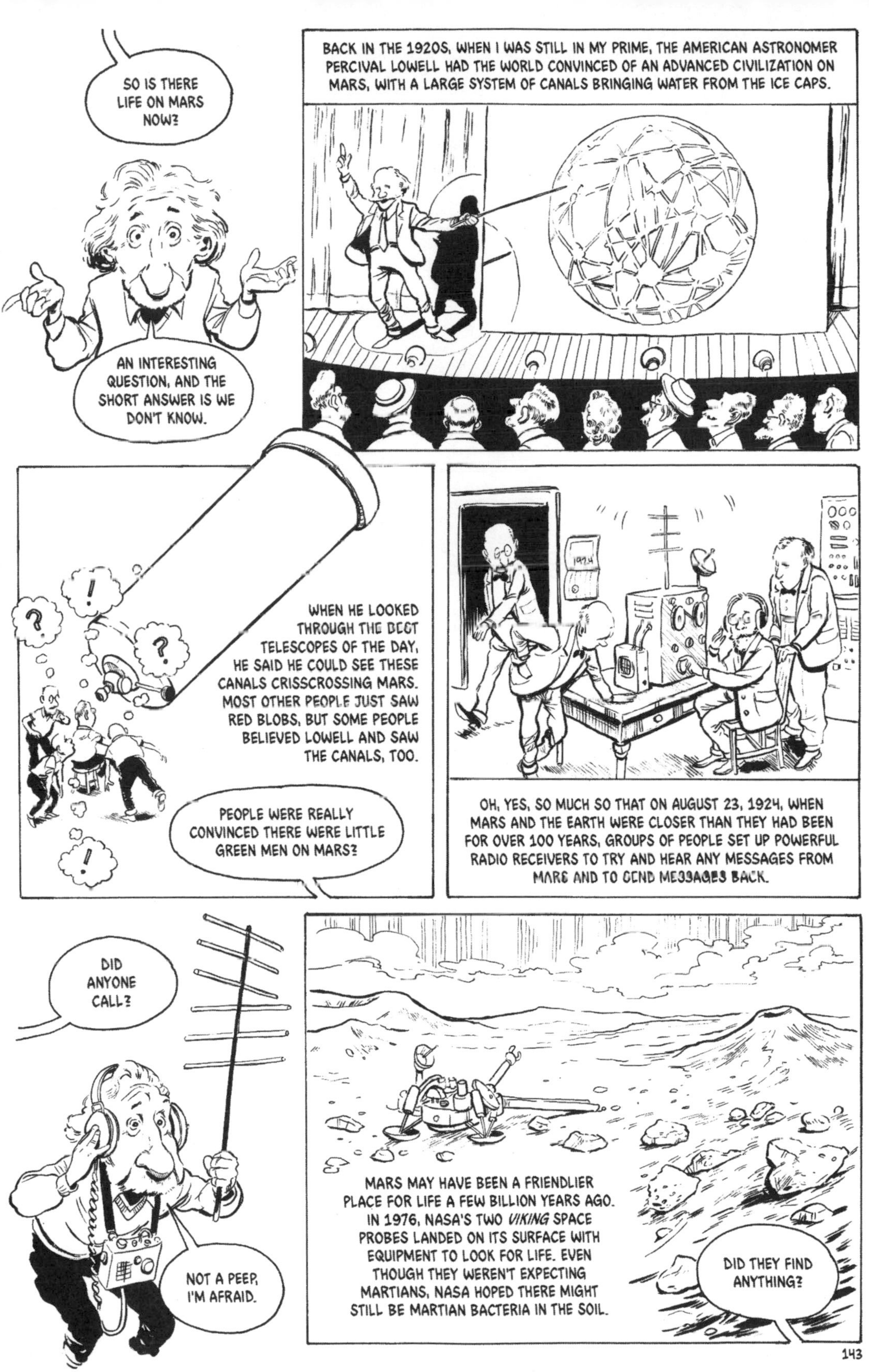
SO IS THERE LIFE ON MARS NOW?
AN INTERESTING QUESTION, AND THE SHORT ANSWER IS WE DON'T KNOW.
BACK IN THE 1920S, WHEN I WAS STILL IN MY PRIME, THE AMERICAN ASTRONOMER PERCIVAL LOWELL HAD THE WORLD CONVINCED OF AN ADVANCED CIVILIZATION ON MARS, WITH A LARGE SYSTEM OF CANALS BRINGING WATER FROM THE ICE CAPS.
WHEN HE LOOKED THROUGH THE BEST TELESCOPES OF THE DAY, HE SAID HE COULD SEE THESE CANALS CRISSCROSSING MARS. MOST OTHER PEOPLE JUST SAW RED BLOBS, BUT SOME PEOPLE BELIEVED LOWELL AND SAW THE CANALS, TOO.
?
!
?
!
PEOPLE WERE REALLY CONVINCED THERE WERE LITTLE GREEN MEN ON MARS?
1924
OH, YES, SO MUCH SO THAT ON AUGUST 23, 1924, WHEN MARS AND THE EARTH WERE CLOSER THAN THEY HAD BEEN FOR OVER 100 YEARS, GROUPS OF PEOPLE SET UP POWERFUL RADIO RECEIVERS TO TRY AND HEAR ANY MESSAGES FROM MARS AND TO SEND MESSAGES BACK.
DID ANYONE CALL?
NOT A PEEP, I'M AFRAID.
MARS MAY HAVE BEEN A FRIENDLIER PLACE FOR LIFE A FEW BILLION YEARS AGO. IN 1976, NASA'S TWO *VIKING* SPACE PROBES LANDED ON ITS SURFACE WITH EQUIPMENT TO LOOK FOR LIFE. EVEN THOUGH THEY WEREN'T EXPECTING MARTIANS, NASA HOPED THERE MIGHT STILL BE MARTIAN BACTERIA IN THE SOIL.
DID THEY FIND ANYTHING?

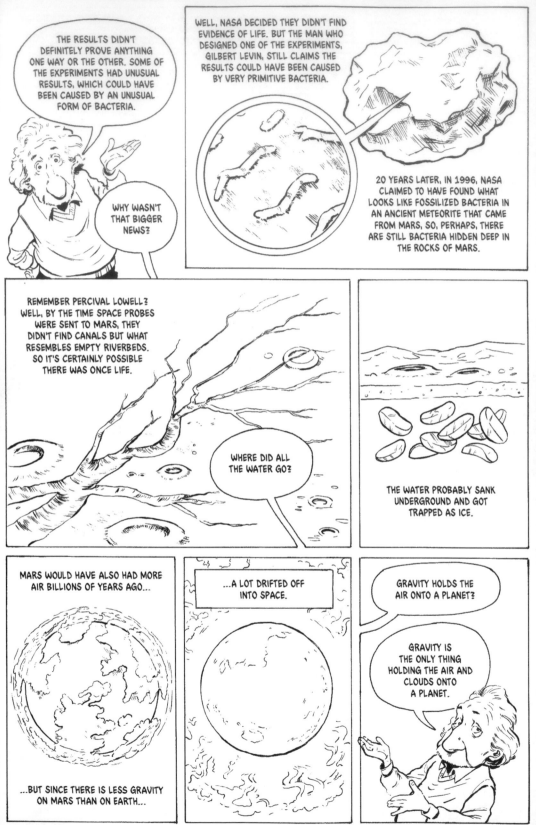

THE RESULTS DIDN'T DEFINITELY PROVE ANYTHING ONE WAY OR THE OTHER. SOME OF THE EXPERIMENTS HAD UNUSUAL RESULTS, WHICH COULD HAVE BEEN CAUSED BY AN UNUSUAL FORM OF BACTERIA.

WHY WASN'T THAT BIGGER NEWS?

WELL, NASA DECIDED THEY DIDN'T FIND EVIDENCE OF LIFE. BUT THE MAN WHO DESIGNED ONE OF THE EXPERIMENTS, GILBERT LEVIN, STILL CLAIMS THE RESULTS COULD HAVE BEEN CAUSED BY VERY PRIMITIVE BACTERIA.

20 YEARS LATER, IN 1996, NASA CLAIMED TO HAVE FOUND WHAT LOOKS LIKE FOSSILIZED BACTERIA IN AN ANCIENT METEORITE THAT CAME FROM MARS, SO, PERHAPS, THERE ARE STILL BACTERIA HIDDEN DEEP IN THE ROCKS OF MARS.

REMEMBER PERCIVAL LOWELL? WELL, BY THE TIME SPACE PROBES WERE SENT TO MARS, THEY DIDN'T FIND CANALS BUT WHAT RESEMBLES EMPTY RIVERBEDS. SO IT'S CERTAINLY POSSIBLE THERE WAS ONCE LIFE.

WHERE DID ALL THE WATER GO?

THE WATER PROBABLY SANK UNDERGROUND AND GOT TRAPPED AS ICE.

MARS WOULD HAVE ALSO HAD MORE AIR BILLIONS OF YEARS AGO...

...BUT SINCE THERE IS LESS GRAVITY ON MARS THAN ON EARTH...

...A LOT DRIFTED OFF INTO SPACE.

GRAVITY HOLDS THE AIR ONTO A PLANET?

GRAVITY IS THE ONLY THING HOLDING THE AIR AND CLOUDS ONTO A PLANET.

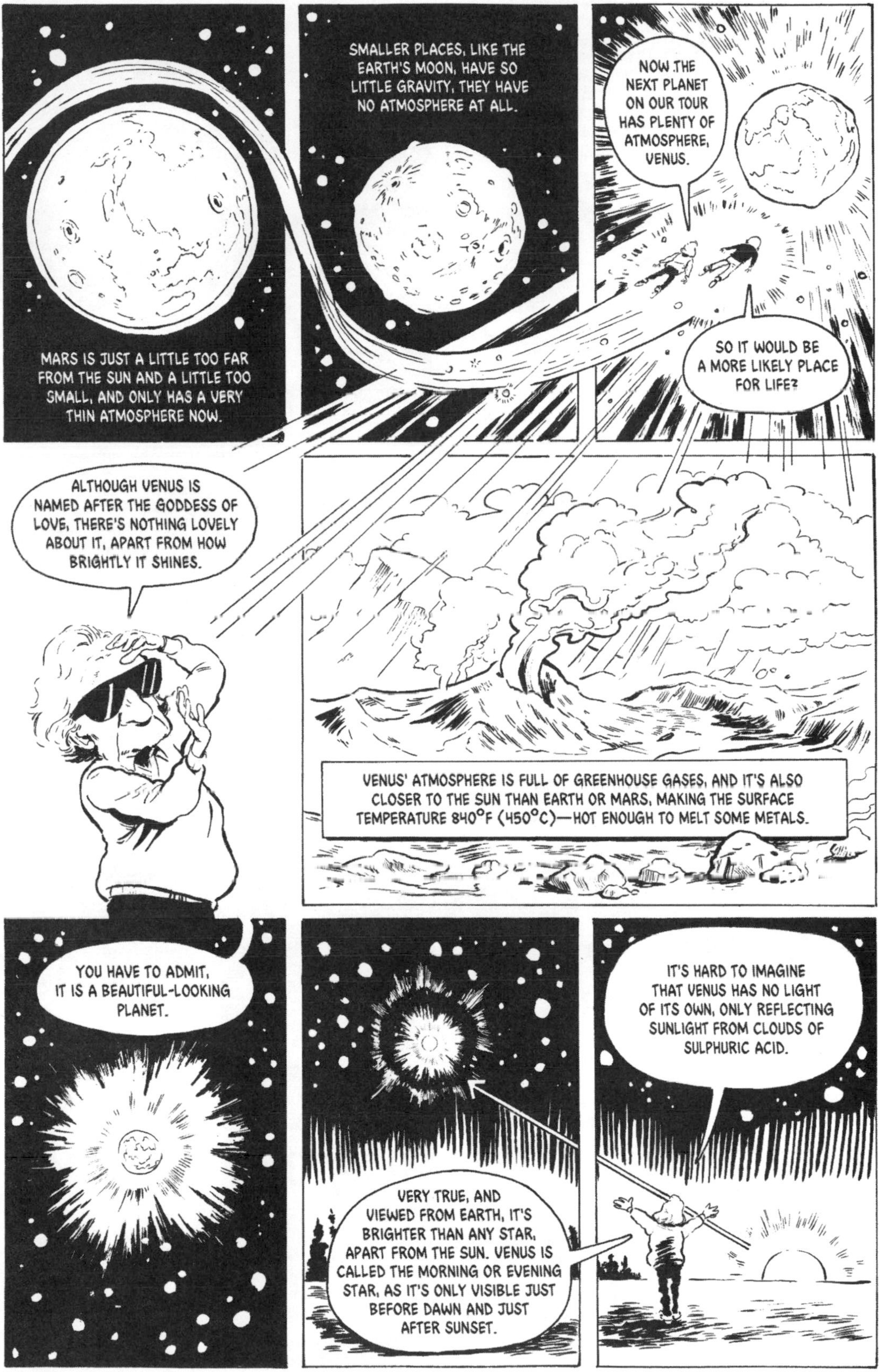

MARS IS JUST A LITTLE TOO FAR FROM THE SUN AND A LITTLE TOO SMALL, AND ONLY HAS A VERY THIN ATMOSPHERE NOW.
SMALLER PLACES, LIKE THE EARTH'S MOON, HAVE SO LITTLE GRAVITY, THEY HAVE NO ATMOSPHERE AT ALL.
NOW THE NEXT PLANET ON OUR TOUR HAS PLENTY OF ATMOSPHERE, VENUS.
SO IT WOULD BE A MORE LIKELY PLACE FOR LIFE?
ALTHOUGH VENUS IS NAMED AFTER THE GODDESS OF LOVE, THERE'S NOTHING LOVELY ABOUT IT, APART FROM HOW BRIGHTLY IT SHINES.
VENUS' ATMOSPHERE IS FULL OF GREENHOUSE GASES, AND IT'S ALSO CLOSER TO THE SUN THAN EARTH OR MARS, MAKING THE SURFACE TEMPERATURE 840°F (450°C)—HOT ENOUGH TO MELT SOME METALS.
YOU HAVE TO ADMIT, IT IS A BEAUTIFUL-LOOKING PLANET.
VERY TRUE, AND VIEWED FROM EARTH, IT'S BRIGHTER THAN ANY STAR, APART FROM THE SUN. VENUS IS CALLED THE MORNING OR EVENING STAR, AS IT'S ONLY VISIBLE JUST BEFORE DAWN AND JUST AFTER SUNSET.
IT'S HARD TO IMAGINE THAT VENUS HAS NO LIGHT OF ITS OWN, ONLY REFLECTING SUNLIGHT FROM CLOUDS OF SULPHURIC ACID.

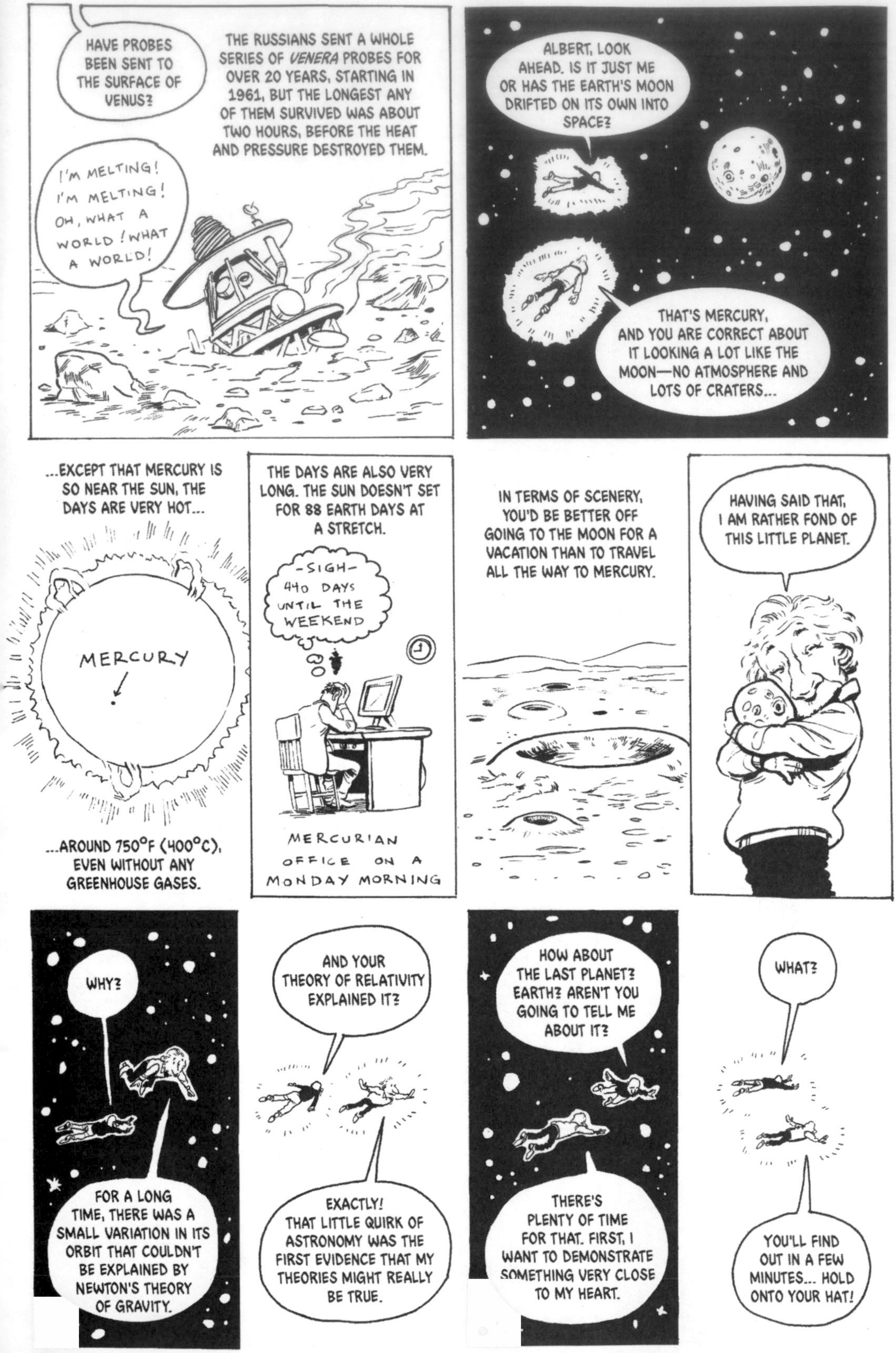
HAVE PROBES BEEN SENT TO THE SURFACE OF VENUS?
THE RUSSIANS SENT A WHOLE SERIES OF *VENERA* PROBES FOR OVER 20 YEARS, STARTING IN 1961, BUT THE LONGEST ANY OF THEM SURVIVED WAS ABOUT TWO HOURS, BEFORE THE HEAT AND PRESSURE DESTROYED THEM.
I'M MELTING! I'M MELTING! OH, WHAT A WORLD! WHAT A WORLD!
ALBERT, LOOK AHEAD. IS IT JUST ME OR HAS THE EARTH'S MOON DRIFTED ON ITS OWN INTO SPACE?
THAT'S MERCURY, AND YOU ARE CORRECT ABOUT IT LOOKING A LOT LIKE THE MOON—NO ATMOSPHERE AND LOTS OF CRATERS...
...EXCEPT THAT MERCURY IS SO NEAR THE SUN, THE DAYS ARE VERY HOT...
MERCURY
...AROUND 750°F (400°C), EVEN WITHOUT ANY GREENHOUSE GASES.
THE DAYS ARE ALSO VERY LONG. THE SUN DOESN'T SET FOR 88 EARTH DAYS AT A STRETCH.
-SIGH- 440 DAYS UNTIL THE WEEKEND
MERCURIAN OFFICE ON A MONDAY MORNING
IN TERMS OF SCENERY, YOU'D BE BETTER OFF GOING TO THE MOON FOR A VACATION THAN TO TRAVEL ALL THE WAY TO MERCURY.
HAVING SAID THAT, I AM RATHER FOND OF THIS LITTLE PLANET.
WHY?
FOR A LONG TIME, THERE WAS A SMALL VARIATION IN ITS ORBIT THAT COULDN'T BE EXPLAINED BY NEWTON'S THEORY OF GRAVITY.
AND YOUR THEORY OF RELATIVITY EXPLAINED IT?
EXACTLY! THAT LITTLE QUIRK OF ASTRONOMY WAS THE FIRST EVIDENCE THAT MY THEORIES MIGHT REALLY BE TRUE.
HOW ABOUT THE LAST PLANET? EARTH? AREN'T YOU GOING TO TELL ME ABOUT IT?
THERE'S PLENTY OF TIME FOR THAT. FIRST, I WANT TO DEMONSTRATE SOMETHING VERY CLOSE TO MY HEART.
WHAT?
YOU'LL FIND OUT IN A FEW MINUTES... HOLD ONTO YOUR HAT!

CHAPTER 27: A CLOSE ENCOUNTER OF THE SOLAR KIND

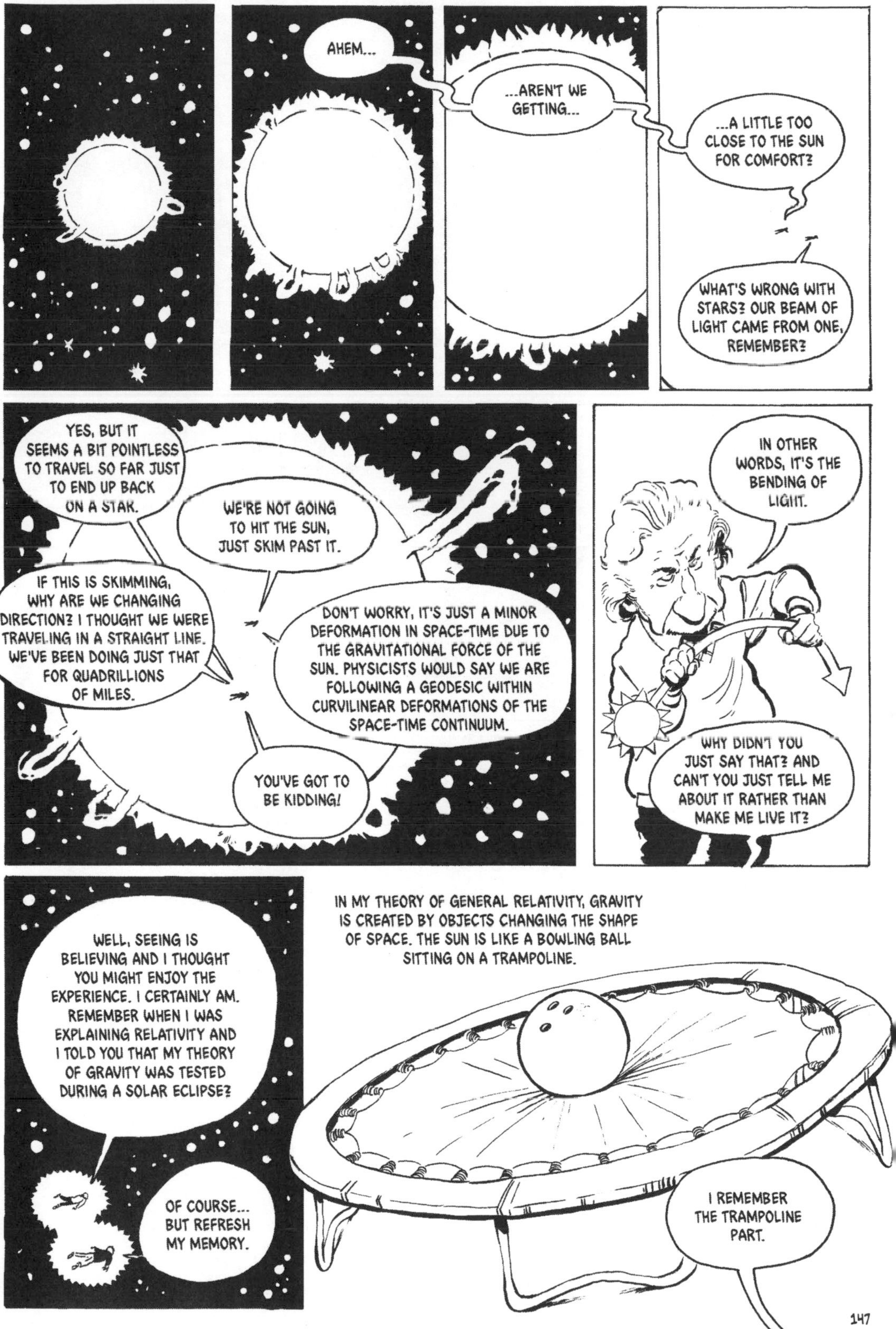

AHEM...
...AREN'T WE GETTING...
...A LITTLE TOO CLOSE TO THE SUN FOR COMFORT?
WHAT'S WRONG WITH STARS? OUR BEAM OF LIGHT CAME FROM ONE, REMEMBER?
YES, BUT IT SEEMS A BIT POINTLESS TO TRAVEL SO FAR JUST TO END UP BACK ON A STAR.
WE'RE NOT GOING TO HIT THE SUN, JUST SKIM PAST IT.
IF THIS IS SKIMMING, WHY ARE WE CHANGING DIRECTION? I THOUGHT WE WERE TRAVELING IN A STRAIGHT LINE. WE'VE BEEN DOING JUST THAT FOR QUADRILLIONS OF MILES.
DON'T WORRY, IT'S JUST A MINOR DEFORMATION IN SPACE-TIME DUE TO THE GRAVITATIONAL FORCE OF THE SUN. PHYSICISTS WOULD SAY WE ARE FOLLOWING A GEODESIC WITHIN CURVILINEAR DEFORMATIONS OF THE SPACE-TIME CONTINUUM.
YOU'VE GOT TO BE KIDDING!
IN OTHER WORDS, IT'S THE BENDING OF LIGHT.
WHY DIDN'T YOU JUST SAY THAT? AND CAN'T YOU JUST TELL ME ABOUT IT RATHER THAN MAKE ME LIVE IT?
WELL, SEEING IS BELIEVING AND I THOUGHT YOU MIGHT ENJOY THE EXPERIENCE. I CERTAINLY AM. REMEMBER WHEN I WAS EXPLAINING RELATIVITY AND I TOLD YOU THAT MY THEORY OF GRAVITY WAS TESTED DURING A SOLAR ECLIPSE?
OF COURSE... BUT REFRESH MY MEMORY.
IN MY THEORY OF GENERAL RELATIVITY, GRAVITY IS CREATED BY OBJECTS CHANGING THE SHAPE OF SPACE. THE SUN IS LIKE A BOWLING BALL SITTING ON A TRAMPOLINE.
I REMEMBER THE TRAMPOLINE PART.

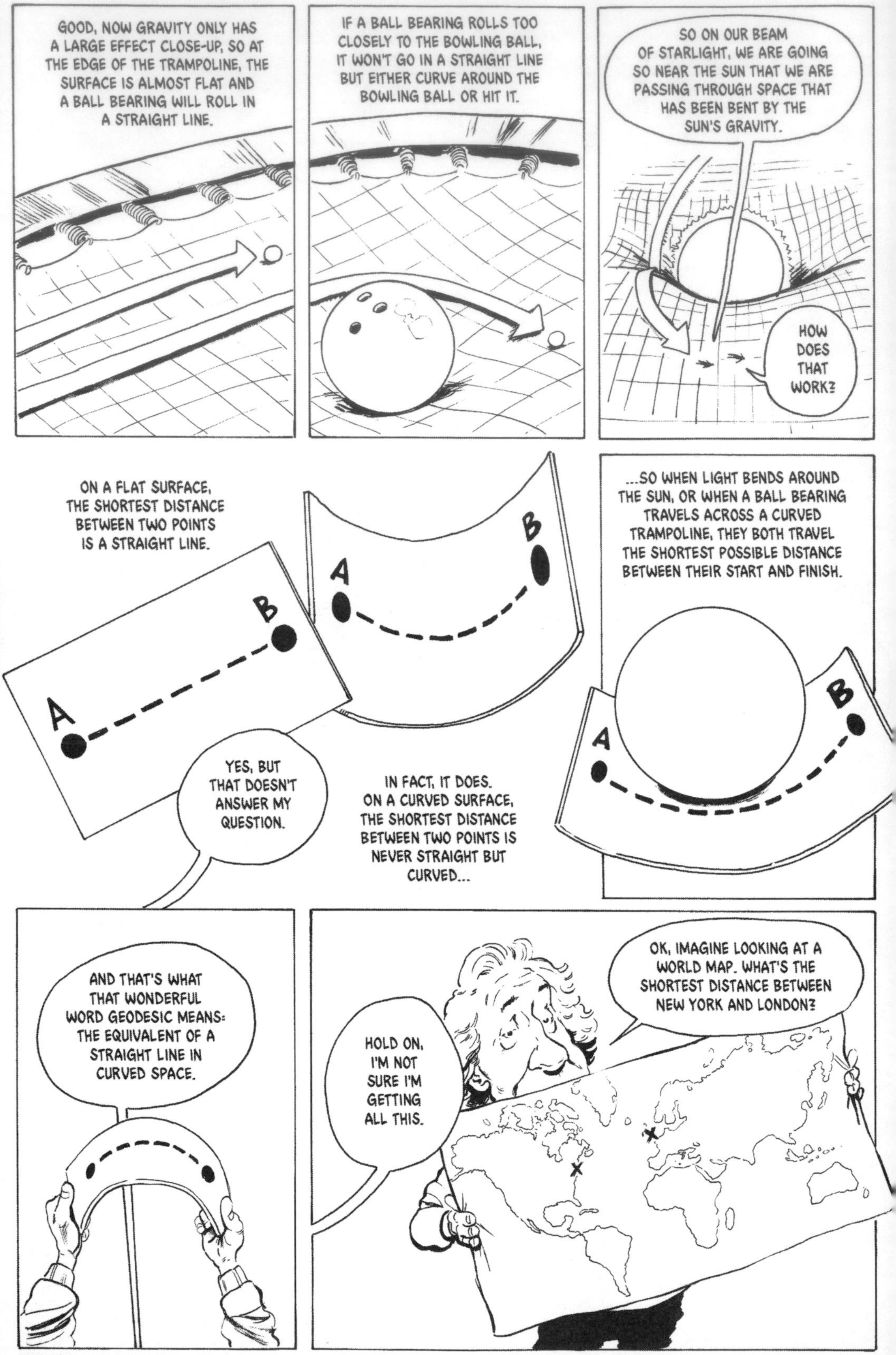
GOOD, NOW GRAVITY ONLY HAS A LARGE EFFECT CLOSE-UP, SO AT THE EDGE OF THE TRAMPOLINE, THE SURFACE IS ALMOST FLAT AND A BALL BEARING WILL ROLL IN A STRAIGHT LINE.
IF A BALL BEARING ROLLS TOO CLOSELY TO THE BOWLING BALL, IT WON'T GO IN A STRAIGHT LINE BUT EITHER CURVE AROUND THE BOWLING BALL OR HIT IT.
SO ON OUR BEAM OF STARLIGHT, WE ARE GOING SO NEAR THE SUN THAT WE ARE PASSING THROUGH SPACE THAT HAS BEEN BENT BY THE SUN'S GRAVITY.
HOW DOES THAT WORK?
ON A FLAT SURFACE, THE SHORTEST DISTANCE BETWEEN TWO POINTS IS A STRAIGHT LINE.
A
B
A
B
YES, BUT THAT DOESN'T ANSWER MY QUESTION.
IN FACT, IT DOES. ON A CURVED SURFACE, THE SHORTEST DISTANCE BETWEEN TWO POINTS IS NEVER STRAIGHT BUT CURVED...
...SO WHEN LIGHT BENDS AROUND THE SUN, OR WHEN A BALL BEARING TRAVELS ACROSS A CURVED TRAMPOLINE, THEY BOTH TRAVEL THE SHORTEST POSSIBLE DISTANCE BETWEEN THEIR START AND FINISH.
A
B
AND THAT'S WHAT THAT WONDERFUL WORD GEODESIC MEANS: THE EQUIVALENT OF A STRAIGHT LINE IN CURVED SPACE.
OK, IMAGINE LOOKING AT A WORLD MAP. WHAT'S THE SHORTEST DISTANCE BETWEEN NEW YORK AND LONDON?
HOLD ON, I'M NOT SURE I'M GETTING ALL THIS.

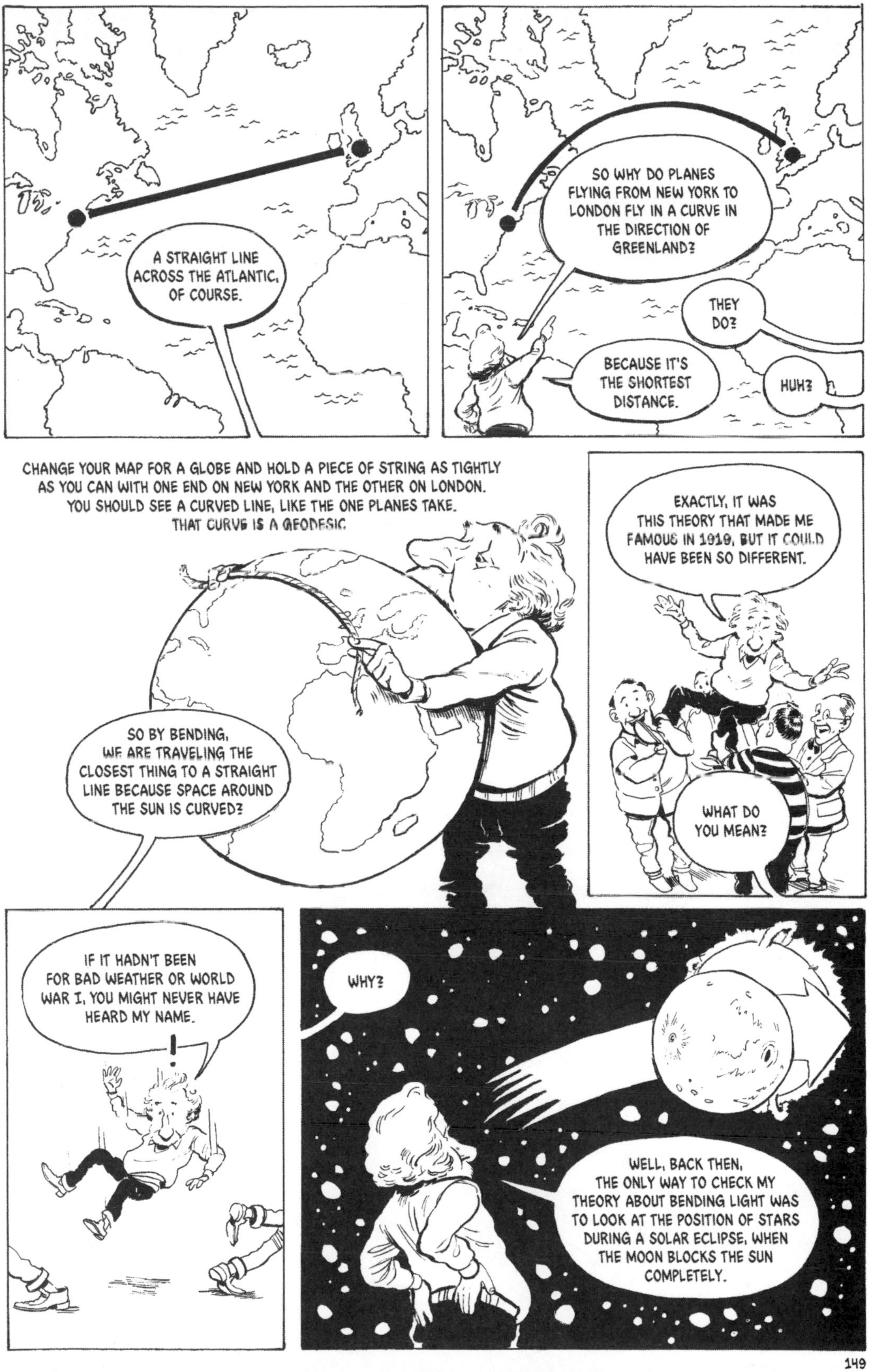
A STRAIGHT LINE ACROSS THE ATLANTIC, OF COURSE.
SO WHY DO PLANES FLYING FROM NEW YORK TO LONDON FLY IN A CURVE IN THE DIRECTION OF GREENLAND?
THEY DO?
BECAUSE IT'S THE SHORTEST DISTANCE.
HUH?
CHANGE YOUR MAP FOR A GLOBE AND HOLD A PIECE OF STRING AS TIGHTLY AS YOU CAN WITH ONE END ON NEW YORK AND THE OTHER ON LONDON. YOU SHOULD SEE A CURVED LINE, LIKE THE ONE PLANES TAKE. THAT CURVE IS A GEODESIC.
SO BY BENDING, WE ARE TRAVELING THE CLOSEST THING TO A STRAIGHT LINE BECAUSE SPACE AROUND THE SUN IS CURVED?
EXACTLY, IT WAS THIS THEORY THAT MADE ME FAMOUS IN 1919, BUT IT COULD HAVE BEEN SO DIFFERENT.
WHAT DO YOU MEAN?
IF IT HADN'T BEEN FOR BAD WEATHER OR WORLD WAR I, YOU MIGHT NEVER HAVE HEARD MY NAME.
!
WHY?
WELL, BACK THEN, THE ONLY WAY TO CHECK MY THEORY ABOUT BENDING LIGHT WAS TO LOOK AT THE POSITION OF STARS DURING A SOLAR ECLIPSE, WHEN THE MOON BLOCKS THE SUN COMPLETELY.

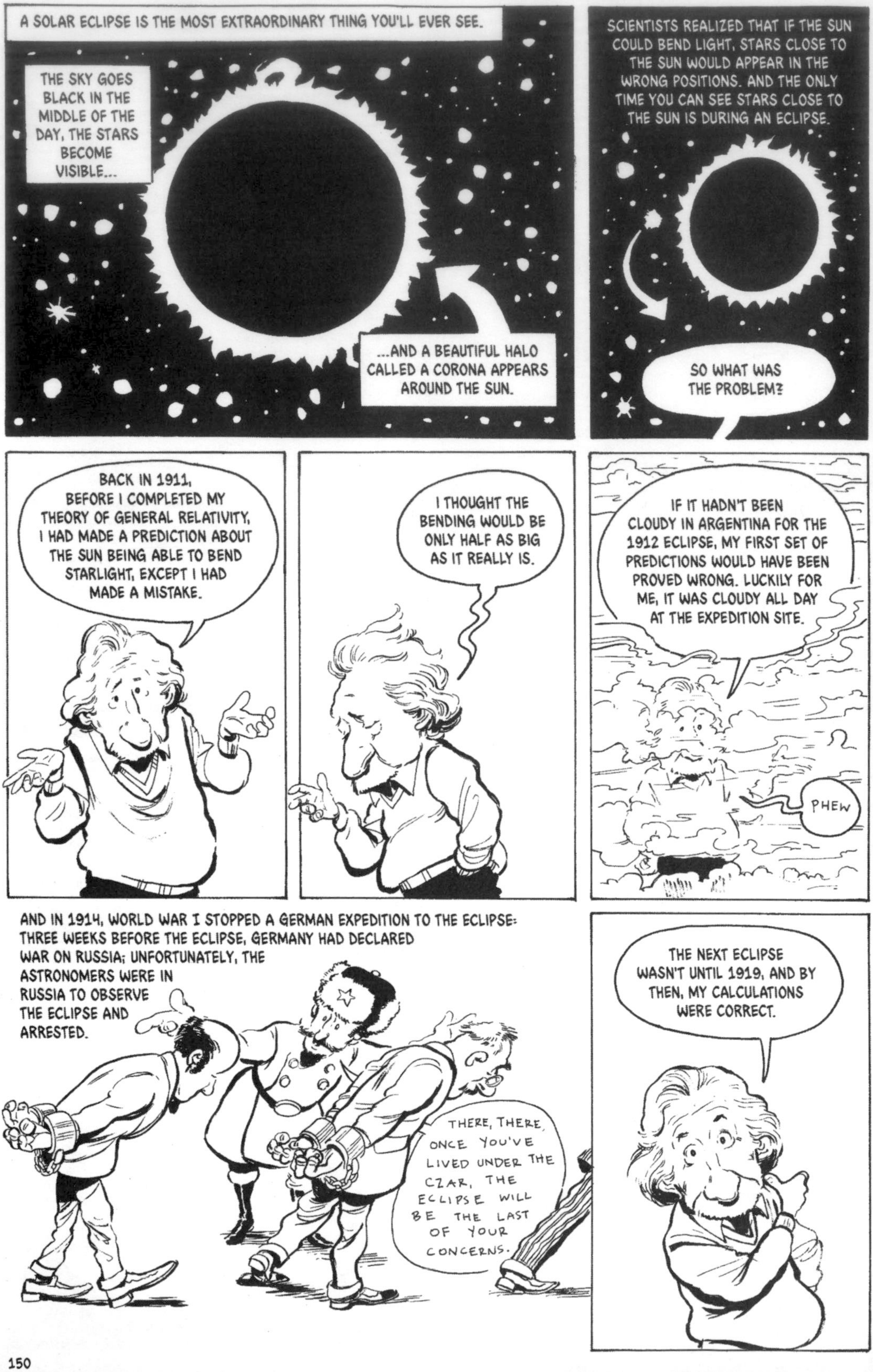
A SOLAR ECLIPSE IS THE MOST EXTRAORDINARY THING YOU'LL EVER SEE.
THE SKY GOES BLACK IN THE MIDDLE OF THE DAY, THE STARS BECOME VISIBLE...
...AND A BEAUTIFUL HALO CALLED A CORONA APPEARS AROUND THE SUN.
SCIENTISTS REALIZED THAT IF THE SUN COULD BEND LIGHT, STARS CLOSE TO THE SUN WOULD APPEAR IN THE WRONG POSITIONS. AND THE ONLY TIME YOU CAN SEE STARS CLOSE TO THE SUN IS DURING AN ECLIPSE.
SO WHAT WAS THE PROBLEM?
BACK IN 1911, BEFORE I COMPLETED MY THEORY OF GENERAL RELATIVITY, I HAD MADE A PREDICTION ABOUT THE SUN BEING ABLE TO BEND STARLIGHT, EXCEPT I HAD MADE A MISTAKE.
I THOUGHT THE BENDING WOULD BE ONLY HALF AS BIG AS IT REALLY IS.
IF IT HADN'T BEEN CLOUDY IN ARGENTINA FOR THE 1912 ECLIPSE, MY FIRST SET OF PREDICTIONS WOULD HAVE BEEN PROVED WRONG. LUCKILY FOR ME, IT WAS CLOUDY ALL DAY AT THE EXPEDITION SITE.
PHEW
AND IN 1914, WORLD WAR I STOPPED A GERMAN EXPEDITION TO THE ECLIPSE: THREE WEEKS BEFORE THE ECLIPSE, GERMANY HAD DECLARED WAR ON RUSSIA; UNFORTUNATELY, THE ASTRONOMERS WERE IN RUSSIA TO OBSERVE THE ECLIPSE AND ARRESTED.
THERE, THERE. ONCE YOU'VE LIVED UNDER THE CZAR, THE ECLIPSE WILL BE THE LAST OF YOUR CONCERNS.
THE NEXT ECLIPSE WASN'T UNTIL 1919, AND BY THEN, MY CALCULATIONS WERE CORRECT.

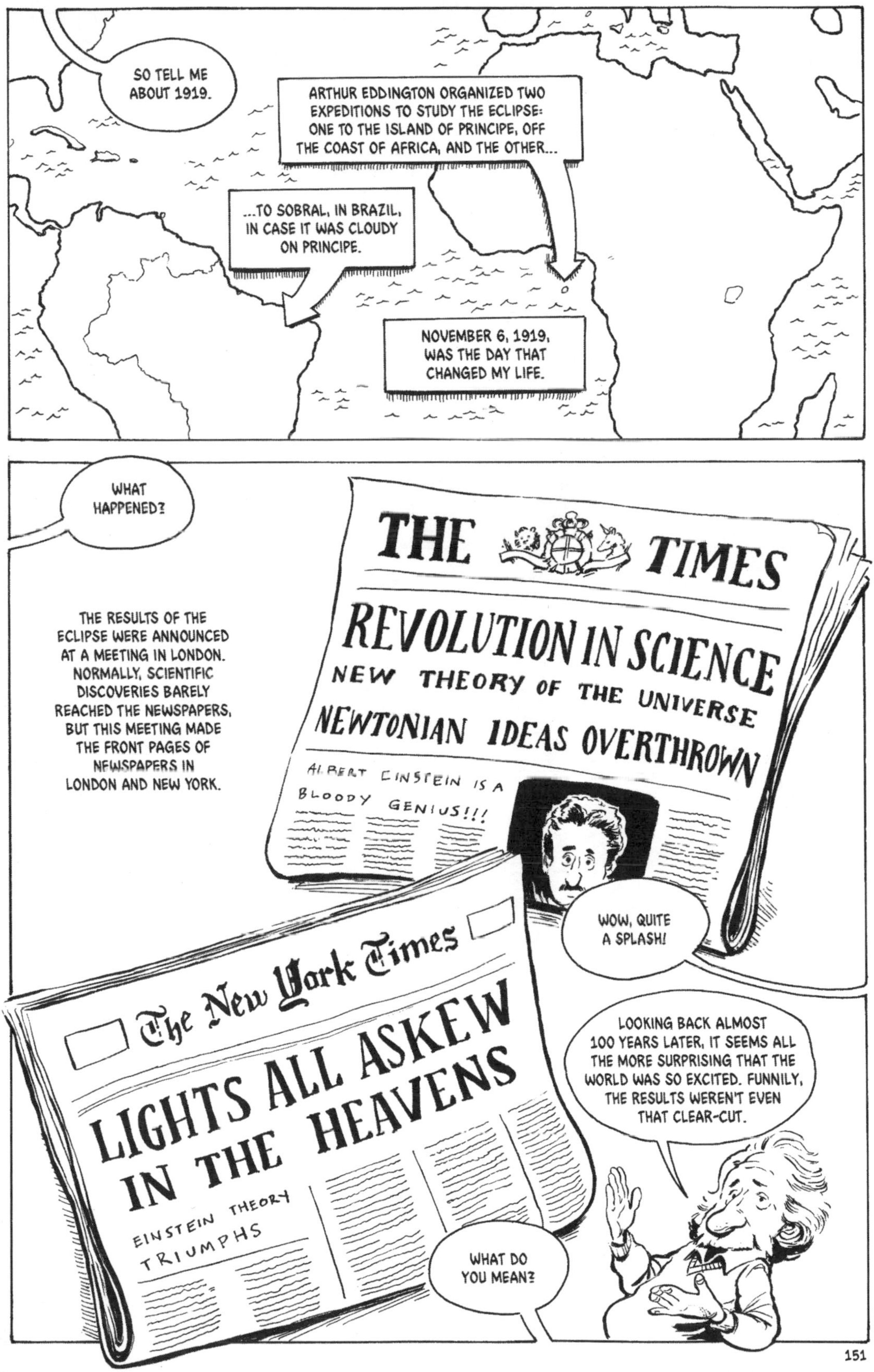
SO TELL ME ABOUT 1919.
ARTHUR EDDINGTON ORGANIZED TWO EXPEDITIONS TO STUDY THE ECLIPSE: ONE TO THE ISLAND OF PRINCIPE, OFF THE COAST OF AFRICA, AND THE OTHER...
...TO SOBRAL, IN BRAZIL, IN CASE IT WAS CLOUDY ON PRINCIPE.
NOVEMBER 6, 1919, WAS THE DAY THAT CHANGED MY LIFE.
WHAT HAPPENED?
THE RESULTS OF THE ECLIPSE WERE ANNOUNCED AT A MEETING IN LONDON. NORMALLY, SCIENTIFIC DISCOVERIES BARELY REACHED THE NEWSPAPERS, BUT THIS MEETING MADE THE FRONT PAGES OF NEWSPAPERS IN LONDON AND NEW YORK.
THE TIMES
REVOLUTION IN SCIENCE
NEW THEORY OF THE UNIVERSE
NEWTONIAN IDEAS OVERTHROWN
ALBERT EINSTEIN IS A BLOODY GENIUS!!!
WOW, QUITE A SPLASH!
The New York Times
LIGHTS ALL ASKEW IN THE HEAVENS
EINSTEIN THEORY TRIUMPHS
LOOKING BACK ALMOST 100 YEARS LATER, IT SEEMS ALL THE MORE SURPRISING THAT THE WORLD WAS SO EXCITED. FUNNILY, THE RESULTS WEREN'T EVEN THAT CLEAR-CUT.
WHAT DO YOU MEAN?

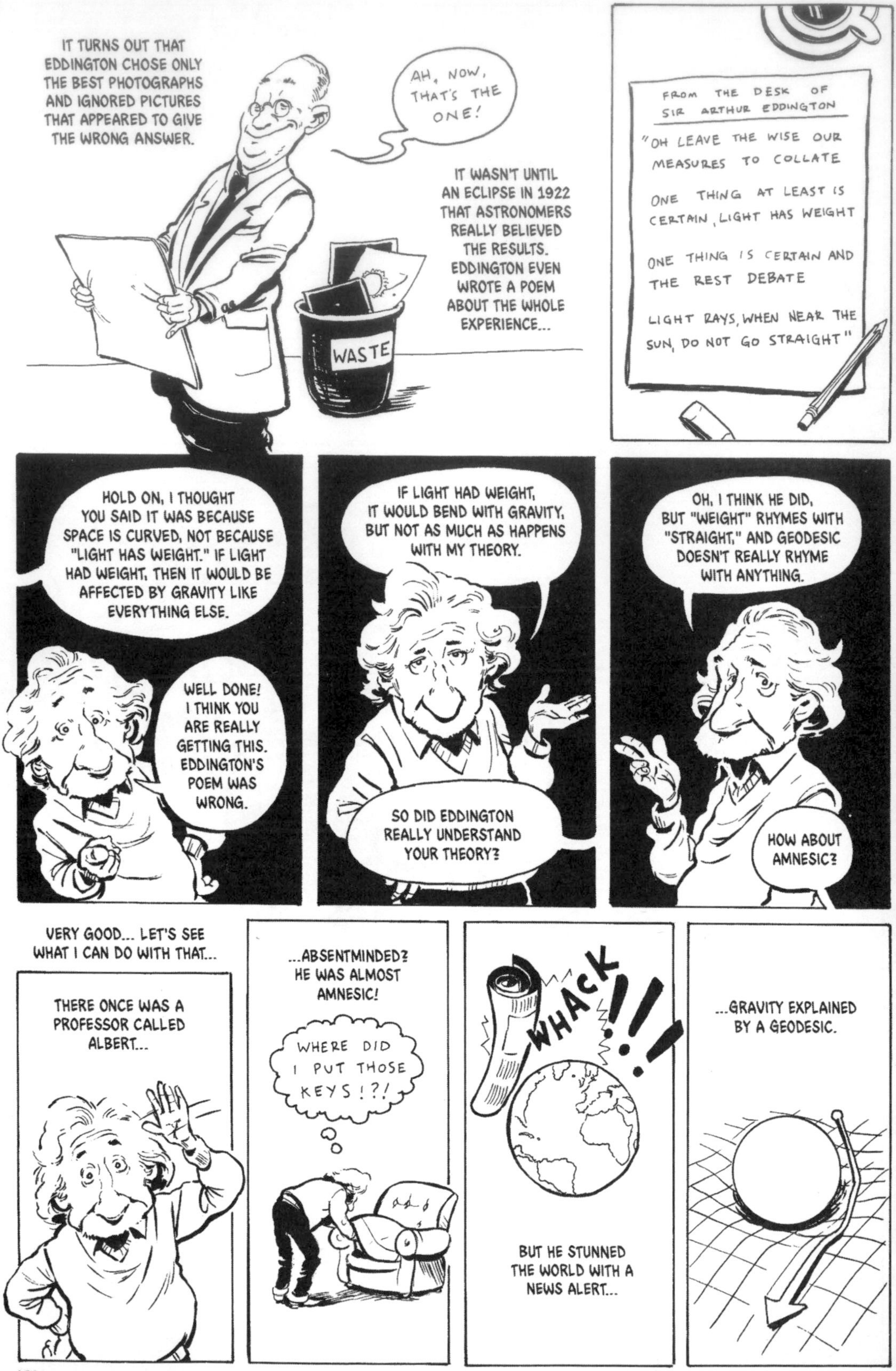
IT TURNS OUT THAT EDDINGTON CHOSE ONLY THE BEST PHOTOGRAPHS AND IGNORED PICTURES THAT APPEARED TO GIVE THE WRONG ANSWER.
AH, NOW, THAT'S THE ONE!
WASTE
IT WASN'T UNTIL AN ECLIPSE IN 1922 THAT ASTRONOMERS REALLY BELIEVED THE RESULTS. EDDINGTON EVEN WROTE A POEM ABOUT THE WHOLE EXPERIENCE...
FROM THE DESK OF SIR ARTHUR EDDINGTON
"OH LEAVE THE WISE OUR MEASURES TO COLLATE
ONE THING AT LEAST IS CERTAIN, LIGHT HAS WEIGHT
ONE THING IS CERTAIN AND THE REST DEBATE
LIGHT RAYS, WHEN NEAR THE SUN, DO NOT GO STRAIGHT"
HOLD ON, I THOUGHT YOU SAID IT WAS BECAUSE SPACE IS CURVED, NOT BECAUSE "LIGHT HAS WEIGHT." IF LIGHT HAD WEIGHT, THEN IT WOULD BE AFFECTED BY GRAVITY LIKE EVERYTHING ELSE.
WELL DONE! I THINK YOU ARE REALLY GETTING THIS. EDDINGTON'S POEM WAS WRONG.
IF LIGHT HAD WEIGHT, IT WOULD BEND WITH GRAVITY, BUT NOT AS MUCH AS HAPPENS WITH MY THEORY.
SO DID EDDINGTON REALLY UNDERSTAND YOUR THEORY?
OH, I THINK HE DID, BUT "WEIGHT" RHYMES WITH "STRAIGHT," AND GEODESIC DOESN'T REALLY RHYME WITH ANYTHING.
HOW ABOUT AMNESIC?
VERY GOOD... LET'S SEE WHAT I CAN DO WITH THAT...
THERE ONCE WAS A PROFESSOR CALLED ALBERT...
...ABSENTMINDED? HE WAS ALMOST AMNESIC!
WHERE DID I PUT THOSE KEYS!?!
WHACK!!!
BUT HE STUNNED THE WORLD WITH A NEWS ALERT...
...GRAVITY EXPLAINED BY A GEODESIC.

CHAPTER 28: WHAT IS LIFE AND WHERE DID IT COME FROM?

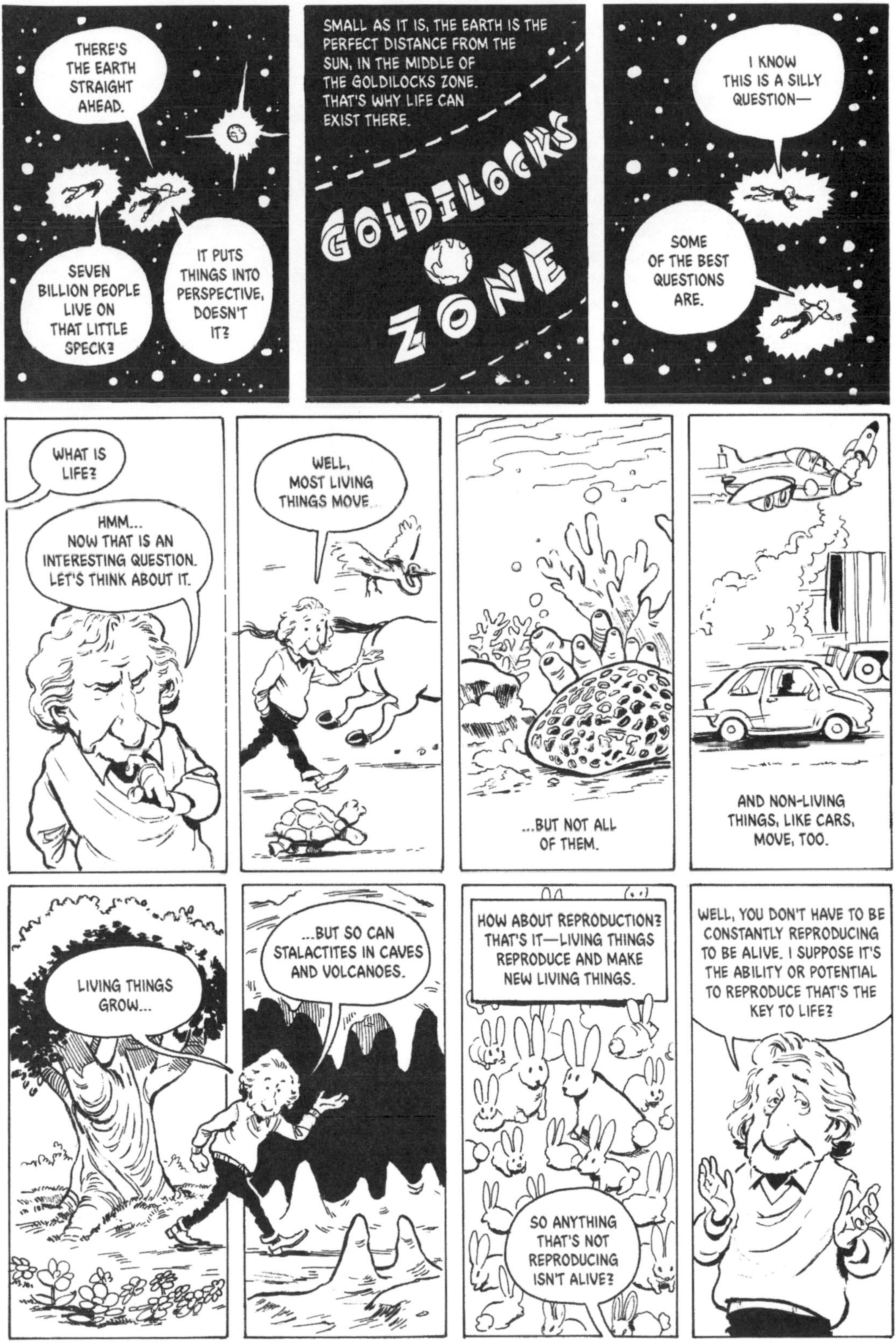
THERE'S THE EARTH STRAIGHT AHEAD.
SEVEN BILLION PEOPLE LIVE ON THAT LITTLE SPECK?
IT PUTS THINGS INTO PERSPECTIVE, DOESN'T IT?
SMALL AS IT IS, THE EARTH IS THE PERFECT DISTANCE FROM THE SUN, IN THE MIDDLE OF THE GOLDILOCKS ZONE. THAT'S WHY LIFE CAN EXIST THERE.
GOLDILOCKS ZONE
I KNOW THIS IS A SILLY QUESTION—
SOME OF THE BEST QUESTIONS ARE.
WHAT IS LIFE?
HMM... NOW THAT IS AN INTERESTING QUESTION. LET'S THINK ABOUT IT.
WELL, MOST LIVING THINGS MOVE...
...BUT NOT ALL OF THEM.
AND NON-LIVING THINGS, LIKE CARS, MOVE, TOO.
LIVING THINGS GROW...
...BUT SO CAN STALACTITES IN CAVES AND VOLCANOES.
HOW ABOUT REPRODUCTION? THAT'S IT—LIVING THINGS REPRODUCE AND MAKE NEW LIVING THINGS.
SO ANYTHING THAT'S NOT REPRODUCING ISN'T ALIVE?
WELL, YOU DON'T HAVE TO BE CONSTANTLY REPRODUCING TO BE ALIVE. I SUPPOSE IT'S THE ABILITY OR POTENTIAL TO REPRODUCE THAT'S THE KEY TO LIFE?

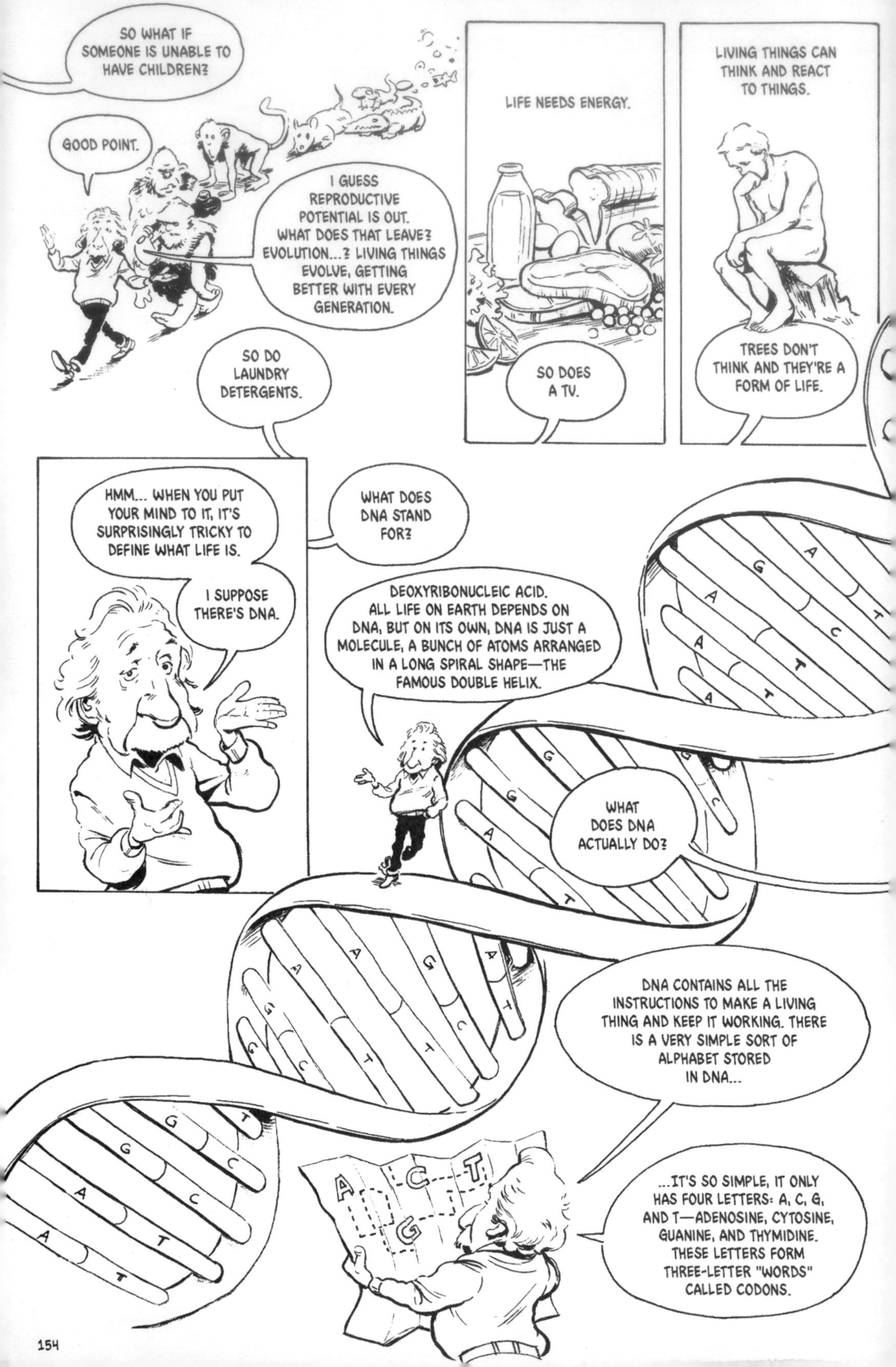

SO WHAT IF SOMEONE IS UNABLE TO HAVE CHILDREN?
GOOD POINT.
I GUESS REPRODUCTIVE POTENTIAL IS OUT. WHAT DOES THAT LEAVE? EVOLUTION...? LIVING THINGS EVOLVE, GETTING BETTER WITH EVERY GENERATION.
SO DO LAUNDRY DETERGENTS.
LIFE NEEDS ENERGY.
SO DOES A TV.
LIVING THINGS CAN THINK AND REACT TO THINGS.
TREES DON'T THINK AND THEY'RE A FORM OF LIFE.
HMM... WHEN YOU PUT YOUR MIND TO IT, IT'S SURPRISINGLY TRICKY TO DEFINE WHAT LIFE IS.
I SUPPOSE THERE'S DNA.
WHAT DOES DNA STAND FOR?
DEOXYRIBONUCLEIC ACID. ALL LIFE ON EARTH DEPENDS ON DNA, BUT ON ITS OWN, DNA IS JUST A MOLECULE, A BUNCH OF ATOMS ARRANGED IN A LONG SPIRAL SHAPE—THE FAMOUS DOUBLE HELIX.
WHAT DOES DNA ACTUALLY DO?
DNA CONTAINS ALL THE INSTRUCTIONS TO MAKE A LIVING THING AND KEEP IT WORKING. THERE IS A VERY SIMPLE SORT OF ALPHABET STORED IN DNA...
A C T G
...IT'S SO SIMPLE, IT ONLY HAS FOUR LETTERS: A, C, G, AND T—ADENOSINE, CYTOSINE, GUANINE, AND THYMIDINE. THESE LETTERS FORM THREE-LETTER "WORDS" CALLED CODONS.

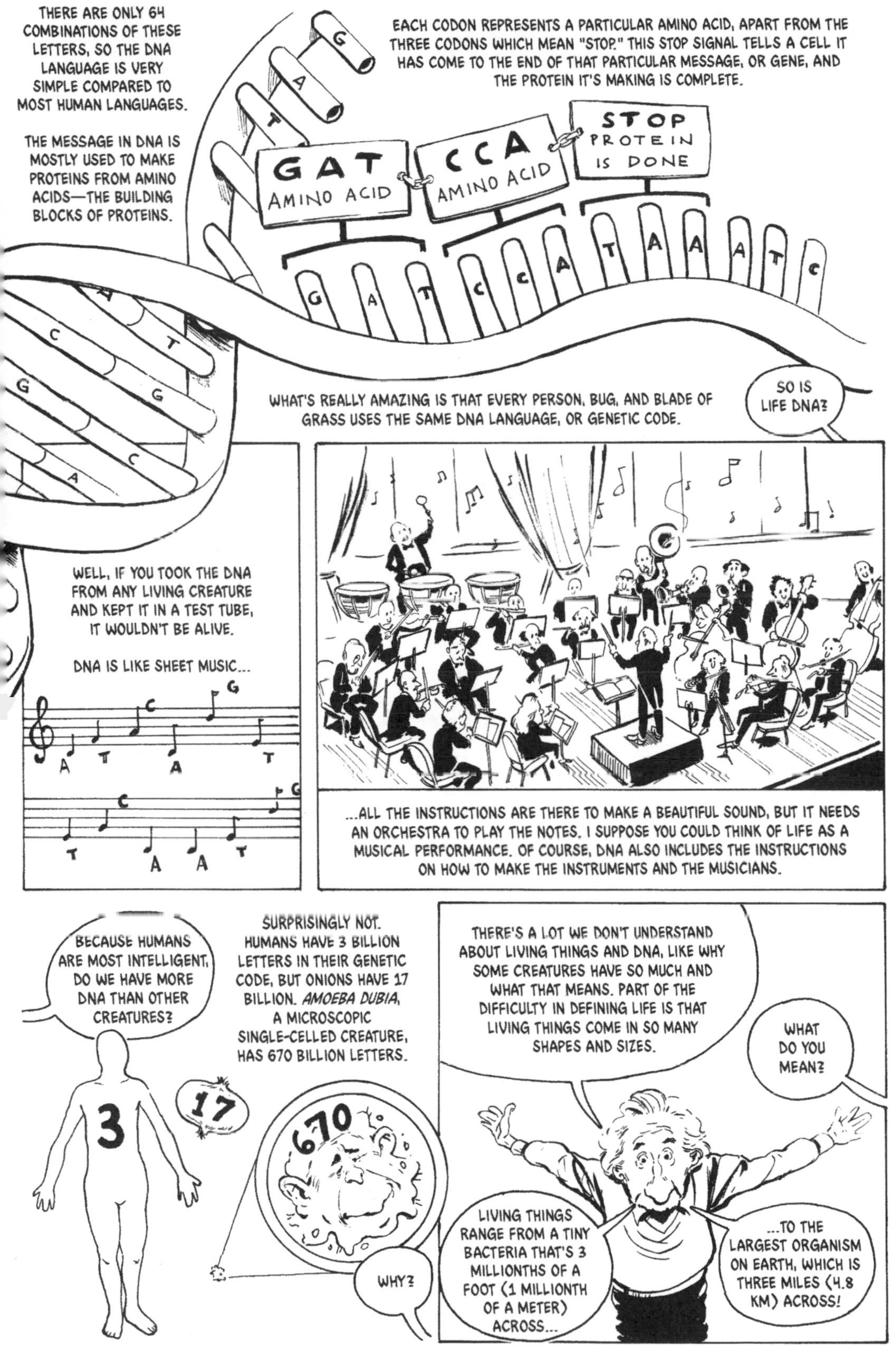

THERE ARE ONLY 64 COMBINATIONS OF THESE LETTERS, SO THE DNA LANGUAGE IS VERY SIMPLE COMPARED TO MOST HUMAN LANGUAGES.
THE MESSAGE IN DNA IS MOSTLY USED TO MAKE PROTEINS FROM AMINO ACIDS—THE BUILDING BLOCKS OF PROTEINS.
EACH CODON REPRESENTS A PARTICULAR AMINO ACID, APART FROM THE THREE CODONS WHICH MEAN "STOP." THIS STOP SIGNAL TELLS A CELL IT HAS COME TO THE END OF THAT PARTICULAR MESSAGE, OR GENE, AND THE PROTEIN IT'S MAKING IS COMPLETE.
GAT AMINO ACID
CCA AMINO ACID
STOP PROTEIN IS DONE
G A T C C A T A A A T C
WHAT'S REALLY AMAZING IS THAT EVERY PERSON, BUG, AND BLADE OF GRASS USES THE SAME DNA LANGUAGE, OR GENETIC CODE.
SO IS LIFE DNA?
WELL, IF YOU TOOK THE DNA FROM ANY LIVING CREATURE AND KEPT IT IN A TEST TUBE, IT WOULDN'T BE ALIVE.
DNA IS LIKE SHEET MUSIC...
A T C A G T
T C A A T G
...ALL THE INSTRUCTIONS ARE THERE TO MAKE A BEAUTIFUL SOUND, BUT IT NEEDS AN ORCHESTRA TO PLAY THE NOTES. I SUPPOSE YOU COULD THINK OF LIFE AS A MUSICAL PERFORMANCE. OF COURSE, DNA ALSO INCLUDES THE INSTRUCTIONS ON HOW TO MAKE THE INSTRUMENTS AND THE MUSICIANS.
BECAUSE HUMANS ARE MOST INTELLIGENT, DO WE HAVE MORE DNA THAN OTHER CREATURES?
SURPRISINGLY NOT. HUMANS HAVE 3 BILLION LETTERS IN THEIR GENETIC CODE, BUT ONIONS HAVE 17 BILLION. *AMOEBA DUBIA*, A MICROSCOPIC SINGLE-CELLED CREATURE, HAS 670 BILLION LETTERS.
3
17
670
WHY?
THERE'S A LOT WE DON'T UNDERSTAND ABOUT LIVING THINGS AND DNA, LIKE WHY SOME CREATURES HAVE SO MUCH AND WHAT THAT MEANS. PART OF THE DIFFICULTY IN DEFINING LIFE IS THAT LIVING THINGS COME IN SO MANY SHAPES AND SIZES.
WHAT DO YOU MEAN?
LIVING THINGS RANGE FROM A TINY BACTERIA THAT'S 3 MILLIONTHS OF A FOOT (1 MILLIONTH OF A METER) ACROSS...
...TO THE LARGEST ORGANISM ON EARTH, WHICH IS THREE MILES (4.8 KM) ACROSS!

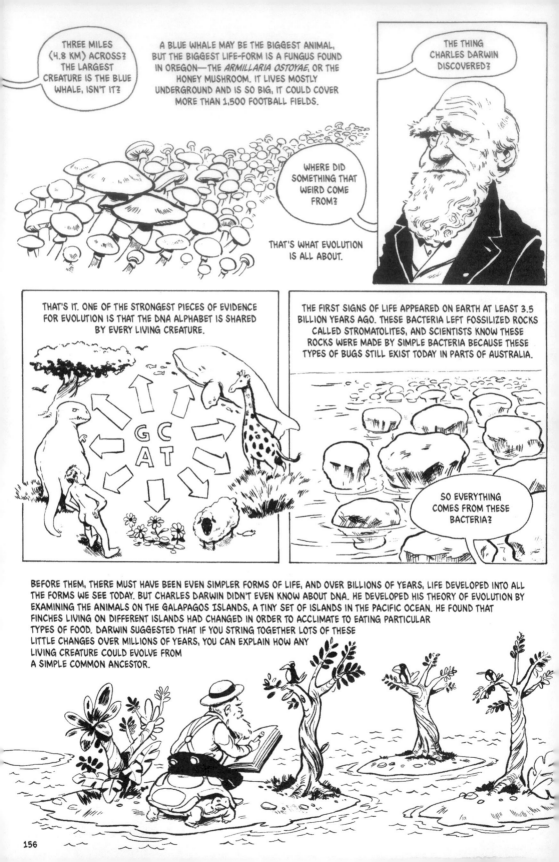

THREE MILES (4.8 KM) ACROSS? THE LARGEST CREATURE IS THE BLUE WHALE, ISN'T IT?

A BLUE WHALE MAY BE THE BIGGEST ANIMAL, BUT THE BIGGEST LIFE-FORM IS A FUNGUS FOUND IN OREGON—THE *ARMILLARIA OSTOYAE*, OR THE HONEY MUSHROOM. IT LIVES MOSTLY UNDERGROUND AND IS SO BIG, IT COULD COVER MORE THAN 1,500 FOOTBALL FIELDS.

THE THING CHARLES DARWIN DISCOVERED?

WHERE DID SOMETHING THAT WEIRD COME FROM?

THAT'S WHAT EVOLUTION IS ALL ABOUT.

THAT'S IT. ONE OF THE STRONGEST PIECES OF EVIDENCE FOR EVOLUTION IS THAT THE DNA ALPHABET IS SHARED BY EVERY LIVING CREATURE.

G C A T

THE FIRST SIGNS OF LIFE APPEARED ON EARTH AT LEAST 3.5 BILLION YEARS AGO. THESE BACTERIA LEFT FOSSILIZED ROCKS CALLED STROMATOLITES, AND SCIENTISTS KNOW THESE ROCKS WERE MADE BY SIMPLE BACTERIA BECAUSE THESE TYPES OF BUGS STILL EXIST TODAY IN PARTS OF AUSTRALIA.

SO EVERYTHING COMES FROM THESE BACTERIA?

BEFORE THEM, THERE MUST HAVE BEEN EVEN SIMPLER FORMS OF LIFE, AND OVER BILLIONS OF YEARS, LIFE DEVELOPED INTO ALL THE FORMS WE SEE TODAY. BUT CHARLES DARWIN DIDN'T EVEN KNOW ABOUT DNA. HE DEVELOPED HIS THEORY OF EVOLUTION BY EXAMINING THE ANIMALS ON THE GALAPAGOS ISLANDS, A TINY SET OF ISLANDS IN THE PACIFIC OCEAN. HE FOUND THAT FINCHES LIVING ON DIFFERENT ISLANDS HAD CHANGED IN ORDER TO ACCLIMATE TO EATING PARTICULAR TYPES OF FOOD. DARWIN SUGGESTED THAT IF YOU STRING TOGETHER LOTS OF THESE LITTLE CHANGES OVER MILLIONS OF YEARS, YOU CAN EXPLAIN HOW ANY LIVING CREATURE COULD EVOLVE FROM A SIMPLE COMMON ANCESTOR.

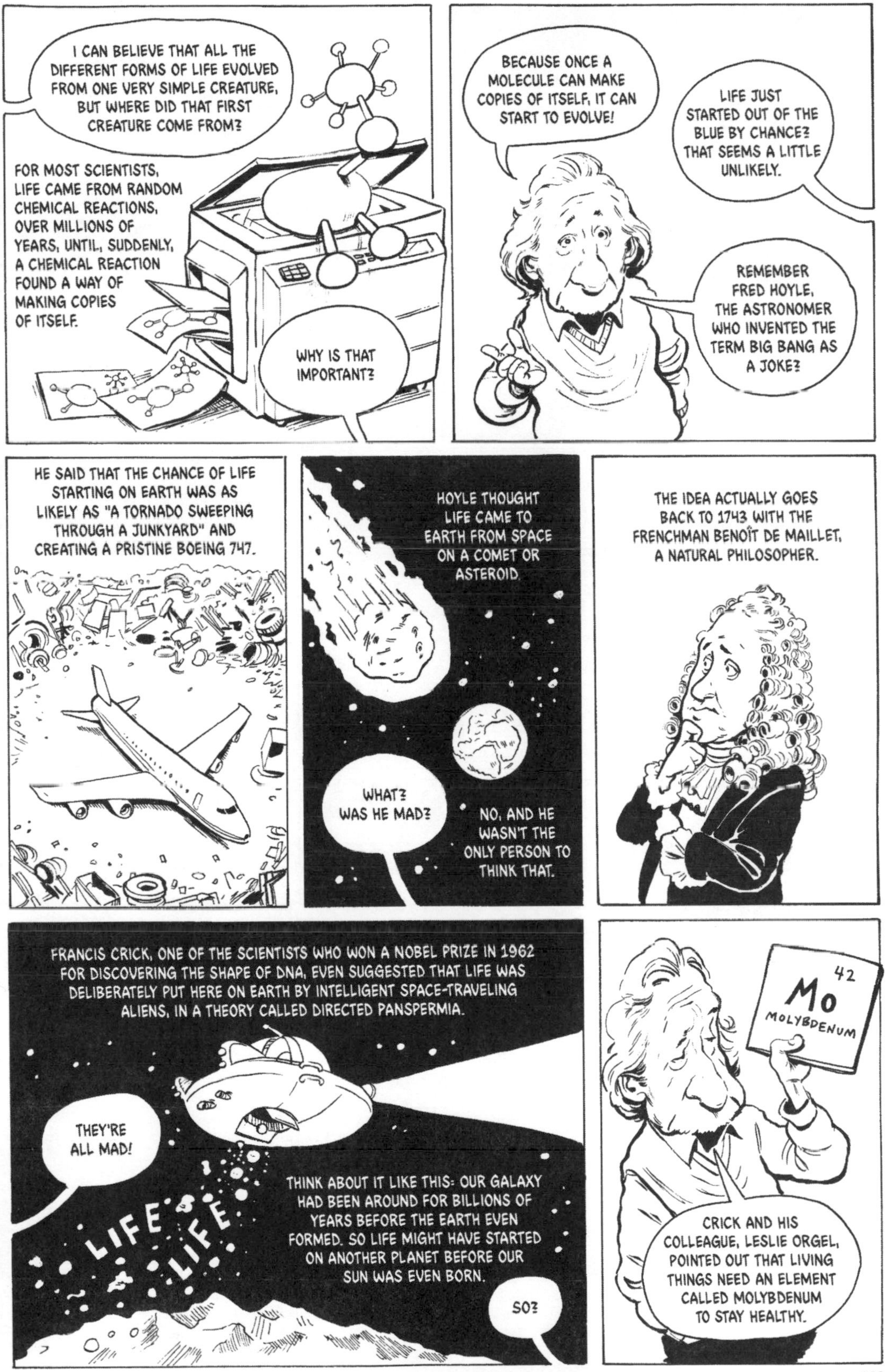
I CAN BELIEVE THAT ALL THE DIFFERENT FORMS OF LIFE EVOLVED FROM ONE VERY SIMPLE CREATURE, BUT WHERE DID THAT FIRST CREATURE COME FROM?
FOR MOST SCIENTISTS, LIFE CAME FROM RANDOM CHEMICAL REACTIONS, OVER MILLIONS OF YEARS, UNTIL, SUDDENLY, A CHEMICAL REACTION FOUND A WAY OF MAKING COPIES OF ITSELF.
WHY IS THAT IMPORTANT?
BECAUSE ONCE A MOLECULE CAN MAKE COPIES OF ITSELF, IT CAN START TO EVOLVE!
LIFE JUST STARTED OUT OF THE BLUE BY CHANCE? THAT SEEMS A LITTLE UNLIKELY.
REMEMBER FRED HOYLE, THE ASTRONOMER WHO INVENTED THE TERM BIG BANG AS A JOKE?
HE SAID THAT THE CHANCE OF LIFE STARTING ON EARTH WAS AS LIKELY AS "A TORNADO SWEEPING THROUGH A JUNKYARD" AND CREATING A PRISTINE BOEING 747.
HOYLE THOUGHT LIFE CAME TO EARTH FROM SPACE ON A COMET OR ASTEROID.
WHAT? WAS HE MAD?
NO, AND HE WASN'T THE ONLY PERSON TO THINK THAT.
THE IDEA ACTUALLY GOES BACK TO 1743 WITH THE FRENCHMAN BENOÎT DE MAILLET, A NATURAL PHILOSOPHER.
FRANCIS CRICK, ONE OF THE SCIENTISTS WHO WON A NOBEL PRIZE IN 1962 FOR DISCOVERING THE SHAPE OF DNA, EVEN SUGGESTED THAT LIFE WAS DELIBERATELY PUT HERE ON EARTH BY INTELLIGENT SPACE-TRAVELING ALIENS, IN A THEORY CALLED DIRECTED PANSPERMIA.
THEY'RE ALL MAD!
LIFE
LIFE
THINK ABOUT IT LIKE THIS: OUR GALAXY HAD BEEN AROUND FOR BILLIONS OF YEARS BEFORE THE EARTH EVEN FORMED. SO LIFE MIGHT HAVE STARTED ON ANOTHER PLANET BEFORE OUR SUN WAS EVEN BORN.
SO?
42
Mo
MOLYBDENUM
CRICK AND HIS COLLEAGUE, LESLIE ORGEL, POINTED OUT THAT LIVING THINGS NEED AN ELEMENT CALLED MOLYBDENUM TO STAY HEALTHY.

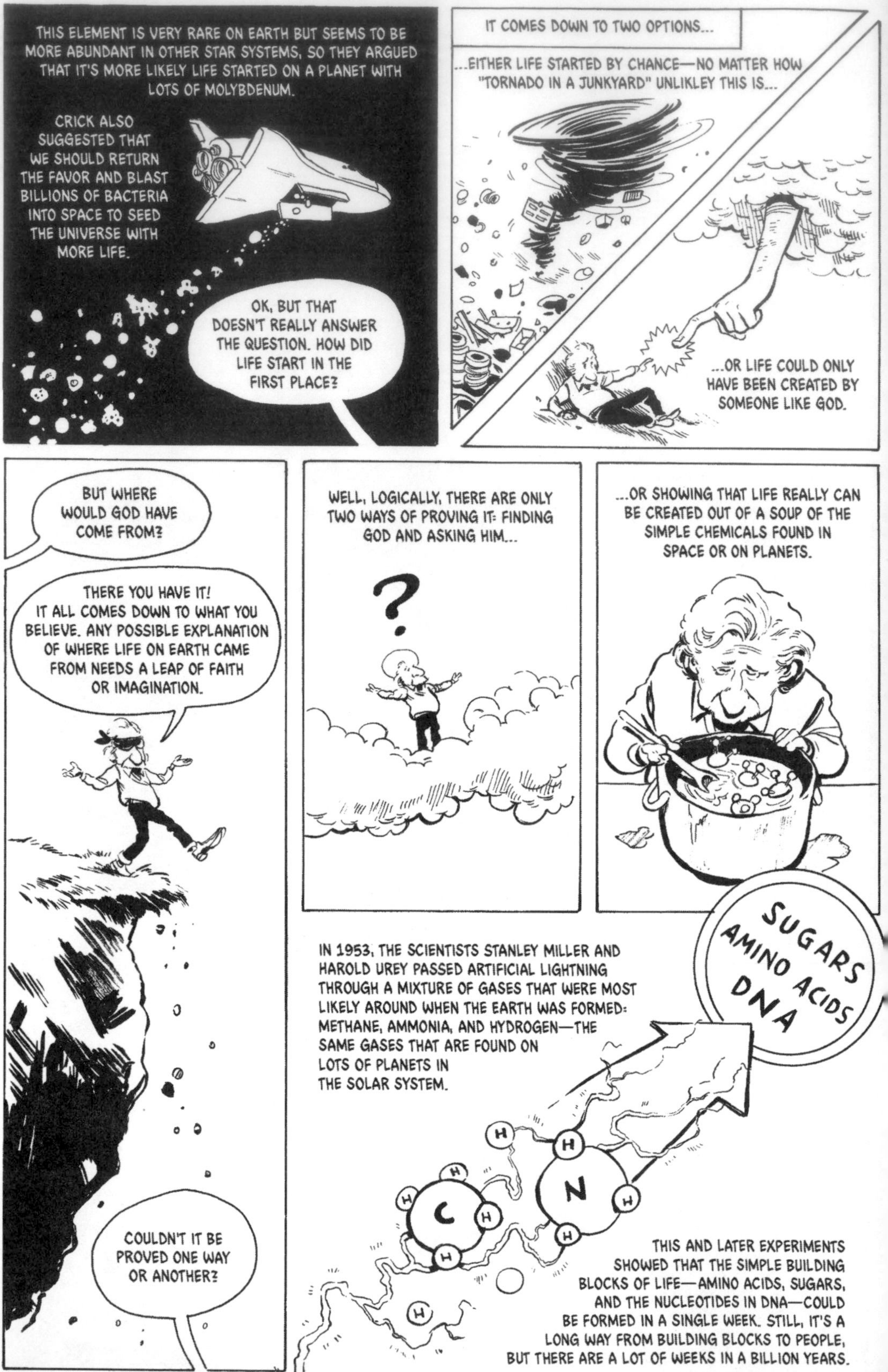
THIS ELEMENT IS VERY RARE ON EARTH BUT SEEMS TO BE MORE ABUNDANT IN OTHER STAR SYSTEMS, SO THEY ARGUED THAT IT'S MORE LIKELY LIFE STARTED ON A PLANET WITH LOTS OF MOLYBDENUM.
CRICK ALSO SUGGESTED THAT WE SHOULD RETURN THE FAVOR AND BLAST BILLIONS OF BACTERIA INTO SPACE TO SEED THE UNIVERSE WITH MORE LIFE.
OK, BUT THAT DOESN'T REALLY ANSWER THE QUESTION. HOW DID LIFE START IN THE FIRST PLACE?
IT COMES DOWN TO TWO OPTIONS...
...EITHER LIFE STARTED BY CHANCE—NO MATTER HOW "TORNADO IN A JUNKYARD" UNLIKLEY THIS IS...
...OR LIFE COULD ONLY HAVE BEEN CREATED BY SOMEONE LIKE GOD.
BUT WHERE WOULD GOD HAVE COME FROM?
THERE YOU HAVE IT! IT ALL COMES DOWN TO WHAT YOU BELIEVE. ANY POSSIBLE EXPLANATION OF WHERE LIFE ON EARTH CAME FROM NEEDS A LEAP OF FAITH OR IMAGINATION.
COULDN'T IT BE PROVED ONE WAY OR ANOTHER?
WELL, LOGICALLY, THERE ARE ONLY TWO WAYS OF PROVING IT: FINDING GOD AND ASKING HIM...
?
...OR SHOWING THAT LIFE REALLY CAN BE CREATED OUT OF A SOUP OF THE SIMPLE CHEMICALS FOUND IN SPACE OR ON PLANETS.
IN 1953, THE SCIENTISTS STANLEY MILLER AND HAROLD UREY PASSED ARTIFICIAL LIGHTNING THROUGH A MIXTURE OF GASES THAT WERE MOST LIKELY AROUND WHEN THE EARTH WAS FORMED: METHANE, AMMONIA, AND HYDROGEN—THE SAME GASES THAT ARE FOUND ON LOTS OF PLANETS IN THE SOLAR SYSTEM.
SUGARS
AMINO ACIDS
DNA
H
C
N
THIS AND LATER EXPERIMENTS SHOWED THAT THE SIMPLE BUILDING BLOCKS OF LIFE—AMINO ACIDS, SUGARS, AND THE NUCLEOTIDES IN DNA—COULD BE FORMED IN A SINGLE WEEK. STILL, IT'S A LONG WAY FROM BUILDING BLOCKS TO PEOPLE, BUT THERE ARE A LOT OF WEEKS IN A BILLION YEARS.

CHAPTER 29: EATING SUNLIGHT: THE IMPORTANCE OF LIGHT FOR LIFE

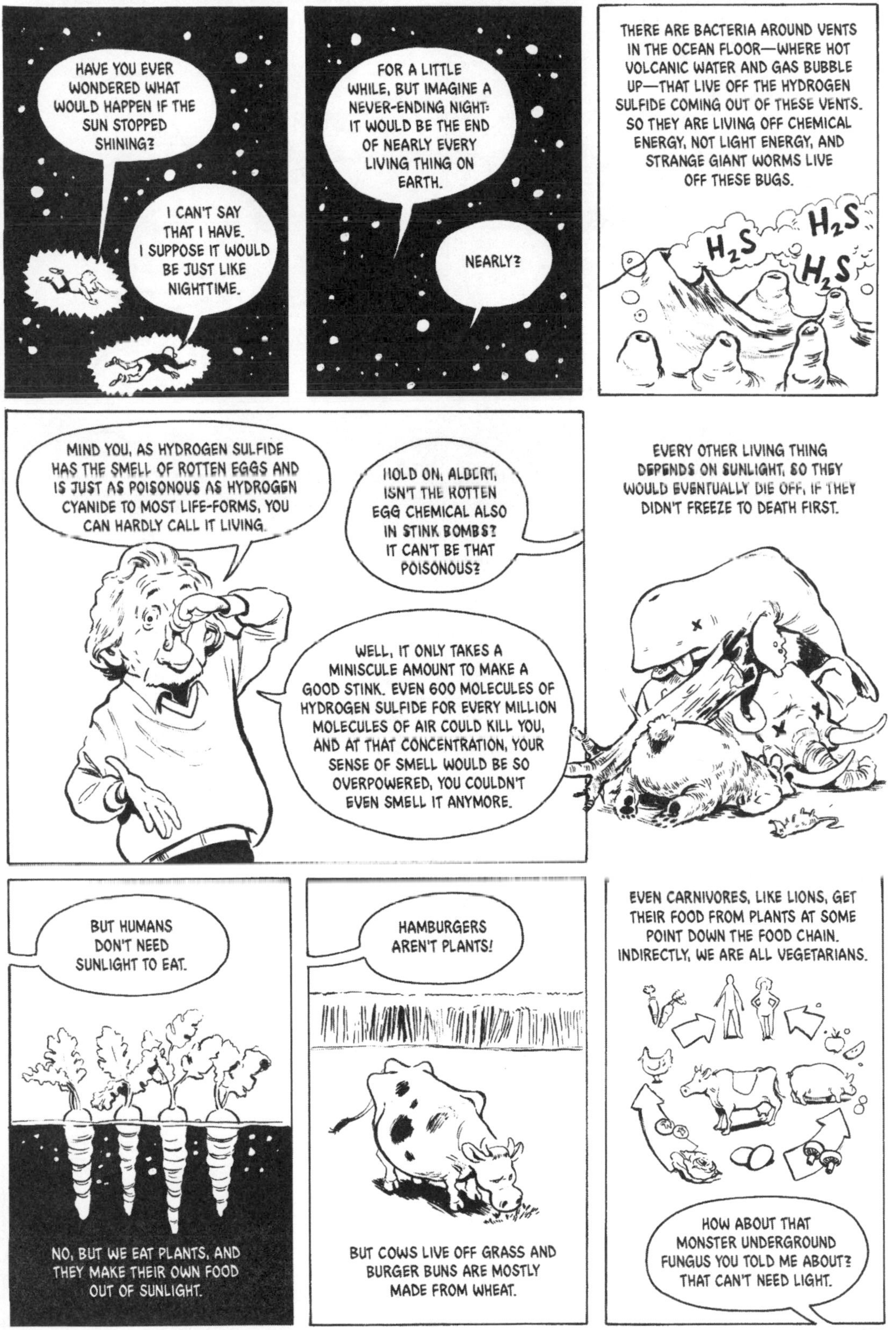
HAVE YOU EVER WONDERED WHAT WOULD HAPPEN IF THE SUN STOPPED SHINING?
I CAN'T SAY THAT I HAVE. I SUPPOSE IT WOULD BE JUST LIKE NIGHTTIME.
FOR A LITTLE WHILE, BUT IMAGINE A NEVER-ENDING NIGHT: IT WOULD BE THE END OF NEARLY EVERY LIVING THING ON EARTH.
NEARLY?
THERE ARE BACTERIA AROUND VENTS IN THE OCEAN FLOOR—WHERE HOT VOLCANIC WATER AND GAS BUBBLE UP—THAT LIVE OFF THE HYDROGEN SULFIDE COMING OUT OF THESE VENTS. SO THEY ARE LIVING OFF CHEMICAL ENERGY, NOT LIGHT ENERGY, AND STRANGE GIANT WORMS LIVE OFF THESE BUGS.
H_2S
H_2S
H_2S
MIND YOU, AS HYDROGEN SULFIDE HAS THE SMELL OF ROTTEN EGGS AND IS JUST AS POISONOUS AS HYDROGEN CYANIDE TO MOST LIFE-FORMS, YOU CAN HARDLY CALL IT LIVING.
HOLD ON, ALBERT, ISN'T THE ROTTEN EGG CHEMICAL ALSO IN STINK BOMBS? IT CAN'T BE THAT POISONOUS?
WELL, IT ONLY TAKES A MINISCULE AMOUNT TO MAKE A GOOD STINK. EVEN 600 MOLECULES OF HYDROGEN SULFIDE FOR EVERY MILLION MOLECULES OF AIR COULD KILL YOU, AND AT THAT CONCENTRATION, YOUR SENSE OF SMELL WOULD BE SO OVERPOWERED, YOU COULDN'T EVEN SMELL IT ANYMORE.
EVERY OTHER LIVING THING DEPENDS ON SUNLIGHT, SO THEY WOULD EVENTUALLY DIE OFF, IF THEY DIDN'T FREEZE TO DEATH FIRST.
BUT HUMANS DON'T NEED SUNLIGHT TO EAT.
NO, BUT WE EAT PLANTS, AND THEY MAKE THEIR OWN FOOD OUT OF SUNLIGHT.
HAMBURGERS AREN'T PLANTS!
BUT COWS LIVE OFF GRASS AND BURGER BUNS ARE MOSTLY MADE FROM WHEAT.
EVEN CARNIVORES, LIKE LIONS, GET THEIR FOOD FROM PLANTS AT SOME POINT DOWN THE FOOD CHAIN. INDIRECTLY, WE ARE ALL VEGETARIANS.
HOW ABOUT THAT MONSTER UNDERGROUND FUNGUS YOU TOLD ME ABOUT? THAT CAN'T NEED LIGHT.

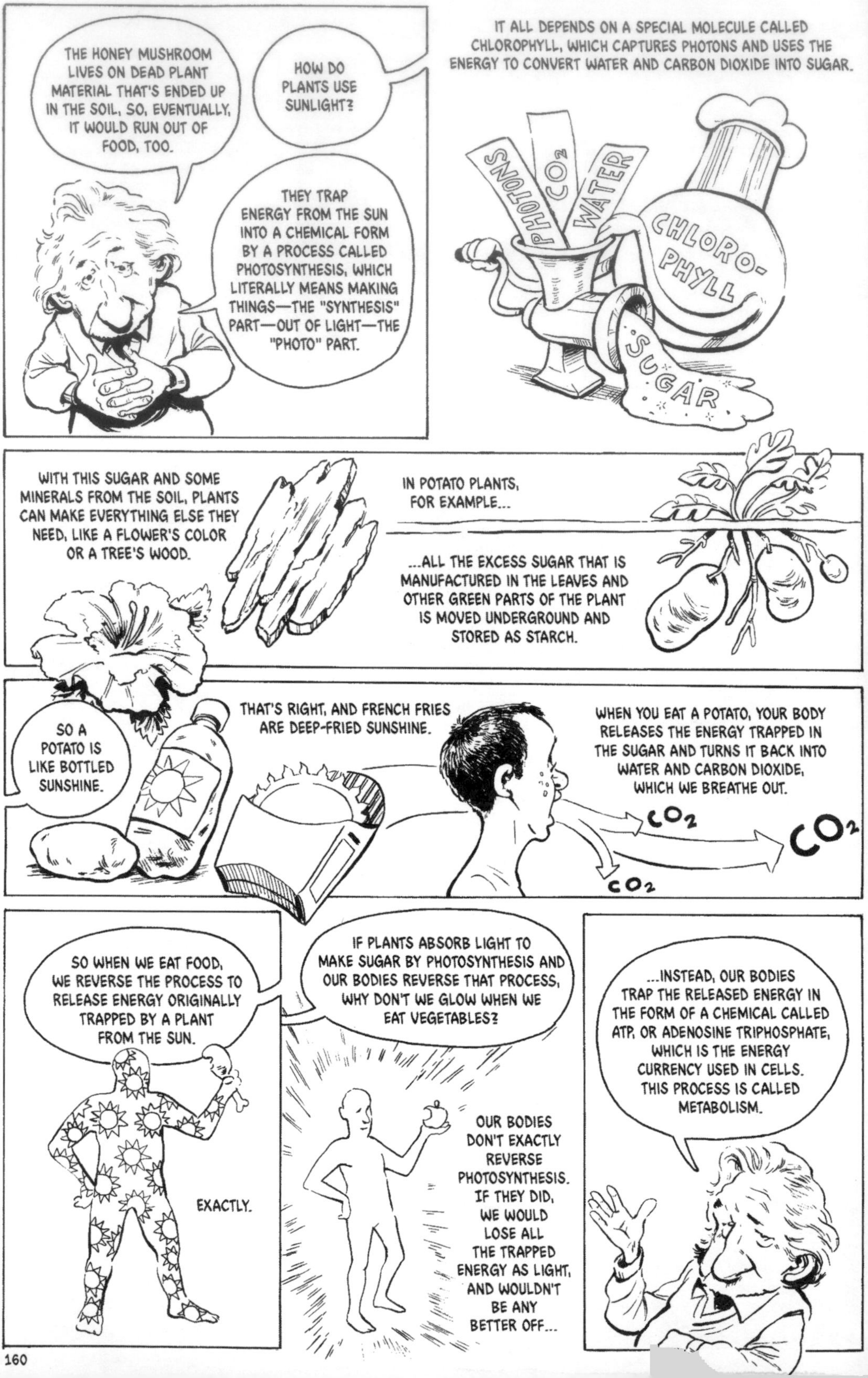
THE HONEY MUSHROOM LIVES ON DEAD PLANT MATERIAL THAT'S ENDED UP IN THE SOIL, SO, EVENTUALLY, IT WOULD RUN OUT OF FOOD, TOO.
HOW DO PLANTS USE SUNLIGHT?
THEY TRAP ENERGY FROM THE SUN INTO A CHEMICAL FORM BY A PROCESS CALLED PHOTOSYNTHESIS, WHICH LITERALLY MEANS MAKING THINGS—THE "SYNTHESIS" PART—OUT OF LIGHT—THE "PHOTO" PART.
IT ALL DEPENDS ON A SPECIAL MOLECULE CALLED CHLOROPHYLL, WHICH CAPTURES PHOTONS AND USES THE ENERGY TO CONVERT WATER AND CARBON DIOXIDE INTO SUGAR.
PHOTONS
CO2
WATER
CHLORO-PHYLL
SUGAR
WITH THIS SUGAR AND SOME MINERALS FROM THE SOIL, PLANTS CAN MAKE EVERYTHING ELSE THEY NEED, LIKE A FLOWER'S COLOR OR A TREE'S WOOD.
IN POTATO PLANTS, FOR EXAMPLE...
...ALL THE EXCESS SUGAR THAT IS MANUFACTURED IN THE LEAVES AND OTHER GREEN PARTS OF THE PLANT IS MOVED UNDERGROUND AND STORED AS STARCH.
SO A POTATO IS LIKE BOTTLED SUNSHINE.
THAT'S RIGHT, AND FRENCH FRIES ARE DEEP-FRIED SUNSHINE.
WHEN YOU EAT A POTATO, YOUR BODY RELEASES THE ENERGY TRAPPED IN THE SUGAR AND TURNS IT BACK INTO WATER AND CARBON DIOXIDE, WHICH WE BREATHE OUT.
CO2
CO2
CO2
SO WHEN WE EAT FOOD, WE REVERSE THE PROCESS TO RELEASE ENERGY ORIGINALLY TRAPPED BY A PLANT FROM THE SUN.
EXACTLY.
IF PLANTS ABSORB LIGHT TO MAKE SUGAR BY PHOTOSYNTHESIS AND OUR BODIES REVERSE THAT PROCESS, WHY DON'T WE GLOW WHEN WE EAT VEGETABLES?
OUR BODIES DON'T EXACTLY REVERSE PHOTOSYNTHESIS. IF THEY DID, WE WOULD LOSE ALL THE TRAPPED ENERGY AS LIGHT, AND WOULDN'T BE ANY BETTER OFF...
...INSTEAD, OUR BODIES TRAP THE RELEASED ENERGY IN THE FORM OF A CHEMICAL CALLED ATP, OR ADENOSINE TRIPHOSPHATE, WHICH IS THE ENERGY CURRENCY USED IN CELLS. THIS PROCESS IS CALLED METABOLISM.

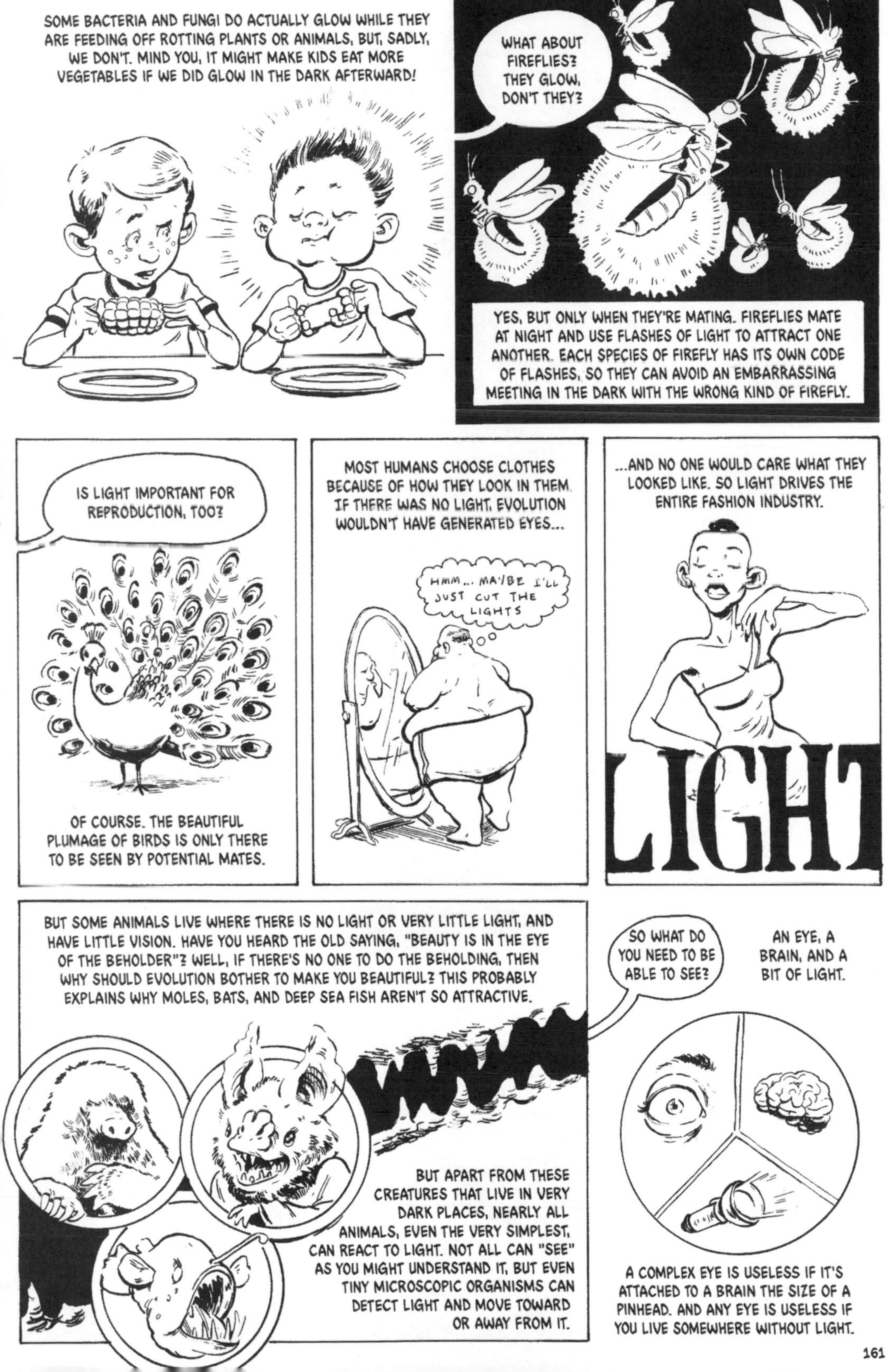
SOME BACTERIA AND FUNGI DO ACTUALLY GLOW WHILE THEY ARE FEEDING OFF ROTTING PLANTS OR ANIMALS, BUT, SADLY, WE DON'T. MIND YOU, IT MIGHT MAKE KIDS EAT MORE VEGETABLES IF WE DID GLOW IN THE DARK AFTERWARD!
WHAT ABOUT FIREFLIES? THEY GLOW, DON'T THEY?
YES, BUT ONLY WHEN THEY'RE MATING. FIREFLIES MATE AT NIGHT AND USE FLASHES OF LIGHT TO ATTRACT ONE ANOTHER. EACH SPECIES OF FIREFLY HAS ITS OWN CODE OF FLASHES, SO THEY CAN AVOID AN EMBARRASSING MEETING IN THE DARK WITH THE WRONG KIND OF FIREFLY.
IS LIGHT IMPORTANT FOR REPRODUCTION, TOO?
OF COURSE. THE BEAUTIFUL PLUMAGE OF BIRDS IS ONLY THERE TO BE SEEN BY POTENTIAL MATES.
MOST HUMANS CHOOSE CLOTHES BECAUSE OF HOW THEY LOOK IN THEM. IF THERE WAS NO LIGHT, EVOLUTION WOULDN'T HAVE GENERATED EYES...
HMM... MA'BE I'LL JUST CUT THE LIGHTS
...AND NO ONE WOULD CARE WHAT THEY LOOKED LIKE. SO LIGHT DRIVES THE ENTIRE FASHION INDUSTRY.
LIGHT
BUT SOME ANIMALS LIVE WHERE THERE IS NO LIGHT OR VERY LITTLE LIGHT, AND HAVE LITTLE VISION. HAVE YOU HEARD THE OLD SAYING, "BEAUTY IS IN THE EYE OF THE BEHOLDER"? WELL, IF THERE'S NO ONE TO DO THE BEHOLDING, THEN WHY SHOULD EVOLUTION BOTHER TO MAKE YOU BEAUTIFUL? THIS PROBABLY EXPLAINS WHY MOLES, BATS, AND DEEP SEA FISH AREN'T SO ATTRACTIVE.
BUT APART FROM THESE CREATURES THAT LIVE IN VERY DARK PLACES, NEARLY ALL ANIMALS, EVEN THE VERY SIMPLEST, CAN REACT TO LIGHT. NOT ALL CAN "SEE" AS YOU MIGHT UNDERSTAND IT, BUT EVEN TINY MICROSCOPIC ORGANISMS CAN DETECT LIGHT AND MOVE TOWARD OR AWAY FROM IT.
SO WHAT DO YOU NEED TO BE ABLE TO SEE?
AN EYE, A BRAIN, AND A BIT OF LIGHT.
A COMPLEX EYE IS USELESS IF IT'S ATTACHED TO A BRAIN THE SIZE OF A PINHEAD. AND ANY EYE IS USELESS IF YOU LIVE SOMEWHERE WITHOUT LIGHT.

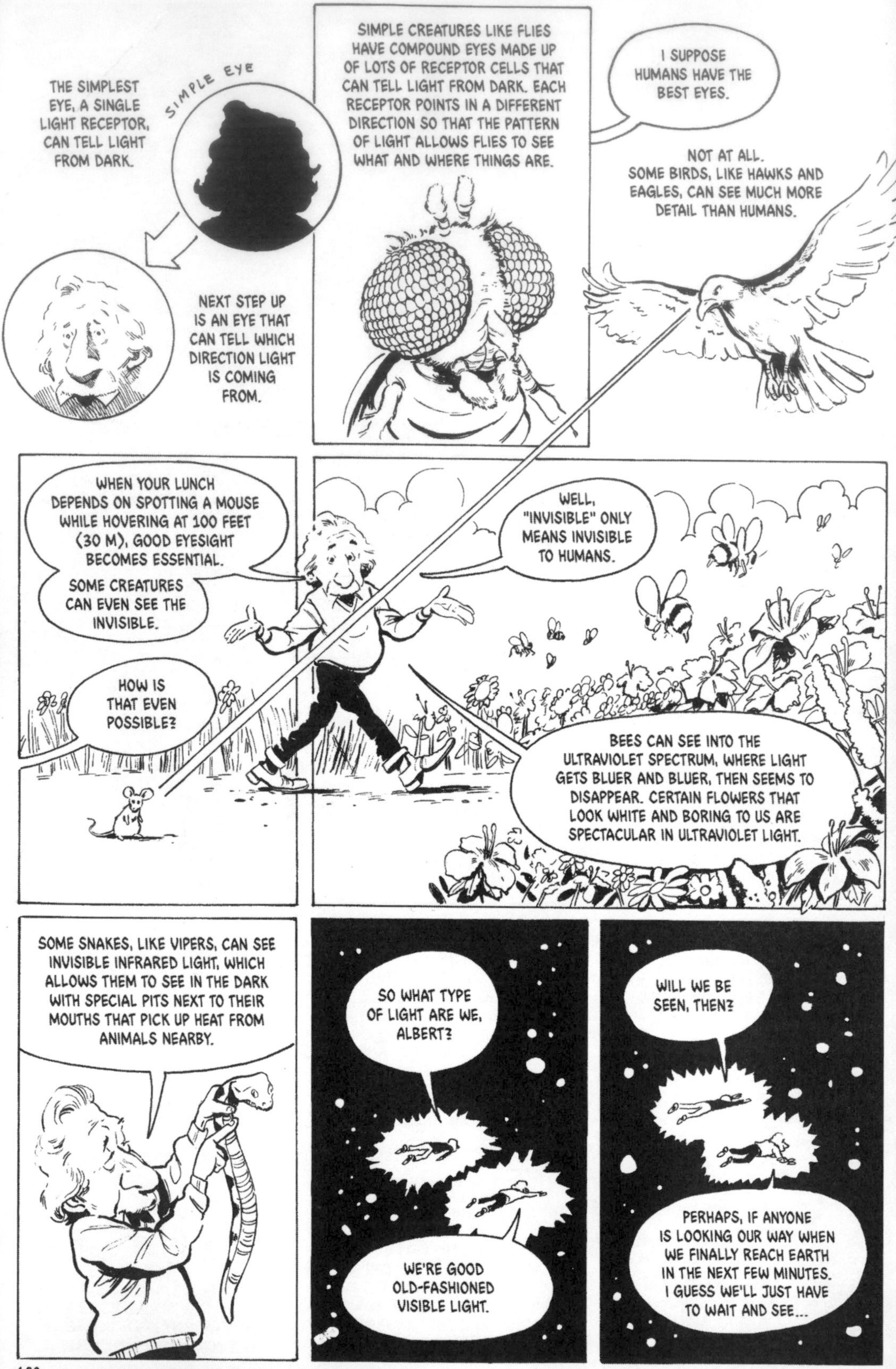
THE SIMPLEST EYE, A SINGLE LIGHT RECEPTOR, CAN TELL LIGHT FROM DARK.
SIMPLE EYE
NEXT STEP UP IS AN EYE THAT CAN TELL WHICH DIRECTION LIGHT IS COMING FROM.
SIMPLE CREATURES LIKE FLIES HAVE COMPOUND EYES MADE UP OF LOTS OF RECEPTOR CELLS THAT CAN TELL LIGHT FROM DARK. EACH RECEPTOR POINTS IN A DIFFERENT DIRECTION SO THAT THE PATTERN OF LIGHT ALLOWS FLIES TO SEE WHAT AND WHERE THINGS ARE.
I SUPPOSE HUMANS HAVE THE BEST EYES.
NOT AT ALL. SOME BIRDS, LIKE HAWKS AND EAGLES, CAN SEE MUCH MORE DETAIL THAN HUMANS.
WHEN YOUR LUNCH DEPENDS ON SPOTTING A MOUSE WHILE HOVERING AT 100 FEET (30 M), GOOD EYESIGHT BECOMES ESSENTIAL.
SOME CREATURES CAN EVEN SEE THE INVISIBLE.
HOW IS THAT EVEN POSSIBLE?
WELL, "INVISIBLE" ONLY MEANS INVISIBLE TO HUMANS.
BEES CAN SEE INTO THE ULTRAVIOLET SPECTRUM, WHERE LIGHT GETS BLUER AND BLUER, THEN SEEMS TO DISAPPEAR. CERTAIN FLOWERS THAT LOOK WHITE AND BORING TO US ARE SPECTACULAR IN ULTRAVIOLET LIGHT.
SOME SNAKES, LIKE VIPERS, CAN SEE INVISIBLE INFRARED LIGHT, WHICH ALLOWS THEM TO SEE IN THE DARK WITH SPECIAL PITS NEXT TO THEIR MOUTHS THAT PICK UP HEAT FROM ANIMALS NEARBY.
SO WHAT TYPE OF LIGHT ARE WE, ALBERT?
WE'RE GOOD OLD-FASHIONED VISIBLE LIGHT.
WILL WE BE SEEN, THEN?
PERHAPS, IF ANYONE IS LOOKING OUR WAY WHEN WE FINALLY REACH EARTH IN THE NEXT FEW MINUTES. I GUESS WE'LL JUST HAVE TO WAIT AND SEE...

CHAPTER 30: ARE WE ALONE? LISTENING FOR EXTRATERRESTRIAL LIFE

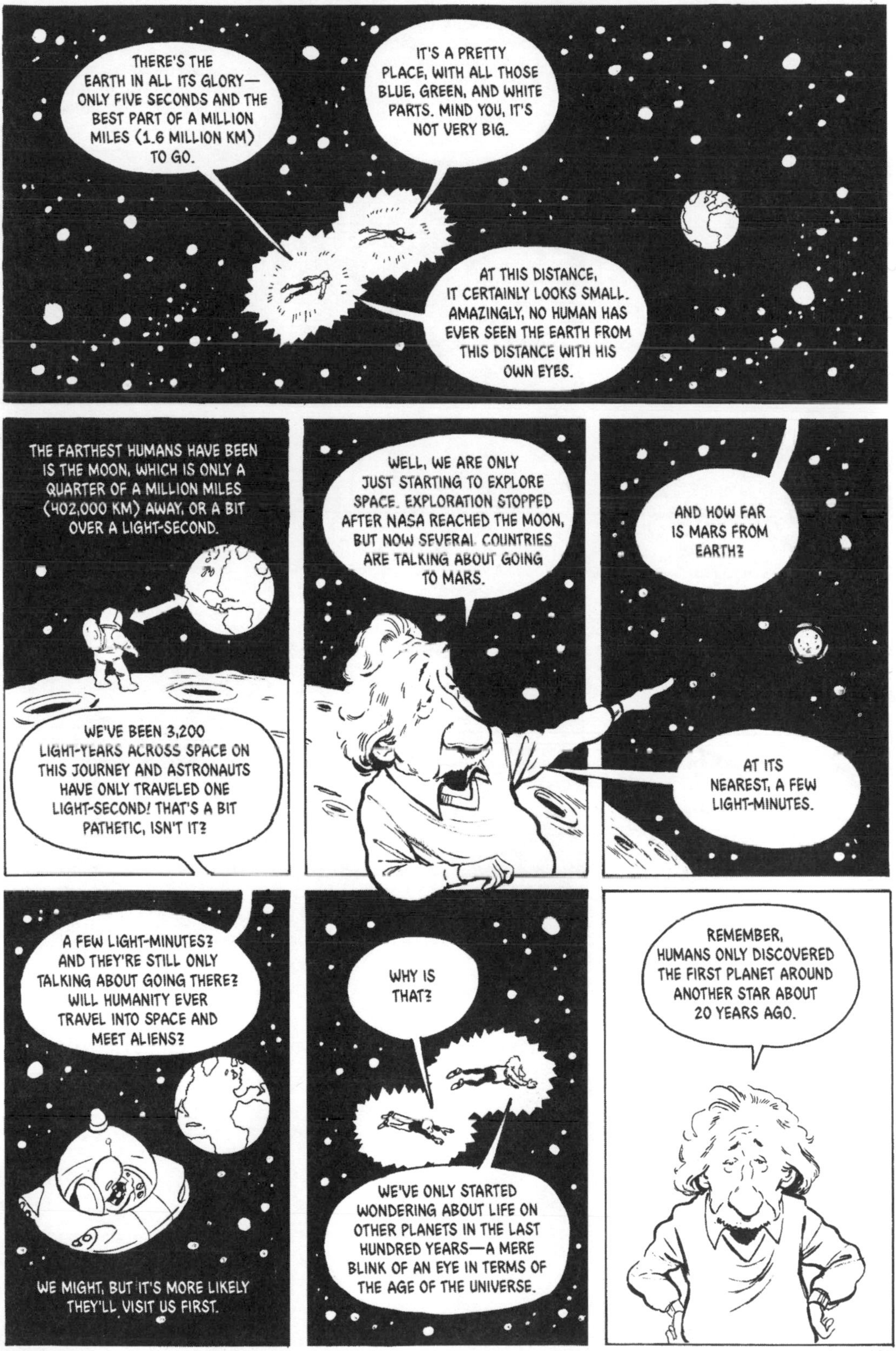
THERE'S THE EARTH IN ALL ITS GLORY—ONLY FIVE SECONDS AND THE BEST PART OF A MILLION MILES (1.6 MILLION KM) TO GO.
IT'S A PRETTY PLACE, WITH ALL THOSE BLUE, GREEN, AND WHITE PARTS. MIND YOU, IT'S NOT VERY BIG.
AT THIS DISTANCE, IT CERTAINLY LOOKS SMALL. AMAZINGLY, NO HUMAN HAS EVER SEEN THE EARTH FROM THIS DISTANCE WITH HIS OWN EYES.
THE FARTHEST HUMANS HAVE BEEN IS THE MOON, WHICH IS ONLY A QUARTER OF A MILLION MILES (402,000 KM) AWAY, OR A BIT OVER A LIGHT-SECOND.
WE'VE BEEN 3,200 LIGHT-YEARS ACROSS SPACE ON THIS JOURNEY AND ASTRONAUTS HAVE ONLY TRAVELED ONE LIGHT-SECOND! THAT'S A BIT PATHETIC, ISN'T IT?
WELL, WE ARE ONLY JUST STARTING TO EXPLORE SPACE. EXPLORATION STOPPED AFTER NASA REACHED THE MOON, BUT NOW SEVERAL COUNTRIES ARE TALKING ABOUT GOING TO MARS.
AND HOW FAR IS MARS FROM EARTH?
AT ITS NEAREST, A FEW LIGHT-MINUTES.
A FEW LIGHT-MINUTES? AND THEY'RE STILL ONLY TALKING ABOUT GOING THERE? WILL HUMANITY EVER TRAVEL INTO SPACE AND MEET ALIENS?
WE MIGHT, BUT IT'S MORE LIKELY THEY'LL VISIT US FIRST.
WHY IS THAT?
WE'VE ONLY STARTED WONDERING ABOUT LIFE ON OTHER PLANETS IN THE LAST HUNDRED YEARS—A MERE BLINK OF AN EYE IN TERMS OF THE AGE OF THE UNIVERSE.
REMEMBER, HUMANS ONLY DISCOVERED THE FIRST PLANET AROUND ANOTHER STAR ABOUT 20 YEARS AGO.

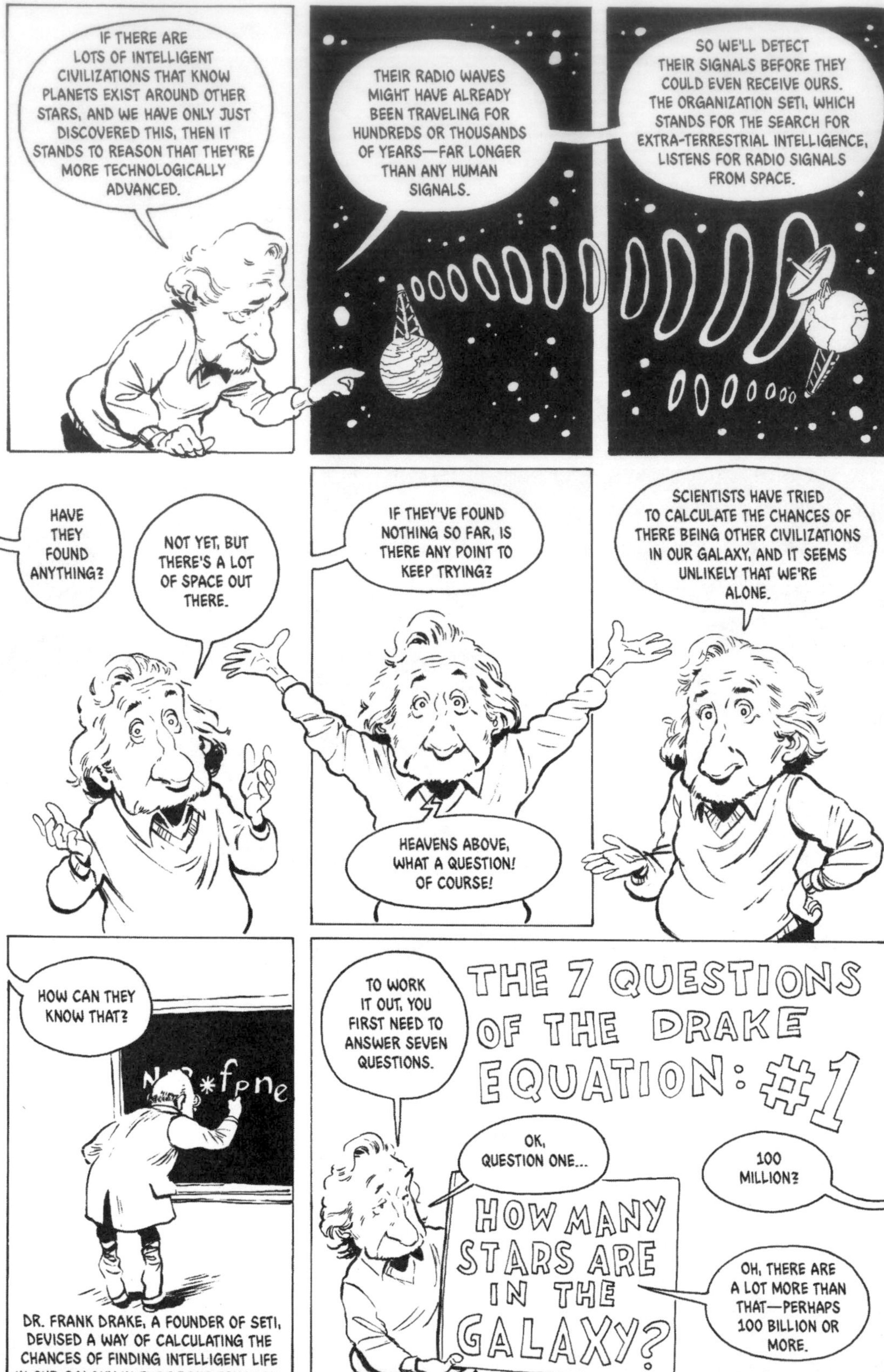
IF THERE ARE LOTS OF INTELLIGENT CIVILIZATIONS THAT KNOW PLANETS EXIST AROUND OTHER STARS, AND WE HAVE ONLY JUST DISCOVERED THIS, THEN IT STANDS TO REASON THAT THEY'RE MORE TECHNOLOGICALLY ADVANCED.
THEIR RADIO WAVES MIGHT HAVE ALREADY BEEN TRAVELING FOR HUNDREDS OR THOUSANDS OF YEARS—FAR LONGER THAN ANY HUMAN SIGNALS.
SO WE'LL DETECT THEIR SIGNALS BEFORE THEY COULD EVEN RECEIVE OURS. THE ORGANIZATION SETI, WHICH STANDS FOR THE SEARCH FOR EXTRA-TERRESTRIAL INTELLIGENCE, LISTENS FOR RADIO SIGNALS FROM SPACE.
HAVE THEY FOUND ANYTHING?
NOT YET, BUT THERE'S A LOT OF SPACE OUT THERE.
IF THEY'VE FOUND NOTHING SO FAR, IS THERE ANY POINT TO KEEP TRYING?
HEAVENS ABOVE, WHAT A QUESTION! OF COURSE!
SCIENTISTS HAVE TRIED TO CALCULATE THE CHANCES OF THERE BEING OTHER CIVILIZATIONS IN OUR GALAXY, AND IT SEEMS UNLIKELY THAT WE'RE ALONE.
HOW CAN THEY KNOW THAT?
DR. FRANK DRAKE, A FOUNDER OF SETI, DEVISED A WAY OF CALCULATING THE CHANCES OF FINDING INTELLIGENT LIFE IN OUR GALAXY IN THE DRAKE EQUATION.
TO WORK IT OUT, YOU FIRST NEED TO ANSWER SEVEN QUESTIONS.
THE 7 QUESTIONS OF THE DRAKE EQUATION: #1
OK, QUESTION ONE...
HOW MANY STARS ARE IN THE GALAXY?
100 MILLION?
OH, THERE ARE A LOT MORE THAN THAT—PERHAPS 100 BILLION OR MORE.

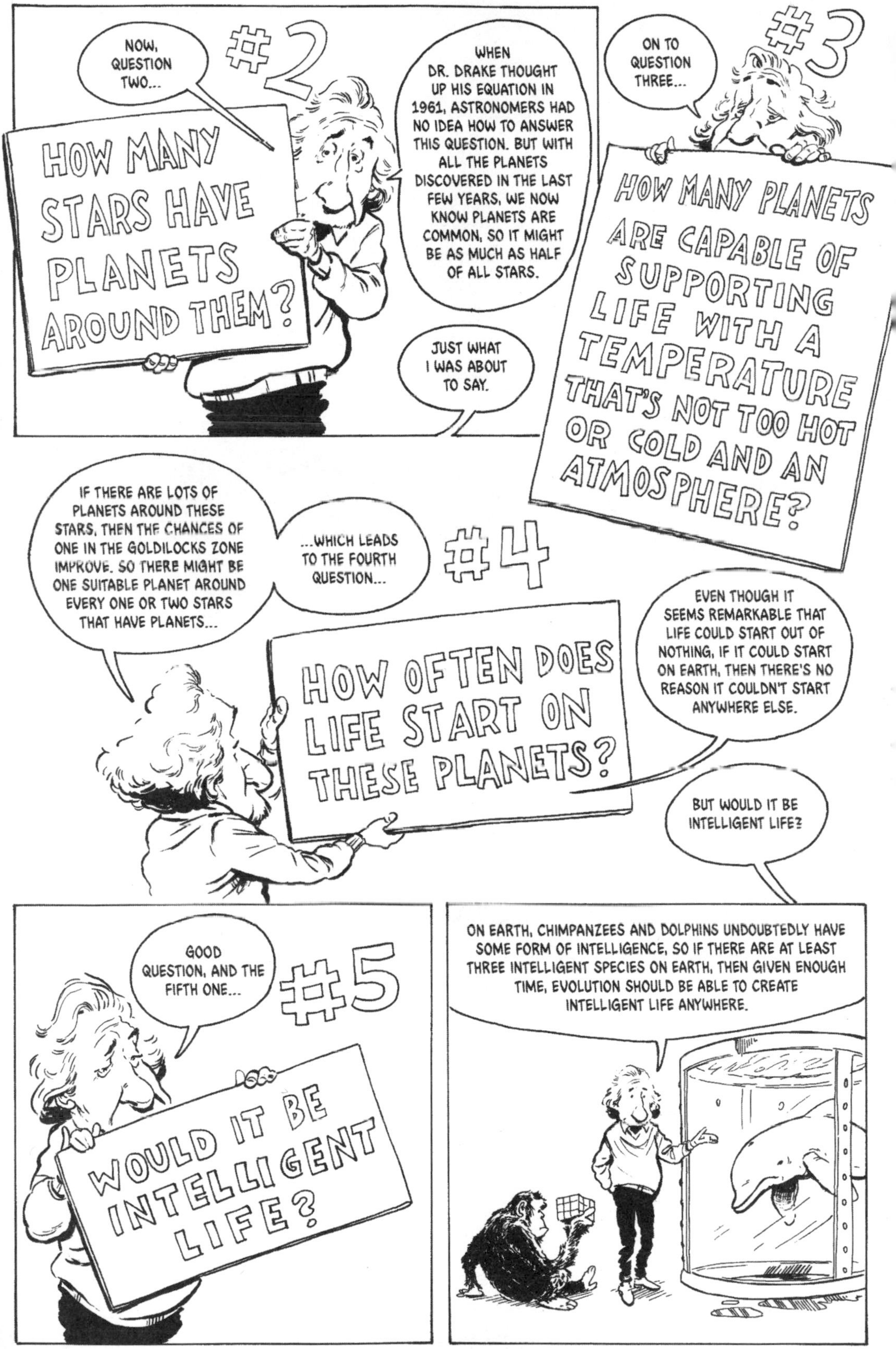

NOW, QUESTION TWO...
#2
HOW MANY STARS HAVE PLANETS AROUND THEM?
WHEN DR. DRAKE THOUGHT UP HIS EQUATION IN 1961, ASTRONOMERS HAD NO IDEA HOW TO ANSWER THIS QUESTION. BUT WITH ALL THE PLANETS DISCOVERED IN THE LAST FEW YEARS, WE NOW KNOW PLANETS ARE COMMON, SO IT MIGHT BE AS MUCH AS HALF OF ALL STARS.
JUST WHAT I WAS ABOUT TO SAY.
ON TO QUESTION THREE...
#3
HOW MANY PLANETS ARE CAPABLE OF SUPPORTING LIFE WITH A TEMPERATURE THAT'S NOT TOO HOT OR COLD AND AN ATMOSPHERE?
IF THERE ARE LOTS OF PLANETS AROUND THESE STARS, THEN THE CHANCES OF ONE IN THE GOLDILOCKS ZONE IMPROVE. SO THERE MIGHT BE ONE SUITABLE PLANET AROUND EVERY ONE OR TWO STARS THAT HAVE PLANETS...
...WHICH LEADS TO THE FOURTH QUESTION...
#4
HOW OFTEN DOES LIFE START ON THESE PLANETS?
EVEN THOUGH IT SEEMS REMARKABLE THAT LIFE COULD START OUT OF NOTHING, IF IT COULD START ON EARTH, THEN THERE'S NO REASON IT COULDN'T START ANYWHERE ELSE.
BUT WOULD IT BE INTELLIGENT LIFE?
GOOD QUESTION, AND THE FIFTH ONE...
#5
WOULD IT BE INTELLIGENT LIFE?
ON EARTH, CHIMPANZEES AND DOLPHINS UNDOUBTEDLY HAVE SOME FORM OF INTELLIGENCE, SO IF THERE ARE AT LEAST THREE INTELLIGENT SPECIES ON EARTH, THEN GIVEN ENOUGH TIME, EVOLUTION SHOULD BE ABLE TO CREATE INTELLIGENT LIFE ANYWHERE.

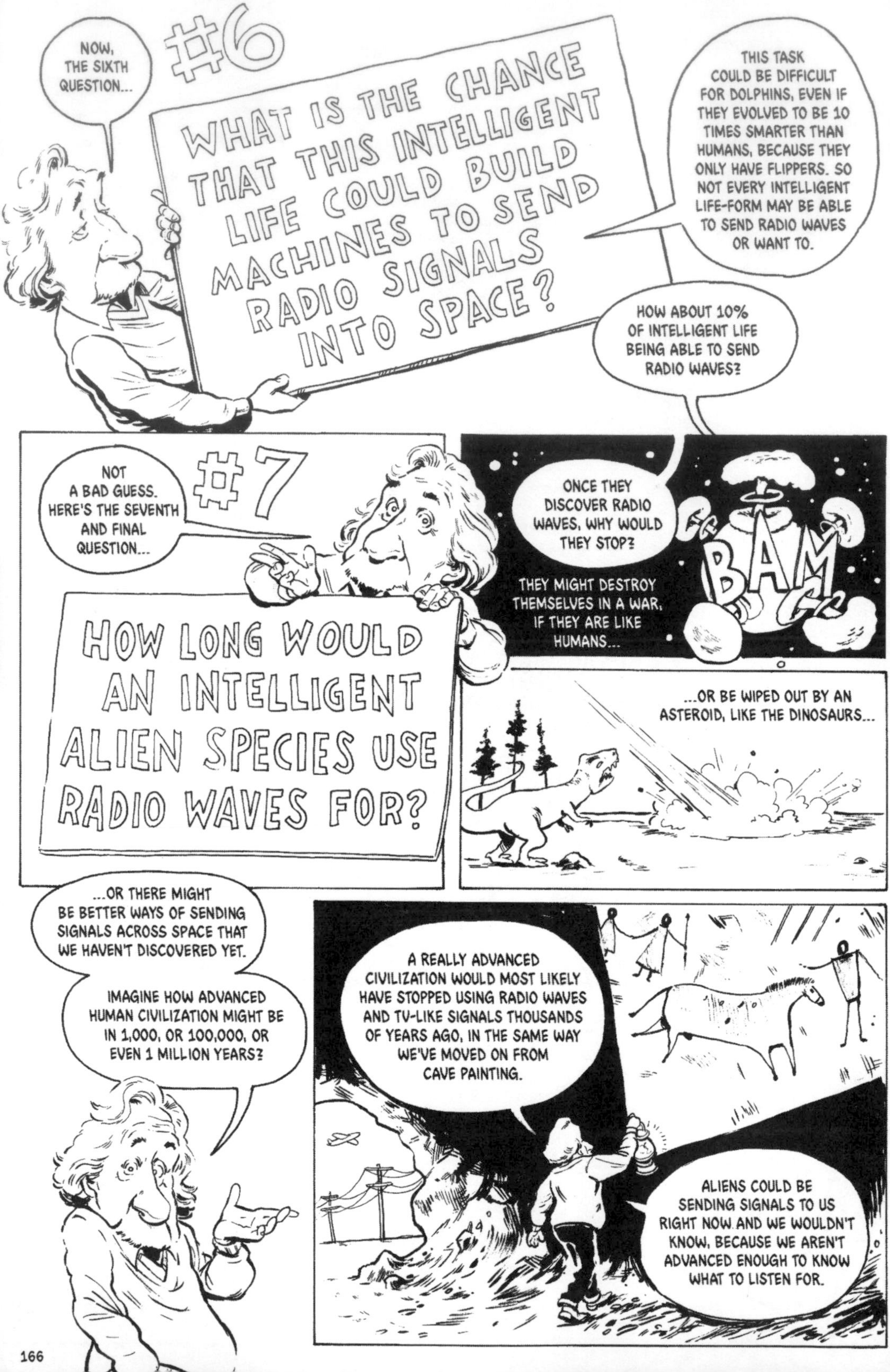
NOW, THE SIXTH QUESTION...
#6
WHAT IS THE CHANCE THAT THIS INTELLIGENT LIFE COULD BUILD MACHINES TO SEND RADIO SIGNALS INTO SPACE?
THIS TASK COULD BE DIFFICULT FOR DOLPHINS, EVEN IF THEY EVOLVED TO BE 10 TIMES SMARTER THAN HUMANS, BECAUSE THEY ONLY HAVE FLIPPERS. SO NOT EVERY INTELLIGENT LIFE-FORM MAY BE ABLE TO SEND RADIO WAVES OR WANT TO.
HOW ABOUT 10% OF INTELLIGENT LIFE BEING ABLE TO SEND RADIO WAVES?
NOT A BAD GUESS. HERE'S THE SEVENTH AND FINAL QUESTION...
#7
HOW LONG WOULD AN INTELLIGENT ALIEN SPECIES USE RADIO WAVES FOR?
ONCE THEY DISCOVER RADIO WAVES, WHY WOULD THEY STOP?
THEY MIGHT DESTROY THEMSELVES IN A WAR, IF THEY ARE LIKE HUMANS...
BAM
...OR BE WIPED OUT BY AN ASTEROID, LIKE THE DINOSAURS...
...OR THERE MIGHT BE BETTER WAYS OF SENDING SIGNALS ACROSS SPACE THAT WE HAVEN'T DISCOVERED YET.
IMAGINE HOW ADVANCED HUMAN CIVILIZATION MIGHT BE IN 1,000, OR 100,000, OR EVEN 1 MILLION YEARS?
A REALLY ADVANCED CIVILIZATION WOULD MOST LIKELY HAVE STOPPED USING RADIO WAVES AND TV-LIKE SIGNALS THOUSANDS OF YEARS AGO, IN THE SAME WAY WE'VE MOVED ON FROM CAVE PAINTING.
ALIENS COULD BE SENDING SIGNALS TO US RIGHT NOW AND WE WOULDN'T KNOW, BECAUSE WE AREN'T ADVANCED ENOUGH TO KNOW WHAT TO LISTEN FOR.

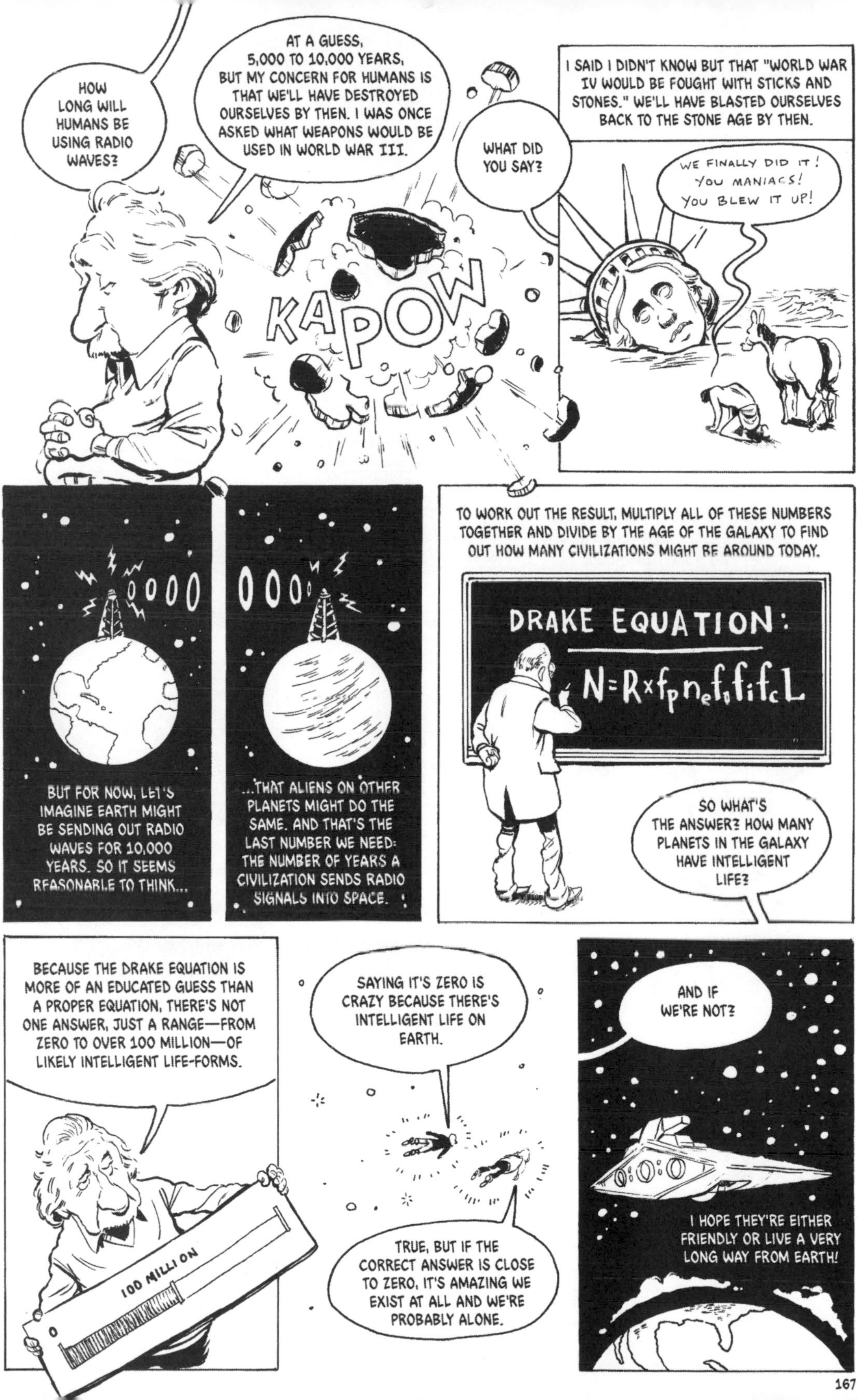
HOW LONG WILL HUMANS BE USING RADIO WAVES?
AT A GUESS, 5,000 TO 10,000 YEARS, BUT MY CONCERN FOR HUMANS IS THAT WE'LL HAVE DESTROYED OURSELVES BY THEN. I WAS ONCE ASKED WHAT WEAPONS WOULD BE USED IN WORLD WAR III.
KAPOW
WHAT DID YOU SAY?
I SAID I DIDN'T KNOW BUT THAT "WORLD WAR IV WOULD BE FOUGHT WITH STICKS AND STONES." WE'LL HAVE BLASTED OURSELVES BACK TO THE STONE AGE BY THEN.
WE FINALLY DID IT! YOU MANIACS! YOU BLEW IT UP!
BUT FOR NOW, LET'S IMAGINE EARTH MIGHT BE SENDING OUT RADIO WAVES FOR 10,000 YEARS. SO IT SEEMS REASONABLE TO THINK...
...THAT ALIENS ON OTHER PLANETS MIGHT DO THE SAME. AND THAT'S THE LAST NUMBER WE NEED: THE NUMBER OF YEARS A CIVILIZATION SENDS RADIO SIGNALS INTO SPACE.
TO WORK OUT THE RESULT, MULTIPLY ALL OF THESE NUMBERS TOGETHER AND DIVIDE BY THE AGE OF THE GALAXY TO FIND OUT HOW MANY CIVILIZATIONS MIGHT BE AROUND TODAY.
DRAKE EQUATION:
$N = R \times f_p n_e f_l f_i f_c L$
SO WHAT'S THE ANSWER? HOW MANY PLANETS IN THE GALAXY HAVE INTELLIGENT LIFE?
BECAUSE THE DRAKE EQUATION IS MORE OF AN EDUCATED GUESS THAN A PROPER EQUATION, THERE'S NOT ONE ANSWER, JUST A RANGE—FROM ZERO TO OVER 100 MILLION—OF LIKELY INTELLIGENT LIFE-FORMS.
100 MILLION
0
SAYING IT'S ZERO IS CRAZY BECAUSE THERE'S INTELLIGENT LIFE ON EARTH.
TRUE, BUT IF THE CORRECT ANSWER IS CLOSE TO ZERO, IT'S AMAZING WE EXIST AT ALL AND WE'RE PROBABLY ALONE.
AND IF WE'RE NOT?
I HOPE THEY'RE EITHER FRIENDLY OR LIVE A VERY LONG WAY FROM EARTH!

CHAPTER 31: WHY GREENHOUSE GASES ARE GOOD FOR THE PLANET

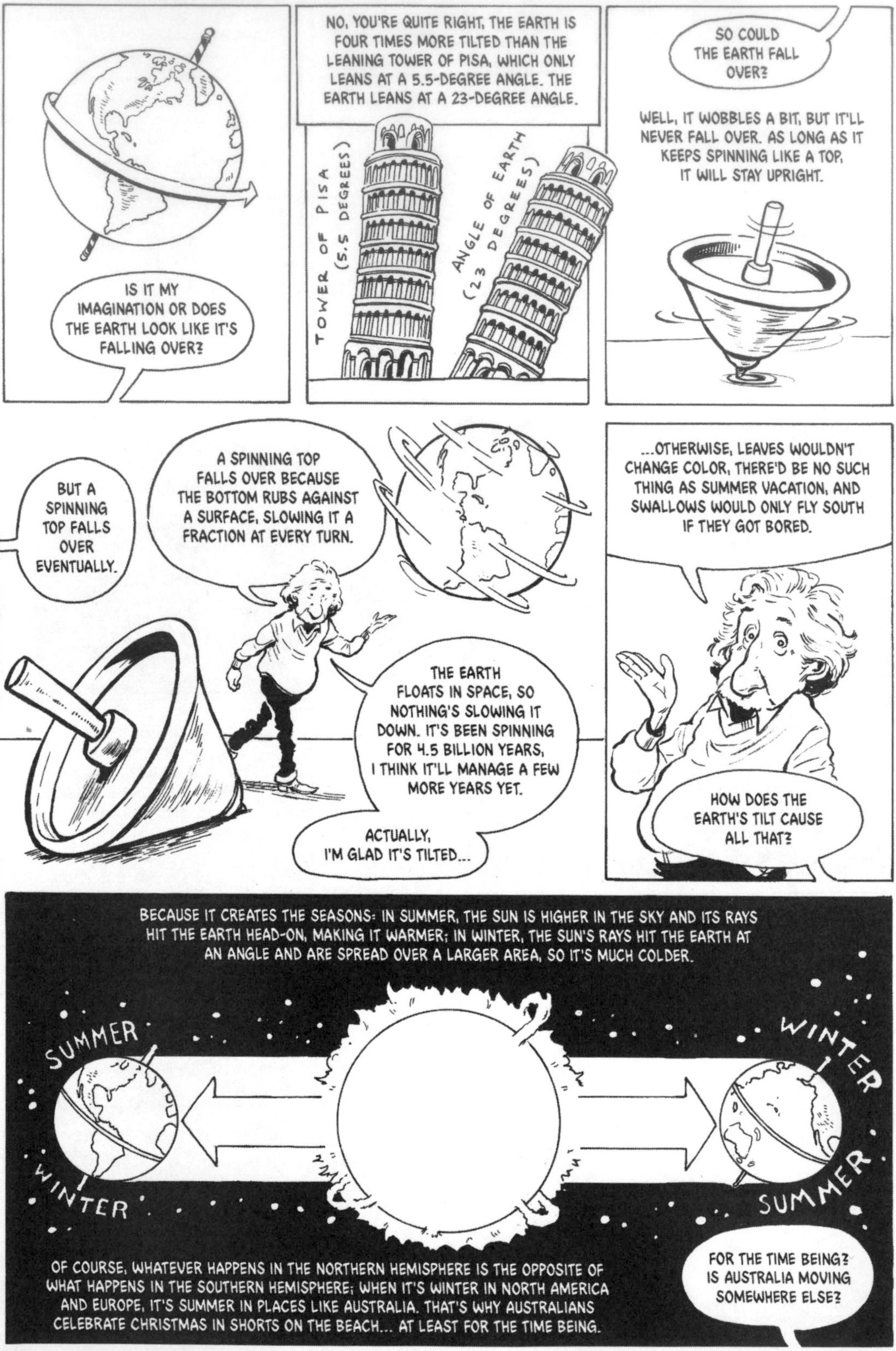
IS IT MY IMAGINATION OR DOES THE EARTH LOOK LIKE IT'S FALLING OVER?
NO, YOU'RE QUITE RIGHT, THE EARTH IS FOUR TIMES MORE TILTED THAN THE LEANING TOWER OF PISA, WHICH ONLY LEANS AT A 5.5-DEGREE ANGLE. THE EARTH LEANS AT A 23-DEGREE ANGLE.
TOWER OF PISA (5.5 DEGREES)
ANGLE OF EARTH (23 DEGREES)
SO COULD THE EARTH FALL OVER?
WELL, IT WOBBLES A BIT, BUT IT'LL NEVER FALL OVER. AS LONG AS IT KEEPS SPINNING LIKE A TOP, IT WILL STAY UPRIGHT.
BUT A SPINNING TOP FALLS OVER EVENTUALLY.
A SPINNING TOP FALLS OVER BECAUSE THE BOTTOM RUBS AGAINST A SURFACE, SLOWING IT A FRACTION AT EVERY TURN.
THE EARTH FLOATS IN SPACE, SO NOTHING'S SLOWING IT DOWN. IT'S BEEN SPINNING FOR 4.5 BILLION YEARS, I THINK IT'LL MANAGE A FEW MORE YEARS YET.
ACTUALLY, I'M GLAD IT'S TILTED...
...OTHERWISE, LEAVES WOULDN'T CHANGE COLOR, THERE'D BE NO SUCH THING AS SUMMER VACATION, AND SWALLOWS WOULD ONLY FLY SOUTH IF THEY GOT BORED.
HOW DOES THE EARTH'S TILT CAUSE ALL THAT?
BECAUSE IT CREATES THE SEASONS: IN SUMMER, THE SUN IS HIGHER IN THE SKY AND ITS RAYS HIT THE EARTH HEAD-ON, MAKING IT WARMER; IN WINTER, THE SUN'S RAYS HIT THE EARTH AT AN ANGLE AND ARE SPREAD OVER A LARGER AREA, SO IT'S MUCH COLDER.
SUMMER
WINTER
WINTER
SUMMER
OF COURSE, WHATEVER HAPPENS IN THE NORTHERN HEMISPHERE IS THE OPPOSITE OF WHAT HAPPENS IN THE SOUTHERN HEMISPHERE; WHEN IT'S WINTER IN NORTH AMERICA AND EUROPE, IT'S SUMMER IN PLACES LIKE AUSTRALIA. THAT'S WHY AUSTRALIANS CELEBRATE CHRISTMAS IN SHORTS ON THE BEACH... AT LEAST FOR THE TIME BEING.
FOR THE TIME BEING? IS AUSTRALIA MOVING SOMEWHERE ELSE?

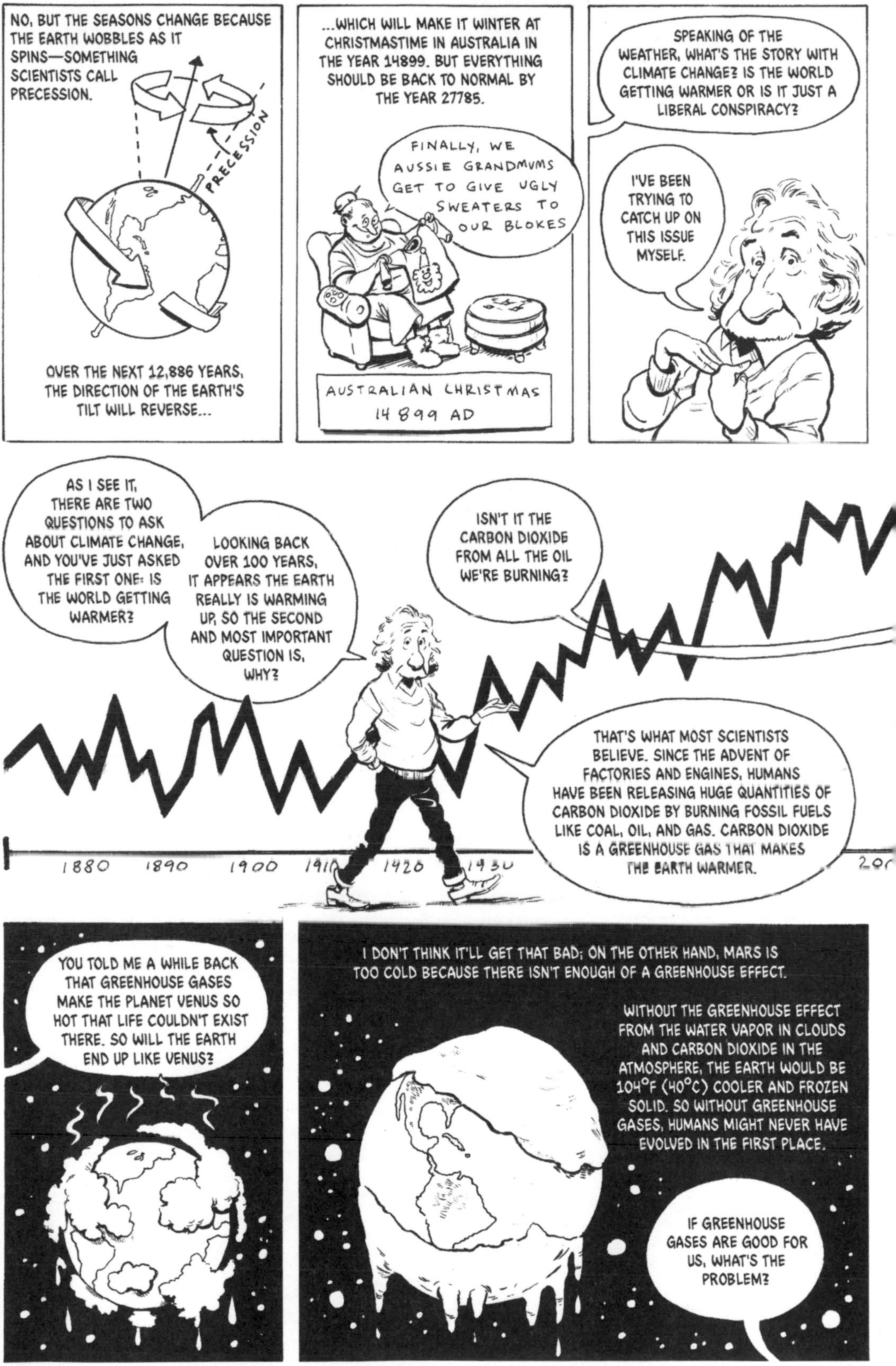
NO, BUT THE SEASONS CHANGE BECAUSE THE EARTH WOBBLES AS IT SPINS—SOMETHING SCIENTISTS CALL PRECESSION.
PRECESSION
OVER THE NEXT 12,886 YEARS, THE DIRECTION OF THE EARTH'S TILT WILL REVERSE...
...WHICH WILL MAKE IT WINTER AT CHRISTMASTIME IN AUSTRALIA IN THE YEAR 14899. BUT EVERYTHING SHOULD BE BACK TO NORMAL BY THE YEAR 27785.
FINALLY, WE AUSSIE GRANDMUMS GET TO GIVE UGLY SWEATERS TO OUR BLOKES
AUSTRALIAN CHRISTMAS 14899 AD
SPEAKING OF THE WEATHER, WHAT'S THE STORY WITH CLIMATE CHANGE? IS THE WORLD GETTING WARMER OR IS IT JUST A LIBERAL CONSPIRACY?
I'VE BEEN TRYING TO CATCH UP ON THIS ISSUE MYSELF.
AS I SEE IT, THERE ARE TWO QUESTIONS TO ASK ABOUT CLIMATE CHANGE, AND YOU'VE JUST ASKED THE FIRST ONE: IS THE WORLD GETTING WARMER?
LOOKING BACK OVER 100 YEARS, IT APPEARS THE EARTH REALLY IS WARMING UP, SO THE SECOND AND MOST IMPORTANT QUESTION IS, WHY?
ISN'T IT THE CARBON DIOXIDE FROM ALL THE OIL WE'RE BURNING?
THAT'S WHAT MOST SCIENTISTS BELIEVE. SINCE THE ADVENT OF FACTORIES AND ENGINES, HUMANS HAVE BEEN RELEASING HUGE QUANTITIES OF CARBON DIOXIDE BY BURNING FOSSIL FUELS LIKE COAL, OIL, AND GAS. CARBON DIOXIDE IS A GREENHOUSE GAS THAT MAKES THE EARTH WARMER.
1880
1890
1900
1920
YOU TOLD ME A WHILE BACK THAT GREENHOUSE GASES MAKE THE PLANET VENUS SO HOT THAT LIFE COULDN'T EXIST THERE. SO WILL THE EARTH END UP LIKE VENUS?
I DON'T THINK IT'LL GET THAT BAD; ON THE OTHER HAND, MARS IS TOO COLD BECAUSE THERE ISN'T ENOUGH OF A GREENHOUSE EFFECT.
WITHOUT THE GREENHOUSE EFFECT FROM THE WATER VAPOR IN CLOUDS AND CARBON DIOXIDE IN THE ATMOSPHERE, THE EARTH WOULD BE 104°F (40°C) COOLER AND FROZEN SOLID. SO WITHOUT GREENHOUSE GASES, HUMANS MIGHT NEVER HAVE EVOLVED IN THE FIRST PLACE.
IF GREENHOUSE GASES ARE GOOD FOR US, WHAT'S THE PROBLEM?

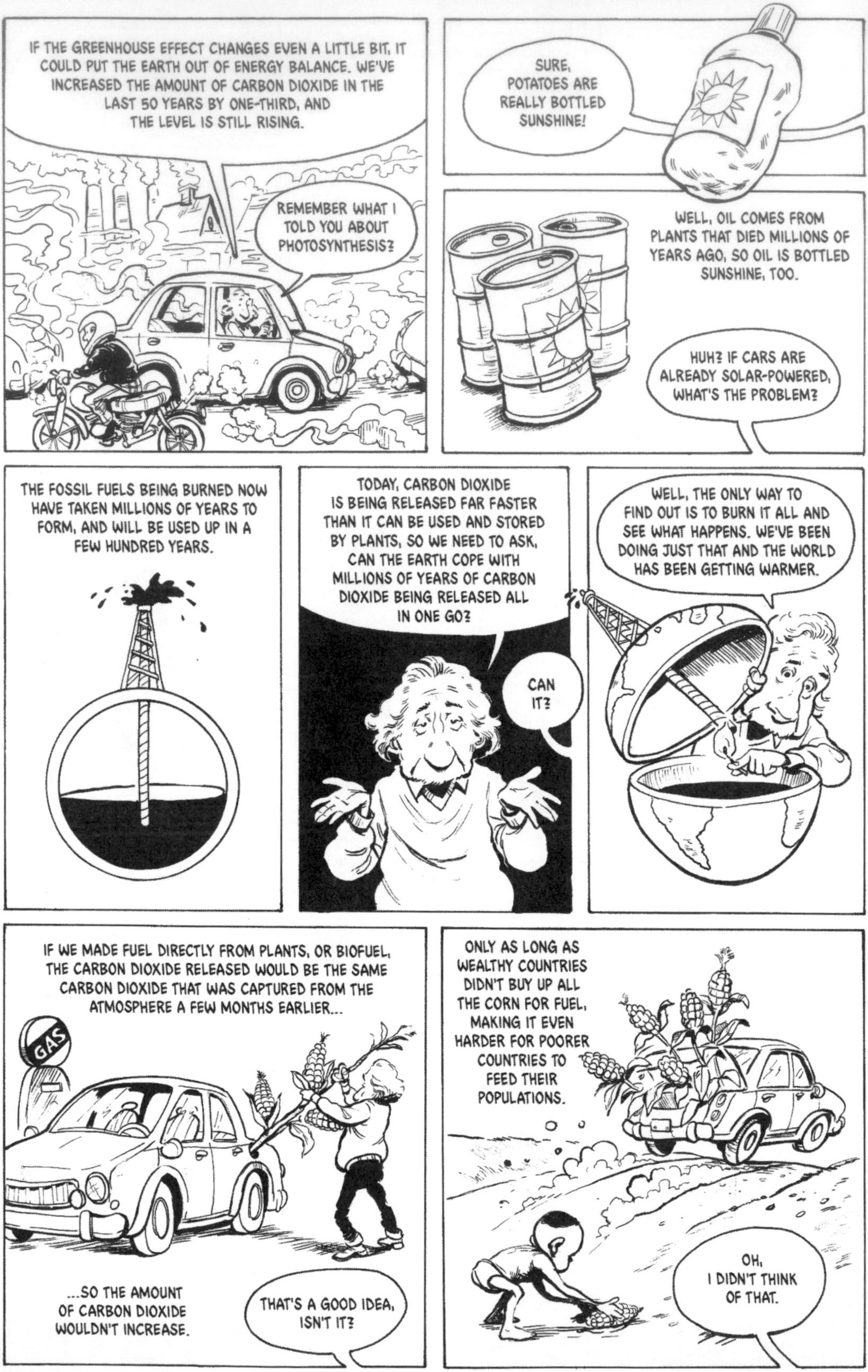
IF THE GREENHOUSE EFFECT CHANGES EVEN A LITTLE BIT, IT COULD PUT THE EARTH OUT OF ENERGY BALANCE. WE'VE INCREASED THE AMOUNT OF CARBON DIOXIDE IN THE LAST 50 YEARS BY ONE-THIRD, AND THE LEVEL IS STILL RISING.
REMEMBER WHAT I TOLD YOU ABOUT PHOTOSYNTHESIS?
SURE, POTATOES ARE REALLY BOTTLED SUNSHINE!
WELL, OIL COMES FROM PLANTS THAT DIED MILLIONS OF YEARS AGO, SO OIL IS BOTTLED SUNSHINE, TOO.
HUH? IF CARS ARE ALREADY SOLAR-POWERED, WHAT'S THE PROBLEM?
THE FOSSIL FUELS BEING BURNED NOW HAVE TAKEN MILLIONS OF YEARS TO FORM, AND WILL BE USED UP IN A FEW HUNDRED YEARS.
TODAY, CARBON DIOXIDE IS BEING RELEASED FAR FASTER THAN IT CAN BE USED AND STORED BY PLANTS, SO WE NEED TO ASK, CAN THE EARTH COPE WITH MILLIONS OF YEARS OF CARBON DIOXIDE BEING RELEASED ALL IN ONE GO?
CAN IT?
WELL, THE ONLY WAY TO FIND OUT IS TO BURN IT ALL AND SEE WHAT HAPPENS. WE'VE BEEN DOING JUST THAT AND THE WORLD HAS BEEN GETTING WARMER.
IF WE MADE FUEL DIRECTLY FROM PLANTS, OR BIOFUEL, THE CARBON DIOXIDE RELEASED WOULD BE THE SAME CARBON DIOXIDE THAT WAS CAPTURED FROM THE ATMOSPHERE A FEW MONTHS EARLIER...
GAS
...SO THE AMOUNT OF CARBON DIOXIDE WOULDN'T INCREASE.
THAT'S A GOOD IDEA, ISN'T IT?
ONLY AS LONG AS WEALTHY COUNTRIES DIDN'T BUY UP ALL THE CORN FOR FUEL, MAKING IT EVEN HARDER FOR POORER COUNTRIES TO FEED THEIR POPULATIONS.
OH, I DIDN'T THINK OF THAT.

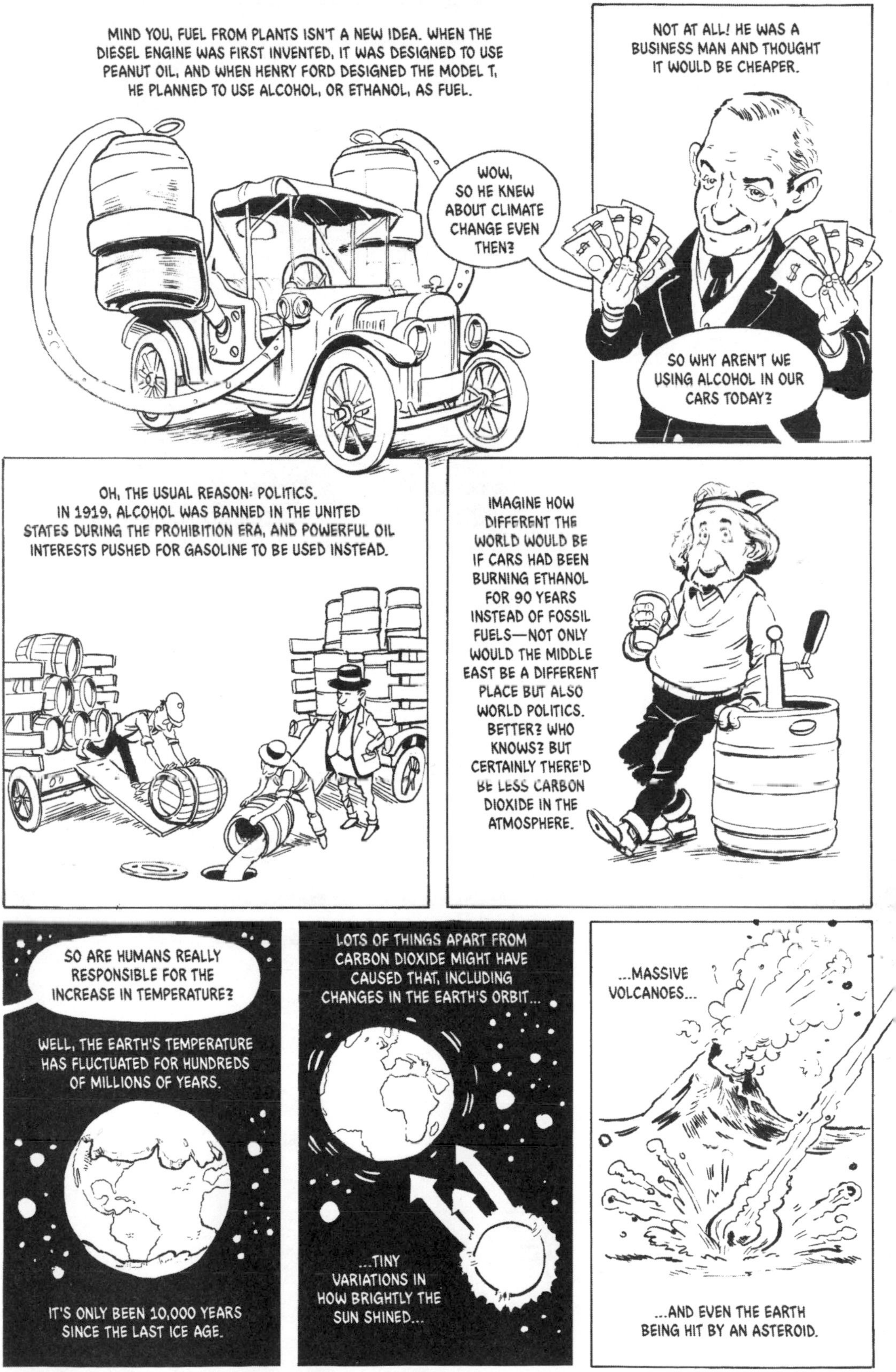

MIND YOU, FUEL FROM PLANTS ISN'T A NEW IDEA. WHEN THE DIESEL ENGINE WAS FIRST INVENTED, IT WAS DESIGNED TO USE PEANUT OIL, AND WHEN HENRY FORD DESIGNED THE MODEL T, HE PLANNED TO USE ALCOHOL, OR ETHANOL, AS FUEL.
WOW, SO HE KNEW ABOUT CLIMATE CHANGE EVEN THEN?
NOT AT ALL! HE WAS A BUSINESS MAN AND THOUGHT IT WOULD BE CHEAPER.
SO WHY AREN'T WE USING ALCOHOL IN OUR CARS TODAY?
OH, THE USUAL REASON: POLITICS. IN 1919, ALCOHOL WAS BANNED IN THE UNITED STATES DURING THE PROHIBITION ERA, AND POWERFUL OIL INTERESTS PUSHED FOR GASOLINE TO BE USED INSTEAD.
IMAGINE HOW DIFFERENT THE WORLD WOULD BE IF CARS HAD BEEN BURNING ETHANOL FOR 90 YEARS INSTEAD OF FOSSIL FUELS—NOT ONLY WOULD THE MIDDLE EAST BE A DIFFERENT PLACE BUT ALSO WORLD POLITICS. BETTER? WHO KNOWS? BUT CERTAINLY THERE'D BE LESS CARBON DIOXIDE IN THE ATMOSPHERE.
SO ARE HUMANS REALLY RESPONSIBLE FOR THE INCREASE IN TEMPERATURE?
WELL, THE EARTH'S TEMPERATURE HAS FLUCTUATED FOR HUNDREDS OF MILLIONS OF YEARS.
IT'S ONLY BEEN 10,000 YEARS SINCE THE LAST ICE AGE.
LOTS OF THINGS APART FROM CARBON DIOXIDE MIGHT HAVE CAUSED THAT, INCLUDING CHANGES IN THE EARTH'S ORBIT...
...TINY VARIATIONS IN HOW BRIGHTLY THE SUN SHINED...
...MASSIVE VOLCANOES...
...AND EVEN THE EARTH BEING HIT BY AN ASTEROID.

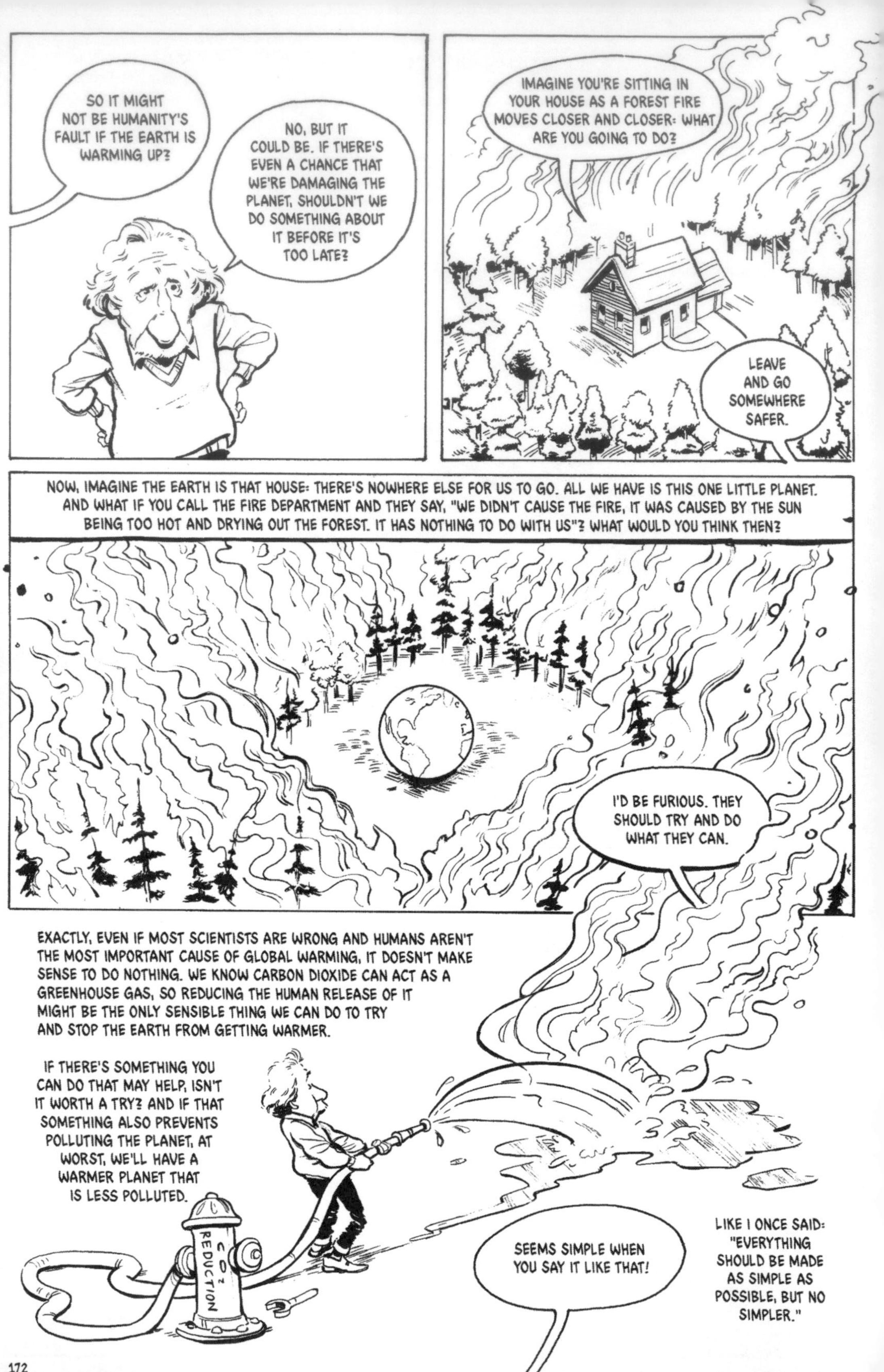
SO IT MIGHT NOT BE HUMANITY'S FAULT IF THE EARTH IS WARMING UP?
NO, BUT IT COULD BE. IF THERE'S EVEN A CHANCE THAT WE'RE DAMAGING THE PLANET, SHOULDN'T WE DO SOMETHING ABOUT IT BEFORE IT'S TOO LATE?
IMAGINE YOU'RE SITTING IN YOUR HOUSE AS A FOREST FIRE MOVES CLOSER AND CLOSER: WHAT ARE YOU GOING TO DO?
LEAVE AND GO SOMEWHERE SAFER.
NOW, IMAGINE THE EARTH IS THAT HOUSE: THERE'S NOWHERE ELSE FOR US TO GO. ALL WE HAVE IS THIS ONE LITTLE PLANET. AND WHAT IF YOU CALL THE FIRE DEPARTMENT AND THEY SAY, "WE DIDN'T CAUSE THE FIRE, IT WAS CAUSED BY THE SUN BEING TOO HOT AND DRYING OUT THE FOREST. IT HAS NOTHING TO DO WITH US"? WHAT WOULD YOU THINK THEN?
I'D BE FURIOUS. THEY SHOULD TRY AND DO WHAT THEY CAN.
EXACTLY, EVEN IF MOST SCIENTISTS ARE WRONG AND HUMANS AREN'T THE MOST IMPORTANT CAUSE OF GLOBAL WARMING, IT DOESN'T MAKE SENSE TO DO NOTHING. WE KNOW CARBON DIOXIDE CAN ACT AS A GREENHOUSE GAS, SO REDUCING THE HUMAN RELEASE OF IT MIGHT BE THE ONLY SENSIBLE THING WE CAN DO TO TRY AND STOP THE EARTH FROM GETTING WARMER.
IF THERE'S SOMETHING YOU CAN DO THAT MAY HELP, ISN'T IT WORTH A TRY? AND IF THAT SOMETHING ALSO PREVENTS POLLUTING THE PLANET, AT WORST, WE'LL HAVE A WARMER PLANET THAT IS LESS POLLUTED.
CO2 REDUCTION
SEEMS SIMPLE WHEN YOU SAY IT LIKE THAT!
LIKE I ONCE SAID: "EVERYTHING SHOULD BE MADE AS SIMPLE AS POSSIBLE, BUT NO SIMPLER."

CHAPTER 32: BLUE SKIES, RAINBOWS, AND THE MEANING OF LIFE

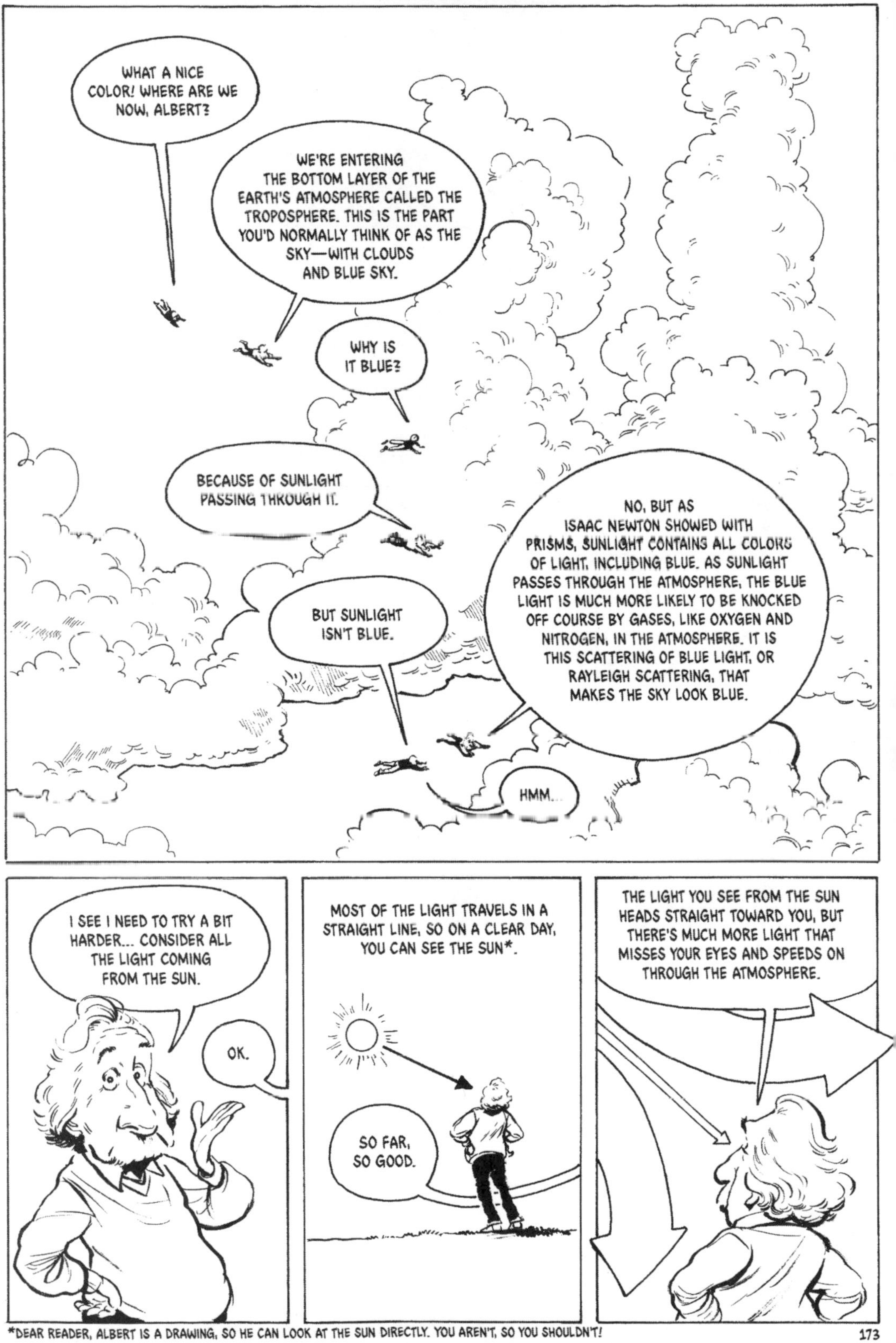
WHAT A NICE COLOR! WHERE ARE WE NOW, ALBERT?
WE'RE ENTERING THE BOTTOM LAYER OF THE EARTH'S ATMOSPHERE CALLED THE TROPOSPHERE. THIS IS THE PART YOU'D NORMALLY THINK OF AS THE SKY—WITH CLOUDS AND BLUE SKY.
WHY IS IT BLUE?
BECAUSE OF SUNLIGHT PASSING THROUGH IT.
BUT SUNLIGHT ISN'T BLUE.
NO, BUT AS ISAAC NEWTON SHOWED WITH PRISMS, SUNLIGHT CONTAINS ALL COLORS OF LIGHT, INCLUDING BLUE. AS SUNLIGHT PASSES THROUGH THE ATMOSPHERE, THE BLUE LIGHT IS MUCH MORE LIKELY TO BE KNOCKED OFF COURSE BY GASES, LIKE OXYGEN AND NITROGEN, IN THE ATMOSPHERE. IT IS THIS SCATTERING OF BLUE LIGHT, OR RAYLEIGH SCATTERING, THAT MAKES THE SKY LOOK BLUE.
HMM...
I SEE I NEED TO TRY A BIT HARDER... CONSIDER ALL THE LIGHT COMING FROM THE SUN.
OK.
MOST OF THE LIGHT TRAVELS IN A STRAIGHT LINE, SO ON A CLEAR DAY, YOU CAN SEE THE SUN*.
SO FAR, SO GOOD.
THE LIGHT YOU SEE FROM THE SUN HEADS STRAIGHT TOWARD YOU, BUT THERE'S MUCH MORE LIGHT THAT MISSES YOUR EYES AND SPEEDS ON THROUGH THE ATMOSPHERE.

*DEAR READER, ALBERT IS A DRAWING, SO HE CAN LOOK AT THE SUN DIRECTLY. YOU AREN'T, SO YOU SHOULDN'T!

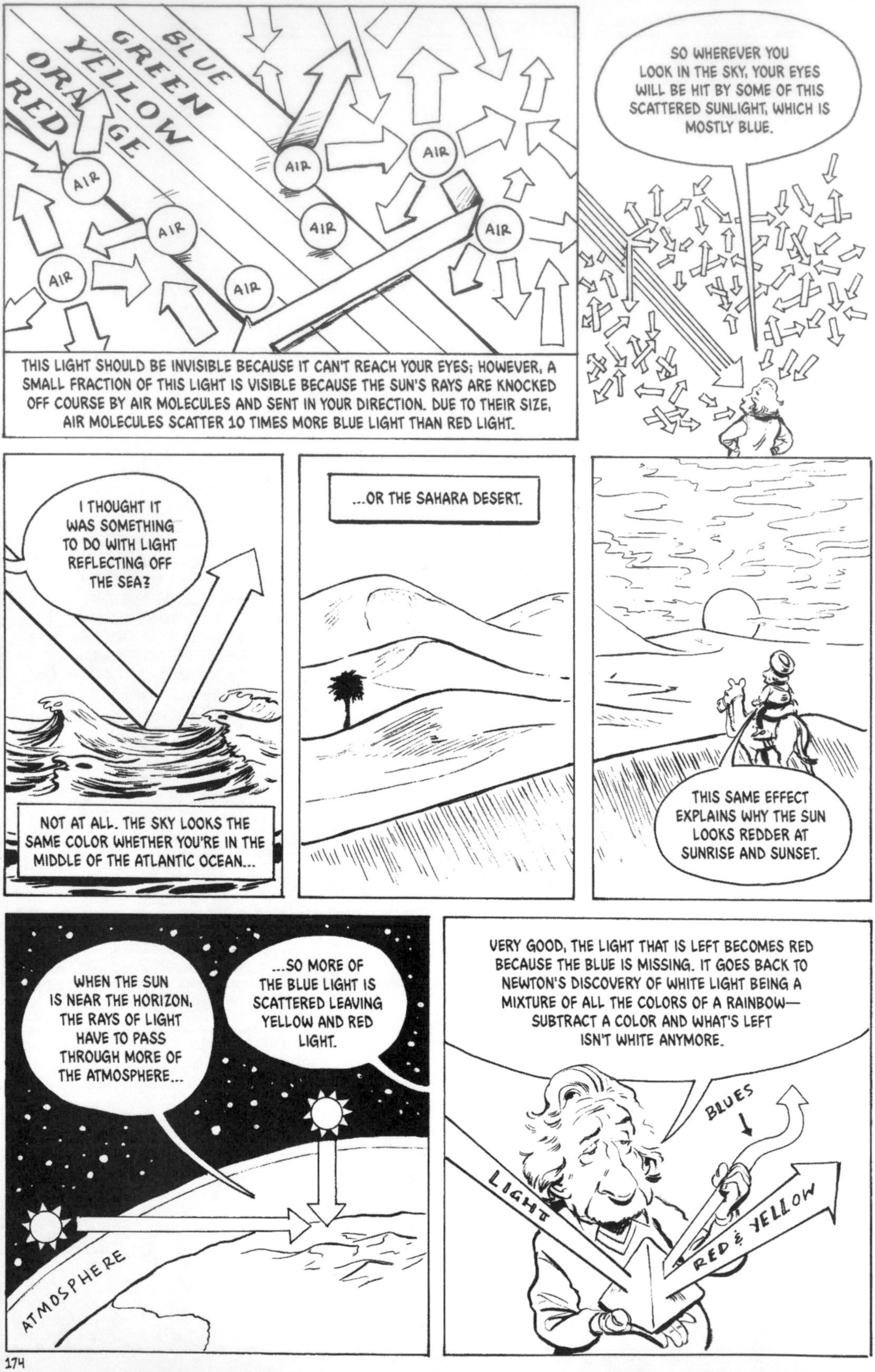
BLUE
GREEN
YELLOW
ORANGE
RED
AIR
THIS LIGHT SHOULD BE INVISIBLE BECAUSE IT CAN'T REACH YOUR EYES; HOWEVER, A SMALL FRACTION OF THIS LIGHT IS VISIBLE BECAUSE THE SUN'S RAYS ARE KNOCKED OFF COURSE BY AIR MOLECULES AND SENT IN YOUR DIRECTION. DUE TO THEIR SIZE, AIR MOLECULES SCATTER 10 TIMES MORE BLUE LIGHT THAN RED LIGHT.
SO WHEREVER YOU LOOK IN THE SKY, YOUR EYES WILL BE HIT BY SOME OF THIS SCATTERED SUNLIGHT, WHICH IS MOSTLY BLUE.
I THOUGHT IT WAS SOMETHING TO DO WITH LIGHT REFLECTING OFF THE SEA?
NOT AT ALL. THE SKY LOOKS THE SAME COLOR WHETHER YOU'RE IN THE MIDDLE OF THE ATLANTIC OCEAN...
...OR THE SAHARA DESERT.
THIS SAME EFFECT EXPLAINS WHY THE SUN LOOKS REDDER AT SUNRISE AND SUNSET.
WHEN THE SUN IS NEAR THE HORIZON, THE RAYS OF LIGHT HAVE TO PASS THROUGH MORE OF THE ATMOSPHERE...
...SO MORE OF THE BLUE LIGHT IS SCATTERED LEAVING YELLOW AND RED LIGHT.
ATMOSPHERE
VERY GOOD, THE LIGHT THAT IS LEFT BECOMES RED BECAUSE THE BLUE IS MISSING. IT GOES BACK TO NEWTON'S DISCOVERY OF WHITE LIGHT BEING A MIXTURE OF ALL THE COLORS OF A RAINBOW— SUBTRACT A COLOR AND WHAT'S LEFT ISN'T WHITE ANYMORE.
BLUES
LIGHT
RED & YELLOW

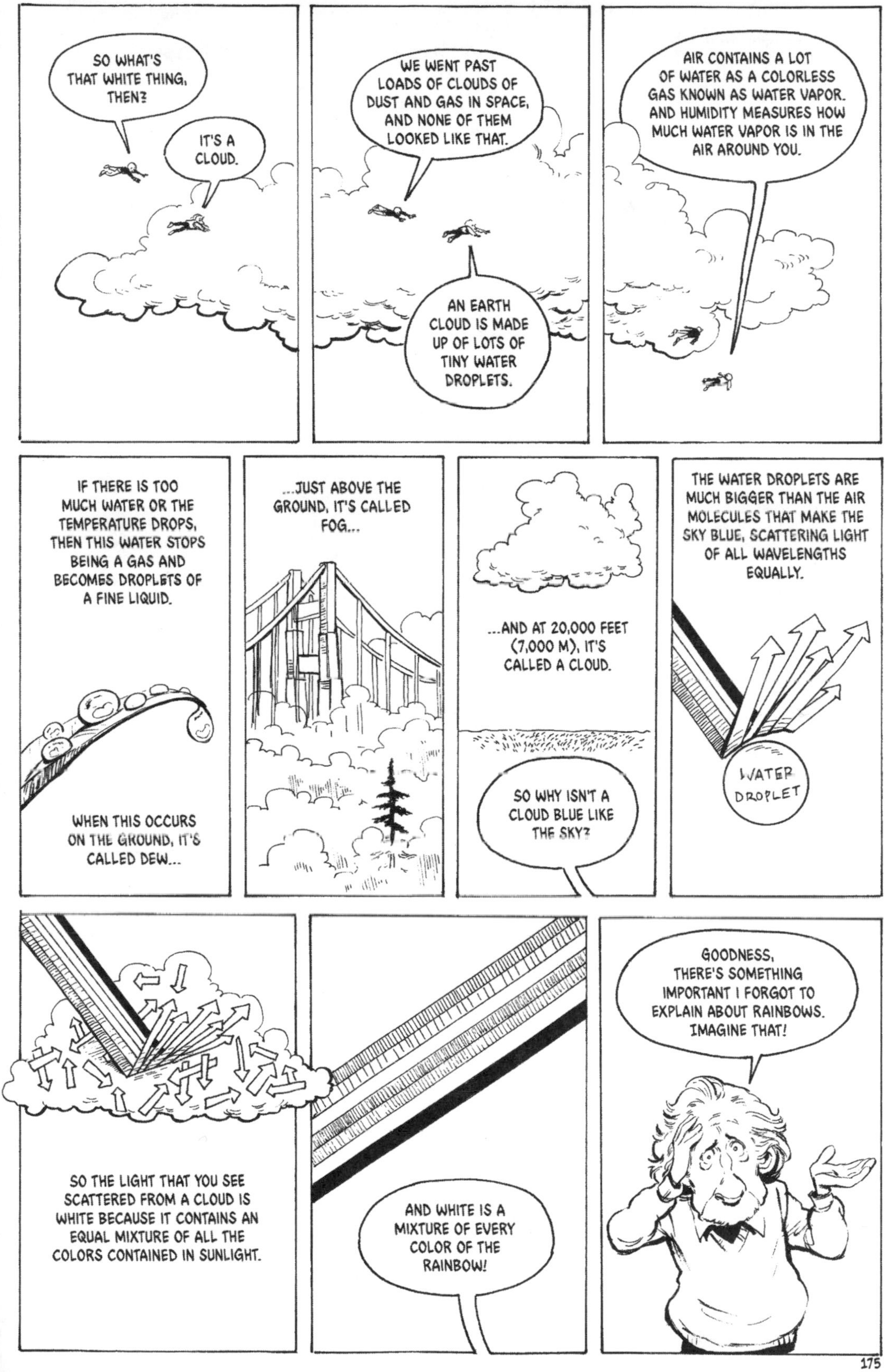
SO WHAT'S THAT WHITE THING, THEN?
IT'S A CLOUD.
WE WENT PAST LOADS OF CLOUDS OF DUST AND GAS IN SPACE, AND NONE OF THEM LOOKED LIKE THAT.
AN EARTH CLOUD IS MADE UP OF LOTS OF TINY WATER DROPLETS.
AIR CONTAINS A LOT OF WATER AS A COLORLESS GAS KNOWN AS WATER VAPOR. AND HUMIDITY MEASURES HOW MUCH WATER VAPOR IS IN THE AIR AROUND YOU.
IF THERE IS TOO MUCH WATER OR THE TEMPERATURE DROPS, THEN THIS WATER STOPS BEING A GAS AND BECOMES DROPLETS OF A FINE LIQUID.
WHEN THIS OCCURS ON THE GROUND, IT'S CALLED DEW...
...JUST ABOVE THE GROUND, IT'S CALLED FOG...
...AND AT 20,000 FEET (7,000 M), IT'S CALLED A CLOUD.
SO WHY ISN'T A CLOUD BLUE LIKE THE SKY?
THE WATER DROPLETS ARE MUCH BIGGER THAN THE AIR MOLECULES THAT MAKE THE SKY BLUE, SCATTERING LIGHT OF ALL WAVELENGTHS EQUALLY.
WATER DROPLET
SO THE LIGHT THAT YOU SEE SCATTERED FROM A CLOUD IS WHITE BECAUSE IT CONTAINS AN EQUAL MIXTURE OF ALL THE COLORS CONTAINED IN SUNLIGHT.
AND WHITE IS A MIXTURE OF EVERY COLOR OF THE RAINBOW!
GOODNESS, THERE'S SOMETHING IMPORTANT I FORGOT TO EXPLAIN ABOUT RAINBOWS. IMAGINE THAT!

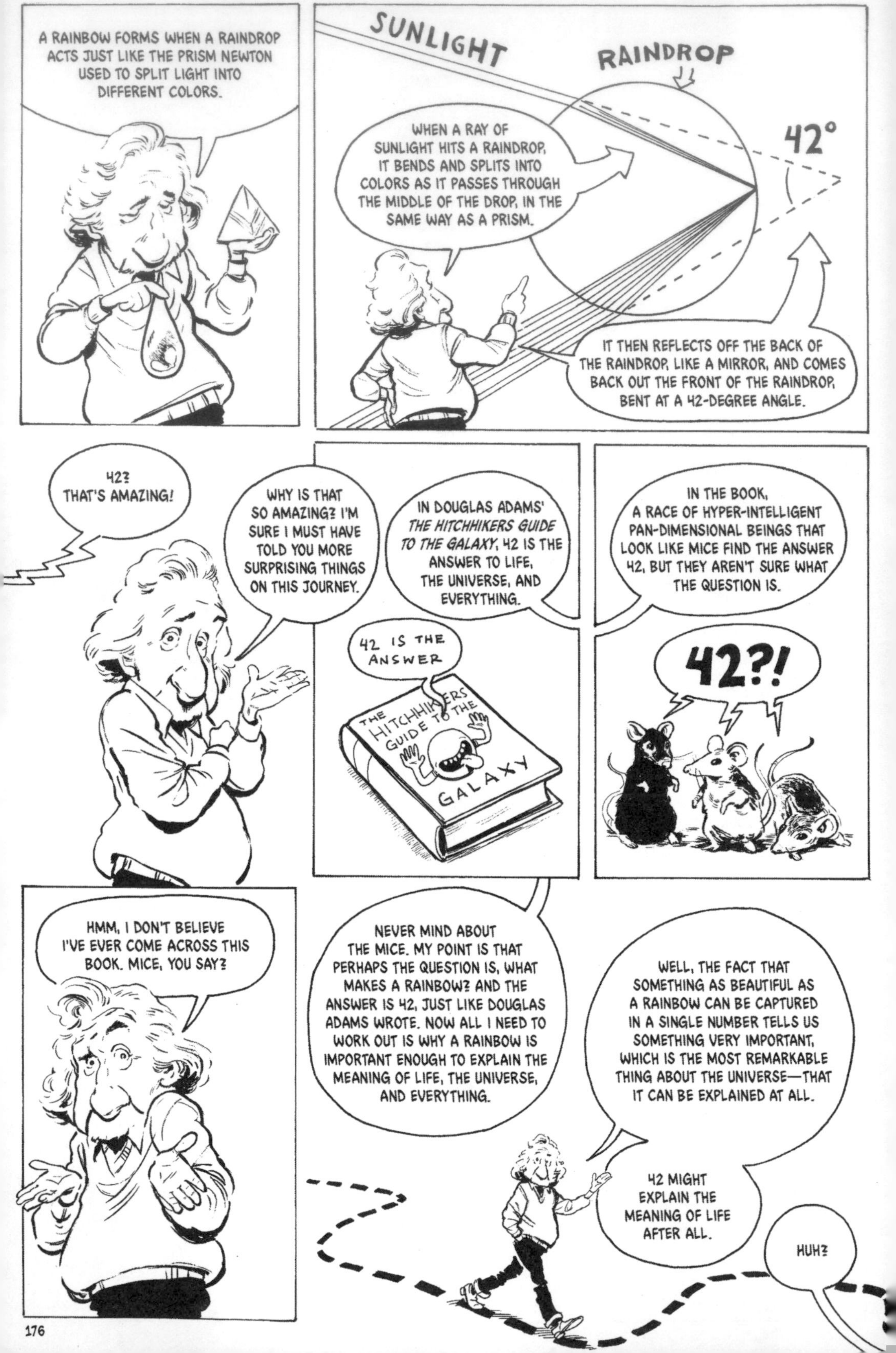
A RAINBOW FORMS WHEN A RAINDROP ACTS JUST LIKE THE PRISM NEWTON USED TO SPLIT LIGHT INTO DIFFERENT COLORS.
SUNLIGHT
RAINDROP
42°
WHEN A RAY OF SUNLIGHT HITS A RAINDROP, IT BENDS AND SPLITS INTO COLORS AS IT PASSES THROUGH THE MIDDLE OF THE DROP, IN THE SAME WAY AS A PRISM.
IT THEN REFLECTS OFF THE BACK OF THE RAINDROP, LIKE A MIRROR, AND COMES BACK OUT THE FRONT OF THE RAINDROP, BENT AT A 42-DEGREE ANGLE.
42? THAT'S AMAZING!
WHY IS THAT SO AMAZING? I'M SURE I MUST HAVE TOLD YOU MORE SURPRISING THINGS ON THIS JOURNEY.
IN DOUGLAS ADAMS' *THE HITCHHIKERS GUIDE TO THE GALAXY*, 42 IS THE ANSWER TO LIFE, THE UNIVERSE, AND EVERYTHING.
42 IS THE ANSWER
THE HITCHHIKERS GUIDE TO THE GALAXY
IN THE BOOK, A RACE OF HYPER-INTELLIGENT PAN-DIMENSIONAL BEINGS THAT LOOK LIKE MICE FIND THE ANSWER 42, BUT THEY AREN'T SURE WHAT THE QUESTION IS.
42?!
HMM, I DON'T BELIEVE I'VE EVER COME ACROSS THIS BOOK. MICE, YOU SAY?
NEVER MIND ABOUT THE MICE. MY POINT IS THAT PERHAPS THE QUESTION IS, WHAT MAKES A RAINBOW? AND THE ANSWER IS 42, JUST LIKE DOUGLAS ADAMS WROTE. NOW ALL I NEED TO WORK OUT IS WHY A RAINBOW IS IMPORTANT ENOUGH TO EXPLAIN THE MEANING OF LIFE, THE UNIVERSE, AND EVERYTHING.
WELL, THE FACT THAT SOMETHING AS BEAUTIFUL AS A RAINBOW CAN BE CAPTURED IN A SINGLE NUMBER TELLS US SOMETHING VERY IMPORTANT, WHICH IS THE MOST REMARKABLE THING ABOUT THE UNIVERSE—THAT IT CAN BE EXPLAINED AT ALL.
42 MIGHT EXPLAIN THE MEANING OF LIFE AFTER ALL.
HUH?

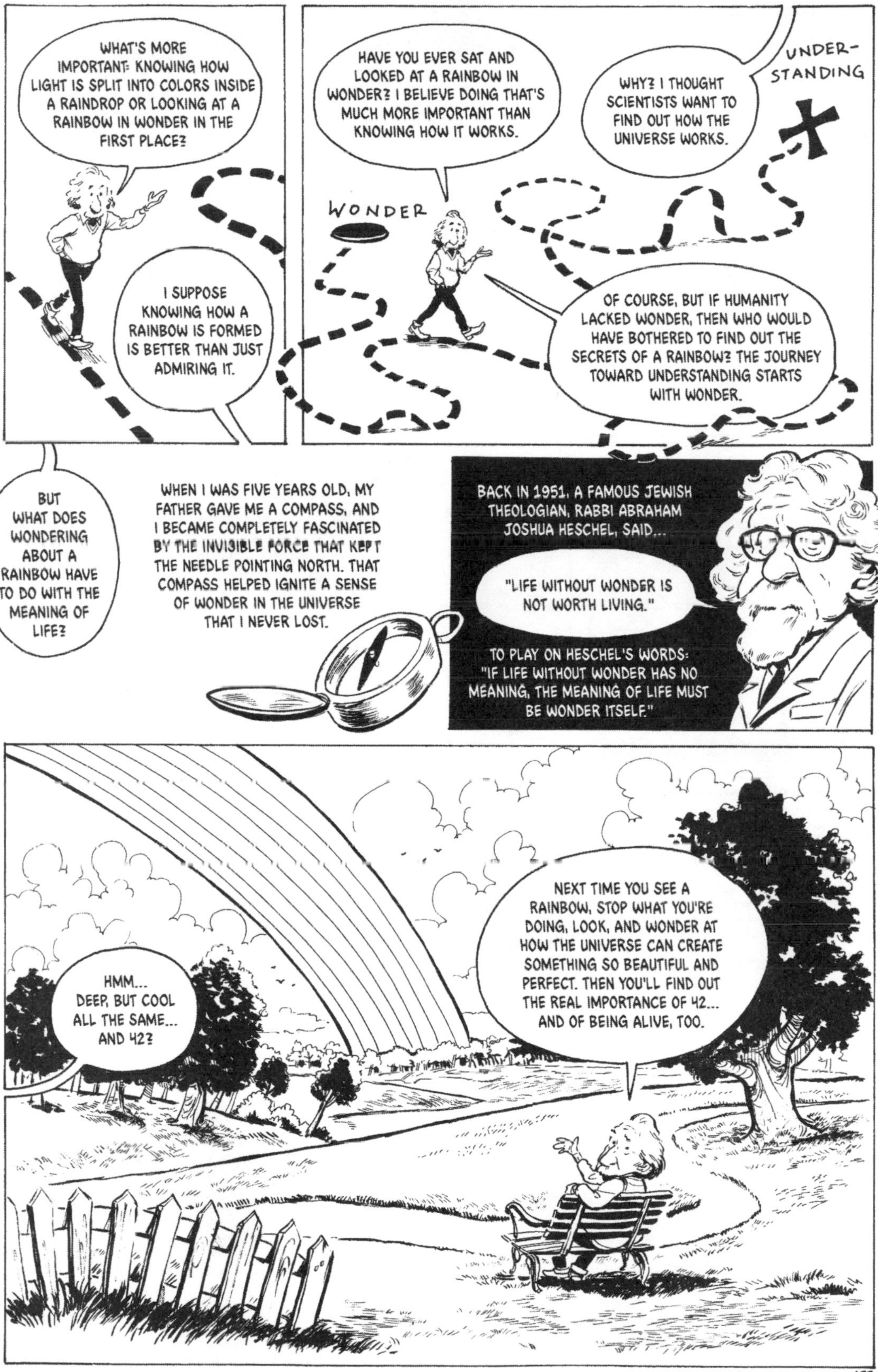

WHAT'S MORE IMPORTANT: KNOWING HOW LIGHT IS SPLIT INTO COLORS INSIDE A RAINDROP OR LOOKING AT A RAINBOW IN WONDER IN THE FIRST PLACE?
I SUPPOSE KNOWING HOW A RAINBOW IS FORMED IS BETTER THAN JUST ADMIRING IT.
HAVE YOU EVER SAT AND LOOKED AT A RAINBOW IN WONDER? I BELIEVE DOING THAT'S MUCH MORE IMPORTANT THAN KNOWING HOW IT WORKS.
WONDER
WHY? I THOUGHT SCIENTISTS WANT TO FIND OUT HOW THE UNIVERSE WORKS.
UNDER-STANDING
OF COURSE, BUT IF HUMANITY LACKED WONDER, THEN WHO WOULD HAVE BOTHERED TO FIND OUT THE SECRETS OF A RAINBOW? THE JOURNEY TOWARD UNDERSTANDING STARTS WITH WONDER.
BUT WHAT DOES WONDERING ABOUT A RAINBOW HAVE TO DO WITH THE MEANING OF LIFE?
WHEN I WAS FIVE YEARS OLD, MY FATHER GAVE ME A COMPASS, AND I BECAME COMPLETELY FASCINATED BY THE INVISIBLE FORCE THAT KEPT THE NEEDLE POINTING NORTH. THAT COMPASS HELPED IGNITE A SENSE OF WONDER IN THE UNIVERSE THAT I NEVER LOST.
BACK IN 1951, A FAMOUS JEWISH THEOLOGIAN, RABBI ABRAHAM JOSHUA HESCHEL, SAID...
"LIFE WITHOUT WONDER IS NOT WORTH LIVING."
TO PLAY ON HESCHEL'S WORDS: "IF LIFE WITHOUT WONDER HAS NO MEANING, THE MEANING OF LIFE MUST BE WONDER ITSELF."
HMM... DEEP, BUT COOL ALL THE SAME... AND 42?
NEXT TIME YOU SEE A RAINBOW, STOP WHAT YOU'RE DOING, LOOK, AND WONDER AT HOW THE UNIVERSE CAN CREATE SOMETHING SO BEAUTIFUL AND PERFECT. THEN YOU'LL FIND OUT THE REAL IMPORTANCE OF 42... AND OF BEING ALIVE, TOO.

CHAPTER 33: MORE BLACK HOLES AND WHY THE SPEED OF LIGHT ISN'T CONSTANT AFTER ALL

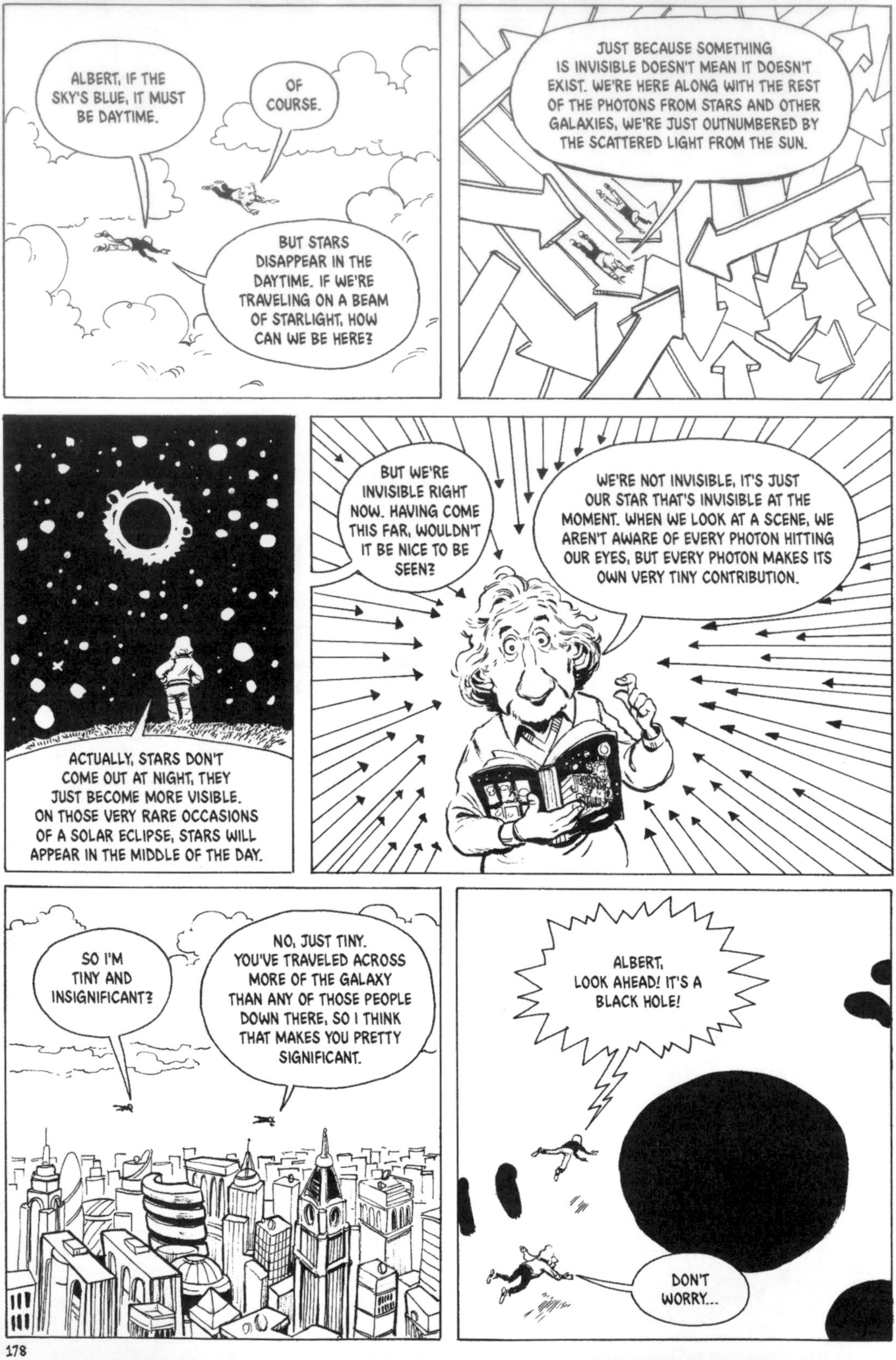

ALBERT, IF THE SKY'S BLUE, IT MUST BE DAYTIME.
OF COURSE.
BUT STARS DISAPPEAR IN THE DAYTIME. IF WE'RE TRAVELING ON A BEAM OF STARLIGHT, HOW CAN WE BE HERE?
JUST BECAUSE SOMETHING IS INVISIBLE DOESN'T MEAN IT DOESN'T EXIST. WE'RE HERE ALONG WITH THE REST OF THE PHOTONS FROM STARS AND OTHER GALAXIES, WE'RE JUST OUTNUMBERED BY THE SCATTERED LIGHT FROM THE SUN.
ACTUALLY, STARS DON'T COME OUT AT NIGHT, THEY JUST BECOME MORE VISIBLE. ON THOSE VERY RARE OCCASIONS OF A SOLAR ECLIPSE, STARS WILL APPEAR IN THE MIDDLE OF THE DAY.
BUT WE'RE INVISIBLE RIGHT NOW. HAVING COME THIS FAR, WOULDN'T IT BE NICE TO BE SEEN?
WE'RE NOT INVISIBLE, IT'S JUST OUR STAR THAT'S INVISIBLE AT THE MOMENT. WHEN WE LOOK AT A SCENE, WE AREN'T AWARE OF EVERY PHOTON HITTING OUR EYES, BUT EVERY PHOTON MAKES ITS OWN VERY TINY CONTRIBUTION.
SO I'M TINY AND INSIGNIFICANT?
NO, JUST TINY. YOU'VE TRAVELED ACROSS MORE OF THE GALAXY THAN ANY OF THOSE PEOPLE DOWN THERE, SO I THINK THAT MAKES YOU PRETTY SIGNIFICANT.
ALBERT, LOOK AHEAD! IT'S A BLACK HOLE!
DON'T WORRY...

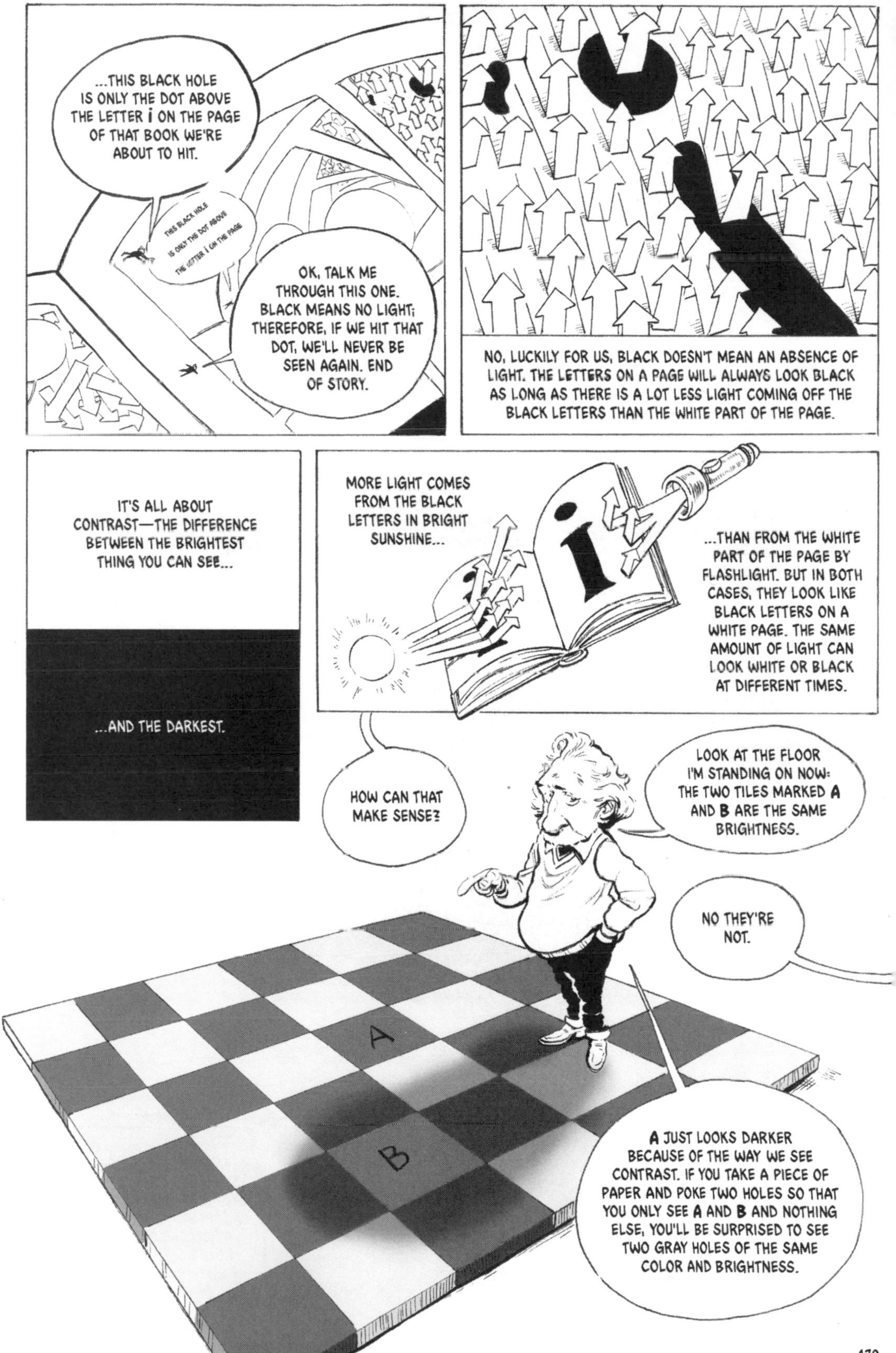
...THIS BLACK HOLE IS ONLY THE DOT ABOVE THE LETTER i ON THE PAGE OF THAT BOOK WE'RE ABOUT TO HIT.
THIS BLACK HOLE IS ONLY THE DOT ABOVE THE LETTER i ON THE PAGE
OK, TALK ME THROUGH THIS ONE. BLACK MEANS NO LIGHT; THEREFORE, IF WE HIT THAT DOT, WE'LL NEVER BE SEEN AGAIN. END OF STORY.
NO, LUCKILY FOR US, BLACK DOESN'T MEAN AN ABSENCE OF LIGHT. THE LETTERS ON A PAGE WILL ALWAYS LOOK BLACK AS LONG AS THERE IS A LOT LESS LIGHT COMING OFF THE BLACK LETTERS THAN THE WHITE PART OF THE PAGE.
IT'S ALL ABOUT CONTRAST—THE DIFFERENCE BETWEEN THE BRIGHTEST THING YOU CAN SEE...
...AND THE DARKEST.
MORE LIGHT COMES FROM THE BLACK LETTERS IN BRIGHT SUNSHINE...
...THAN FROM THE WHITE PART OF THE PAGE BY FLASHLIGHT. BUT IN BOTH CASES, THEY LOOK LIKE BLACK LETTERS ON A WHITE PAGE. THE SAME AMOUNT OF LIGHT CAN LOOK WHITE OR BLACK AT DIFFERENT TIMES.
HOW CAN THAT MAKE SENSE?
LOOK AT THE FLOOR I'M STANDING ON NOW: THE TWO TILES MARKED A AND B ARE THE SAME BRIGHTNESS.
NO THEY'RE NOT.
A
B
A JUST LOOKS DARKER BECAUSE OF THE WAY WE SEE CONTRAST. IF YOU TAKE A PIECE OF PAPER AND POKE TWO HOLES SO THAT YOU ONLY SEE A AND B AND NOTHING ELSE, YOU'LL BE SURPRISED TO SEE TWO GRAY HOLES OF THE SAME COLOR AND BRIGHTNESS.

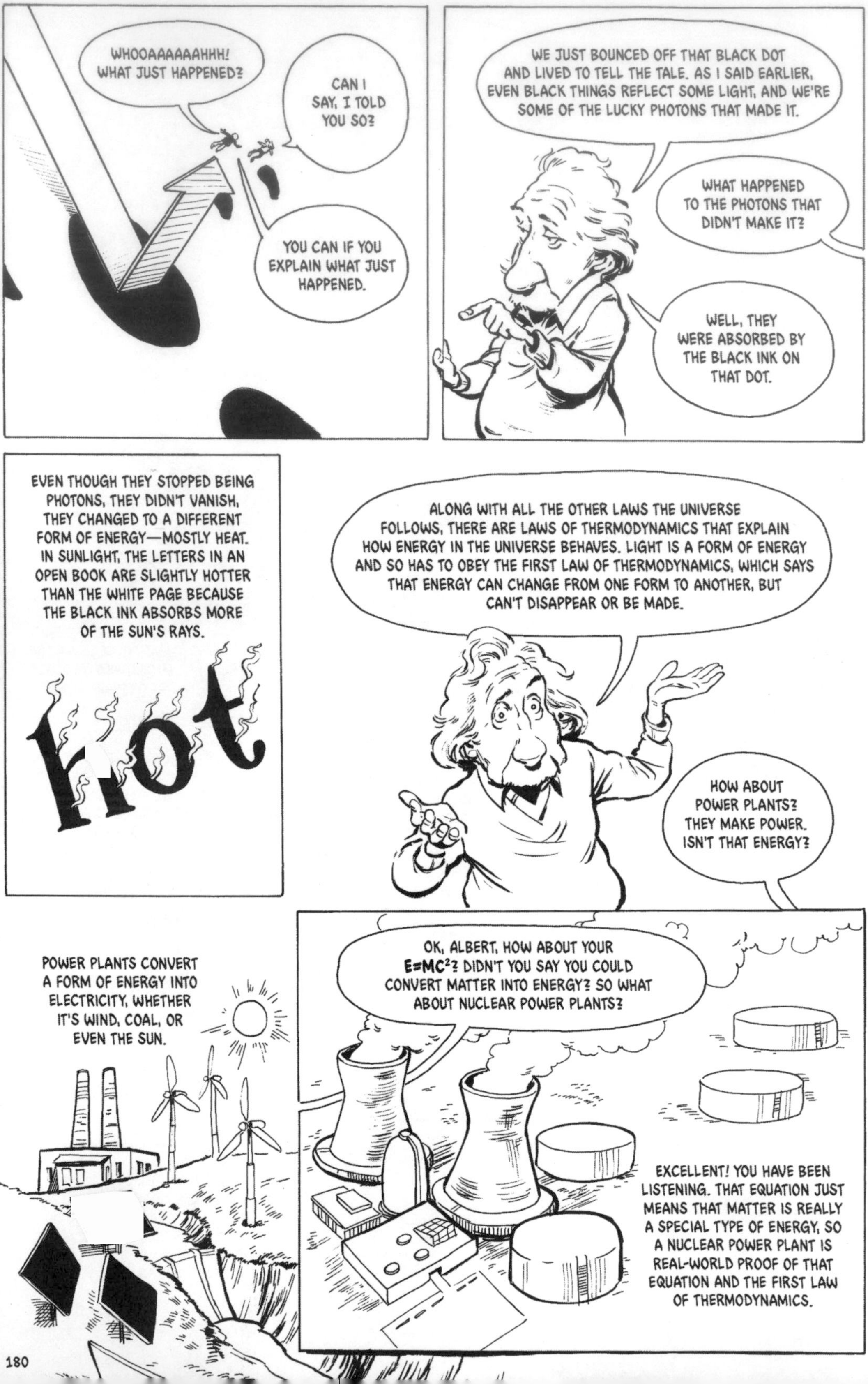
WHOOAAAAAAHHH! WHAT JUST HAPPENED?
CAN I SAY, I TOLD YOU SO?
YOU CAN IF YOU EXPLAIN WHAT JUST HAPPENED.
WE JUST BOUNCED OFF THAT BLACK DOT AND LIVED TO TELL THE TALE. AS I SAID EARLIER, EVEN BLACK THINGS REFLECT SOME LIGHT, AND WE'RE SOME OF THE LUCKY PHOTONS THAT MADE IT.
WHAT HAPPENED TO THE PHOTONS THAT DIDN'T MAKE IT?
WELL, THEY WERE ABSORBED BY THE BLACK INK ON THAT DOT.
EVEN THOUGH THEY STOPPED BEING PHOTONS, THEY DIDN'T VANISH, THEY CHANGED TO A DIFFERENT FORM OF ENERGY—MOSTLY HEAT. IN SUNLIGHT, THE LETTERS IN AN OPEN BOOK ARE SLIGHTLY HOTTER THAN THE WHITE PAGE BECAUSE THE BLACK INK ABSORBS MORE OF THE SUN'S RAYS.
hot
ALONG WITH ALL THE OTHER LAWS THE UNIVERSE FOLLOWS, THERE ARE LAWS OF THERMODYNAMICS THAT EXPLAIN HOW ENERGY IN THE UNIVERSE BEHAVES. LIGHT IS A FORM OF ENERGY AND SO HAS TO OBEY THE FIRST LAW OF THERMODYNAMICS, WHICH SAYS THAT ENERGY CAN CHANGE FROM ONE FORM TO ANOTHER, BUT CAN'T DISAPPEAR OR BE MADE.
HOW ABOUT POWER PLANTS? THEY MAKE POWER. ISN'T THAT ENERGY?
POWER PLANTS CONVERT A FORM OF ENERGY INTO ELECTRICITY, WHETHER IT'S WIND, COAL, OR EVEN THE SUN.
OK, ALBERT, HOW ABOUT YOUR E=MC²? DIDN'T YOU SAY YOU COULD CONVERT MATTER INTO ENERGY? SO WHAT ABOUT NUCLEAR POWER PLANTS?
EXCELLENT! YOU HAVE BEEN LISTENING. THAT EQUATION JUST MEANS THAT MATTER IS REALLY A SPECIAL TYPE OF ENERGY, SO A NUCLEAR POWER PLANT IS REAL-WORLD PROOF OF THAT EQUATION AND THE FIRST LAW OF THERMODYNAMICS.

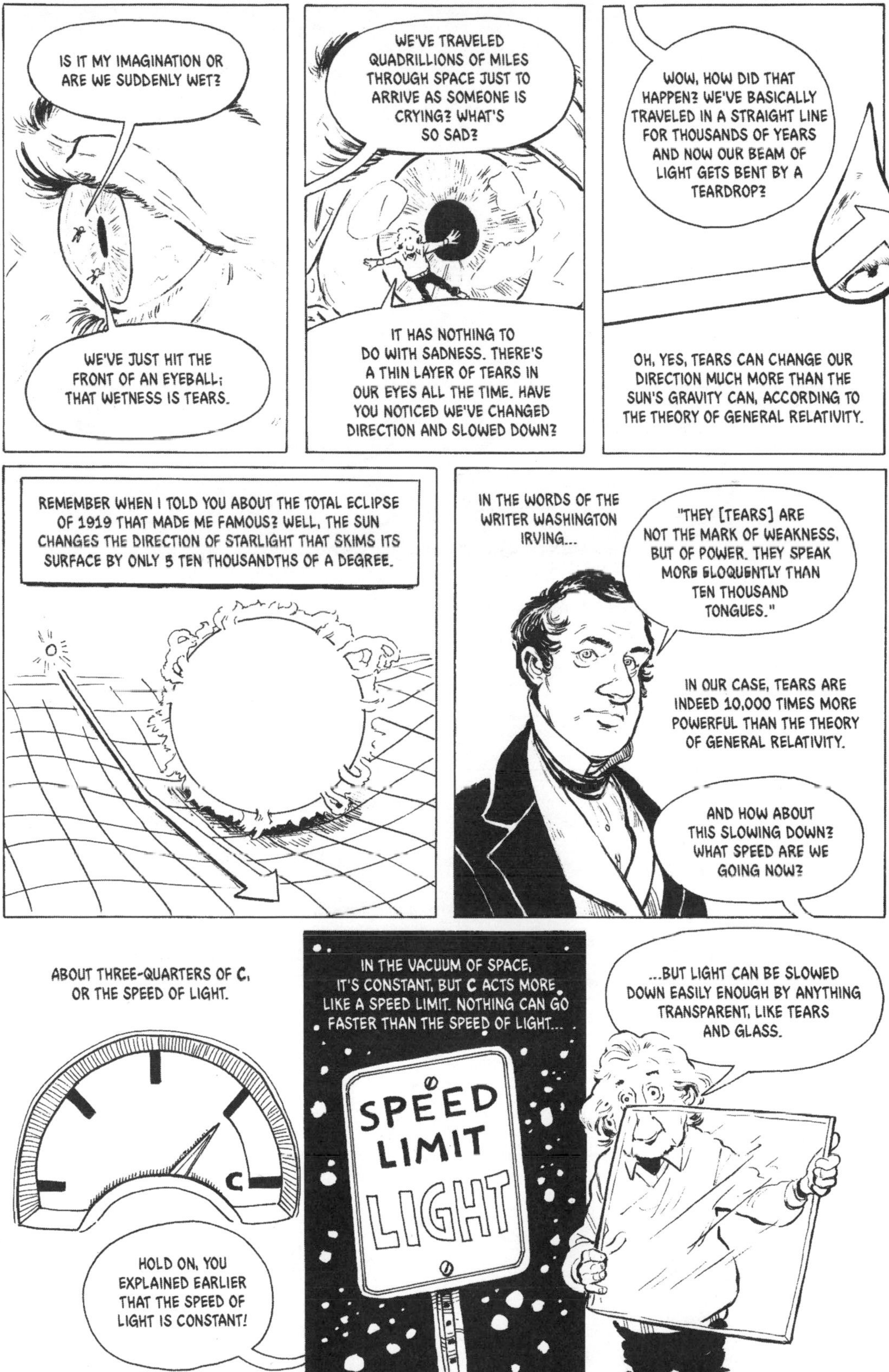
IS IT MY IMAGINATION OR ARE WE SUDDENLY WET?
WE'VE JUST HIT THE FRONT OF AN EYEBALL; THAT WETNESS IS TEARS.
WE'VE TRAVELED QUADRILLIONS OF MILES THROUGH SPACE JUST TO ARRIVE AS SOMEONE IS CRYING? WHAT'S SO SAD?
IT HAS NOTHING TO DO WITH SADNESS. THERE'S A THIN LAYER OF TEARS IN OUR EYES ALL THE TIME. HAVE YOU NOTICED WE'VE CHANGED DIRECTION AND SLOWED DOWN?
WOW, HOW DID THAT HAPPEN? WE'VE BASICALLY TRAVELED IN A STRAIGHT LINE FOR THOUSANDS OF YEARS AND NOW OUR BEAM OF LIGHT GETS BENT BY A TEARDROP?
OH, YES, TEARS CAN CHANGE OUR DIRECTION MUCH MORE THAN THE SUN'S GRAVITY CAN, ACCORDING TO THE THEORY OF GENERAL RELATIVITY.
REMEMBER WHEN I TOLD YOU ABOUT THE TOTAL ECLIPSE OF 1919 THAT MADE ME FAMOUS? WELL, THE SUN CHANGES THE DIRECTION OF STARLIGHT THAT SKIMS ITS SURFACE BY ONLY 5 TEN THOUSANDTHS OF A DEGREE.
IN THE WORDS OF THE WRITER WASHINGTON IRVING...
"THEY [TEARS] ARE NOT THE MARK OF WEAKNESS, BUT OF POWER. THEY SPEAK MORE ELOQUENTLY THAN TEN THOUSAND TONGUES."
IN OUR CASE, TEARS ARE INDEED 10,000 TIMES MORE POWERFUL THAN THE THEORY OF GENERAL RELATIVITY.
AND HOW ABOUT THIS SLOWING DOWN? WHAT SPEED ARE WE GOING NOW?
ABOUT THREE-QUARTERS OF C, OR THE SPEED OF LIGHT.
C
HOLD ON, YOU EXPLAINED EARLIER THAT THE SPEED OF LIGHT IS CONSTANT!
IN THE VACUUM OF SPACE, IT'S CONSTANT, BUT C ACTS MORE LIKE A SPEED LIMIT. NOTHING CAN GO FASTER THAN THE SPEED OF LIGHT...
SPEED LIMIT LIGHT
...BUT LIGHT CAN BE SLOWED DOWN EASILY ENOUGH BY ANYTHING TRANSPARENT, LIKE TEARS AND GLASS.

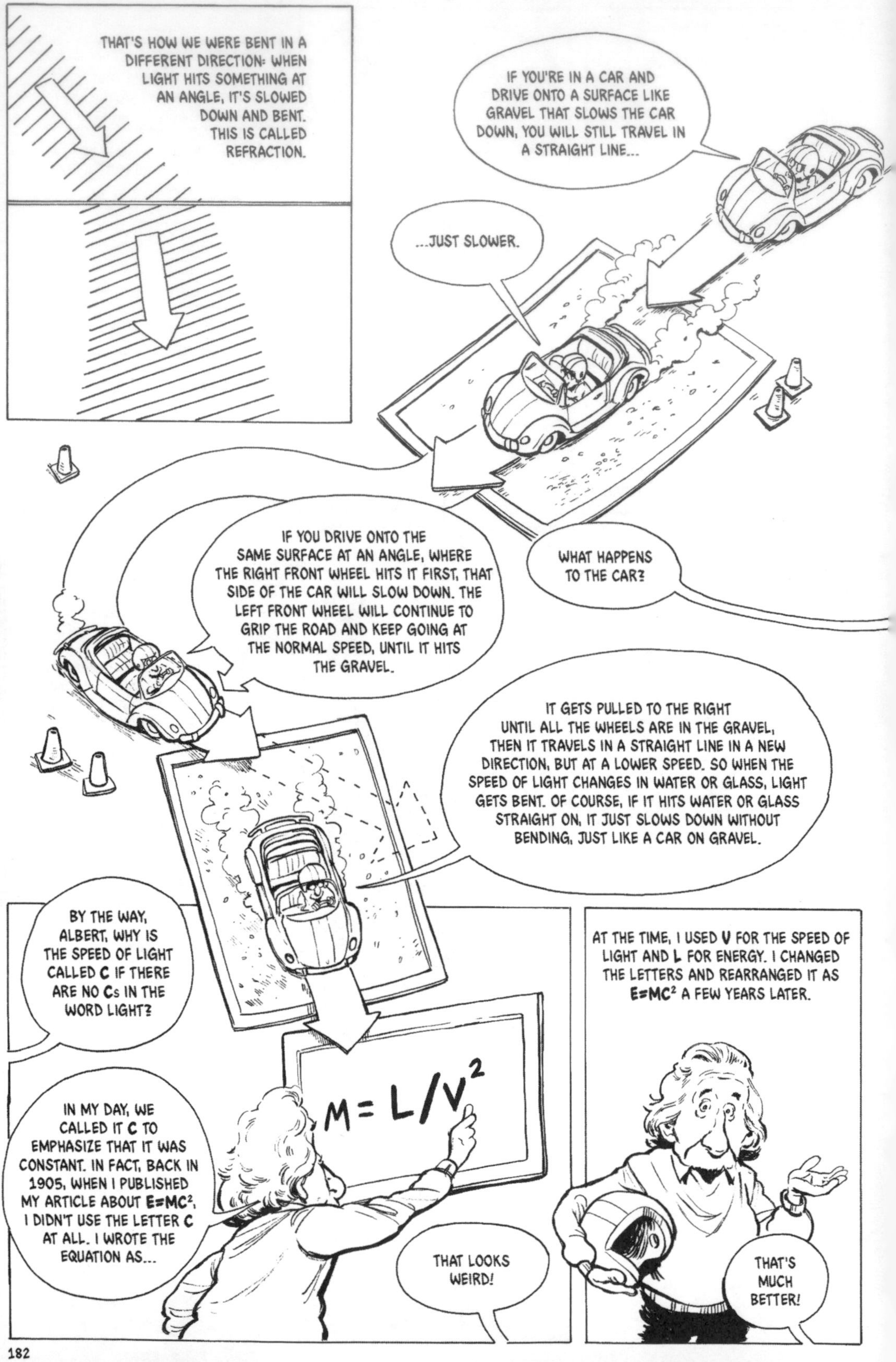
THAT'S HOW WE WERE BENT IN A DIFFERENT DIRECTION: WHEN LIGHT HITS SOMETHING AT AN ANGLE, IT'S SLOWED DOWN AND BENT. THIS IS CALLED REFRACTION.
IF YOU'RE IN A CAR AND DRIVE ONTO A SURFACE LIKE GRAVEL THAT SLOWS THE CAR DOWN, YOU WILL STILL TRAVEL IN A STRAIGHT LINE...
...JUST SLOWER.
WHAT HAPPENS TO THE CAR?
IF YOU DRIVE ONTO THE SAME SURFACE AT AN ANGLE, WHERE THE RIGHT FRONT WHEEL HITS IT FIRST, THAT SIDE OF THE CAR WILL SLOW DOWN. THE LEFT FRONT WHEEL WILL CONTINUE TO GRIP THE ROAD AND KEEP GOING AT THE NORMAL SPEED, UNTIL IT HITS THE GRAVEL.
IT GETS PULLED TO THE RIGHT UNTIL ALL THE WHEELS ARE IN THE GRAVEL, THEN IT TRAVELS IN A STRAIGHT LINE IN A NEW DIRECTION, BUT AT A LOWER SPEED. SO WHEN THE SPEED OF LIGHT CHANGES IN WATER OR GLASS, LIGHT GETS BENT. OF COURSE, IF IT HITS WATER OR GLASS STRAIGHT ON, IT JUST SLOWS DOWN WITHOUT BENDING, JUST LIKE A CAR ON GRAVEL.
BY THE WAY, ALBERT, WHY IS THE SPEED OF LIGHT CALLED C IF THERE ARE NO Cs IN THE WORD LIGHT?
IN MY DAY, WE CALLED IT C TO EMPHASIZE THAT IT WAS CONSTANT. IN FACT, BACK IN 1905, WHEN I PUBLISHED MY ARTICLE ABOUT E=MC², I DIDN'T USE THE LETTER C AT ALL. I WROTE THE EQUATION AS...
M = L/V²
THAT LOOKS WEIRD!
AT THE TIME, I USED V FOR THE SPEED OF LIGHT AND L FOR ENERGY. I CHANGED THE LETTERS AND REARRANGED IT AS E=MC² A FEW YEARS LATER.
THAT'S MUCH BETTER!

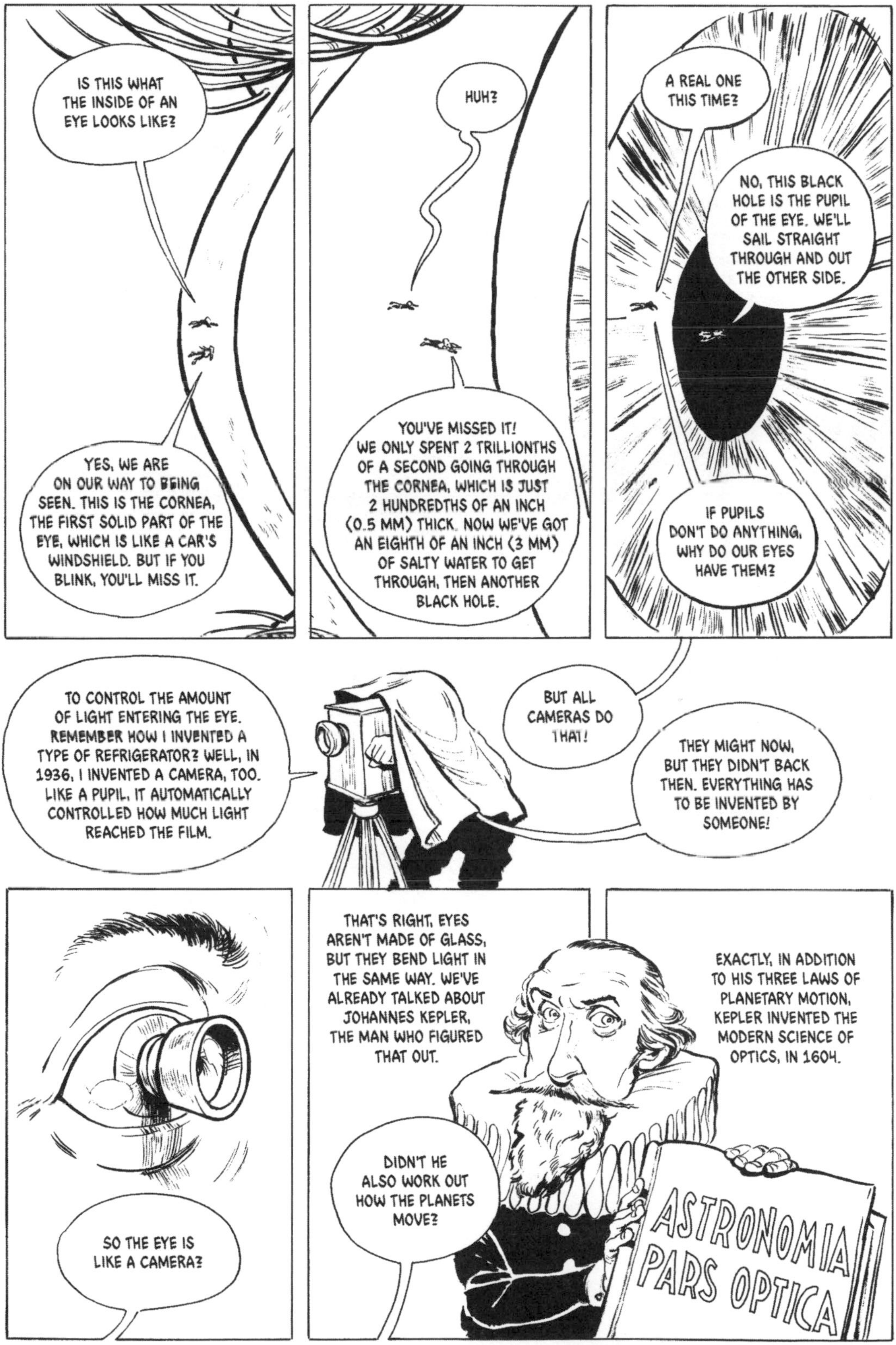
IS THIS WHAT THE INSIDE OF AN EYE LOOKS LIKE?
YES, WE ARE ON OUR WAY TO BEING SEEN. THIS IS THE CORNEA, THE FIRST SOLID PART OF THE EYE, WHICH IS LIKE A CAR'S WINDSHIELD. BUT IF YOU BLINK, YOU'LL MISS IT.
HUH?
YOU'VE MISSED IT! WE ONLY SPENT 2 TRILLIONTHS OF A SECOND GOING THROUGH THE CORNEA, WHICH IS JUST 2 HUNDREDTHS OF AN INCH (0.5 MM) THICK. NOW WE'VE GOT AN EIGHTH OF AN INCH (3 MM) OF SALTY WATER TO GET THROUGH, THEN ANOTHER BLACK HOLE.
A REAL ONE THIS TIME?
NO, THIS BLACK HOLE IS THE PUPIL OF THE EYE. WE'LL SAIL STRAIGHT THROUGH AND OUT THE OTHER SIDE.
IF PUPILS DON'T DO ANYTHING, WHY DO OUR EYES HAVE THEM?
TO CONTROL THE AMOUNT OF LIGHT ENTERING THE EYE. REMEMBER HOW I INVENTED A TYPE OF REFRIGERATOR? WELL, IN 1936, I INVENTED A CAMERA, TOO. LIKE A PUPIL, IT AUTOMATICALLY CONTROLLED HOW MUCH LIGHT REACHED THE FILM.
BUT ALL CAMERAS DO THAT!
THEY MIGHT NOW, BUT THEY DIDN'T BACK THEN. EVERYTHING HAS TO BE INVENTED BY SOMEONE!
SO THE EYE IS LIKE A CAMERA?
THAT'S RIGHT, EYES AREN'T MADE OF GLASS, BUT THEY BEND LIGHT IN THE SAME WAY. WE'VE ALREADY TALKED ABOUT JOHANNES KEPLER, THE MAN WHO FIGURED THAT OUT.
DIDN'T HE ALSO WORK OUT HOW THE PLANETS MOVE?
EXACTLY. IN ADDITION TO HIS THREE LAWS OF PLANETARY MOTION, KEPLER INVENTED THE MODERN SCIENCE OF OPTICS, IN 1604.
ASTRONOMIA PARS OPTICA

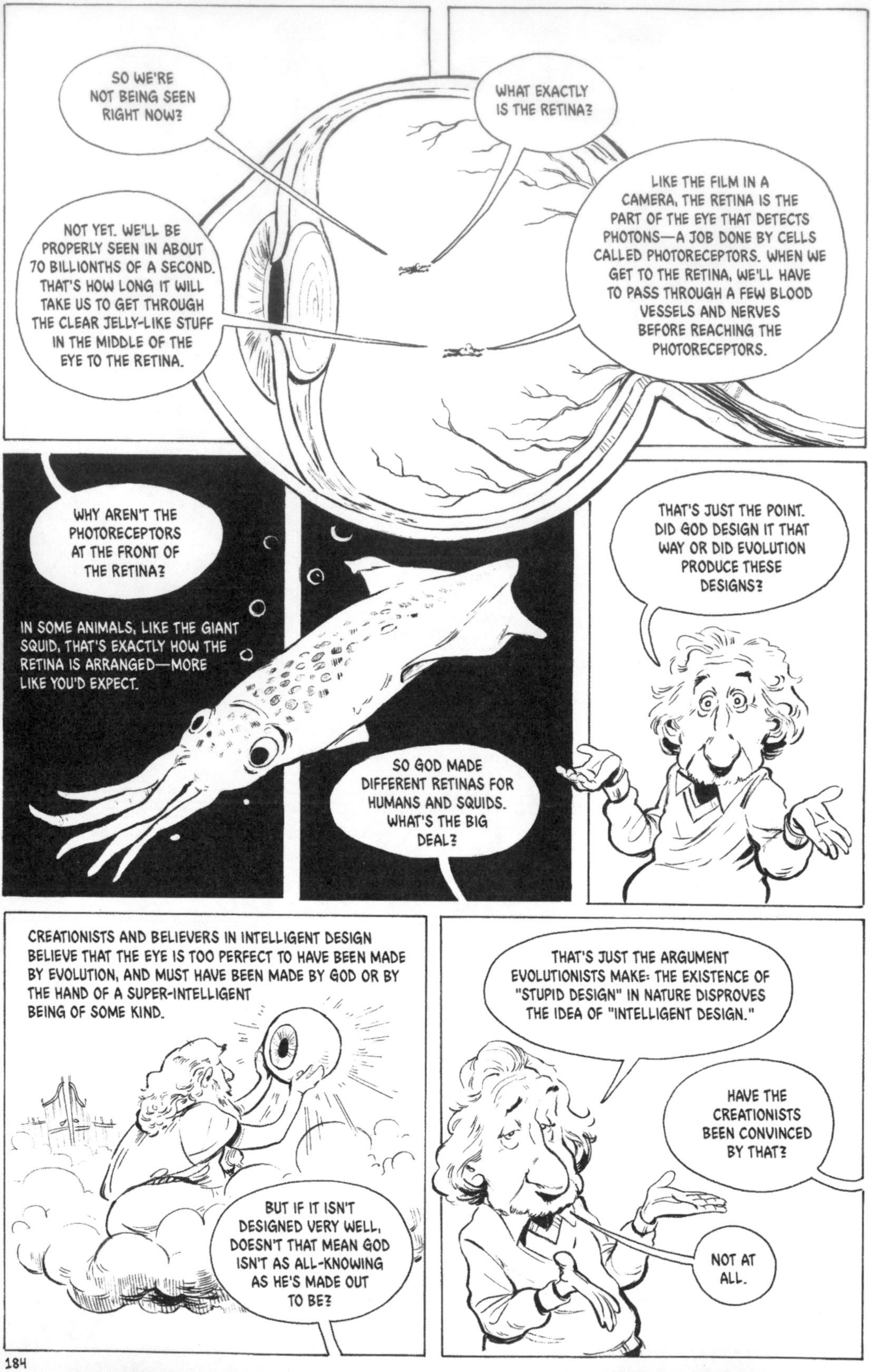
SO WE'RE NOT BEING SEEN RIGHT NOW?
NOT YET. WE'LL BE PROPERLY SEEN IN ABOUT 70 BILLIONTHS OF A SECOND. THAT'S HOW LONG IT WILL TAKE US TO GET THROUGH THE CLEAR JELLY-LIKE STUFF IN THE MIDDLE OF THE EYE TO THE RETINA.
WHAT EXACTLY IS THE RETINA?
LIKE THE FILM IN A CAMERA, THE RETINA IS THE PART OF THE EYE THAT DETECTS PHOTONS—A JOB DONE BY CELLS CALLED PHOTORECEPTORS. WHEN WE GET TO THE RETINA, WE'LL HAVE TO PASS THROUGH A FEW BLOOD VESSELS AND NERVES BEFORE REACHING THE PHOTORECEPTORS.
WHY AREN'T THE PHOTORECEPTORS AT THE FRONT OF THE RETINA?
IN SOME ANIMALS, LIKE THE GIANT SQUID, THAT'S EXACTLY HOW THE RETINA IS ARRANGED—MORE LIKE YOU'D EXPECT.
SO GOD MADE DIFFERENT RETINAS FOR HUMANS AND SQUIDS. WHAT'S THE BIG DEAL?
THAT'S JUST THE POINT. DID GOD DESIGN IT THAT WAY OR DID EVOLUTION PRODUCE THESE DESIGNS?
CREATIONISTS AND BELIEVERS IN INTELLIGENT DESIGN BELIEVE THAT THE EYE IS TOO PERFECT TO HAVE BEEN MADE BY EVOLUTION, AND MUST HAVE BEEN MADE BY GOD OR BY THE HAND OF A SUPER-INTELLIGENT BEING OF SOME KIND.
BUT IF IT ISN'T DESIGNED VERY WELL, DOESN'T THAT MEAN GOD ISN'T AS ALL-KNOWING AS HE'S MADE OUT TO BE?
THAT'S JUST THE ARGUMENT EVOLUTIONISTS MAKE: THE EXISTENCE OF "STUPID DESIGN" IN NATURE DISPROVES THE IDEA OF "INTELLIGENT DESIGN."
HAVE THE CREATIONISTS BEEN CONVINCED BY THAT?
NOT AT ALL.

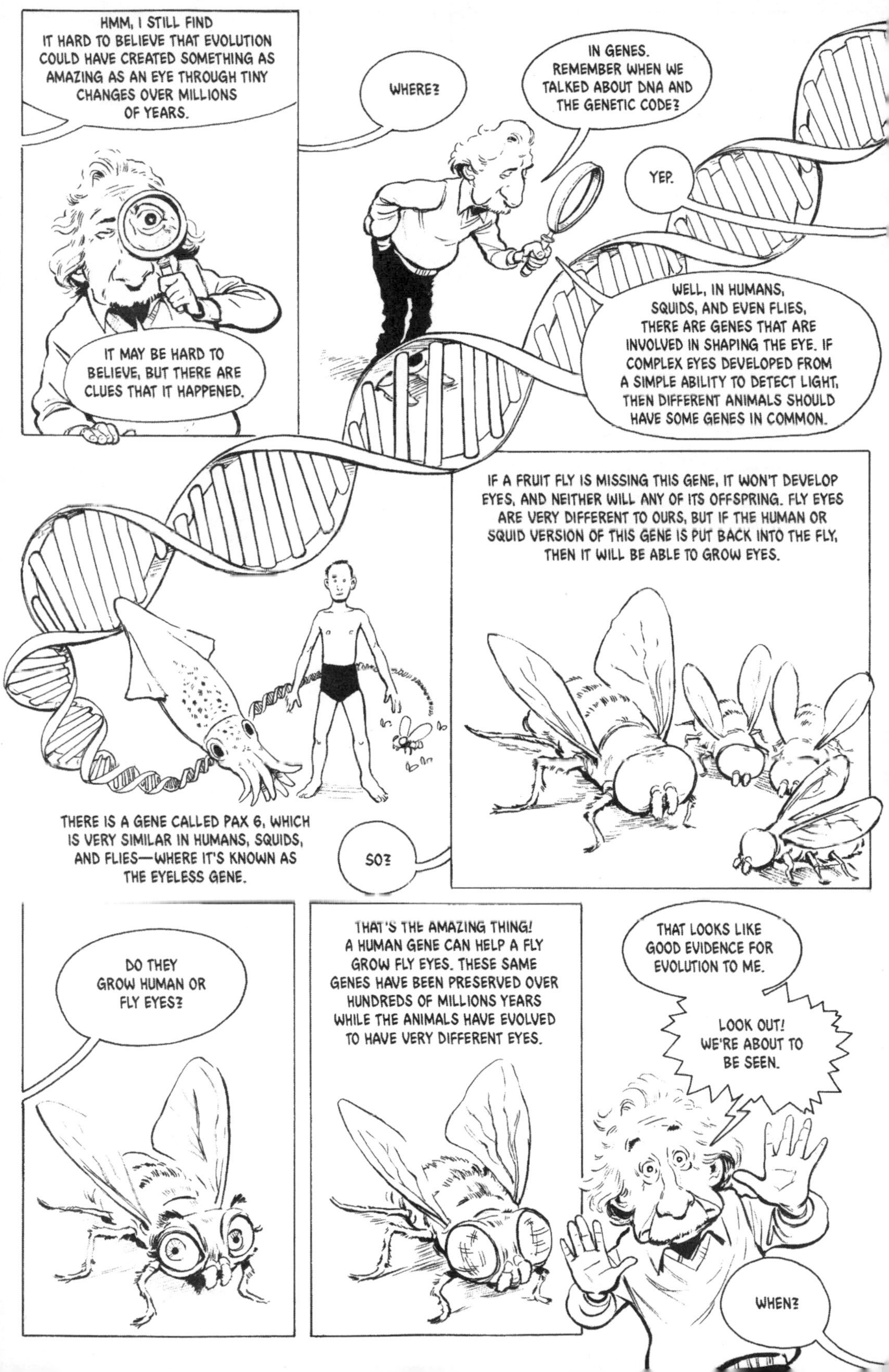
HMM, I STILL FIND IT HARD TO BELIEVE THAT EVOLUTION COULD HAVE CREATED SOMETHING AS AMAZING AS AN EYE THROUGH TINY CHANGES OVER MILLIONS OF YEARS.
IT MAY BE HARD TO BELIEVE, BUT THERE ARE CLUES THAT IT HAPPENED.
WHERE?
IN GENES. REMEMBER WHEN WE TALKED ABOUT DNA AND THE GENETIC CODE?
YEP.
WELL, IN HUMANS, SQUIDS, AND EVEN FLIES, THERE ARE GENES THAT ARE INVOLVED IN SHAPING THE EYE. IF COMPLEX EYES DEVELOPED FROM A SIMPLE ABILITY TO DETECT LIGHT, THEN DIFFERENT ANIMALS SHOULD HAVE SOME GENES IN COMMON.
THERE IS A GENE CALLED PAX 6, WHICH IS VERY SIMILAR IN HUMANS, SQUIDS, AND FLIES—WHERE IT'S KNOWN AS THE EYELESS GENE.
SO?
IF A FRUIT FLY IS MISSING THIS GENE, IT WON'T DEVELOP EYES, AND NEITHER WILL ANY OF ITS OFFSPRING. FLY EYES ARE VERY DIFFERENT TO OURS, BUT IF THE HUMAN OR SQUID VERSION OF THIS GENE IS PUT BACK INTO THE FLY, THEN IT WILL BE ABLE TO GROW EYES.
DO THEY GROW HUMAN OR FLY EYES?
THAT'S THE AMAZING THING! A HUMAN GENE CAN HELP A FLY GROW FLY EYES. THESE SAME GENES HAVE BEEN PRESERVED OVER HUNDREDS OF MILLIONS YEARS WHILE THE ANIMALS HAVE EVOLVED TO HAVE VERY DIFFERENT EYES.
THAT LOOKS LIKE GOOD EVIDENCE FOR EVOLUTION TO ME.
LOOK OUT! WE'RE ABOUT TO BE SEEN.
WHEN?

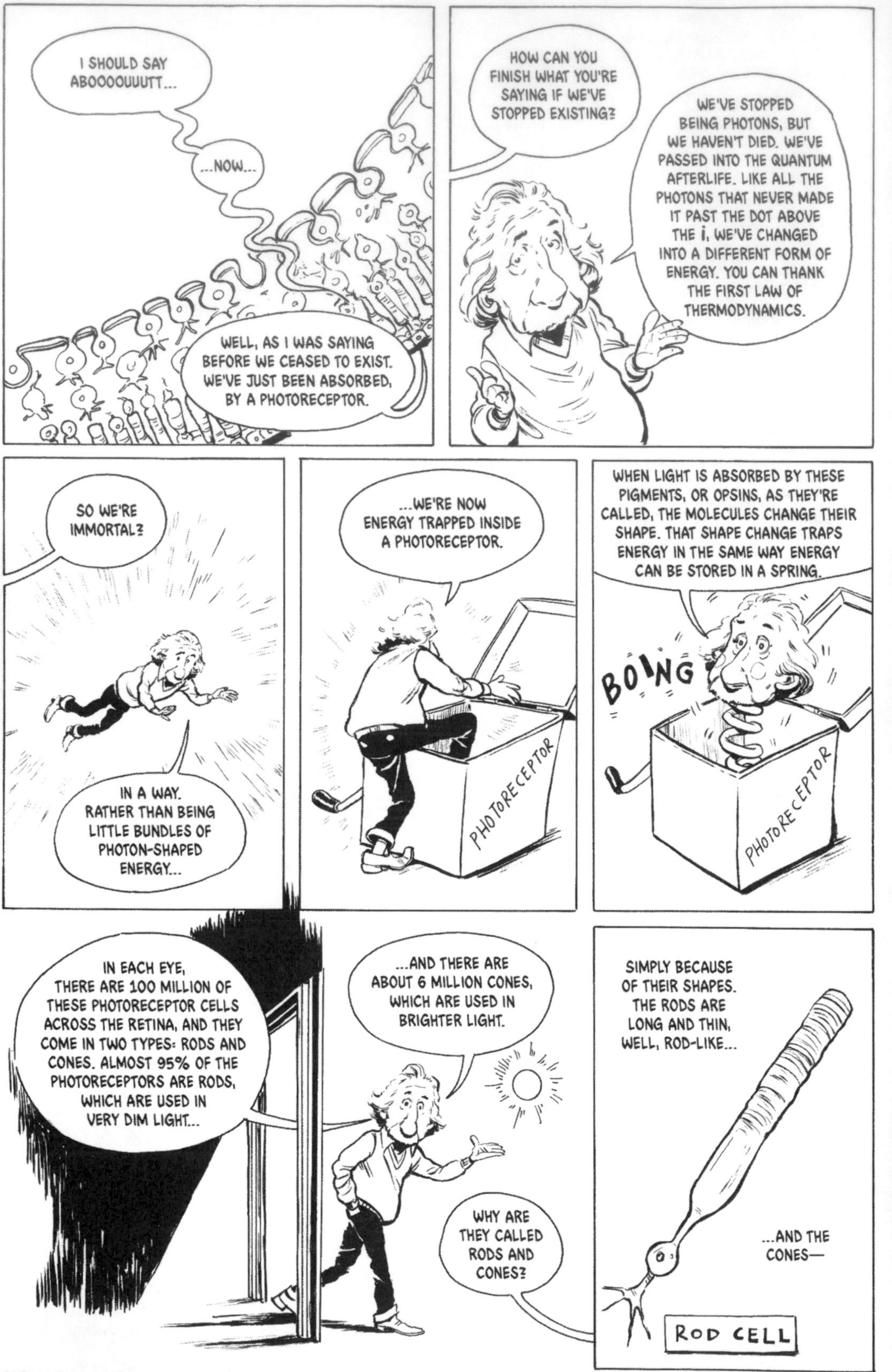
I SHOULD SAY ABOOOOUUUTT...
...NOW...
WELL, AS I WAS SAYING BEFORE WE CEASED TO EXIST. WE'VE JUST BEEN ABSORBED, BY A PHOTORECEPTOR.
HOW CAN YOU FINISH WHAT YOU'RE SAYING IF WE'VE STOPPED EXISTING?
WE'VE STOPPED BEING PHOTONS, BUT WE HAVEN'T DIED. WE'VE PASSED INTO THE QUANTUM AFTERLIFE. LIKE ALL THE PHOTONS THAT NEVER MADE IT PAST THE DOT ABOVE THE i, WE'VE CHANGED INTO A DIFFERENT FORM OF ENERGY. YOU CAN THANK THE FIRST LAW OF THERMODYNAMICS.
SO WE'RE IMMORTAL?
IN A WAY. RATHER THAN BEING LITTLE BUNDLES OF PHOTON-SHAPED ENERGY...
...WE'RE NOW ENERGY TRAPPED INSIDE A PHOTORECEPTOR.
PHOTORECEPTOR
WHEN LIGHT IS ABSORBED BY THESE PIGMENTS, OR OPSINS, AS THEY'RE CALLED, THE MOLECULES CHANGE THEIR SHAPE. THAT SHAPE CHANGE TRAPS ENERGY IN THE SAME WAY ENERGY CAN BE STORED IN A SPRING.
BOING
PHOTORECEPTOR
IN EACH EYE, THERE ARE 100 MILLION OF THESE PHOTORECEPTOR CELLS ACROSS THE RETINA, AND THEY COME IN TWO TYPES: RODS AND CONES. ALMOST 95% OF THE PHOTORECEPTORS ARE RODS, WHICH ARE USED IN VERY DIM LIGHT...
...AND THERE ARE ABOUT 6 MILLION CONES, WHICH ARE USED IN BRIGHTER LIGHT.
WHY ARE THEY CALLED RODS AND CONES?
SIMPLY BECAUSE OF THEIR SHAPES. THE RODS ARE LONG AND THIN, WELL, ROD-LIKE...
...AND THE CONES—
ROD CELL

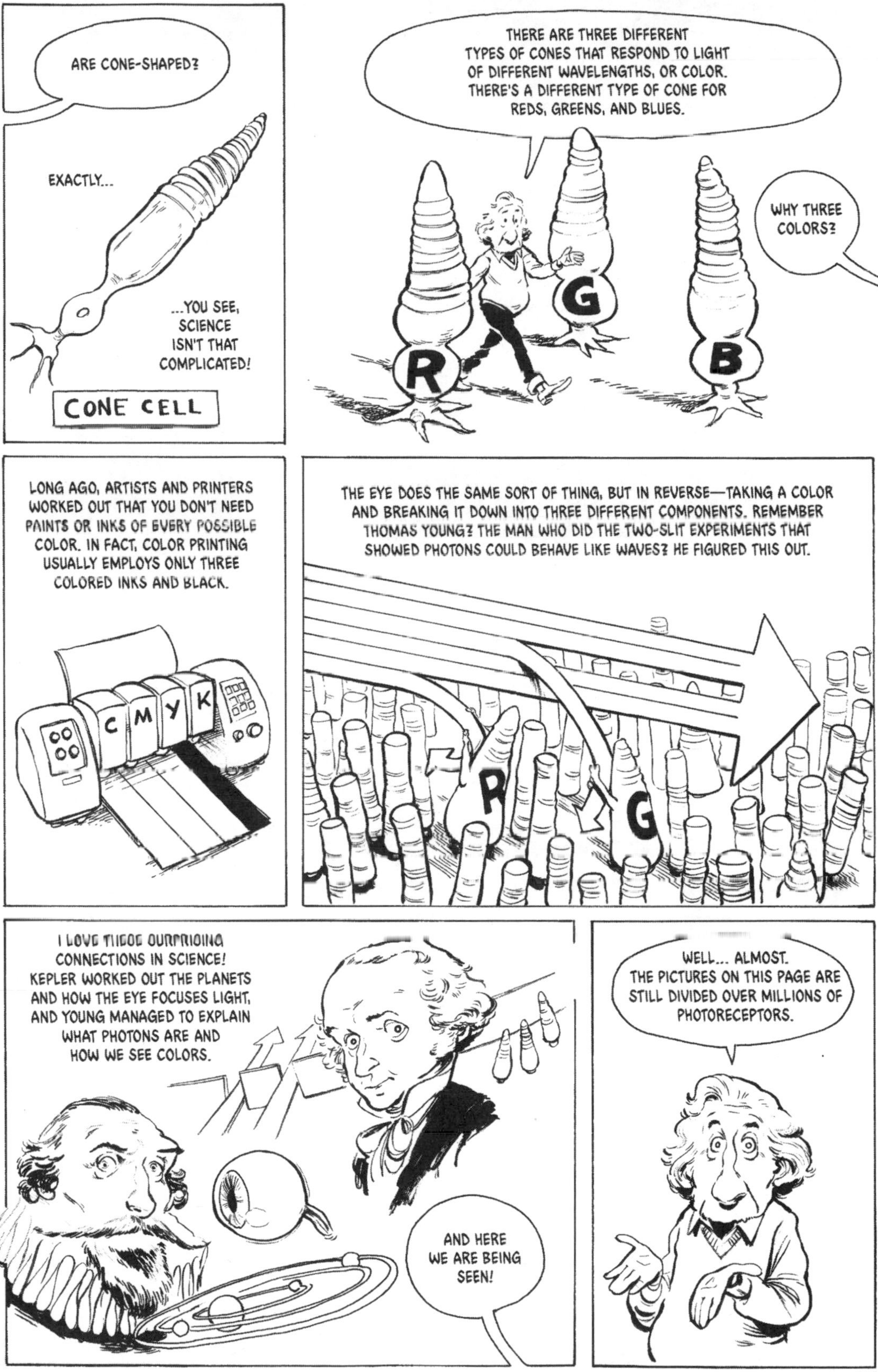
ARE CONE-SHAPED?
EXACTLY...
...YOU SEE, SCIENCE ISN'T THAT COMPLICATED!
CONE CELL
THERE ARE THREE DIFFERENT TYPES OF CONES THAT RESPOND TO LIGHT OF DIFFERENT WAVELENGTHS, OR COLOR. THERE'S A DIFFERENT TYPE OF CONE FOR REDS, GREENS, AND BLUES.
R
G
B
WHY THREE COLORS?
LONG AGO, ARTISTS AND PRINTERS WORKED OUT THAT YOU DON'T NEED PAINTS OR INKS OF EVERY POSSIBLE COLOR. IN FACT, COLOR PRINTING USUALLY EMPLOYS ONLY THREE COLORED INKS AND BLACK.
C M Y K
THE EYE DOES THE SAME SORT OF THING, BUT IN REVERSE—TAKING A COLOR AND BREAKING IT DOWN INTO THREE DIFFERENT COMPONENTS. REMEMBER THOMAS YOUNG? THE MAN WHO DID THE TWO-SLIT EXPERIMENTS THAT SHOWED PHOTONS COULD BEHAVE LIKE WAVES? HE FIGURED THIS OUT.
R
G
I LOVE THESE SURPRISING CONNECTIONS IN SCIENCE! KEPLER WORKED OUT THE PLANETS AND HOW THE EYE FOCUSES LIGHT, AND YOUNG MANAGED TO EXPLAIN WHAT PHOTONS ARE AND HOW WE SEE COLORS.
AND HERE WE ARE BEING SEEN!
WELL... ALMOST. THE PICTURES ON THIS PAGE ARE STILL DIVIDED OVER MILLIONS OF PHOTORECEPTORS.

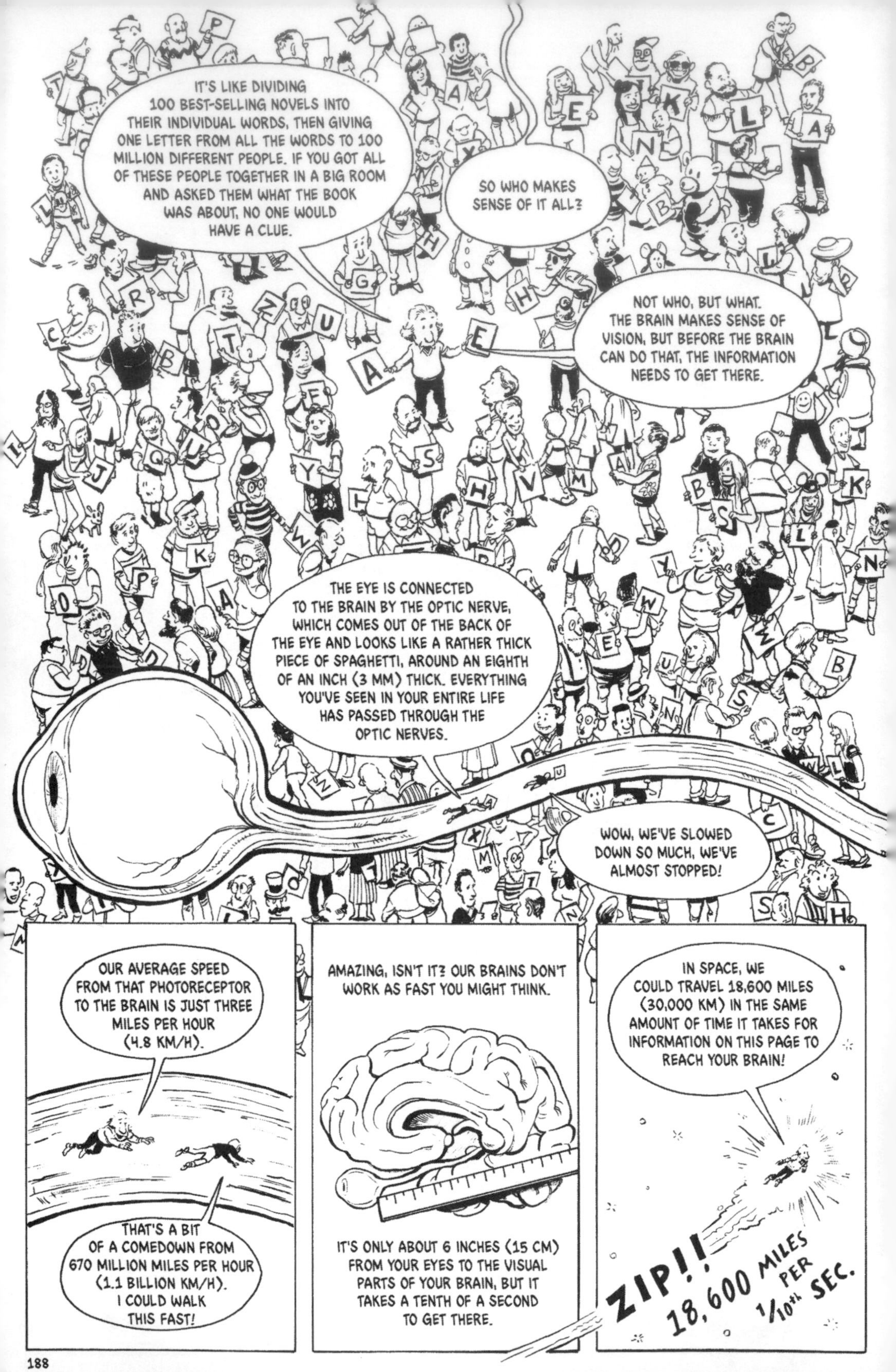
IT'S LIKE DIVIDING 100 BEST-SELLING NOVELS INTO THEIR INDIVIDUAL WORDS, THEN GIVING ONE LETTER FROM ALL THE WORDS TO 100 MILLION DIFFERENT PEOPLE. IF YOU GOT ALL OF THESE PEOPLE TOGETHER IN A BIG ROOM AND ASKED THEM WHAT THE BOOK WAS ABOUT, NO ONE WOULD HAVE A CLUE.
SO WHO MAKES SENSE OF IT ALL?
NOT WHO, BUT WHAT. THE BRAIN MAKES SENSE OF VISION, BUT BEFORE THE BRAIN CAN DO THAT, THE INFORMATION NEEDS TO GET THERE.
THE EYE IS CONNECTED TO THE BRAIN BY THE OPTIC NERVE, WHICH COMES OUT OF THE BACK OF THE EYE AND LOOKS LIKE A RATHER THICK PIECE OF SPAGHETTI, AROUND AN EIGHTH OF AN INCH (3 MM) THICK. EVERYTHING YOU'VE SEEN IN YOUR ENTIRE LIFE HAS PASSED THROUGH THE OPTIC NERVES.
WOW, WE'VE SLOWED DOWN SO MUCH, WE'VE ALMOST STOPPED!
OUR AVERAGE SPEED FROM THAT PHOTORECEPTOR TO THE BRAIN IS JUST THREE MILES PER HOUR (4.8 KM/H).
THAT'S A BIT OF A COMEDOWN FROM 670 MILLION MILES PER HOUR (1.1 BILLION KM/H). I COULD WALK THIS FAST!
AMAZING, ISN'T IT? OUR BRAINS DON'T WORK AS FAST YOU MIGHT THINK.
IT'S ONLY ABOUT 6 INCHES (15 CM) FROM YOUR EYES TO THE VISUAL PARTS OF YOUR BRAIN, BUT IT TAKES A TENTH OF A SECOND TO GET THERE.
IN SPACE, WE COULD TRAVEL 18,600 MILES (30,000 KM) IN THE SAME AMOUNT OF TIME IT TAKES FOR INFORMATION ON THIS PAGE TO REACH YOUR BRAIN!
ZIP!!
18,600 MILES PER 1/10th SEC.

CHAPTER 35: WHAT HAPPENS IN YOUR BRAIN AND WHAT HAPPENED TO ALBERT EINSTEIN'S BRAIN

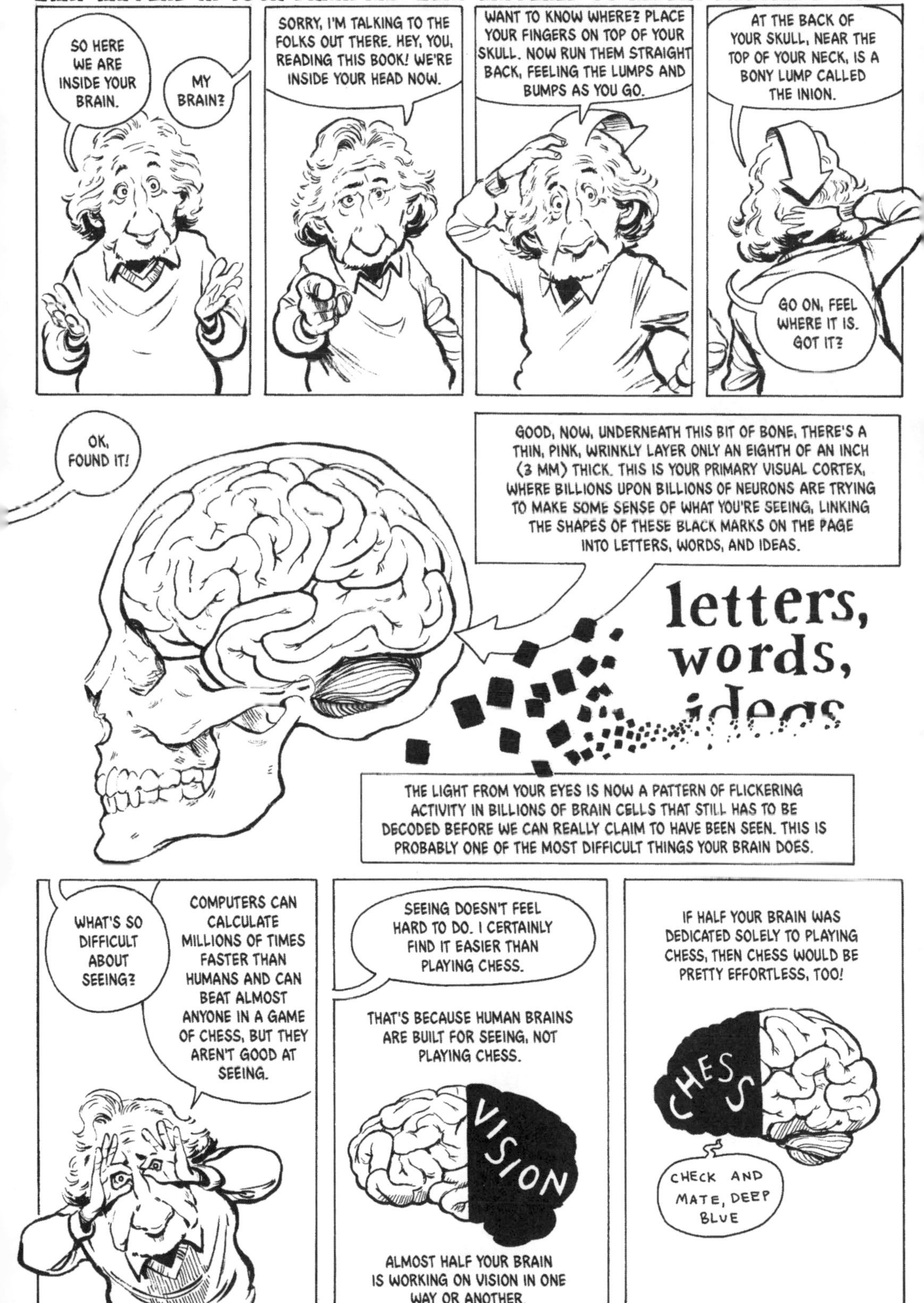

SO HERE WE ARE INSIDE YOUR BRAIN.
MY BRAIN?
SORRY, I'M TALKING TO THE FOLKS OUT THERE. HEY, YOU, READING THIS BOOK! WE'RE INSIDE YOUR HEAD NOW.
WANT TO KNOW WHERE? PLACE YOUR FINGERS ON TOP OF YOUR SKULL. NOW RUN THEM STRAIGHT BACK, FEELING THE LUMPS AND BUMPS AS YOU GO.
AT THE BACK OF YOUR SKULL, NEAR THE TOP OF YOUR NECK, IS A BONY LUMP CALLED THE INION.
GO ON, FEEL WHERE IT IS. GOT IT?
OK, FOUND IT!
GOOD, NOW, UNDERNEATH THIS BIT OF BONE, THERE'S A THIN, PINK, WRINKLY LAYER ONLY AN EIGHTH OF AN INCH (3 MM) THICK. THIS IS YOUR PRIMARY VISUAL CORTEX, WHERE BILLIONS UPON BILLIONS OF NEURONS ARE TRYING TO MAKE SOME SENSE OF WHAT YOU'RE SEEING, LINKING THE SHAPES OF THESE BLACK MARKS ON THE PAGE INTO LETTERS, WORDS, AND IDEAS.
letters,
words,
THE LIGHT FROM YOUR EYES IS NOW A PATTERN OF FLICKERING ACTIVITY IN BILLIONS OF BRAIN CELLS THAT STILL HAS TO BE DECODED BEFORE WE CAN REALLY CLAIM TO HAVE BEEN SEEN. THIS IS PROBABLY ONE OF THE MOST DIFFICULT THINGS YOUR BRAIN DOES.
WHAT'S SO DIFFICULT ABOUT SEEING?
COMPUTERS CAN CALCULATE MILLIONS OF TIMES FASTER THAN HUMANS AND CAN BEAT ALMOST ANYONE IN A GAME OF CHESS, BUT THEY AREN'T GOOD AT SEEING.
SEEING DOESN'T FEEL HARD TO DO. I CERTAINLY FIND IT EASIER THAN PLAYING CHESS.
THAT'S BECAUSE HUMAN BRAINS ARE BUILT FOR SEEING, NOT PLAYING CHESS.
VISION
ALMOST HALF YOUR BRAIN IS WORKING ON VISION IN ONE WAY OR ANOTHER.
IF HALF YOUR BRAIN WAS DEDICATED SOLELY TO PLAYING CHESS, THEN CHESS WOULD BE PRETTY EFFORTLESS, TOO!
CHESS
CHECK AND MATE, DEEP BLUE

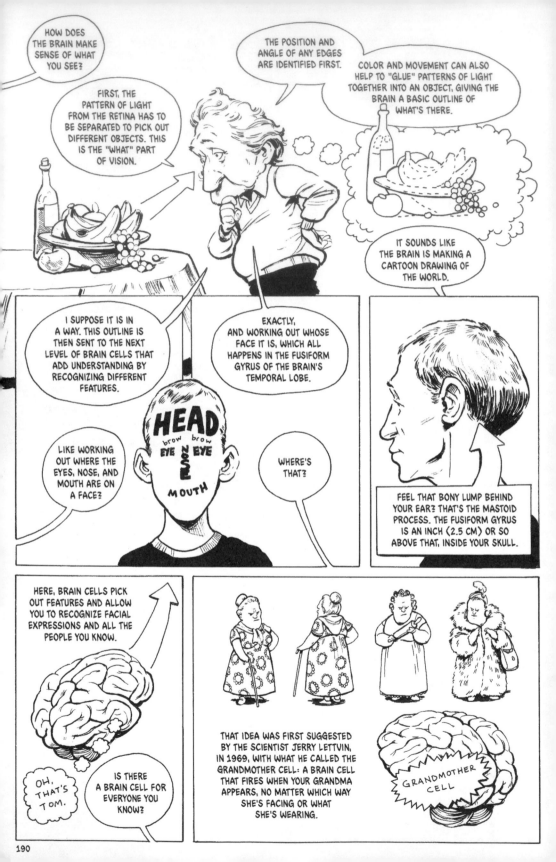

HOW DOES THE BRAIN MAKE SENSE OF WHAT YOU SEE?

FIRST, THE PATTERN OF LIGHT FROM THE RETINA HAS TO BE SEPARATED TO PICK OUT DIFFERENT OBJECTS. THIS IS THE "WHAT" PART OF VISION.

THE POSITION AND ANGLE OF ANY EDGES ARE IDENTIFIED FIRST.

COLOR AND MOVEMENT CAN ALSO HELP TO "GLUE" PATTERNS OF LIGHT TOGETHER INTO AN OBJECT, GIVING THE BRAIN A BASIC OUTLINE OF WHAT'S THERE.

IT SOUNDS LIKE THE BRAIN IS MAKING A CARTOON DRAWING OF THE WORLD.

I SUPPOSE IT IS IN A WAY. THIS OUTLINE IS THEN SENT TO THE NEXT LEVEL OF BRAIN CELLS THAT ADD UNDERSTANDING BY RECOGNIZING DIFFERENT FEATURES.

EXACTLY, AND WORKING OUT WHOSE FACE IT IS, WHICH ALL HAPPENS IN THE FUSIFORM GYRUS OF THE BRAIN'S TEMPORAL LOBE.

LIKE WORKING OUT WHERE THE EYES, NOSE, AND MOUTH ARE ON A FACE?

HEAD
brow brow
EYE NOSE EYE
MOUTH

WHERE'S THAT?

FEEL THAT BONY LUMP BEHIND YOUR EAR? THAT'S THE MASTOID PROCESS. THE FUSIFORM GYRUS IS AN INCH (2.5 CM) OR SO ABOVE THAT, INSIDE YOUR SKULL.

HERE, BRAIN CELLS PICK OUT FEATURES AND ALLOW YOU TO RECOGNIZE FACIAL EXPRESSIONS AND ALL THE PEOPLE YOU KNOW.

OH, THAT'S TOM.

IS THERE A BRAIN CELL FOR EVERYONE YOU KNOW?

THAT IDEA WAS FIRST SUGGESTED BY THE SCIENTIST JERRY LETTVIN, IN 1969, WITH WHAT HE CALLED THE GRANDMOTHER CELL: A BRAIN CELL THAT FIRES WHEN YOUR GRANDMA APPEARS, NO MATTER WHICH WAY SHE'S FACING OR WHAT SHE'S WEARING.

GRANDMOTHER CELL

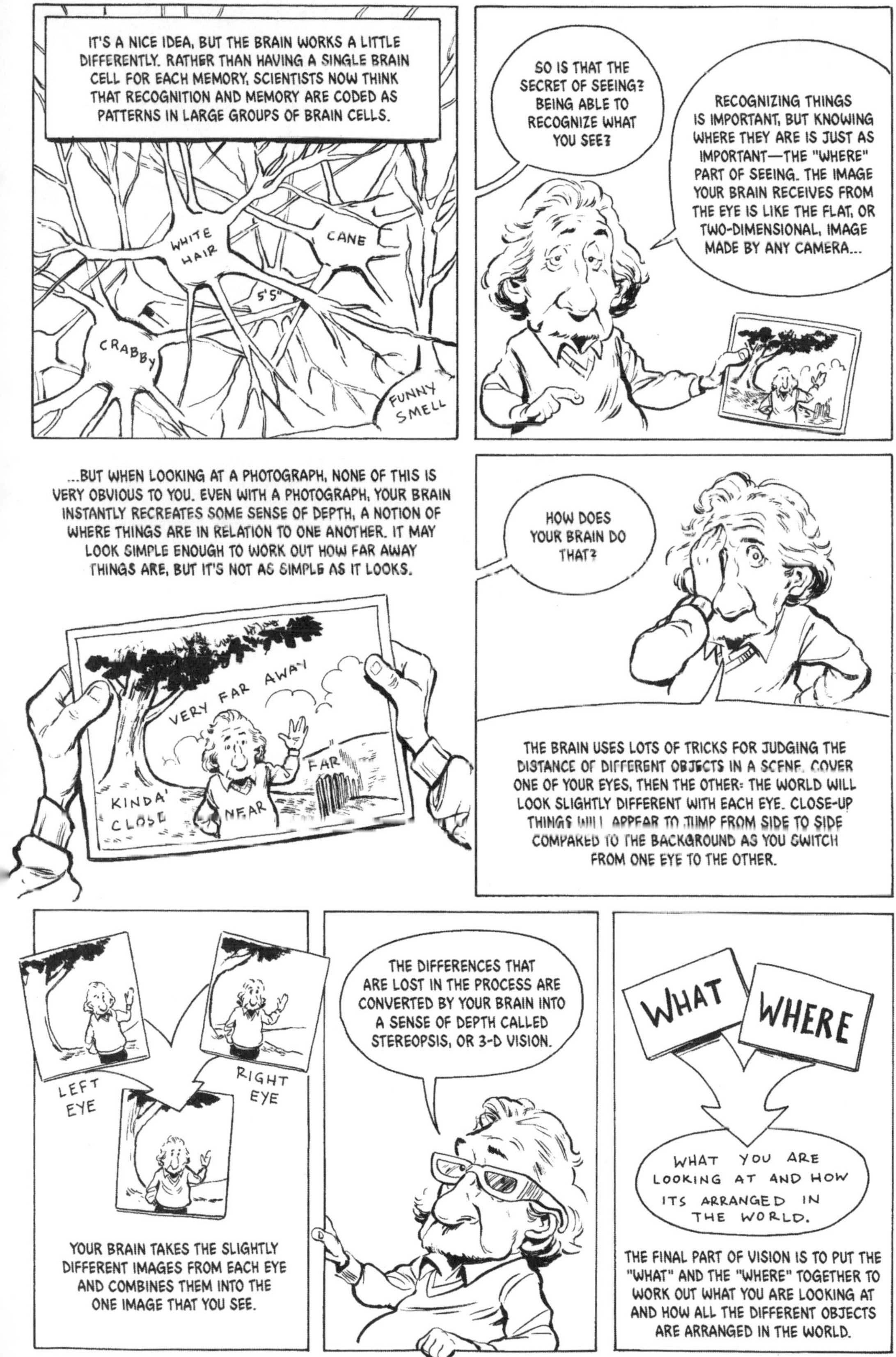

IT'S A NICE IDEA, BUT THE BRAIN WORKS A LITTLE DIFFERENTLY. RATHER THAN HAVING A SINGLE BRAIN CELL FOR EACH MEMORY, SCIENTISTS NOW THINK THAT RECOGNITION AND MEMORY ARE CODED AS PATTERNS IN LARGE GROUPS OF BRAIN CELLS.
WHITE HAIR
CANE
5'5"
CRABBY
FUNNY SMELL
SO IS THAT THE SECRET OF SEEING? BEING ABLE TO RECOGNIZE WHAT YOU SEE?
RECOGNIZING THINGS IS IMPORTANT, BUT KNOWING WHERE THEY ARE IS JUST AS IMPORTANT—THE "WHERE" PART OF SEEING. THE IMAGE YOUR BRAIN RECEIVES FROM THE EYE IS LIKE THE FLAT, OR TWO-DIMENSIONAL, IMAGE MADE BY ANY CAMERA...
...BUT WHEN LOOKING AT A PHOTOGRAPH, NONE OF THIS IS VERY OBVIOUS TO YOU. EVEN WITH A PHOTOGRAPH, YOUR BRAIN INSTANTLY RECREATES SOME SENSE OF DEPTH, A NOTION OF WHERE THINGS ARE IN RELATION TO ONE ANOTHER. IT MAY LOOK SIMPLE ENOUGH TO WORK OUT HOW FAR AWAY THINGS ARE, BUT IT'S NOT AS SIMPLE AS IT LOOKS.
VERY FAR AWAY
FAR
KINDA' CLOSE
NEAR
HOW DOES YOUR BRAIN DO THAT?
THE BRAIN USES LOTS OF TRICKS FOR JUDGING THE DISTANCE OF DIFFERENT OBJECTS IN A SCENE. COVER ONE OF YOUR EYES, THEN THE OTHER: THE WORLD WILL LOOK SLIGHTLY DIFFERENT WITH EACH EYE. CLOSE-UP THINGS WILL APPEAR TO JUMP FROM SIDE TO SIDE COMPARED TO THE BACKGROUND AS YOU SWITCH FROM ONE EYE TO THE OTHER.
LEFT EYE
RIGHT EYE
YOUR BRAIN TAKES THE SLIGHTLY DIFFERENT IMAGES FROM EACH EYE AND COMBINES THEM INTO THE ONE IMAGE THAT YOU SEE.
THE DIFFERENCES THAT ARE LOST IN THE PROCESS ARE CONVERTED BY YOUR BRAIN INTO A SENSE OF DEPTH CALLED STEREOPSIS, OR 3-D VISION.
WHAT
WHERE
WHAT YOU ARE LOOKING AT AND HOW ITS ARRANGED IN THE WORLD.
THE FINAL PART OF VISION IS TO PUT THE "WHAT" AND THE "WHERE" TOGETHER TO WORK OUT WHAT YOU ARE LOOKING AT AND HOW ALL THE DIFFERENT OBJECTS ARE ARRANGED IN THE WORLD.

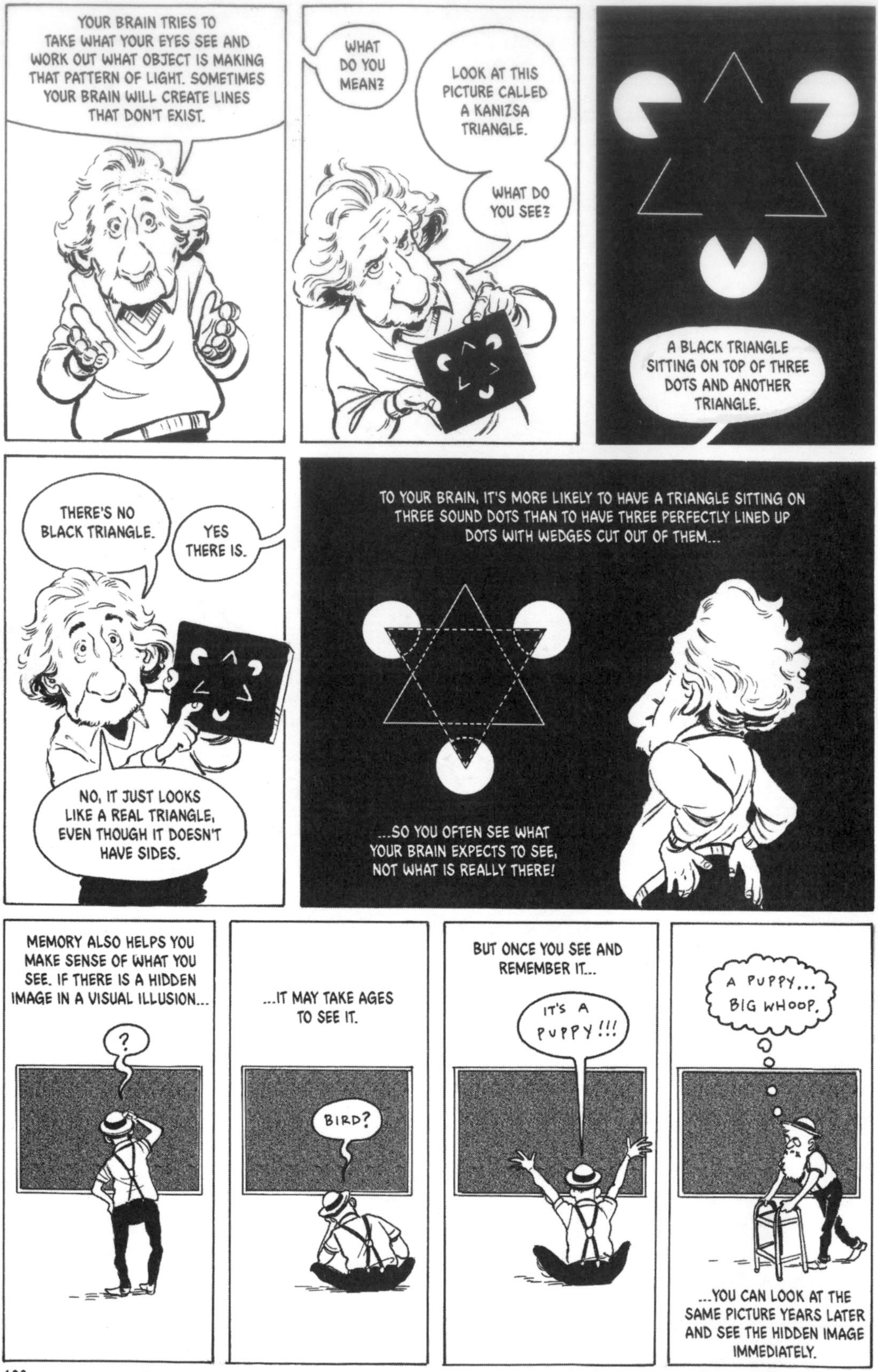
YOUR BRAIN TRIES TO TAKE WHAT YOUR EYES SEE AND WORK OUT WHAT OBJECT IS MAKING THAT PATTERN OF LIGHT. SOMETIMES YOUR BRAIN WILL CREATE LINES THAT DON'T EXIST.
WHAT DO YOU MEAN?
LOOK AT THIS PICTURE CALLED A KANIZSA TRIANGLE.
WHAT DO YOU SEE?
A BLACK TRIANGLE SITTING ON TOP OF THREE DOTS AND ANOTHER TRIANGLE.
THERE'S NO BLACK TRIANGLE.
YES THERE IS.
NO, IT JUST LOOKS LIKE A REAL TRIANGLE, EVEN THOUGH IT DOESN'T HAVE SIDES.
TO YOUR BRAIN, IT'S MORE LIKELY TO HAVE A TRIANGLE SITTING ON THREE SOUND DOTS THAN TO HAVE THREE PERFECTLY LINED UP DOTS WITH WEDGES CUT OUT OF THEM...
...SO YOU OFTEN SEE WHAT YOUR BRAIN EXPECTS TO SEE, NOT WHAT IS REALLY THERE!
MEMORY ALSO HELPS YOU MAKE SENSE OF WHAT YOU SEE. IF THERE IS A HIDDEN IMAGE IN A VISUAL ILLUSION...
?
...IT MAY TAKE AGES TO SEE IT.
BIRD?
BUT ONCE YOU SEE AND REMEMBER IT...
IT'S A PUPPY!!!
A PUPPY... BIG WHOOP.
...YOU CAN LOOK AT THE SAME PICTURE YEARS LATER AND SEE THE HIDDEN IMAGE IMMEDIATELY.

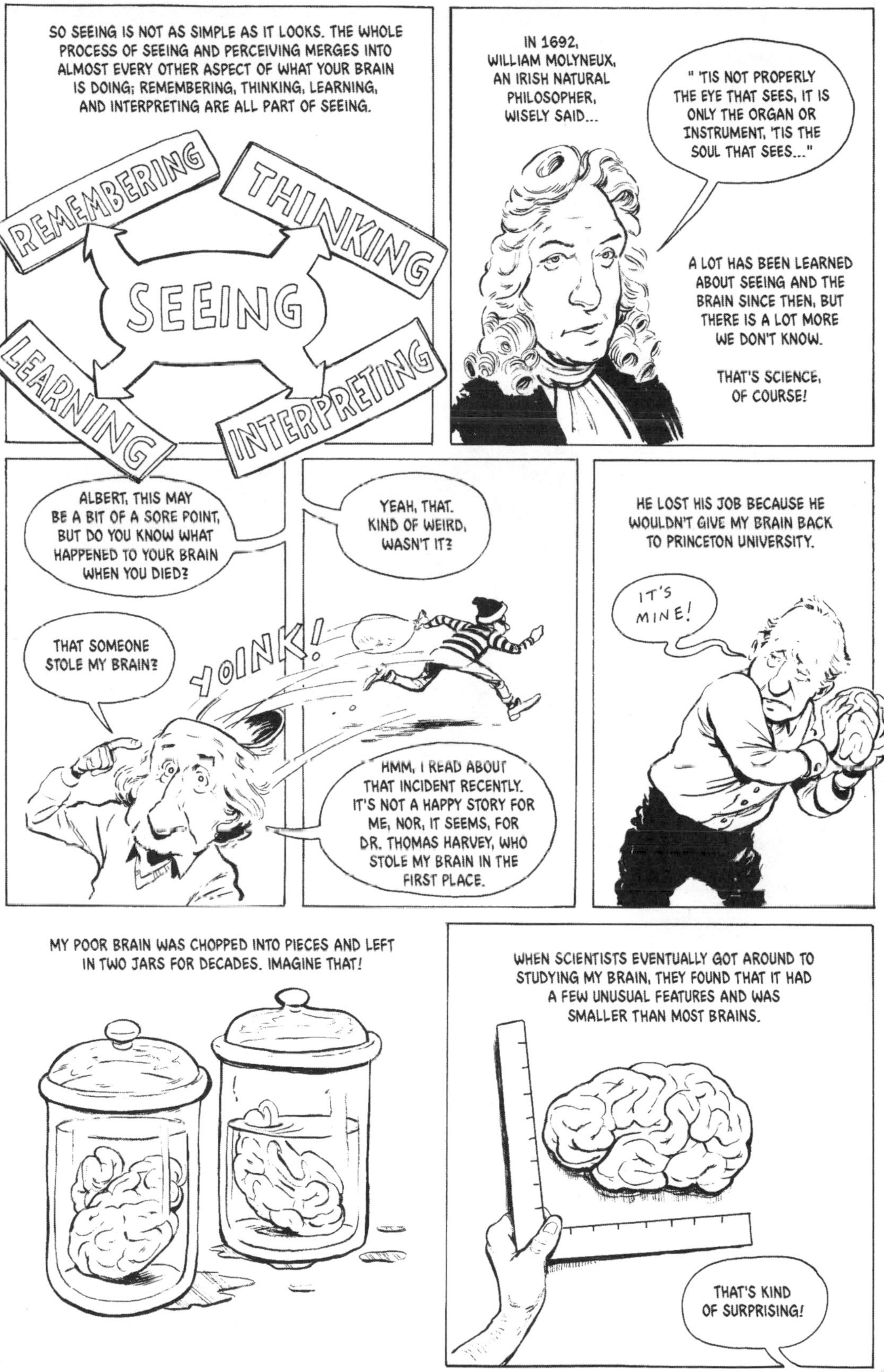

SO SEEING IS NOT AS SIMPLE AS IT LOOKS. THE WHOLE PROCESS OF SEEING AND PERCEIVING MERGES INTO ALMOST EVERY OTHER ASPECT OF WHAT YOUR BRAIN IS DOING; REMEMBERING, THINKING, LEARNING, AND INTERPRETING ARE ALL PART OF SEEING.
REMEMBERING
THINKING
SEEING
LEARNING
INTERPRETING
IN 1692, WILLIAM MOLYNEUX, AN IRISH NATURAL PHILOSOPHER, WISELY SAID...
" 'TIS NOT PROPERLY THE EYE THAT SEES, IT IS ONLY THE ORGAN OR INSTRUMENT, 'TIS THE SOUL THAT SEES..."
A LOT HAS BEEN LEARNED ABOUT SEEING AND THE BRAIN SINCE THEN, BUT THERE IS A LOT MORE WE DON'T KNOW.
THAT'S SCIENCE, OF COURSE!
ALBERT, THIS MAY BE A BIT OF A SORE POINT, BUT DO YOU KNOW WHAT HAPPENED TO YOUR BRAIN WHEN YOU DIED?
YEAH, THAT. KIND OF WEIRD, WASN'T IT?
THAT SOMEONE STOLE MY BRAIN?
YOINK!
HMM, I READ ABOUT THAT INCIDENT RECENTLY. IT'S NOT A HAPPY STORY FOR ME, NOR, IT SEEMS, FOR DR. THOMAS HARVEY, WHO STOLE MY BRAIN IN THE FIRST PLACE.
HE LOST HIS JOB BECAUSE HE WOULDN'T GIVE MY BRAIN BACK TO PRINCETON UNIVERSITY.
IT'S MINE!
MY POOR BRAIN WAS CHOPPED INTO PIECES AND LEFT IN TWO JARS FOR DECADES. IMAGINE THAT!
WHEN SCIENTISTS EVENTUALLY GOT AROUND TO STUDYING MY BRAIN, THEY FOUND THAT IT HAD A FEW UNUSUAL FEATURES AND WAS SMALLER THAN MOST BRAINS.
THAT'S KIND OF SURPRISING!

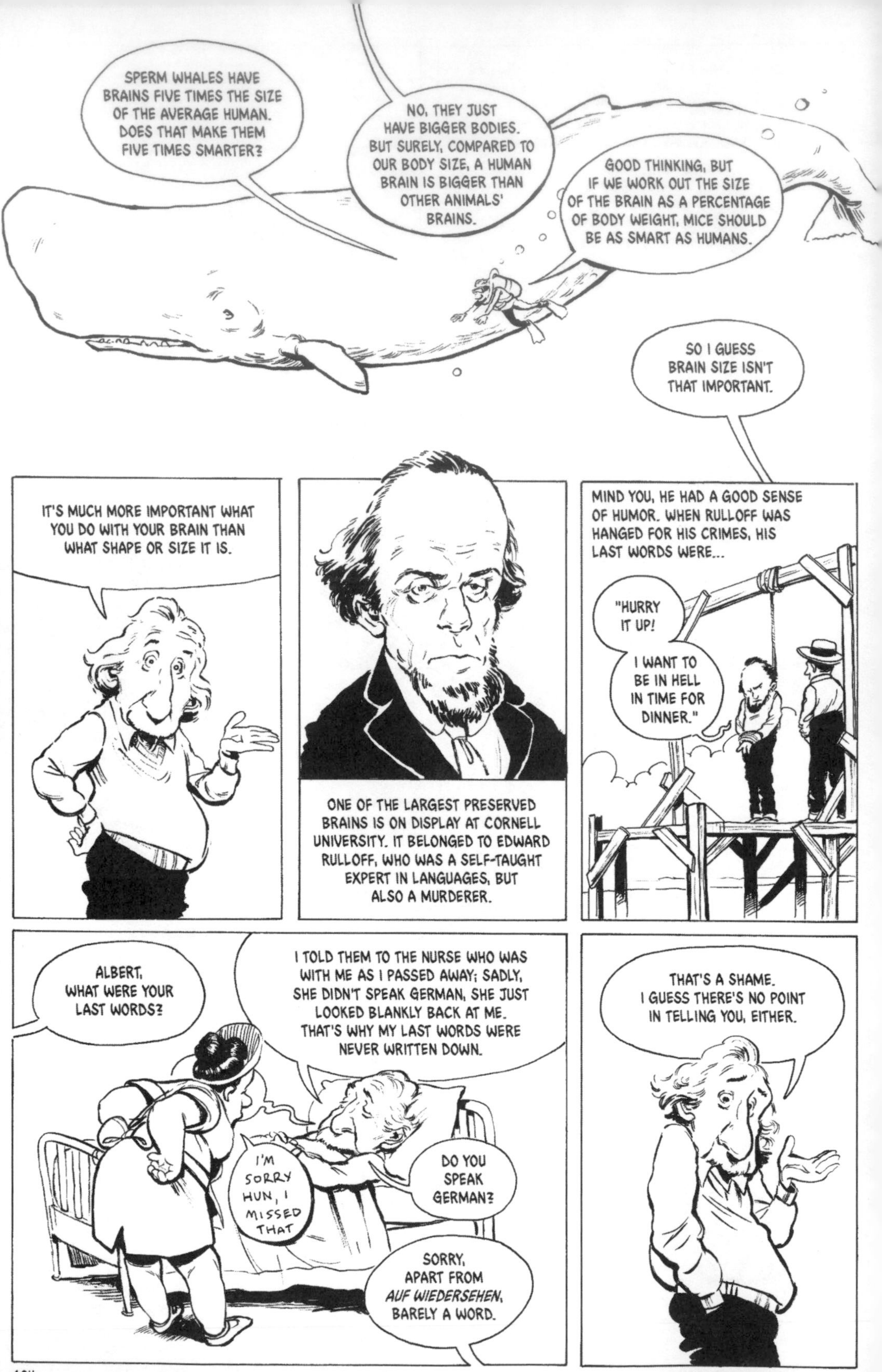
SPERM WHALES HAVE BRAINS FIVE TIMES THE SIZE OF THE AVERAGE HUMAN. DOES THAT MAKE THEM FIVE TIMES SMARTER?
NO, THEY JUST HAVE BIGGER BODIES. BUT SURELY, COMPARED TO OUR BODY SIZE, A HUMAN BRAIN IS BIGGER THAN OTHER ANIMALS' BRAINS.
GOOD THINKING, BUT IF WE WORK OUT THE SIZE OF THE BRAIN AS A PERCENTAGE OF BODY WEIGHT, MICE SHOULD BE AS SMART AS HUMANS.
SO I GUESS BRAIN SIZE ISN'T THAT IMPORTANT.
IT'S MUCH MORE IMPORTANT WHAT YOU DO WITH YOUR BRAIN THAN WHAT SHAPE OR SIZE IT IS.
ONE OF THE LARGEST PRESERVED BRAINS IS ON DISPLAY AT CORNELL UNIVERSITY. IT BELONGED TO EDWARD RULLOFF, WHO WAS A SELF-TAUGHT EXPERT IN LANGUAGES, BUT ALSO A MURDERER.
MIND YOU, HE HAD A GOOD SENSE OF HUMOR. WHEN RULLOFF WAS HANGED FOR HIS CRIMES, HIS LAST WORDS WERE...
"HURRY IT UP!
I WANT TO BE IN HELL IN TIME FOR DINNER."
ALBERT, WHAT WERE YOUR LAST WORDS?
I TOLD THEM TO THE NURSE WHO WAS WITH ME AS I PASSED AWAY; SADLY, SHE DIDN'T SPEAK GERMAN, SHE JUST LOOKED BLANKLY BACK AT ME. THAT'S WHY MY LAST WORDS WERE NEVER WRITTEN DOWN.
I'M SORRY HUN, I MISSED THAT
DO YOU SPEAK GERMAN?
SORRY, APART FROM *AUF WIEDERSEHEN*, BARELY A WORD.
THAT'S A SHAME. I GUESS THERE'S NO POINT IN TELLING YOU, EITHER.

CHAPTER 36: ALBERT EINSTEIN'S LOST SECRET REVEALED: THE ONE THING THAT CAN TRAVEL FASTER THAN LIGHT

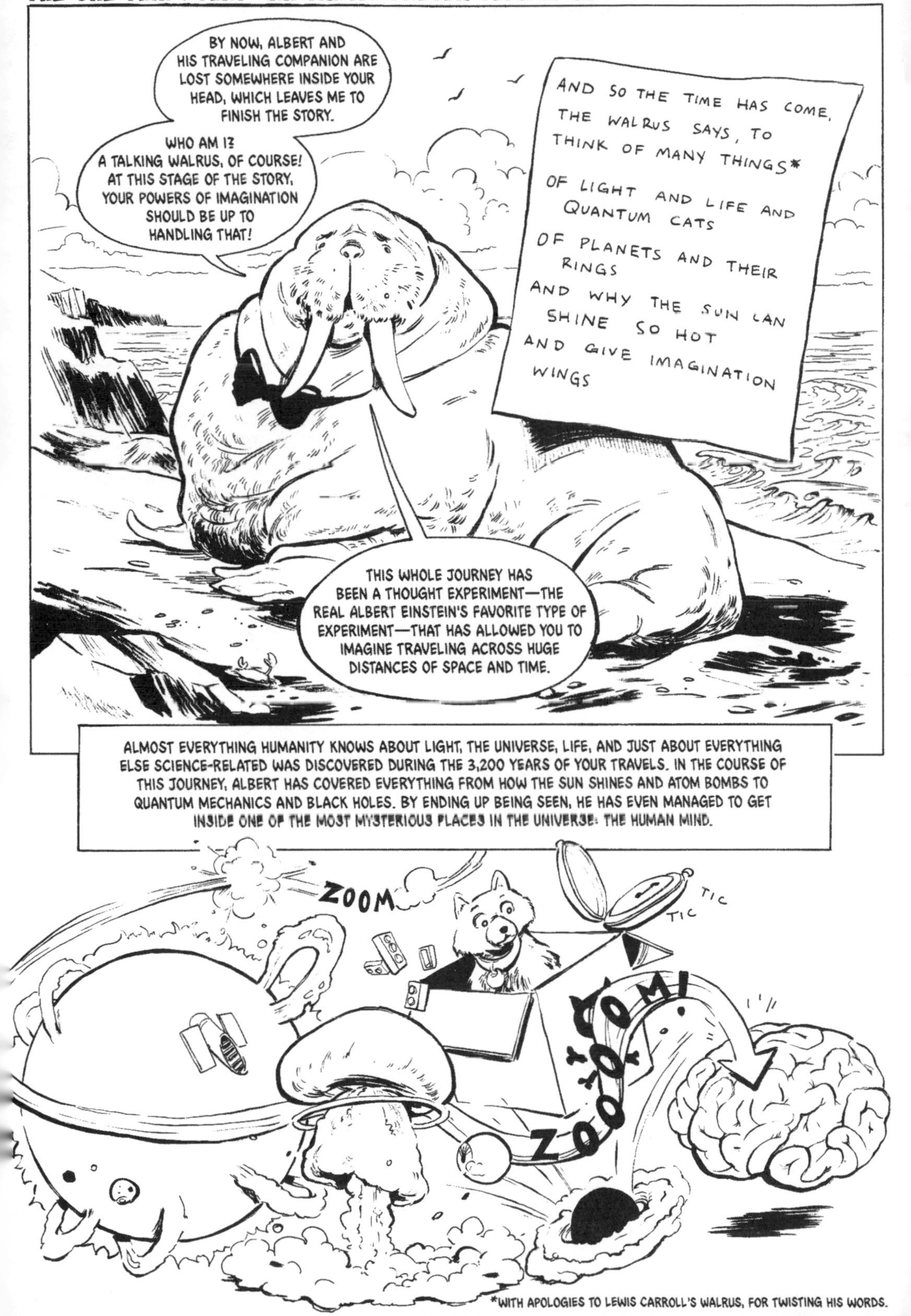
BY NOW, ALBERT AND HIS TRAVELING COMPANION ARE LOST SOMEWHERE INSIDE YOUR HEAD, WHICH LEAVES ME TO FINISH THE STORY.
WHO AM I? A TALKING WALRUS, OF COURSE! AT THIS STAGE OF THE STORY, YOUR POWERS OF IMAGINATION SHOULD BE UP TO HANDLING THAT!
AND SO THE TIME HAS COME, THE WALRUS SAYS, TO THINK OF MANY THINGS*
OF LIGHT AND LIFE AND QUANTUM CATS
OF PLANETS AND THEIR RINGS
AND WHY THE SUN CAN SHINE SO HOT
AND GIVE IMAGINATION WINGS
THIS WHOLE JOURNEY HAS BEEN A THOUGHT EXPERIMENT—THE REAL ALBERT EINSTEIN'S FAVORITE TYPE OF EXPERIMENT—THAT HAS ALLOWED YOU TO IMAGINE TRAVELING ACROSS HUGE DISTANCES OF SPACE AND TIME.
ALMOST EVERYTHING HUMANITY KNOWS ABOUT LIGHT, THE UNIVERSE, LIFE, AND JUST ABOUT EVERYTHING ELSE SCIENCE-RELATED WAS DISCOVERED DURING THE 3,200 YEARS OF YOUR TRAVELS. IN THE COURSE OF THIS JOURNEY, ALBERT HAS COVERED EVERYTHING FROM HOW THE SUN SHINES AND ATOM BOMBS TO QUANTUM MECHANICS AND BLACK HOLES. BY ENDING UP BEING SEEN, HE HAS EVEN MANAGED TO GET INSIDE ONE OF THE MOST MYSTERIOUS PLACES IN THE UNIVERSE: THE HUMAN MIND.
ZOOM
TIC
TIC
ZOOOOOOM!

*WITH APOLOGIES TO LEWIS CARROLL'S WALRUS, FOR TWISTING HIS WORDS.

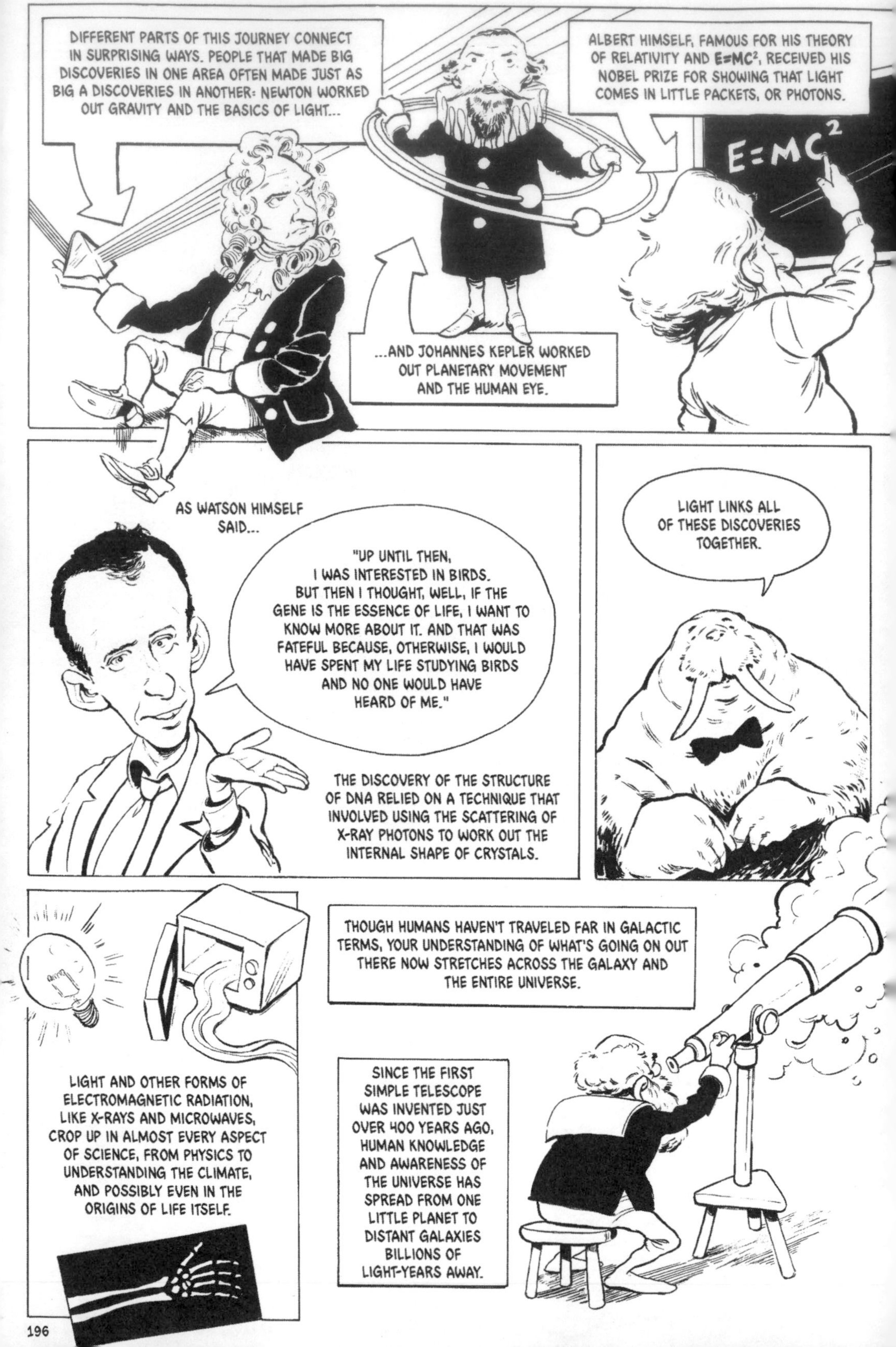

DIFFERENT PARTS OF THIS JOURNEY CONNECT IN SURPRISING WAYS. PEOPLE THAT MADE BIG DISCOVERIES IN ONE AREA OFTEN MADE JUST AS BIG A DISCOVERIES IN ANOTHER: NEWTON WORKED OUT GRAVITY AND THE BASICS OF LIGHT...
...AND JOHANNES KEPLER WORKED OUT PLANETARY MOVEMENT AND THE HUMAN EYE.
ALBERT HIMSELF, FAMOUS FOR HIS THEORY OF RELATIVITY AND E=MC², RECEIVED HIS NOBEL PRIZE FOR SHOWING THAT LIGHT COMES IN LITTLE PACKETS, OR PHOTONS.
E=MC²
AS WATSON HIMSELF SAID...
"UP UNTIL THEN, I WAS INTERESTED IN BIRDS. BUT THEN I THOUGHT, WELL, IF THE GENE IS THE ESSENCE OF LIFE, I WANT TO KNOW MORE ABOUT IT. AND THAT WAS FATEFUL BECAUSE, OTHERWISE, I WOULD HAVE SPENT MY LIFE STUDYING BIRDS AND NO ONE WOULD HAVE HEARD OF ME."
THE DISCOVERY OF THE STRUCTURE OF DNA RELIED ON A TECHNIQUE THAT INVOLVED USING THE SCATTERING OF X-RAY PHOTONS TO WORK OUT THE INTERNAL SHAPE OF CRYSTALS.
LIGHT LINKS ALL OF THESE DISCOVERIES TOGETHER.
LIGHT AND OTHER FORMS OF ELECTROMAGNETIC RADIATION, LIKE X-RAYS AND MICROWAVES, CROP UP IN ALMOST EVERY ASPECT OF SCIENCE, FROM PHYSICS TO UNDERSTANDING THE CLIMATE, AND POSSIBLY EVEN IN THE ORIGINS OF LIFE ITSELF.
THOUGH HUMANS HAVEN'T TRAVELED FAR IN GALACTIC TERMS, YOUR UNDERSTANDING OF WHAT'S GOING ON OUT THERE NOW STRETCHES ACROSS THE GALAXY AND THE ENTIRE UNIVERSE.
SINCE THE FIRST SIMPLE TELESCOPE WAS INVENTED JUST OVER 400 YEARS AGO, HUMAN KNOWLEDGE AND AWARENESS OF THE UNIVERSE HAS SPREAD FROM ONE LITTLE PLANET TO DISTANT GALAXIES BILLIONS OF LIGHT-YEARS AWAY.

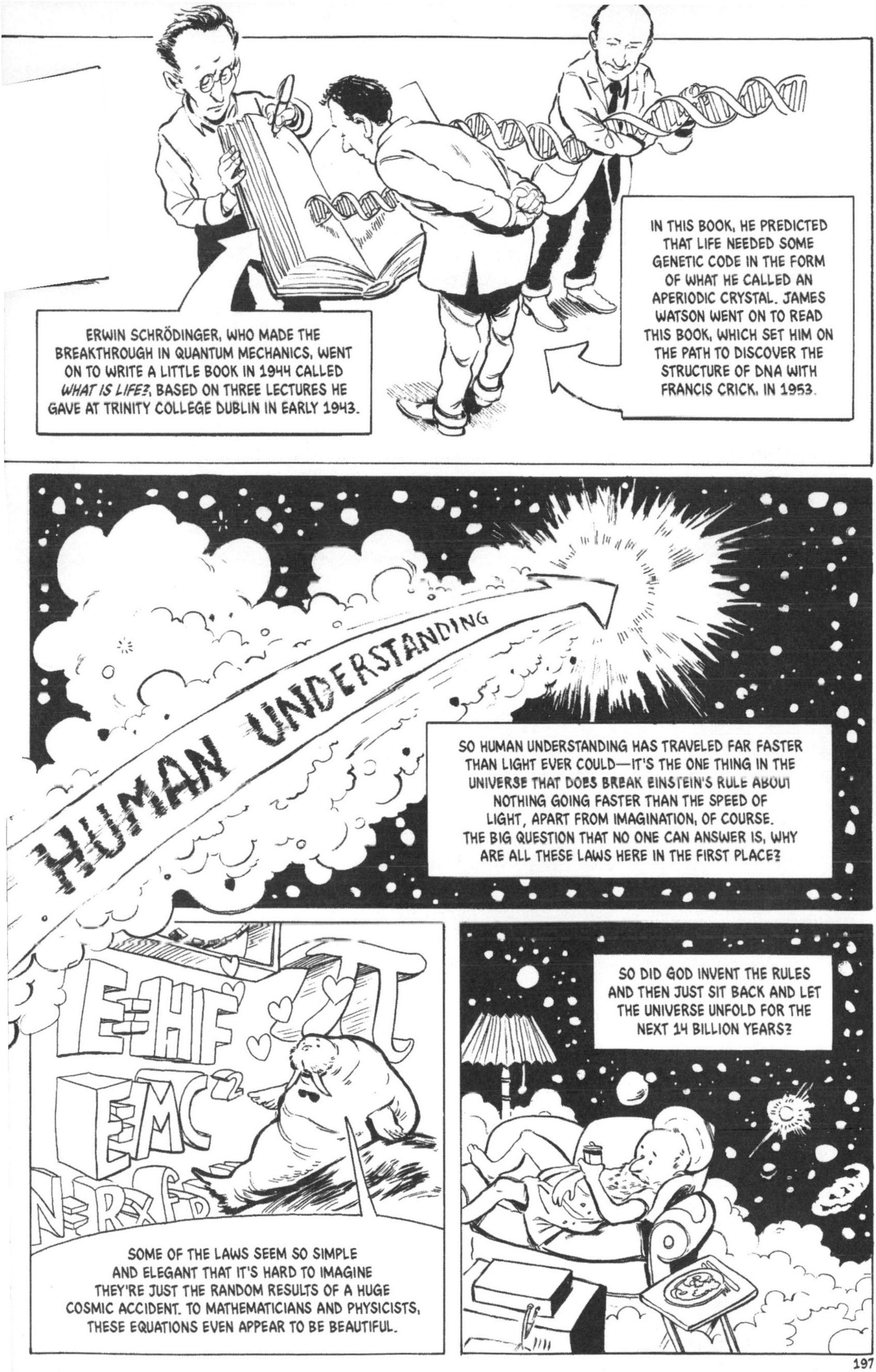
ERWIN SCHRÖDINGER, WHO MADE THE BREAKTHROUGH IN QUANTUM MECHANICS, WENT ON TO WRITE A LITTLE BOOK IN 1944 CALLED *WHAT IS LIFE?*, BASED ON THREE LECTURES HE GAVE AT TRINITY COLLEGE DUBLIN IN EARLY 1943.
IN THIS BOOK, HE PREDICTED THAT LIFE NEEDED SOME GENETIC CODE IN THE FORM OF WHAT HE CALLED AN APERIODIC CRYSTAL. JAMES WATSON WENT ON TO READ THIS BOOK, WHICH SET HIM ON THE PATH TO DISCOVER THE STRUCTURE OF DNA WITH FRANCIS CRICK, IN 1953.
HUMAN UNDERSTANDING
SO HUMAN UNDERSTANDING HAS TRAVELED FAR FASTER THAN LIGHT EVER COULD—IT'S THE ONE THING IN THE UNIVERSE THAT DOES BREAK EINSTEIN'S RULE ABOUT NOTHING GOING FASTER THAN THE SPEED OF LIGHT, APART FROM IMAGINATION, OF COURSE. THE BIG QUESTION THAT NO ONE CAN ANSWER IS, WHY ARE ALL THESE LAWS HERE IN THE FIRST PLACE?
E=HF
E=MC²
SOME OF THE LAWS SEEM SO SIMPLE AND ELEGANT THAT IT'S HARD TO IMAGINE THEY'RE JUST THE RANDOM RESULTS OF A HUGE COSMIC ACCIDENT. TO MATHEMATICIANS AND PHYSICISTS, THESE EQUATIONS EVEN APPEAR TO BE BEAUTIFUL.
SO DID GOD INVENT THE RULES AND THEN JUST SIT BACK AND LET THE UNIVERSE UNFOLD FOR THE NEXT 14 BILLION YEARS?

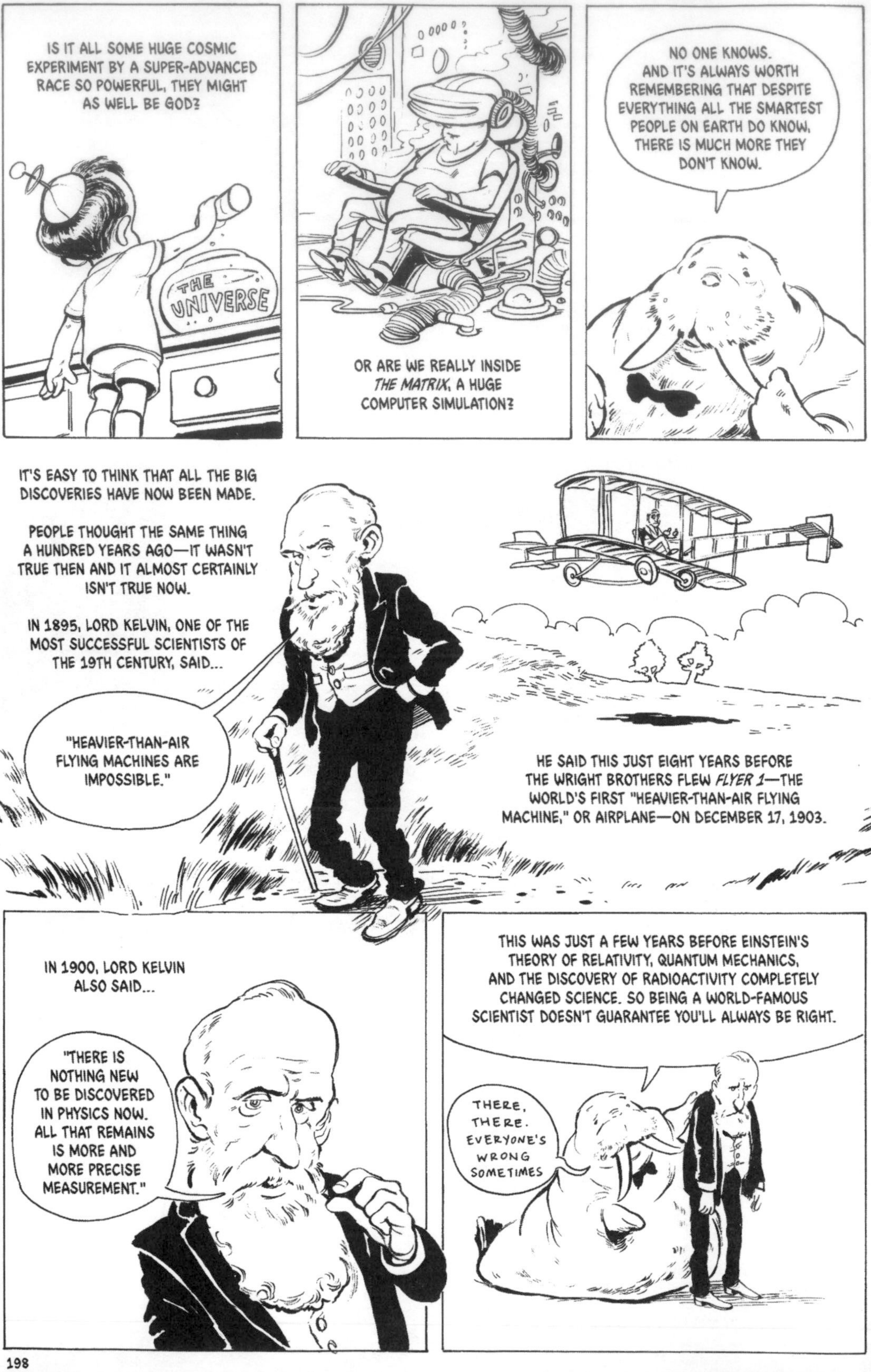
IS IT ALL SOME HUGE COSMIC EXPERIMENT BY A SUPER-ADVANCED RACE SO POWERFUL, THEY MIGHT AS WELL BE GOD?
THE UNIVERSE
OR ARE WE REALLY INSIDE *THE MATRIX*, A HUGE COMPUTER SIMULATION?
NO ONE KNOWS. AND IT'S ALWAYS WORTH REMEMBERING THAT DESPITE EVERYTHING ALL THE SMARTEST PEOPLE ON EARTH DO KNOW, THERE IS MUCH MORE THEY DON'T KNOW.
IT'S EASY TO THINK THAT ALL THE BIG DISCOVERIES HAVE NOW BEEN MADE.
PEOPLE THOUGHT THE SAME THING A HUNDRED YEARS AGO—IT WASN'T TRUE THEN AND IT ALMOST CERTAINLY ISN'T TRUE NOW.
IN 1895, LORD KELVIN, ONE OF THE MOST SUCCESSFUL SCIENTISTS OF THE 19TH CENTURY, SAID...
"HEAVIER-THAN-AIR FLYING MACHINES ARE IMPOSSIBLE."
HE SAID THIS JUST EIGHT YEARS BEFORE THE WRIGHT BROTHERS FLEW *FLYER 1*—THE WORLD'S FIRST "HEAVIER-THAN-AIR FLYING MACHINE," OR AIRPLANE—ON DECEMBER 17, 1903.
IN 1900, LORD KELVIN ALSO SAID...
"THERE IS NOTHING NEW TO BE DISCOVERED IN PHYSICS NOW. ALL THAT REMAINS IS MORE AND MORE PRECISE MEASUREMENT."
THIS WAS JUST A FEW YEARS BEFORE EINSTEIN'S THEORY OF RELATIVITY, QUANTUM MECHANICS, AND THE DISCOVERY OF RADIOACTIVITY COMPLETELY CHANGED SCIENCE. SO BEING A WORLD-FAMOUS SCIENTIST DOESN'T GUARANTEE YOU'LL ALWAYS BE RIGHT.
THERE, THERE. EVERYONE'S WRONG SOMETIMES

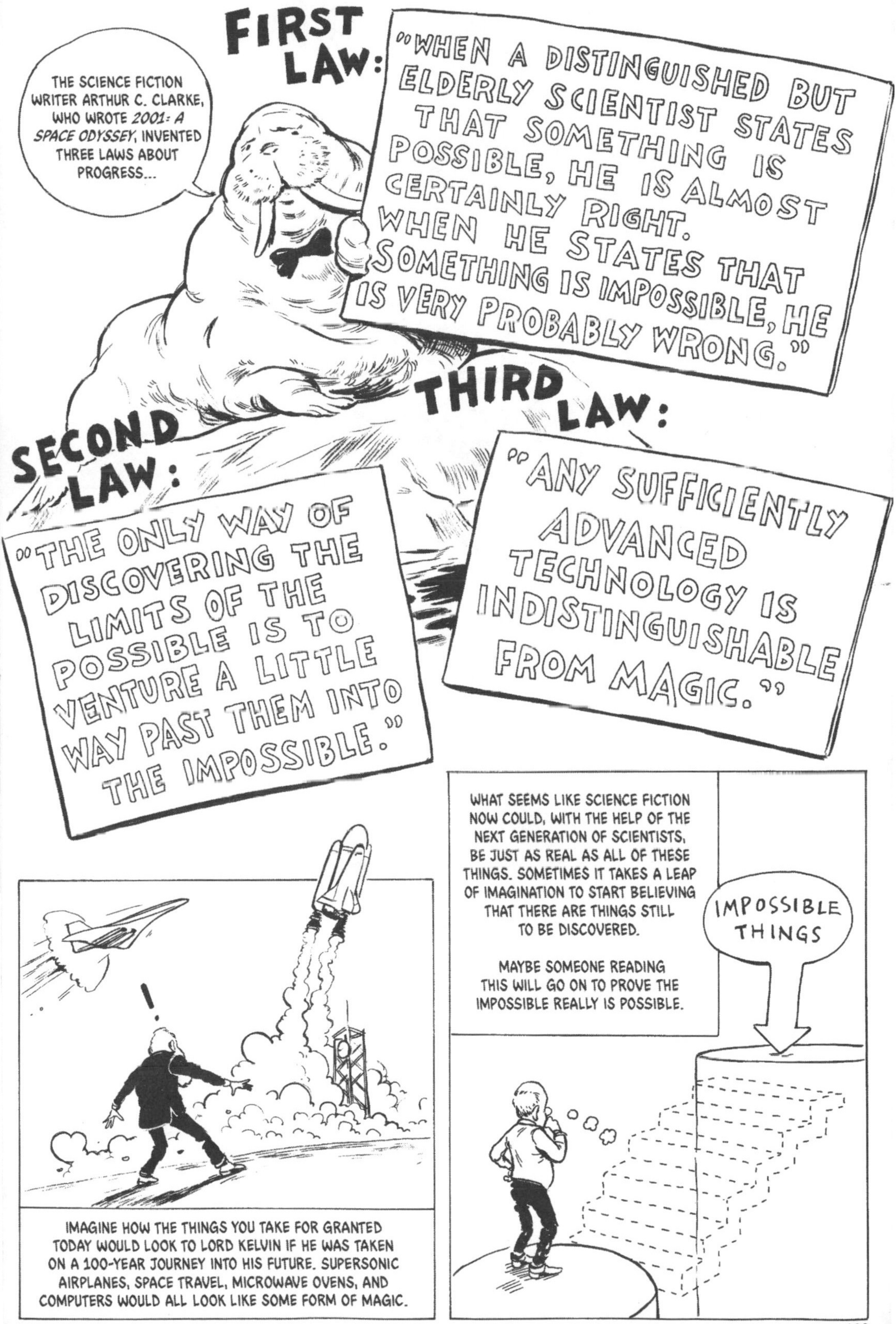
FIRST LAW:
THE SCIENCE FICTION WRITER ARTHUR C. CLARKE, WHO WROTE *2001: A SPACE ODYSSEY*, INVENTED THREE LAWS ABOUT PROGRESS...
"WHEN A DISTINGUISHED BUT ELDERLY SCIENTIST STATES THAT SOMETHING IS POSSIBLE, HE IS ALMOST CERTAINLY RIGHT. WHEN HE STATES THAT SOMETHING IS IMPOSSIBLE, HE IS VERY PROBABLY WRONG."
SECOND LAW:
"THE ONLY WAY OF DISCOVERING THE LIMITS OF THE POSSIBLE IS TO VENTURE A LITTLE WAY PAST THEM INTO THE IMPOSSIBLE."
THIRD LAW:
"ANY SUFFICIENTLY ADVANCED TECHNOLOGY IS INDISTINGUISHABLE FROM MAGIC."
!
IMAGINE HOW THE THINGS YOU TAKE FOR GRANTED TODAY WOULD LOOK TO LORD KELVIN IF HE WAS TAKEN ON A 100-YEAR JOURNEY INTO HIS FUTURE. SUPERSONIC AIRPLANES, SPACE TRAVEL, MICROWAVE OVENS, AND COMPUTERS WOULD ALL LOOK LIKE SOME FORM OF MAGIC.
WHAT SEEMS LIKE SCIENCE FICTION NOW COULD, WITH THE HELP OF THE NEXT GENERATION OF SCIENTISTS, BE JUST AS REAL AS ALL OF THESE THINGS. SOMETIMES IT TAKES A LEAP OF IMAGINATION TO START BELIEVING THAT THERE ARE THINGS STILL TO BE DISCOVERED.
MAYBE SOMEONE READING THIS WILL GO ON TO PROVE THE IMPOSSIBLE REALLY IS POSSIBLE.
IMPOSSIBLE THINGS

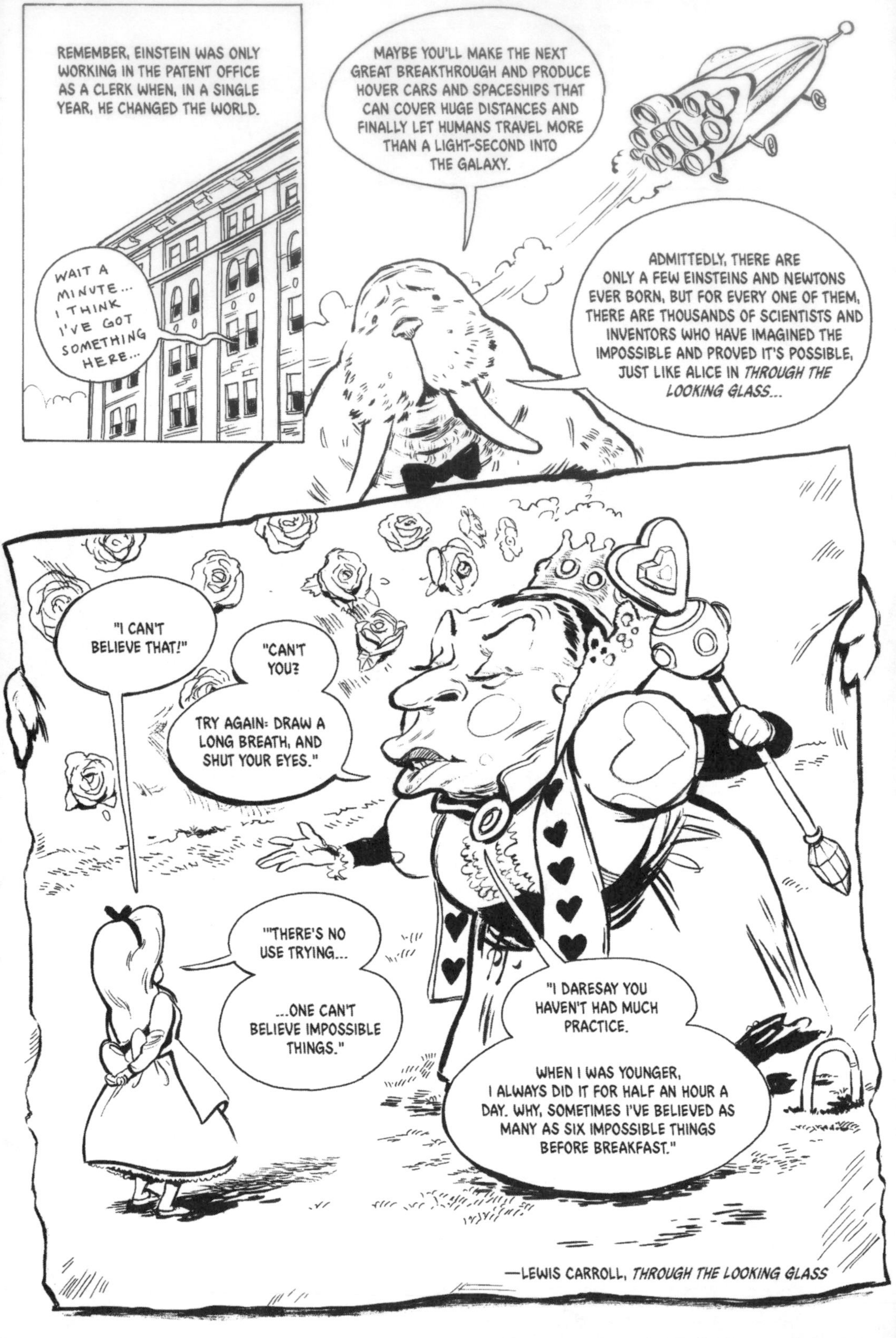

REMEMBER, EINSTEIN WAS ONLY WORKING IN THE PATENT OFFICE AS A CLERK WHEN, IN A SINGLE YEAR, HE CHANGED THE WORLD.
WAIT A MINUTE... I THINK I'VE GOT SOMETHING HERE...
MAYBE YOU'LL MAKE THE NEXT GREAT BREAKTHROUGH AND PRODUCE HOVER CARS AND SPACESHIPS THAT CAN COVER HUGE DISTANCES AND FINALLY LET HUMANS TRAVEL MORE THAN A LIGHT-SECOND INTO THE GALAXY.
ADMITTEDLY, THERE ARE ONLY A FEW EINSTEINS AND NEWTONS EVER BORN, BUT FOR EVERY ONE OF THEM, THERE ARE THOUSANDS OF SCIENTISTS AND INVENTORS WHO HAVE IMAGINED THE IMPOSSIBLE AND PROVED IT'S POSSIBLE, JUST LIKE ALICE IN *THROUGH THE LOOKING GLASS*...
"I CAN'T BELIEVE THAT!"
"CAN'T YOU?
TRY AGAIN: DRAW A LONG BREATH, AND SHUT YOUR EYES."
"'THERE'S NO USE TRYING...
...ONE CAN'T BELIEVE IMPOSSIBLE THINGS."
"I DARESAY YOU HAVEN'T HAD MUCH PRACTICE.
WHEN I WAS YOUNGER, I ALWAYS DID IT FOR HALF AN HOUR A DAY. WHY, SOMETIMES I'VE BELIEVED AS MANY AS SIX IMPOSSIBLE THINGS BEFORE BREAKFAST."
—LEWIS CARROLL, *THROUGH THE LOOKING GLASS*

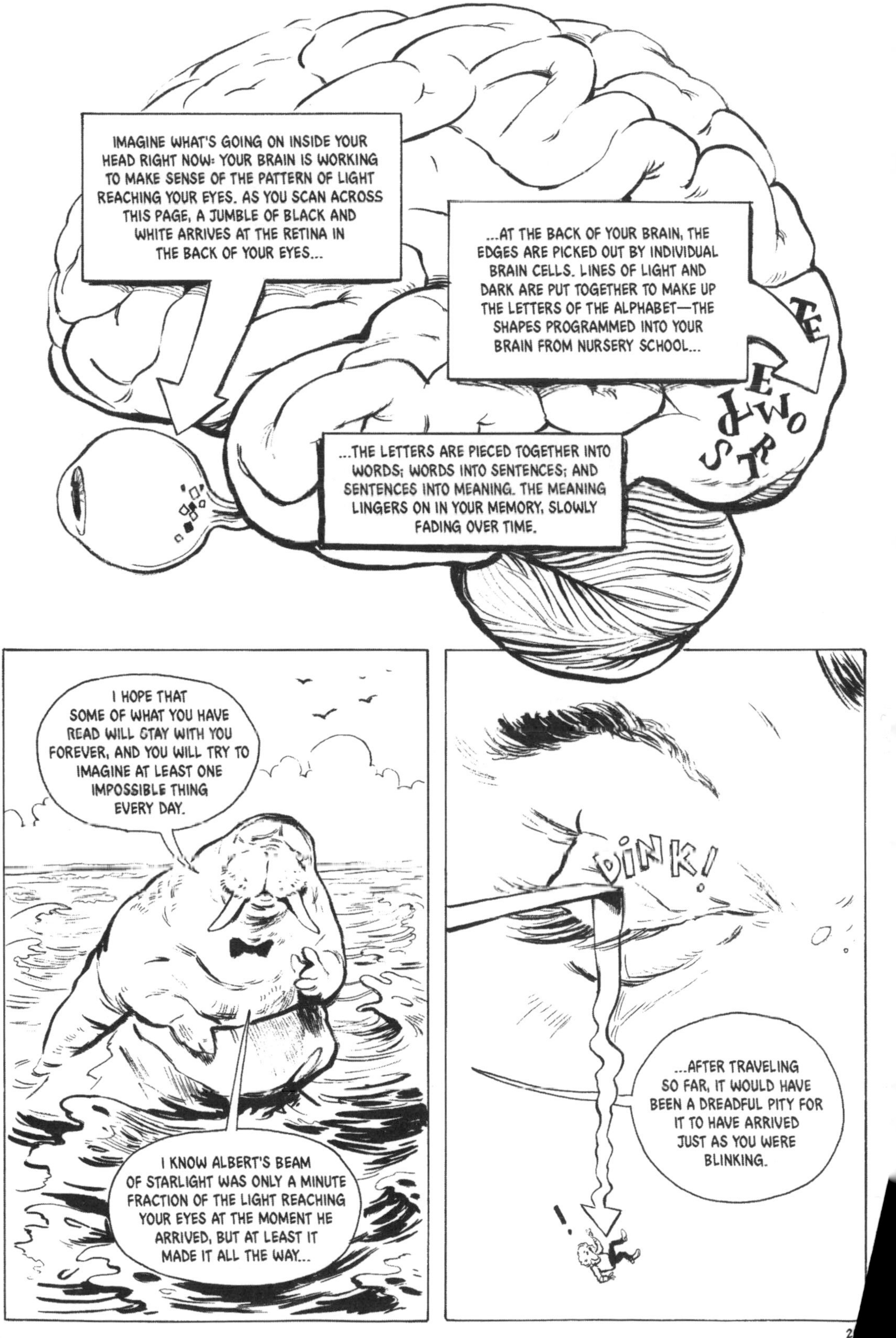
IMAGINE WHAT'S GOING ON INSIDE YOUR HEAD RIGHT NOW: YOUR BRAIN IS WORKING TO MAKE SENSE OF THE PATTERN OF LIGHT REACHING YOUR EYES. AS YOU SCAN ACROSS THIS PAGE, A JUMBLE OF BLACK AND WHITE ARRIVES AT THE RETINA IN THE BACK OF YOUR EYES...
...AT THE BACK OF YOUR BRAIN, THE EDGES ARE PICKED OUT BY INDIVIDUAL BRAIN CELLS. LINES OF LIGHT AND DARK ARE PUT TOGETHER TO MAKE UP THE LETTERS OF THE ALPHABET—THE SHAPES PROGRAMMED INTO YOUR BRAIN FROM NURSERY SCHOOL...
...THE LETTERS ARE PIECED TOGETHER INTO WORDS; WORDS INTO SENTENCES; AND SENTENCES INTO MEANING. THE MEANING LINGERS ON IN YOUR MEMORY, SLOWLY FADING OVER TIME.
I HOPE THAT SOME OF WHAT YOU HAVE READ WILL STAY WITH YOU FOREVER, AND YOU WILL TRY TO IMAGINE AT LEAST ONE IMPOSSIBLE THING EVERY DAY.
I KNOW ALBERT'S BEAM OF STARLIGHT WAS ONLY A MINUTE FRACTION OF THE LIGHT REACHING YOUR EYES AT THE MOMENT HE ARRIVED, BUT AT LEAST IT MADE IT ALL THE WAY...
DINK!
...AFTER TRAVELING SO FAR, IT WOULD HAVE BEEN A DREADFUL PITY FOR IT TO HAVE ARRIVED JUST AS YOU WERE BLINKING.
!

INDEX

ACKNOWLEDGMENTS

IAN:

LIKE ALBERT'S JOURNEY, THIS PROJECT HAS TAKEN A WHILE TO COME TO FRUITION. ALONG THE WAY, I HAVE BEEN HELPED BY ALL THE GREAT FEEDBACK AND COMMENTS FROM READERS OF THE JOURNEY BY STARLIGHT BLOG. I WOULD PARTICULARLY LIKE TO THANK MY AGENT, BRANDI BOWLES (FOUNDRY LITERARY + MEDIA), FOR MAKING CONTACT AND CONVINCING ME THAT JOURNEY BY STARLIGHT WOULD MAKE A GREAT GRAPHIC NOVEL.

AS MY DRAWING ABILITY IS AT THE LEVEL OF STICK-MEN, THIS SEEMED DAUNTING UNTIL BRANDI TRACKED DOWN THE BRILLIANT BRITT SPENCER, WHO BROUGHT THE STORY TO LIFE GRAPHICALLY. HIS DEDICATION IN COMPLETING THIS BOOK MADE HIM SOLDIER ON THROUGH THE PAIN OF A BROKEN HAND. ALL GREAT ART IS BORN OUT OF SUFFERING, AND BRITT'S WONDERFUL DRAWINGS ARE A TESTAMENT TO THAT!

THE FINAL STEP OF TURNING THIS JOURNEY INTO A PHYSICAL BOOK CAME ABOUT THANKS TO ERIN CANNING, OUR EDITOR AT ONE PEACE BOOKS. I WANT TO REALLY THANK HER FOR HER GREAT INSIGHT IN DECIDING TO PUBLISH *JOURNEY BY STARLIGHT*, HER EDITORIAL GUIDANCE, AND FOR MAKING THE WHOLE PROCESS SO ENJOYABLE.

ON A MORE PERSONAL NOTE, THE PERSON WHO HAS SUPPORTED AND ENCOURAGED ME THROUGHOUT ALL THIS IS THE LIGHT OF MY LIFE, JEAN. SHE HAS MADE SURE I HAVE REMAINED SANE, HUMBLE, ENTHUSED, AND LOVED. WHO COULD ASK FOR MORE?

BRITT:

OH, BOY, THIS WAS A BIG'N. UNDERTAKING A PROJECT SUCH AS THIS REQUIRED A LOT OF BLOOD, SWEAT, AND, UNFORTUNATELY, EVEN SOME TEARS. BUT IT ALSO NEEDED A LOT OF HELP FROM OTHERS. WHERE DO I BEGIN? I RECKON AT THE BEGINNING...

THANKS TO BRANDI BOWLES FOR RECOGNIZING THE POTENTIAL OF THIS BOOK AND BRINGING IAN AND I TOGETHER FOR THIS COLLABORATION.

THANKS TO IAN FOR WRITING A MARVELOUS BOOK. INITIALLY, WHEN I WAS PRESENTED WITH THE IDEA OF ILLUSTRATING AN ENTIRE GRAPHIC NOVEL ABOUT SCIENCE, I HAD RESERVATIONS. I ASKED, "WHY ME? WHAT DO I KNOW ABOUT SCIENCE?!" AS IT TURNS OUT, I DIDN'T NEED TO KNOW ANYTHING BECAUSE IAN WRITES IN SUCH A PALATABLE WAY THAT IT MAKES COMPREHENSION OF QUANTUM MECHANICS SOMEHOW EASY. IT MADE MY CONTRIBUTION, IN PART, FEEL LIKE MERE DONKEY-WORK.

THANKS TO ERIN CANNING FOR PUTTING UP WITH THE TROUBLING SCHEDULE CHANGES THAT A BROKEN HAND MADE NECESSARY (THIS "THOUGHT EXPERIMENT" OF OURS GOT A LITTLE ROUGH AT POINTS).

BIG THANKS TO MY TEAM OF DEDICATED AND TALENTED INTERNS, ALLIE JACHIMOWICZ, EMILY SPENCER, AND LOMAHO KRETZMANN. I'D PROBABLY STILL BE INKING IF IT WEREN'T FOR ALL THEIR HELP.

AND LASTLY, THANKS TO A NETWORK OF GREAT FRIENDS AND FAMILY THAT HELPED ME GET THROUGH THE PROCESS WITHOUT LOSING MY MIND COMPLETELY.